Improving the Understanding, Diagnostics, and Prediction of Precipitation

Improving the Understanding, Diagnostics, and Prediction of Precipitation

Editors

Zuohao Cao
Huaqing Cai
Xiaofan Li

MDPI • Basel • Beijing • Wuhan • Barcelona • Belgrade • Manchester • Tokyo • Cluj • Tianjin

MDPI

Editors
Zuohao Cao
Meteorological Research Division
Environment and Climate Change Canada
Toronto
Canada

Huaqing Cai
Army Research Directorate
The U.S. Army Combat Capabilities Development Command Army Research Laboratory
White Sands Missile Range
United States

Xiaofan Li
Department of Atmospheric Sciences
Zhejiang University
Hangzhou
China

Editorial Office
MDPI
St. Alban-Anlage 66
4052 Basel, Switzerland

This is a reprint of articles from the Special Issue published online in the open access journal *Atmosphere* (ISSN 2073-4433) (available at: www.mdpi.com/journal/atmosphere/special_issues/Prediction_Precipitation).

For citation purposes, cite each article independently as indicated on the article page online and as indicated below:

LastName, A.A.; LastName, B.B.; LastName, C.C. Article Title. *Journal Name* **Year**, *Volume Number*, Page Range.

ISBN 978-3-0365-7607-7 (Hbk)
ISBN 978-3-0365-7606-0 (PDF)

Contents

Preface to "Improving the Understanding, Diagnostics, and Prediction of Precipitation"

This reprint is a collection of the papers published in the Special Issue of *Atmosphere* entitled "Improving the Understanding, Diagnostics, and Prediction of Precipitation". It provides atmospheric researchers and operational meteorologists with an update on recent research and its applications on precipitation prediction.

Heavy precipitation remains one of the least understood meteorological phenomena in research and operational communities due to its involvement of multi-scale dynamic and thermodynamic processes associated with precipitating weather systems. This Special Issue aims at advancing the knowledge of these processes, systems, and their interactions; building a bridge between the academic and the operational communities in order to improve the accuracy of numerical weather prediction (NWP) in precipitation forecasting, especially heavy precipitation associated with high-impact weather. These goals can be achieved by developing innovative theories, diagnostic methods, numerical approaches, and verification techniques. Insightful diagnoses are usually expected to provide a guidance on why an NWP model makes a right or wrong prediction. Given the challenges in precipitation forecast, an alternative approach is required to anticipate large-scale environments favorable for development of heavy precipitation.

As a result of rigorous peer reviews, 14 papers have been accepted for publication in this Special Issue. These articles cover topics in (a) data assimilation, such as assimilation of ground-based microwave radiometers [p. 1] and combined techniques of data assimilation [p. 15]; (b) microphysical parameterizations in NWP models [p. 41]; (c) analog ensemble post-processing [p. 61]; (d) deep learning-based short-term intensive rainfall forecast [p. 85], nowcasting [p. 105], and monthly forecast [p. 121]; (e) trend and projection of the long-term spatial-temporal precipitation changes [p. 145]; (f) intercomparison of satellite-based and X-band radar rainfall products [p. 167]; (g) verification of various analyzed precipitation data with observations [p. 191]; (h) climatic patterns of Meiyu and its associated circulations [p. 207]; (i) composite analysis of warm-sector heavy rainfall and its association with large-scale circulations, pre-storm environments, and mesoscale convective systems [p. 221]; and (j) precipitation diurnal cycle [p. 245] and precipitation recycling and moisture sources [p. 263]. It should be noted that all viewpoints in the published papers merely represent those authors'viewpoints, and, certainly, they do not represent our and our organization's viewpoints.

Finally, we would like to take this opportunity to thank all authors for their contributions to this Special Issue, reviewers for their time and efforts to improve the quality of the Special Issue, and the Atmosphere Editorial Office for their prompt assistance.

Conflicts of Interest: The editors declare no conflicts of interest.

Zuohao Cao, Huaqing Cai, and Xiaofan Li
Editors

atmosphere

MDPI

Article

Assimilation of Ground-Based Microwave Radiometer on Heavy Rainfall Forecast in Beijing

Yajie Qi [1,2], Shuiyong Fan [1,*], Bai Li [3], Jiajia Mao [3] and Dawei Lin [2]

1 Institute of Urban Meteorology, China Meteorological Administration, Beijing 100089, China; yjqi@ium.cn
2 Key Laboratory for Cloud Physics, China Meteorological Administration, Beijing 100081, China; lindw@cma.gov.cn
3 Meteorological Observation Center, China Meteorological Administration, Beijing 100081, China; Libai@cma.gov.cn (B.L.); maojj@cma.gov.cn (J.M.)
* Correspondence: syfan@ium.cn

Abstract: Ground-based microwave radiometers (MWRPS) can provide continuous atmospheric temperature and relative humidity profiles for a weather prediction model. We investigated the impact of assimilation of ground-based microwave radiometers based on the rapid-refresh multiscale analysis and prediction system-short term (RMAPS-ST). In this study, five MWRP-retrieved profiles were assimilated for the precipitation enhancement that occurred in Beijing on 21 May 2020. To evaluate the influence of their assimilation, two experiments with and without the MWRPS assimilation were set. Compared to the control experiment, which only assimilated conventional observations and radar data, the MWRPS experiment, which assimilated conventional observations, the ground-based microwave radiometer profiles and the radar data, had a positive impact on the forecasts of the RMAPS-ST. The results show that in comparison with the control test, the MWRPS experiment reproduced the heat island phenomenon in the observation better. The MWRPS assimilation reduced the bias and RMSE of two-meter temperature and two-meter specific humidity forecasting in the 0–12 h of the forecast range. Furthermore, assimilating the MWRPS improved both the distribution and the intensity of the hourly rainfall forecast, as compared with that of the control experiment, with observations that predicted the process of the precipitation enhancement in the urban area of Beijing.

Keywords: heavy rainfall; ground-based microwave radiometer; heat island effect

Citation: Qi, Y.; Fan, S.; Li, B.; Mao, J.; Lin, D. Assimilation of Ground-Based Microwave Radiometer on Heavy Rainfall Forecast in Beijing. *Atmosphere* **2022**, *13*, 74. https://doi.org/10.3390/atmos13010074

Academic Editors: Zuohao Cao, Huaqing Cai and Xiaofan Li

Received: 26 November 2021
Accepted: 28 December 2021
Published: 31 December 2021

1. Introduction

The temporal variation and spatial distribution of meteorological elements represent the state of the atmosphere in the troposphere, and the vertical distribution and variation of meteorological elements are very important for simulating and predicting atmospheric movement in numerical weather prediction models, as the World Meteorological Organization guidance for numerical weather prediction applications has highlighted. Although satellites can provide data in the upper troposphere, it is particularly difficult to observe the lower few kilometers of the atmosphere due to poor sampling [1]. Compared to satellites, radiosondes have a better vertical resolution on atmospheric profiles [2,3]. However, they cannot provide continuous monitoring data since their data are usually available at an interval of 12 h [4]. A ground-based microwave radiometer is a meteorological observation instrument using remote sensing technology. Its secondary products can detect temperature profile, humidity profile, and other elements [5–8], and can conduct continuous observation of vertical changes of meteorological elements within a certain precision range. It can provide high time resolution information of the atmospheric motion state, close the observational gap in the lower troposphere, and help to improve the ability and accuracy of weather forecasts. The profiling capability of the ground-based microwave radiometer has proven to be valuable in the lower troposphere.

Researchers have paid great attention to the performance of the microwave radiometer and how to widely use it in operational systems. Its products have been widely used in many fields such as air pollution monitoring, site climate analysis, and water vapor analysis [9–12]. At the same time, the ground-based microwave radiometer can provide a continuous, high-resolution, and stable observation of temperature and humidity profiles, which can effectively make up for the shortage of atmospheric information obtained by conventional sounding due to the long observation interval, and it can better meet the observation requirements of high-resolution NWP systems [12]. However, unlike radar and GPSZTD data, which have even become operational at some national weather centers, the assimilation of ground-based microwave radiometers to numerical models is still in its infancy [13–17].

For example, assimilating the temperature and humidity profile from a single MWR station showed better a forecast of winter fog using the Fifth-Generation Pennsylvania State University/National Center for Atmospheric Research Mesoscale Model [18,19]. A winter storm case was simulated by an observing system simulation experiment assimilating simulated MWRs, which demonstrated that the impact was positive on the temperature and humidity forecast [20,21]. The mesoscale prediction system Arome-WMed was used to assimilate the profiles retrieved by multiple ground-based microwave radiometers, and the results showed that the skill of the precipitation forecast was improved slightly [12]. The Weather Research and Forecasting (WRF) model was used to assimilate MWR temperature and humidity profiles for simulating a rainstorm event that occurred in Beijing, China, and the results showed that the assimilation of MWR data had a positive impact on the distribution and intensity of rainfall [16]. The rapid-refresh multiscale analysis and prediction system-short term (RMAPS-ST) was used to assimilate MWRPS temperature and relative humidity profiles in Beijing for a precipitation bifurcation case, the results showed that the assimilation of MWRPS improved the precipitation forecast in terms of distribution and the intensity [17].

Previous studies have all shown the promising impacts of assimilating ground-based microwave radiometers into the numerical model, though the results show different impacts on forecasts. However, the assimilating of ground-based microwave radiometer data to the regional operational forecast over North China is rare. In particular, the urban heat island effect, coupled with other factors, increases the difficulty and uncertainty of precipitation forecast accuracy in Beijing. Reproducing observed urban effects can help increase precipitation forecast accuracy in Beijing. Therefore, weather forecasting can provide better information to meet the public demand, especially in the urban area of Beijing [22,23].

In this study, five temperature and humidity profiles, retrieved by ground-based microwave radiometers, were assimilated into the rapid-refresh multi-scale analysis and prediction system-short term (RMAPS-ST). We evaluated the impact of the ground-based microwave radiometer data on the analyses and forecasts of a case of heavy rainfall in Beijing. Two assimilation experiments were carried out in this study. Combined with comparative analysis, we explored the impact of ground-based microwave radiometer data on the precipitation forecast in Beijing. In this study, we aimed to improve urban weather forecast in North China, while providing better information to help the public with their daily activities.

The outline of this article is as follows. Section 2 presents the data and methods used in the study, including the characteristics of ground-based microwave radiometer data and the experimental setting. In Section 3, the impact of assimilated ground-based microwave radiometer data on both the prediction of radar composite reflectivity and hourly rainfall evolution are compared for the control and MWRPS experiments. Section 4 discusses the diagnosis for this heavy rainfall event with and without the data assimilation of the ground-based microwave radiometers. Finally, Section 5 summarizes the conclusions.

2. Data and Methods

2.1. The Heavy Precipitation Case

In the present study, we took the heavy rainfall process in Beijing on 21 May 2020 as an example to investigate whether the assimilation of ground-based microwave radiometer data could improve the precipitation forecast. The evolution of the radar echo in this process is shown in Figure 1. The observation shows that there are two echo bands and that the echo that affected this precipitation process in Beijing moved from the northwest to the southeast at 0500 UTC (Coordinated Universal Time) on 21 May 2020; the belt-shaped echo emerged in Inner Mongolia, followed by convection. The strong echo belt moved rapidly to the southeast, reaching Beijing at 0700 UTC, and strengthening at 0800 UTC when it approached the urban area of Beijing. Within the urban area, the echo strengthened. The echo continued to move outside of Beijing at 1000 UTC.

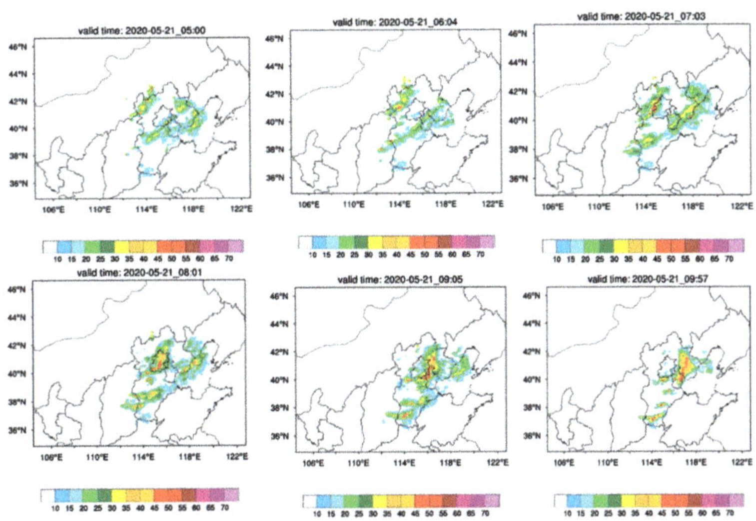

Figure 1. Radar composite reflectivity (CREF) evolution for 05 UTC–10 UTC on 21 May 2020.

For the echo band that affected this precipitation process in Beijing, a northeast and southwest belt-shaped rainfall emerged in Inner Mongolia at 0400 UTC on 21 May 2020. It is shown that the observed precipitation first occurred in the western mountainous area of Beijing at 0700 UTC with the echo moving from northwest to southeast, and then expanding eastward. By 0800 UTC, the precipitation system had reached the urban area of Beijing, and the rainfall intensity exceeded 20 mm/1 h (Figure 2). The heavy rainfall center moved eastward to Tianjin after 0900 UTC, when the echo moved outside of Beijing. The weather process was characterized by the strengthened radar echo in the urban area of Beijing, during which heavy precipitation occurred in the urban area of Beijing as it moved from the northwest to the southeast (Figures 1 and 2).

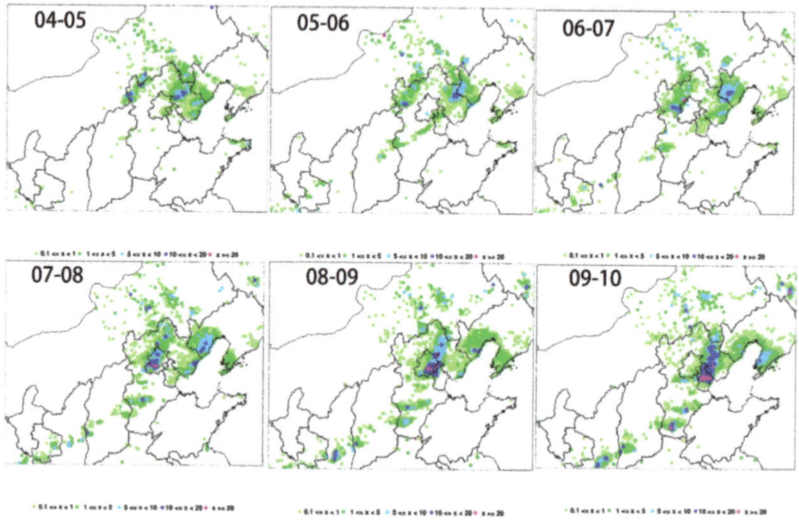

Figure 2. The evolution of hourly accumulated precipitation from AWS observation from 0500 UTC to 1000 UTC on 21 May 2020 in North China.

2.2. *Microwave Radiometer Observations*

In the metropolitan observation experiments, seven microwave radiometers were deployed in Beijing. Two microwave radiometers in the southern suburbs and in Shangdianzi Village were lost due to equipment problems, while the remaining five microwave radiometers deployed in Xiayunling Village, Yanqing District, Haidian District, Huairou District, and Pinggu District were available. The level-2 products from the five ground-based microwave radiometers in Beijing were obtained using inversion software from the microwave radiometer manufacturers, including the temperature and relative humidity profiles. The continuous observations of the temperature and humidity profiles used a high temporal resolution, at a high frequency rate up to two minutes.

Figure 3 shows the vertical distribution of retrieved temperature and humidity profiles at five stations, as well as their evolutions over time. It reveals that microwave radiometers overcome the spatial and temporal shortcomings of conventional observation in temporal resolution. Prior to the three-dimension variation assimilation, the temperature profile and relative humidity profile retrieved by microwave radiometers at heights of 0–10 km are processed. Since precipitation has a great impact on the temperature and relative humidity retrieved by the microwave radiometer, the observed data from the radiometer during precipitation should be prudently dealt with. In this study, we set the data at the corresponding time of precipitation as a missing value. In addition, since the temperature and humidity profiles retrieved by microwave radiometers have high vertical resolutions at heights of 0–10 km, the reference atmospheric pressure at each height layer from 0 to 10 km is calculated according to Zhang et al. (2006) [24].

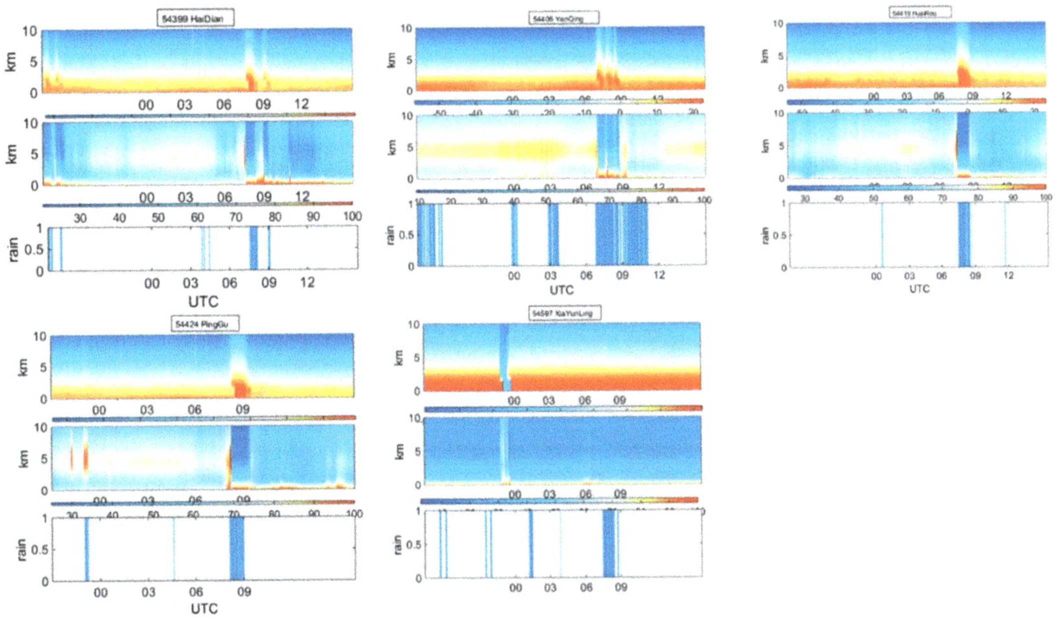

Figure 3. Time series of temperature and humidity profiles retrieved from five ground-based microwave radiometers and the prediction of rain at corresponding times on 21 May 2020. The blue bar indicates the time of precipitation in the MWRPS observations. Altitudes are given in kilometers above ground level.

2.3. Experiment Design

In this study, the experiment was carried out using the RMAPS-ST numerical forecast model, which is a short-term forecasting subsystem of a new generation of RMAPS developed by the Institute of Urban Meteorology, CMA, Beijing. It is based on the previous generation of the North China rapid-refresh cyclic assimilation and forecast system and has been in operation since May 2017 [17,25–28]. The RMAPS-ST features double nesting, a nine-kilometer-resolution outermost D01 area with 649 $\times$ 500 grid points covering the whole of China, and a three-kilometer-resolution inner D02 area with the innermost 550 $\times$ 424 grid points in the simulated area covering North China. The parameterized schemes of the main physical process of the experiment included a new Thompson cloud microphysics scheme, a Noah land surface scheme, a Yonsei University (YSU) boundary layer scheme [29], the global parameterization of the Rapid Radiative Transfer Model (RRTMG) scheme, and short wave and long wave radiation schemes [30,31]. In this paper, ECMWF medium-range forecast ($0.25° \times 0.25°$) was selected to provide the initial field and side boundary conditions for the model. The observed data, including the conventional data and radar data, underwent quality control before inputting into the assimilation system. The radar data assimilation was performed, including radial velocity and reflectivity. The Weather Research and Forecasting model system and the three-dimensional variational data assimilation system (3DVar) were used to assimilate observations due to low calculation cost, small resource occupation, and high efficiency. The solution of 3DVar can be interpreted as obtaining the minimization of the objective function. Based on the optimization theory, the optimal solution was obtained using an iterative descent algorithm. Specifically, the optimal state of the atmosphere was estimated by using both the background field and observed values, thus the statistical optimal analysis was obtained. When U and V are used as dynamic control variables, the correlation between variables is smaller than that when traditional flow function and potential function control variables are used, which satisfies the assumption of variational data assimilation. Such an algorithm is more conducive to the description of medium and small-scale systems [32]. Background error covariance was calculated by the National Meteorological Center (NMC) method [33].

The control experiment assimilated conventional observation data and Beijing–Tianjin–Hebei weather radar data. Data assimilated by the RMAPS-ST data assimilation system included observations from different types of conventional observations to improve the analysis. In Figure 4, the aircraft meteorological data relay, synoptic, sounding, oceanographic buoys, ship-based observations, and wind profile radar observations are shown. Radar data, including radial velocity and reflectivity, were mainly assimilated into Domain 2 of the RMAPS system. Based on the control experiment, the MWRPS experiment adds the data of five microwave radiometers in Beijing. The data distribution is shown in Figure 3. In the next section, we compare the forecast differences between the control experiment and the MWRPS experiment to verify the impact of MWRPS assimilation.

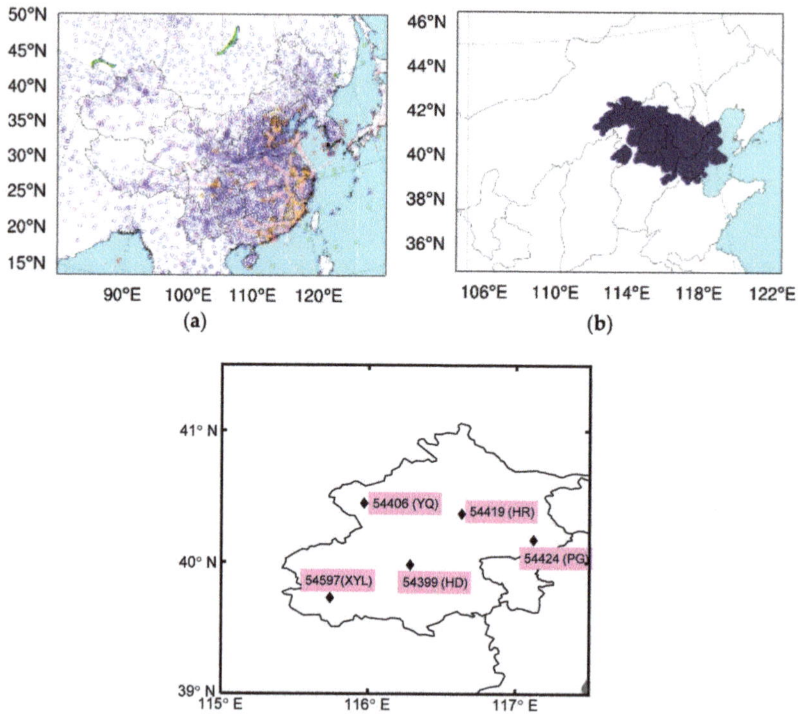

(c)

Figure 4. (**a**) The distribution of radiosonde launch sites are shown as purple solid circles; oceanographic buoys, wind profile radar observations, and ship-based observations are shown as red circles, orange circles, and green circles; synoptic and aircraft meteorological data relay are shown as dark blue circles and pink circles in Domain 1 of RMAPS-ST. (**b**) The radar locations are represented as dark blue solid circles in Domain 2 of RMAPS-ST. (**c**) Locations of MWRPS sites are represented as black diamonds in Beijing.

3. Results

3.1. Impact of Ground-Based Microwave Radiometer Data Assimilation on the Rainfall Prediction

In this section, we compare the forecast results after assimilation. For the enhancement of the belt-shaped echo in the urban area of Beijing, the radar reflectivity simulations from the two experiments are compared during the period from 0700 UTC to 1000 UTC on 21 May 2020. Figure 5 shows the composite radar reflectivity simulated by the MWRPS and control experiments. The top side shows the MWRPS simulations, whereas the bottom side shows the control simulations.

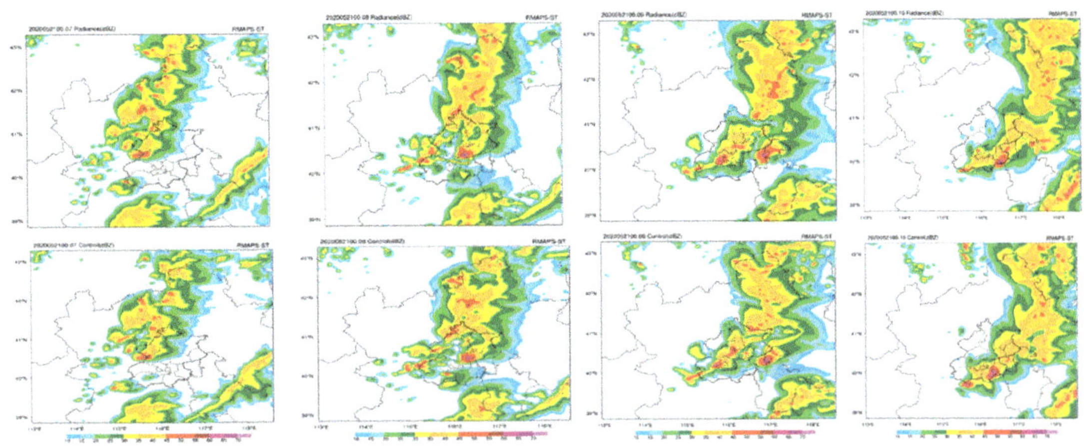

Figure 5. Composite reflectivity during the period from 0700 UTC to 1000 UTC on 21 May 2020 simulated by MWRPS test and control test from top to bottom (Beijing-Tianjin-Hebei region shown).

The zone of the radar echo usually corresponds to the distribution of a convection cell, and the intensity of radar reflectivity corresponds to the intensity of a convection cell. Compared with the control experiment, the MWRPS experiment had better prediction ability for simulating the observed belt-shaped convection enhancement at 0800 UTC: the MWRPS experiment was able to produce better simulations in both the location and intensity of the convection cells when the system impacted on the urban area of Beijing. The MWRPS experiment was also better able to simulate the dissipation process of the band echo during eastward movement. It reproduced the location of the observed band echo better when it moved outside of Beijing at 1000 UTC, while the simulated band echo in the control experiment was still in Beijing at 1000 UTC (Figure 5).

To examine the improvement in the precipitation forecast, the observed and forecasted 1 h accumulated precipitation in Domain 2 during 21 May 2020 are shown in Figure 6. The rain gauge observations from the ground stations are shown in the first column for evaluating the rainfall forecasts. The corresponding forecast results simulated by the MWRPS and control experiments are shown in the second column and the third column, respectively. Compared to the observations, there was a noticeable under-prediction for the precipitation in the control experiment in terms of both location and intensity at 0800 UTC. MWRPS forecasts improved the prediction for the corresponding precipitation intensity at the main center of the heavy rainfall, despite underestimating the range of precipitation. The precipitation intensities predicted by the MWRPS test are the same as the observations (>20 mm.h^{-1}). The spatial patterns of the 1 h accumulated precipitation simulated by the MWRPS experiment agreed with the observations better than those in the control experiment at 0900 UTC, especially for the location of the heavy rainfall's main center. The precipitation simulation in the control and MWRPS experiments are evaluated by TS (threat score) [34]. TS is the ratio of correct prediction times to the total number of events, representing the accuracy of the rainfall prediction. The numerical range is 0–1. The closer the TS value is to 1, the better the prediction is. The results showed that the assimilation of MWRPS provided an advantage in predicting rainfall, especially for larger precipitation events (>10 mm.h^{-1}). At 1000 UTC, the main differences in heavy rainfall between the two experiments is the location of the heavy rainfall, the MWRPS test shows a result consistent with the observations: the center of the precipitation moved outside of Beijing. In contrast, the center of heavy rainfall, as simulated by the control test, was still in Beijing (Figure 6).

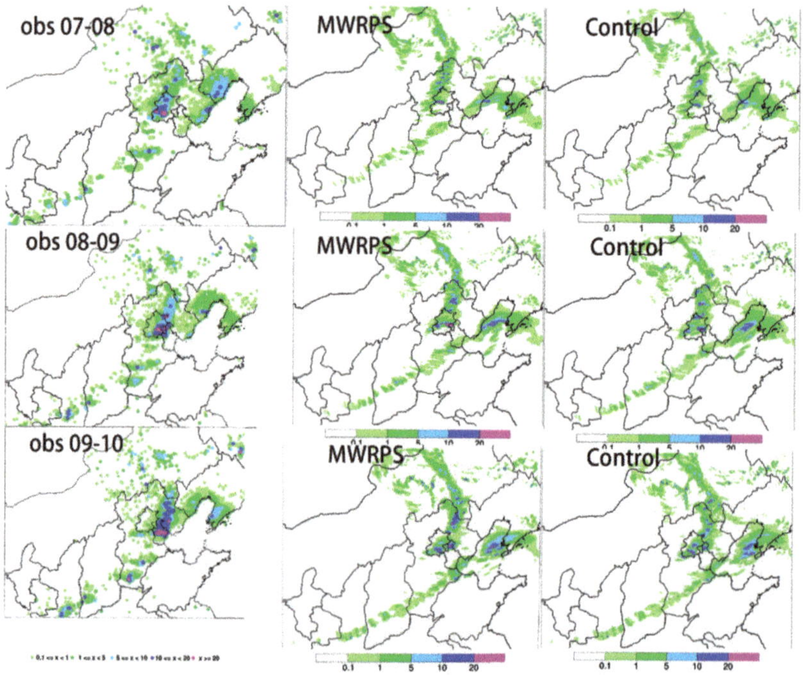

Figure 6. One-hour accumulated precipitation during the periods 0700 UTC–0800 UTC, 0800 UTC–0900 UTC, and 0900 UTC–1000 UTC on 21 May, 2020: observation (**left** column), D02 forecasting in MWRPS experiment (**middle** column), and D02 forecasting in Control experiment (**right** column).

Overall, the composite reflectivity and the 1 h accumulated precipitation from the observations and the control and MWRPS experiments were compared. We found that the assimilation of ground-based microwave radiometers increased the scope of heavy rainfall in MWRPS, which better agreed with the observations in spatial distribution patterns, as compared to the control experiment.

3.2. *Impact of Ground-Based Microwave Radiometer Data Assimilation on Meteorological Element Prediction before Urban Rainfall*

The urbanization process was significant in the Beijing–Tianjin–Hebei megalopolis: due to the joint effects of topography and urban thermal circulation [35], precipitation in Beijing is unique and complex [36–38]. Previous studies have suggested that the intensity of the urban heat island before the rainfall began could project the thermodynamic impact of the underlying urban surface on the rainfall process. The heat island intensity prior to the start of the rainfall determines the kinds of urban effects which will impact on the rainfall. Under the effect of a strong heat island which precedes the rainfall, the precipitation is concentrated in urban areas since the thermal effect of the urban land surface prevails, increasing the intensity of the convective system after it has moved to the urban area. Under the effect of weak heat island, precipitation bifurcation takes place, meaning that precipitation is mainly distributed upwind of the city and on both sides of the city since urban dynamics prevails [35,39,40].

Figure 7 shows the spatial distribution of the observed two-meter air temperature at 0700 UTC before the echo moved to the urban area of Beijing. The black rectangular box was selected to identify the heat island intensity and perform statistical analysis in accordance with Zhang et al. (2017) and Qi et al. (2021) [17,35]. This region includes urban and suburban areas in Beijing, which reported no precipitation at 0700 UTC. It can be seen that before the belt-shaped echo moved to the urban area of Beijing, there was an obvious difference between the urban temperature in the Fifth Ring Road and the temperature in its surrounding area. The heat island intensity in Beijing is strong [35].

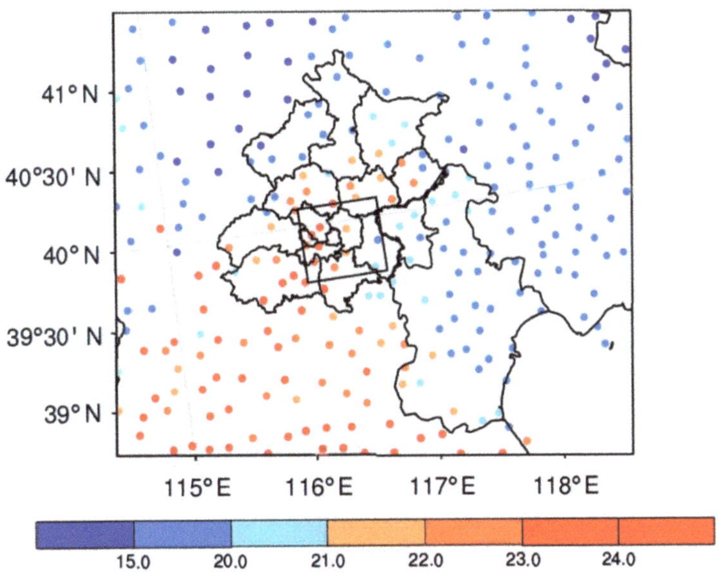

Figure 7. Spatial distribution of 2 m air temperature at 0700UTC on 21 May 2020. The black rectangular box indicates the study area where temperatures prior to the start of rainfall are statistically analyzed (116.12°–116.79° E, 39.65°–40.115° N).

As indicated by the spatial distribution of two-meter temperature bias at 0700 UTC simulated by the control and MWRPS experiments (Figure 8a,b), the temperature deviation predicted by the control test in and around Beijing's Fifth Ring Road (shown by the black rectangular frame) is larger than that predicted by the MWRPS test. The control test overestimates the observed heat island intensity in the urban area within the Fifth Ring Road. The assimilation of the ground-based microwave radiometers corrected the warm bias in this area, and the modified indexes of the two-meter air temperature are all negative, which agrees better with the observations (Figure 8c).

Compared with the control experiment, the MWRPS experiment improves the forecast of the two-meter temperature in Beijing and better reproduces the observed heat island phenomenon. Thus, the observed rainfall enhancement can be better simulated in the MWRPS experiment, mainly due to the effect of the urban surface in Beijing on rainfall since its thermo-dynamic prevails under urban heat island condition.

Additionally, the spatial characteristics of forecast of the two-meter temperature and two-meter specific humidity distribution from the two experiments were also evaluated against observations using the statistical metrics of mean bias and the root mean square error (RMSE). The closer the bias value is to 0 and the smaller the RMSE value is, the better the prediction is. Figure 9 shows the bias and RMSE of the average two-meter temperature and two-meter specific humidity over the black rectangular boxes, including urban and suburban areas in Beijing.

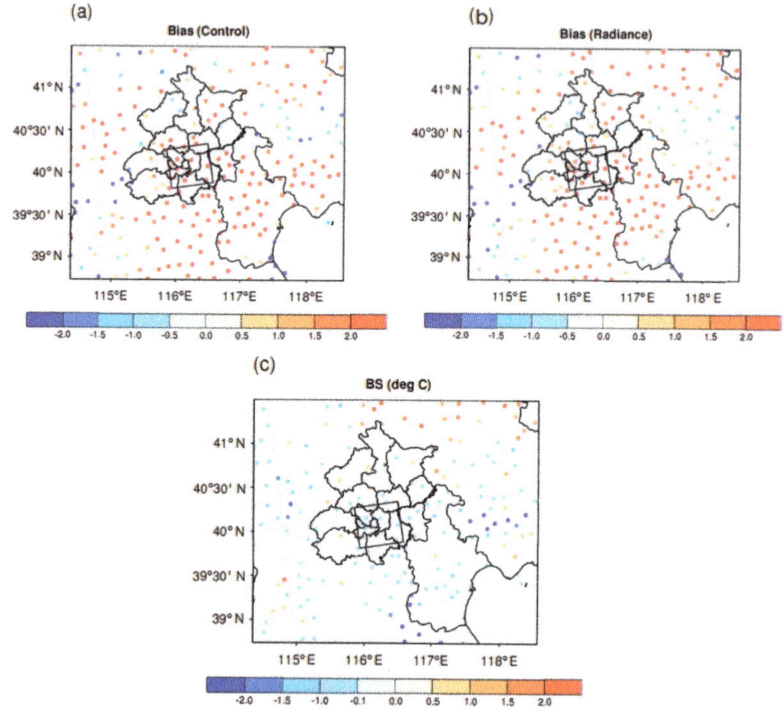

Figure 8. (**a**,**b**) The bias of 2 m temperature (unit: °C) from control experiment and the bias of 2 m temperature from MWRPS experiment; (**c**) improving index (unit: °C) with the assimilation of ground-based microwave radiometers at 0700 UTC on 21 May 2020.

Both experiments overestimate the observed two-meter temperature with a positive bias. Both the bias and RMSE in the MWRPS test are smaller than those in the control test. The two-meter temperature forecast indicates a better forecast performance for a longer period of time. The two-meter specific humidity is drier than the observation in both the control and MWRPS tests. The MWRPS test effectively improves for the first 8 h. To sum up, the assimilation of the ground-based microwave radiometers solved the warm bias for the two-meter temperature and dry bias for the two-meter specific humidity, thus laying a good foundation for the simulated urban precipitation enhancement process.

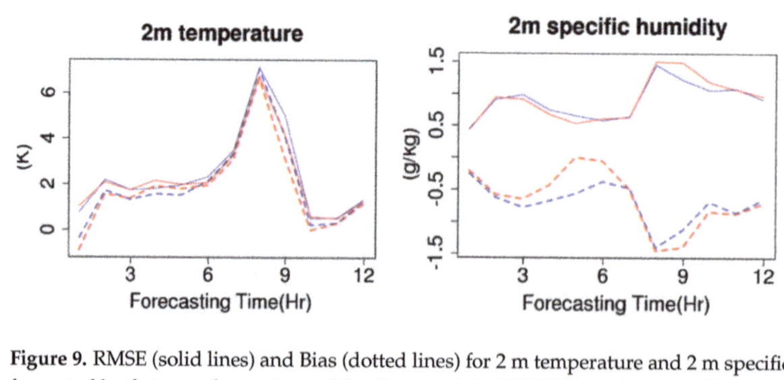

Figure 9. RMSE (solid lines) and Bias (dotted lines) for 2 m temperature and 2 m specific humidity forecasted by the control experiment (blue lines) and the MWRPS experiment (red lines) in relation to observations for the black rectangular box.

4. Discussion

Figure 10 shows the radial vertical profile of temperature and vertical velocity along 40°N in the boundary layer of the Fifth Ring Road and its surrounding areas in the forecast field from two experiments at 0800 UTC on 21 May 2020. It can be seen that the temperature in the lower atmosphere simulated by the control experiment is larger than that simulated by the MWRPS experiment. However, the observed heat island effect cannot be reproduced in the control experiment because of the small temperature difference between the urban and suburban areas. For the MWRPS experiment, a clear heat island was found. Compared to the control experiment, a clear updraft with a larger vertical velocity was also found, producing stronger updrafts in the urban area, which led to a rainfall forecast. This urban ascending motion is strengthened under the strong heat island effect, which promotes the emergence of updrafts. This suggests that the assimilation of ground-based microwave radiometer observations in Beijing, which reproduced the heat island and the intensified local updraft in the urban area in Beijing, is consequently able to improve predictions of rainfall enhancements in urban areas.

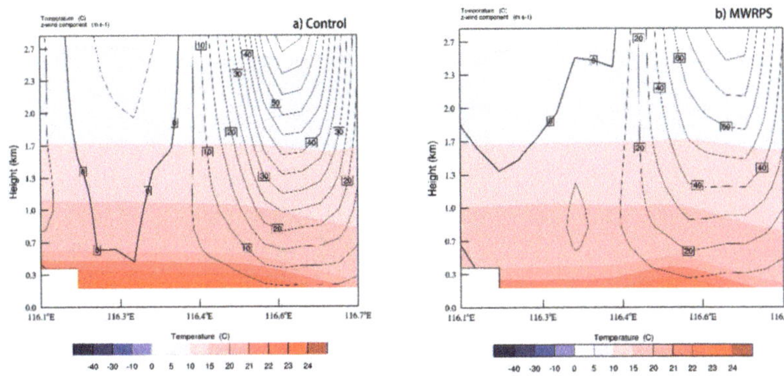

Figure 10. The cross sections of the vertical profiles along 40° N from control (**a**) and MWRPS (**b**) forecast field at 0800 UTC on 21 May 2020, in which the vertical velocity (unit: 10^{-1} m/s) is represented as the contour, and the temperature (unit: °C) is represented as shaded.

5. Conclusions

In view of the case of heavy rainfall in Beijing on 21 May 2020, the RMASPS-ST was used to explore whether the data assimilation of the ground-based microwave radiometers with high spatial and temporal resolution in Beijing could improve the weather forecast. Two experiments—to gather control and MWRPS —were conducted in this study. The simulation results for this case of heavy rainfall with and without the assimilation of MWRPS data were verified. The experimental results show that the assimilation of ground-based microwave radiometers in Beijing did improve the prediction of precipitation and echo and better predicted the rainfall enhancement process. The main conclusions are as follows:

(1) The RMAPS-ST model system can provide a good simulation of the selected rainfall case, by assimilating the MWRPS data in Beijing. It can clearly reproduce the observed urban heat island of the main urban area in Beijing prior to the start of this rainfall, thus reproducing the forecast of precipitation enhancement in the urban area. Compared with the control experiment, the simulated precipitation and radar reflectivity are closer in the MWRPS experiment to the observation.

(2) After the data from the ground-based microwave radiometers are assimilated, the observed weak heat island phenomenon is better reproduced. The simulated surface temperature distribution in Beijing is also closer to the observation prior to the start of the rainfall in the urban area. The model not only clearly improves the forecast

of precipitation distribution, but also makes the precipitation intensity prediction closer to the actual situation and accurately predicts the enhancement process of the belt-shaped echo and the precipitation in the urban area of Beijing.

(3) The heavy rainfall process in Beijing on 21 May 2020 shows that the assimilation of the ground-based microwave radiometer can improve the numerical forecast, contributing to improving the precipitation simulation in the urban area of Beijing, indicating a bright prospect for applications in numerical models. This rainfall event can also help us understand the impact of urban space on the rainfall system, considering urban heat island conditions.

These conclusions are based on heavy rainfall, which occurred in the urban area of Beijing. Further investigation into the impact of the ground-based microwave radiometer on weather forecasts will continue. More cases will be investigated to study the application of the assimilation of ground-based microwave radiometer data in the future. In this way, we will gain a more insightful understanding of the impact of assimilation of ground-based microwave radiometer data on forecasts.

Author Contributions: Conceptualization, B.L., Y.Q. and S.F.; methodology, Y.Q. and S.F.; software, Y.Q.; validation, Y.Q.; formal analysis, Y.Q. and J.M.; computing resources, Y.Q.; writing—original draft preparation, Y.Q.; writing—review and editing, S.F., Y.Q. and D.L.; supervision, S.F.; funding acquisition, S.F. and Y.Q. All authors have read and agreed to the published version of the manuscript.

Funding: This research was funded by the National Key Technologies Research and Development Program of China (2017YFC1501704), National Natural Science Foundation of China (42005124), and the Key Laboratory for Cloud Physics of China Meteorological Administration LCP/CMA (2020Z007).

Data Availability Statement: The ground-based microwave radiometer data are be available from Meteorological Observation Center of China Meteorological Administration.

Acknowledgments: The authors appreciate the help of the RMAPS-ST group, including code support and the sharing of valuable data. We acknowledge the Meteorological Observation Center of China Meteorological Administration for the ground-based microwave radiometer data.

Conflicts of Interest: The authors declare no conflict of interest.

References

1. Illingworth, A.J.; Cimini, D.; Haefele, A.; Haeffelin, M.; Hervo, M.; Kotthaus, S.; Löhnert, U.; Martinet, P.; Mattis, I.; O'Connor, E.J.; et al. How can existing ground-based profiling instruments improve European weather forecasts? *Bull. Am. Meteorol. Soc.* **2019**, *100*, 605–619. [CrossRef]
2. Boylan, P.; Wang, J.; Cohn, S.A.; Fetzer, E.; Maddy, E.S.; Wong, S. Validation of AIRS version 6 temperature profiles and surface-based inversions over Antarctica using Concordiasi dropsonde data. *J. Geophys. Res. Atmos.* **2015**, *120*, 992–1007. [CrossRef]
3. Sun, Q.; Vihma, T.; Jonassen, M.O.; Zhang, Z. Impact of assimilation of radiosonde and UAV observations from the Southern Ocean in the Polar WRF Model. *Adv. Atmos. Sci.* **2020**, *37*, 441–454. [CrossRef]
4. Rüfenacht, R.; Haefele, A.; Pospichal, B.; Cimini, D.; Bircher-Adrot, S.; Turp, M.; Sugier, J. EUMETNET opens to microwave radiometers for operational thermodynamical profiling in Europe. *Bull. Atmos. Sci. Technol.* **2021**, *2*, 1–5. [CrossRef]
5. Cimini, D.; Hewison, T.; Martin, L.; Güldner, J.; Gaffard, C.; Marzano, F.S. Temperature and humidity profile retrievals from ground-based microwave radiometers during TUC. *Meteorol. Z.* **2006**, *15*, 45–56. [CrossRef]
6. Martinet, P.; Cimini, D.; De Angelis, F.; Canut, G.; Unger, V.; Guillot, R.; Tzanos, D.; Paci, A. Combining ground-based mi-crowave radiometer and the AROME convective scale model through 1DVAR retrievals in complex terrain: An Alpine valley case study. *Atmos. Meas. Tech.* **2017**, *10*, 3385–3402. [CrossRef]
7. Crewell, S.; Lohnert, U. Accuracy of boundary layer temperature profiles retrieved with multifrequency multiangle microwave radiometry, IEEE T. *Geosci. Remote* **2007**, *45*, 2195–2201. [CrossRef]
8. Löhnert, U.; Maier, O. Operational profiling of temperature using ground-based microwave radiometry at Payerne: Prospects and challenges. *Atmos. Meas. Tech.* **2012**, *5*, 1121–1134. [CrossRef]
9. Qiu, J.; Chen, H.; Wang, P.; Liu, Y.; Xia, X. Recent progress in atmospheric observation research in China. *Adv. Atmos. Sci.* **2007**, *24*, 940–953. [CrossRef]

10. Cimini, D.; Campos, E.; Ware, R.; Albers, S.; Giuliani, G.; Oreamuno, J.; Joe, P.; Koch, S.E.; Cober, S.; Westwate, E. Thermo-dynamic atmospheric profiling during the 2010 Winter Olympics using ground-based microwave radiometry. *IEEE Trans. Geosci. Remote Sens.* **2011**, *49*, 4959–4969. [CrossRef]
11. Cimini, D.; Nelson, M.; Güldner, J.; Ware, R. Forecast indices from a ground-based microwave radiometer for operational meteorology. *Atmos. Meas. Tech.* **2015**, *8*, 315–333. [CrossRef]
12. Caumont, O.; Cimini, D.; Löhnert, U.; Alados-Arboledas, L.; Bleisch, R.; Buffa, F.; Ferrario, M.E.; Haefele, A.; Huet, T.; Madonna, F.; et al. Assimilation of humidity and temperature observations retrieved from ground-based microwave radiometers into a convective-scale NWP model. *Q. J. R. Meteorol. Soc.* **2016**, *142*, 2692–2704. [CrossRef]
13. Yan, X.; Ducrocq, V.; Poli, P.; Hakam, M.; Jaubert, G.; Walpersdorf, A. Impact of GPS zenith delay assimilation on convective-scale prediction of Mediterranean heavy rainfall. *J. Geophys. Res. Space Phys.* **2009**, *114*, D03104. [CrossRef]
14. Sun, J. Convective-scale assimilation of radar data: Progress and challenges. *Q. J. R. Meteorol. Soc.* **2005**, *131*, 3439–3463. [CrossRef]
15. Stephan, K.; Klink, S.; Schraff, C. Assimilation of radar-derived rain rates into the convective-scale model COSMO-DE at DWD. *Q. J. R. Meteorol. Soc.* **2008**, *134*, 1315–1326. [CrossRef]
16. Wenying, H.; Chen, H.; Li, J. Influence of assimilating ground-based microwave radiometer data into the WRF model on pre-cipitation. *Atmos. Ocean. Sci. Lett.* **2020**, *13*, 107–112.
17. Qi, Y.; Fan, S.; Mao, J.; Li, B.; Guo, C.; Zhang, S. Impact of assimilating ground-based microwave radiometer data on the precipitation bifurcation forecast: A case study in Beijing. *Atmosphere* **2021**, *12*, 551. [CrossRef]
18. Grell, G.A.; Dudhia, J.; Stauffer, D.R. *A Description of the Fifth-Generation Penn State/NCAR Mesoscale Model (MM5)*; University Corporation for Atmospheric Research: Boulder, CO, USA, 1994. [CrossRef]
19. Vandenberghe, F.; Ware, R. Four-dimensional variational assimilation of ground-based microwave observations during a winter fog event. In Proceedings of the International Workshop on GPS Meteorology, International Symposium on Atmospheric Sensing with GPS, Tsukuba, Japan, 14–17 January 2003.
20. Otkin, J.A.; Hartung, D.C.; Turner, D.; Petersen, R.A.; Feltz, W.F.; Janzon, E. Assimilation of surface-based boundary layer profiler observations during a cool-season weather event using an observing system simulation experiment. Part I: Analysis impact. *Mon. Weather Rev.* **2011**, *139*, 2309–2326. [CrossRef]
21. Hartung, D.C.; Otkin, J.; Petersen, R.A.; Turner, D.; Feltz, W.F. Assimilation of surface-based boundary layer profiler observations during a cool-season weather event using an observing system simulation experiment. Part II: Forecast assessment. *Mon. Weather Rev.* **2011**, *139*, 2327–2346. [CrossRef]
22. Miao, S.; Chen, F.; Li, Q.; Fan, S. Impacts of urban processes and urbanization on summer precipitation: A case study of heavy rainfall in Beijing on 1 August 2006. *J. Appl. Meteorol. Clim.* **2011**, *50*, 806–825. [CrossRef]
23. Chen, F.; Dudhia, J. Coupling an advanced land-surface/ hydrology model with the Penn State/NCAR MM5 modeling system. Part I: Model description and implementation. *Mon. Weather Rev.* **2001**, *129*, 569–585. [CrossRef]
24. Zhang, L.; Ni, Y.Q. Four-dimensional variational data assimilation of radar radial velocity observations. *Chin. J. Atmos. Sci.* **2006**, *30*, 433–440.
25. Xie, Y.; Shi, J.; Fan, S.; Chen, M.; Dou, Y.; Ji, D. Impact of radiance data assimilation on the prediction of heavy rainfall in RMAPS: A case study. *Remote Sens.* **2018**, *10*, 1380. [CrossRef]
26. Xie, Y.; Chen, M.; Shi, J.; Fan, S.; He, J.; Dou, Y. Impacts of assimilating ATMS radiances on heavy rainfall forecast in RMAPS-ST. *Remote Sens.* **2020**, *12*, 1147. [CrossRef]
27. Fan, S.; Wang, H.L.; Chen, M.; Gao, H. Study of the data assimilation of radar reflectivity with the WRF 3DVar. *Acta Meteor. Sin.* **2013**, *71*, 527–537.
28. Qi, Y.J.; Chen, M.; Zhong, J.Q.; Fan, S.Y.; Liu, R.T.; Guo, C.W. Effect evaluation of short-term forecast of surface meteorological elements by using RMAPS-ST coupled urban canopy model. *J. Arid. Meteorol.* **2020**, *38*, 859–868.
29. Hong, S.-Y.; Noh, Y.; Dudhia, J. A new vertical diffusion package with an explicit treatment of entrainment processes. *Mon. Weather Rev.* **2006**, *134*, 2318–2341. [CrossRef]
30. Pincus, R.; Mlawer, E.J.; Oreopoulos, L.; Ackerman, A.S.; Baek, S.; Brath, M.; Buelhler, S.A.; Cady-Pereira, K.; Cole, J.; Dufresne, J.-L.; et al. Radiative flux and forcing parameterization error in aerosol-free clear skies. *Geophys. Res. Lett.* **2015**, *42*, 5485–5492. [CrossRef]
31. Mlawer, E.J.; Taubman, S.J.; Brown, P.D.; Iacono, M.J.; Clough, S.A. Radiative transfer for inhomogeneous atmospheres: RRTM, a validated correlated-k model for the longwave. *J. Geophys. Res. Atmos.* **1997**, *102*, 16663–16682. [CrossRef]
32. Sun, J.; Wang, H.; Tong, W.; Zhang, Y.; Lin, C.-Y.; Xu, D. Comparison of the impacts of momentum control variables on high-resolution variational data assimilation and precipitation forecasting. *Mon. Weather Rev.* **2015**, *144*, 149–169. [CrossRef]
33. Parrish, D.F.; Derber, J.C. The National Meteorological Center's Spectral Statistical-Interpolation analysis system. *Mon. Weather Rev.* **1992**, *120*, 1747–1763. [CrossRef]
34. Schaefer, J.T. The critical success index as an indicator of warning skill. *Weather Forecast.* **1990**, *5*, 570–575. [CrossRef]
35. Zhang, Y.; Miao, S.; Dai, Y.; Bornstein, R. Numerical simulation of urban land surface effects on summer convective rainfall under different UHI intensity in Beijing. *J. Geophys. Res. Atmos.* **2017**, *122*, 7851–7868. [CrossRef]
36. Wang, Y.; Zhao, X.; Zuo, L.; Zhang, Z.; Wang, X.; Yi, L.; Liu, F.; Xu, J. Spatial differentiation of land use and landscape pattern changes in the Beijing–Tianjin–Hebei area. *Sustainability* **2020**, *12*, 3040. [CrossRef]

37. Zhang, S.; Huang, G.; Qi, Y.; Jia, G. Impact of urbanization on summer rainfall in Beijing–Tianjin–Hebei metropolis under different climate backgrounds. *Theor. Appl. Clim.* **2017**, *133*, 1093–1106. [CrossRef]
38. Lin, D.W.; Bueh, C.L.; Xie, Z.W. A study on the coupling relationships among Pacific Sea surface temperature and summer rainfalls over North China and India. *Chin. J. Atmos. Sci.* **2018**, *42*, 1175–1190.
39. Bornstein, R.D. Establishment of meso-met modeling case studies to evaluate the relative roles of urban dynamics and aerosols on summer thunderstorms: A proposal. In Proceedings of the 18th Conference on Planned and Inadvertent Weather Modifi-cation, and Third Symposium on Aerosol–Cloud–Climate Interactions, San Jose, CA, USA, 27 January 2011.
40. Dou, J.; Wang, Y.; Bornstein, R.; Miao, S. Observed spatial characteristics of Beijing urban climate impacts on summer thunderstorms. *J. Appl. Meteorol. Clim.* **2015**, *54*, 94–105. [CrossRef]

atmosphere

MDPI

Article

Combining vLAPS and Nudging Data Assimilation

Brian P. Reen [1,*], Huaqing Cai [2], Robert E. Dumais, Jr. [2], Yuanfu Xie [3,†], Steve Albers [3,4,‡] and John W. Raby [2]

1 U.S. Army DEVCOM Army Research Laboratory, Adelphi, MD 20783, USA
2 U.S. Army DEVCOM Army Research Laboratory, White Sands Missile Range, NM 88002, USA; huaqing.cai.civ@army.mil (H.C.); robert.e.dumais.civ@army.mil (R.E.D.J.); john.w.raby2.civ@army.mil (J.W.R.)
3 Earth Systems Research Laboratory, National Oceanic and Atmospheric Administration, Boulder, CO 80305, USA; xieyf@gbamwf.com (Y.X.); Steven.Albers@colostate.edu (S.A.)
4 Cooperative Institute for Research in the Atmosphere, Boulder, CO 80305, USA
* Correspondence: brian.p.reen.civ@army.mil
† Current Affiliation: Guangdong-Hong Kong-Macau Greater Bay Area Weather Research Center for Monitoring Warning and Forecasting (Shenzhen Institute of Meteorological Innovation), Futian District, Shenzhen 518040, China.
‡ Current Affiliation: Spire Global, Boulder, CO 80301, USA.

Abstract: The combination of techniques that incorporate observational data may improve numerical weather prediction forecasts; thus, in this study, the methodology and potential value of one such combination were investigated. A series of experiments on a single case day was used to explore a 3DVAR-based technique (the variational version of the Local Analysis and Prediction System; vLAPS) in combination with Newtonian relaxation (observation and analysis nudging) for simulating moist convection in the Advanced Research version of the Weather Research and Forecasting model. Experiments were carried out with various combinations of vLAPS and nudging for a series of forecast start times. A limited subjective analysis of reflectivity suggested all experiments generally performed similarly in reproducing the overall convective structures. Objective verification indicated that applying vLAPS analyses without nudging performs best during the 0–2 h forecast in terms of placement of moist convection but worst in the 3–5 h forecast and quickly develops the most substantial overforecast bias. The analyses used for analysis nudging were at much finer temporal and spatial scales than usually used in pre-forecast analysis nudging, and the results suggest that further research is needed on how to best apply analysis nudging of analyses at these scales.

Keywords: variational assimilation; nudging; Newtonian relaxation; 3DVAR

Citation: Reen, B.P.; Cai, H.; Dumais, R.E., Jr.; Xie, Y.; Albers, S.; Raby, J.W. Combining vLAPS and Nudging Data Assimilation. *Atmosphere* **2022**, *13*, 127. https://doi.org/10.3390/atmos13010127

Academic Editor: Stephan Havemann

Received: 16 November 2021
Accepted: 7 January 2022
Published: 13 January 2022

1. Introduction

A variety of techniques to incorporate observations have been used in numerical weather prediction models to allow observations to improve the model forecasts. One example is 3D variational analysis (3DVAR; e.g., [1,2]), and another is Newtonian relaxation (nudging; e.g., [3,4]).

The 3DVAR technique has the potential to determine an optimal analysis by combining observations with a background field. However, since an optimal analysis requires that the background error covariance and the observation error covariance be perfectly known, in practice, the analysis will not be optimal. Background error covariance may be estimated by using model forecasts from a series of past times to find a climatological value, from multiple model forecasts of the current time (i.e., an ensemble) to find case-dependent values, or from a combination of these two methods (e.g., [5]). However, these techniques are difficult to apply when creating on-demand forecasts for an area where equivalently configured forecasts (e.g., horizontal resolution) have not been run in the past and there are insufficient computational resources to carry out an ensemble matching the horizontal grid spacing of the on-demand forecasts (although coarser-resolution global ensembles can be used to estimate the case-dependent background error covariance; [5]).

A 3DVAR analysis is commonly applied at a single time at the beginning of a model simulation and can lack physical consistency. Applying the analysis at a single time could result in noise as the model adjusts to a solution consistent with its equations. Additionally, model fields not part of the 3DVAR analyses may not be consistent with the fields in the 3DVAR analyses and may pull the model solution away from the 3DVAR analyses. These potential issues might be mitigated by applying the increments indicated by a 3DVAR analysis over multiple model time steps using incremental analysis updating (e.g., [6,7]) or by nudging towards the 3DVAR analysis. When applying a single analysis, observations made at times other than the analysis time cannot easily be applied at their valid time. While there are methods to mitigate this issue (e.g., first guess at appropriate time, [8]), it is a challenge to know the actual state of the model at the valid time of the observation and to apply the observation at its valid time.

Newtonian relaxation [9,10], also known as nudging, adds nonphysical terms to the tendency terms of the model over a period of time to gradually nudge the model toward analyses or observations. Since the nonphysical terms added by nudging should be smaller than the physical terms in the governing equations, nudging can adjust the model's state variables while maintaining an improved physical consistency compared to inserting modified fields directly at a single time. Nudging allows observations to be applied over a time period centered on their individual valid times, and assimilating analyses valid at multiple times allows observations to be applied to the analysis closest to the valid time of the observation. Fields that are nudged towards must be prognostic variables of the numerical weather prediction model, and thus observations of non-prognostic fields must be converted to prognostic fields in order to be used with nudging. Additionally, nudging is not usually applied in a way that fully accounts for case-specific background error covariances.

Several techniques have combined nudging with another technique. Shaw et al. [11] initialized WRF with a 3DVAR analysis and then used observation nudging and found that the combination appeared to perform better than either solution individually. Liu et al. [12] nudged towards 3DVAR analyses of radar-derived fields while also applying observation nudging. Lei et al. [13–15] combined nudging and the ensemble Kalman filter (EnKF) and applied it first in simplified models and then in the Advanced Research version of the Weather Research and Forecasting model (WRF-ARW; [16]). Information from the ensemble was used to determine how the influence of the observations was spread and how observations of one variable affect another variable. They found the hybrid technique performed better than EnKF for wind direction and better than observation nudging in temperature and relative humidity. Lei et al. [15] also found that noise levels in the hybrid technique were much lower than the EnKF simulation. Liu et al. [17] reported on a hybrid nudging–ensemble system that uses an ensemble to determine the strength at which observation nudging is applied. Lei and Hacker [18] used a Lorenz model to test observation nudging, EnKF, and the nudging–EnKF hybrid developed by Lei et al. [13–15]. They found that in the Lorenz model, the nudging–EnKF hybrid they tested could not simultaneously outperform both nudging and EnKF.

This study demonstrates the combination of a scheme involving 3DVAR (the variational version of the Local Analysis and Prediction System; vLAPS; [19]) and nudging for a single case day with strong convection. This combination takes advantage of vLAPS analysis with the best fit of the observations and nudging with improvement in the physical consistency among the analysis states. Verification of the WRF reflectivity forecasts suggests that further research is needed to improve analysis nudging of high-temporospatial resolution analyses. The experiments described here were previously described in a report [20]; however, that report does not include the objective verification included here. Nonetheless, since this study reports on a subset of the experiments in Reen et al. [20], there are similarities between the text of that document and this study.

This study is unique in its use of analysis nudging to incorporate vLAPS analyses, and there appears to be little past work on assimilating any 3DVAR analyses using analysis nudging. Liu et al. [12] assimilated 3DVAR analyses using analysis nudging but used solely radar observations as the basis for the 3DVAR analyses compared to the much broader set of observations included in the 3DVAR analyses used in our study. While Shaw et al. [11] used 3DVAR and observation nudging, they did not use analysis nudging to apply the 3DVAR analyses, as is done in this study. Analysis nudging is usually used for model forecasts with a horizontal grid spacing much coarser than the 1 km spacing used here (e.g., [21,22]) since analyses are not usually available to nudge towards at such a high resolution. Huo et al. [23] applied what appears to be analysis nudging to assimilate radar-derived fields at a fairly high resolution (3 km grid spacing), but one that is somewhat coarser than the 1 km grid spacing in this study. As the current study is limited to a single case day and evaluation of the radar reflectivity field, it is intended as a limited demonstration of the combination of a 3DVAR-based technique (vLAPS) and nudging. One motivating factor in this investigation was to explore assimilation techniques that may be able to provide value for forward-deployed on-demand nowcasting using limited computational capability.

2. Materials and Methods

2.1. WRF-ARW

The Advanced Research version of the Weather Research and Forecasting model (WRF-ARW) V3.6.1 [16] was used with 56 vertical layers over a 1 km-spaced 801 × 801 horizontal grid centered over Oklahoma (Figure 1). The Mellor–Yamada–Janjić scheme (MYJ; [24]) was used to parameterize the atmospheric boundary layer (ABL) with the background turbulent kinetic energy and atmospheric boundary layer depth calculation altered as in Lee et al. [25] and Reen et al. [26]. The Thompson microphysics parameterization [27], the Rapid Radiative Transfer Model longwave scheme [28], the Dudhia shortwave scheme [29], and the Noah land surface model [30] were used.

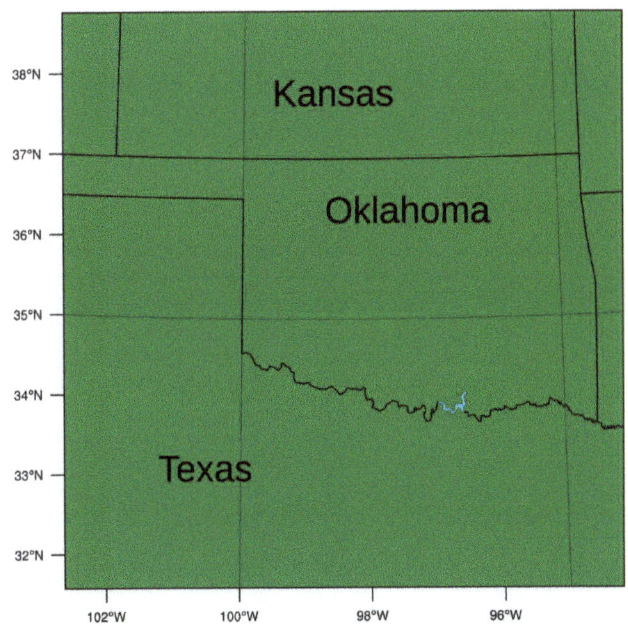

Figure 1. Areal extent of the Advanced Research version of the Weather Research and Forecasting model (WRF-ARW) 1 km domain. Adapted from Reen et al. [20].

2.2. vLAPS

LAPS [31] uses a modified Barnes analysis [32] with some 1D variational components [33], while vLAPS changes to a multiscale 3DVAR scheme for specific fields (temperature, pressure, winds, and humidity) [19]. The vLAPS 3DVAR scheme is based on the Space and Time Multiscale Analysis System [34] and is multiscale spatially and temporally. The analysis was performed here four times with the coarsest analysis at 16 km horizontally and 50 hPa vertically, and subsequent analyses divided the resolution in half through to the final analysis, which was 2 km horizontally and 25 hPa vertically. The first two analyses used a 30 min time window, whereas the final two analyses used a 15 min time window. The multiscale technique allows observations to spread broadly in data-sparse areas while retaining fine-scale features in more data-rich areas and makes the technique less dependent on the accuracy of the background error covariance. For wind, the control variables were the u- and v-wind components [35] since Xie et al. [34,36] demonstrated that the alternative of the stream function and velocity potential introduce numerical errors and noise. The cloud analysis [33] combined Geostationary Operational Environmental Satellite (GOES) infrared and visible data, radar data, surface observations, and model first-guess fields to construct a cloud fraction. The cloud fraction was used to determine hydrometeor fields and vertical velocity, with the latter affecting the 3D wind. Additionally, the cloud analysis was used as a constraint in determining the vLAPS relative humidity field.

Analyses were created every 15 min using vLAPS. The background field for these analyses was taken from the 3 km-horizontal grid spacing High-Resolution Rapid Refresh (HRRR; [37]); specifically, we used the output from the HRRR 15 UTC cycle on 20 May 2013 (i.e., the integration of HRRR that starts at 15 UTC and provides forecasts at a 15 min temporal spacing). Only the 15 UTC cycle of HRRR was used here to approximate conditions in the application driving this research, namely, use in regions where cycles of the numerical weather prediction model used for initial conditions are available much less frequently than hourly. Each vLAPS analysis used the forecast hour of the 15 UTC HRRR cycle corresponding to the vLAPS analysis time (e.g., the 18 UTC vLAPS analysis used the 3 h forecast of the 15 UTC HRRR cycle). Sources of observations used to create the analyses included Meteorological Assimilation Data Ingest System (MADIS; https://madis.noaa.gov) surface, mesonet, profiler, radiosonde, and Aircraft Communications Addressing and Reporting System (ACARS) observations; pilot observations; WSR-88D radars; and GOES. In addition to the quality control applied by the MADIS data provider, MADIS data were also quality controlled by comparing them to the HRRR fields used as the background.

2.3. Nudging

Nudging [9,10,38] adds a term to the model's tendency equations based on the difference between the current model value and the value from an analysis (analysis nudging; e.g., [39]) or an observation (observation nudging; [40]). This difference is known as the innovation and is multiplied by weighting factors to create the nonphysical addition to the tendency equation. In this study, some experiments applied analysis nudging using vLAPS analysis, and some experiments applied observation nudging of MADIS observations. For brevity, the details of nudging are simplified in the following discussion (see Reen et al. [20] for additional details).

The vLAPS analyses were used by analysis nudging to modify the potential temperature, water vapor mixing ratio, and u- and v-wind components. Within the ABL (atmospheric boundary layer), only the surface analysis was used to calculate analysis nudging terms; at the first level above the ABL, both the surface and above-surface analyses contributed; and above this, only the above-surface analyses contributed. The innovation calculated at the surface (by comparing the vLAPS surface analysis to the WRF surface values) was applied throughout the ABL for the potential temperature and wind. For water vapor, the innovation for each level in the ABL was calculated by comparing the surface analysis to the model value at that level. Innovations above the ABL were calculated by comparing the above-surface analysis to the model value at that level.

WRF-ARW analysis nudging linearly interpolates between analyses, which, in this case, were vLAPS analyses available every 15 min. The assimilation period was 3 h, and thus vLAPS analyses from 13 different times were applied. The analysis valid at the end of the assimilation period was applied with a linearly decreasing weight over the 15 min following the end of the assimilation period. The released version of WRF-ARW V3.6.1 does not properly execute this rampdown period, and thus we modified the code to fix this (the fix was provided to the WRF-ARW maintainers and is included in WRF-ARW starting with V3.8). The strength at which analysis nudging was applied was 3×10^{-4} s^{-1}. Analysis (and observation) nudging is designed to gradually modify the model solution while allowing the physical tendency terms to dominate the equations so that the physical consistency of the atmosphere is maintained; this should mitigate any potential issues with overfitting.

Observation nudging used MADIS surface, mesonet, profiler, radiosonde, and ACARS observations (ACARS and profiler observations were inadvertently omitted for part of the simulation). In addition to using the quality control carried out by the data provider, to account for the data quality issues that may be more prevalent in mesonet datasets (e.g., siting issues), mesonet observations were filtered using use/reject lists designed for the Real-Time Mesoscale Analysis [41]. The Obsgrid program [42] was also used to apply quality control. It compared the observations against the 15 UTC cycle of the 3 km HRRR model on 20 May 2013 (i.e., the HRRR forecast that creates forecasts with valid times every 15 min starting at 15 UTC) and against nearby observations. A non-standard version of the WRF Preprocessing System (WPS) program Ungrib was used to vertically interpolate the HRRR field to additional levels to facilitate quality control of single-level above-surface observations (ACARS), and to allow more vertical structure to be retained in multilevel observations (the capability is available in standard Ungrib starting with WPS V3.9).

As with analysis nudging, one must specify the strength of the observation nudging (6×10^{-4} s^{-1} was used here), but additional specification is needed in regard to the vertical, horizontal, and temporal weighting since, unlike the analysis, one does not have one value per model grid cell to nudge towards. Vertical weighting depends on observation type, with the innovation from surface observations being spread throughout the ABL during convective conditions, innovations from single-level above-surface observations being applied in a 100 hPa range, and innovations from multilevel observations being interpolated to model layers within the vertical range of the observation. A 30 km radius of influence was used for surface observations, but terrain differences limited the spreading further. For above-surface observations, the radius of influence increased from 60 km near the surface to 120 km at 500 hPa. Surface observations were used for a 2 h time period centered on the observation valid time, and above-surface observations were used for 3 h time periods. At the end of the assimilation period, observation nudging was ramped down to zero in 1 h (no observations valid during the rampdown were assimilated). The modification of water vapor observation nudging to prevent excessive drying described in [43] was applied. When observation nudging spreads the innovation from a location where the model is too moist to a nearby location that is much drier, this can result in unrealistic drying of the location the innovation is spread to. To mitigate this issue, the modification limits the magnitude of negative water vapor innovations being applied in locations drier than the location where the innovation is calculated.

2.4. Case Description

The day investigated here (20 May 2013) had strong convection in the southern Great Plains region. The base reflectivity (composited over relevant individual radars) of the observed radar (Figure 2) indicated echoes <35 dBZ at 1200, 1500, and 1800 UTC, with most echoes generally along an approximately southwest–northeast-oriented line in the northwest of the domain at 1200 and 1500 UTC, but moving to the north central portion of the domain by 1800 UTC. However, after 1800 UTC, a strong line of southwest-to-northeast-oriented convection developed ahead of the previous echoes and moved eastward. Multiple tornadoes were observed this day, with one that was determined to be an EF5 on the ground approximately from 1956 to 2035 UTC starting in Newcastle, Oklahoma, and traveling through Moore, Oklahoma [44–46]. Severe hail was also observed in multiple locations.

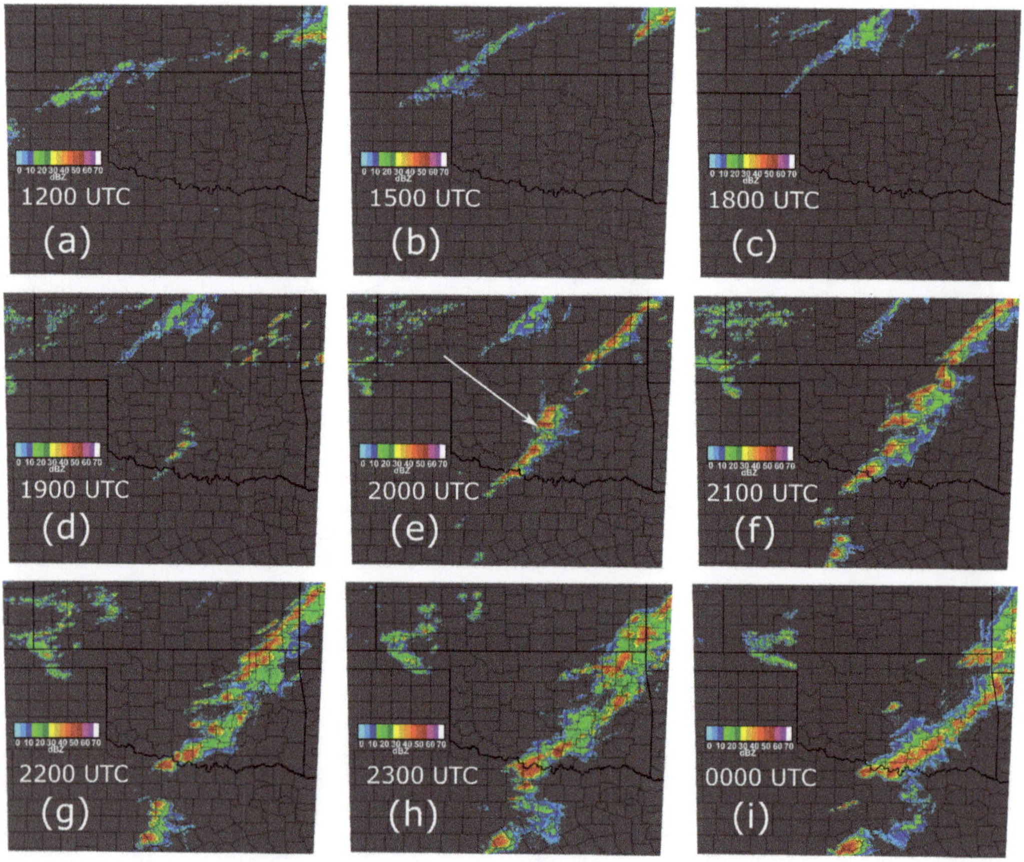

Figure 2. Observed composite base reflectivity for 20 May 2013 at (**a**) 1200 UTC, (**b**) 1500 UTC, (**c**) 1800 UTC, (**d**) 1900 UTC, (**e**) 2000 UTC, (**f**) 2100 UTC, (**g**) 2200 UTC, and (**h**) 2300 UTC, and for 21 May 2013 at (**i**) 0000 UTC. The white arrow in panel e shows the approximate location of the Moore tornado at that time. Imagery from the Iowa Environmental Mesonet (http://mesonet.agron.iastate.edu/GIS/apps/rview/warnings.phtml; accessed on 10 November 2021).

Various modeling studies have been performed on this case. Hanley et al. [47] used the United Kingdom Met Office's Unified Model with 4.4/2.2/0.5/0.2/0.1 km horizontal grid spacings without data assimilation and found convection initiating in the Oklahoma City area at approximately the right time, but the tornado-like vortices in their finest grids occurred approximately 2.5 h later than the Moore tornado. Zhang et al. [48,49] used WRF-ARW to look at predictability in this case. Zhang et al. [48], using 27/9/3/1 km horizontal grid spacings, found that temporal shifting of initial conditions generally temporally shifts convection but, in some cases, does not because lateral boundary conditions control convective initiation. Zhang et al. [49] used a 1 km ensemble with perturbations smaller than the current observational network could resolve and found that the ensemble members all produce a line of storms, but that details of individual storms differed. Snook et al. [50] used 500 m-horizontal grid spacing Advanced Regional Prediction System (ARPS) simulations with assimilation of radar and surface observations every 5 min using EnKF and found skill in predicting hail.

2.5. Experiment Design

The experiments in Table 1 and Figure 3 were used to investigate the potential value of combining vLAPS (2.2) and nudging (2.3) in WRF-ARW (2.1) for the case day of 20 May 2013 (2.4). The names of the experiments begin with the source of the initial conditions (HRRR or vLAPS), followed by the length of the pre-forecast in hours (0 or 3), and finally an O is added if observation nudging was applied, and an A is added if analysis nudging was applied. The pre-forecast refers to the period at the beginning of the model integration during which the model is assumed to be coming into dynamic balance, and during which observations may be assimilated. Hydrometeors are included in the initial conditions provided by both HRRR and vLAPS. Boundary conditions are based on the output from the 15 UTC HRRR cycle. Assimilation of observations or analyses valid during the pre-forecast time may extend into the beginning of the free forecast, but observations or analyses valid during the free forecast should not be assimilated. However, the time window over which observations are included in vLAPS analyses extends 15 min after the valid time of the analysis, and thus observations during the first 15 min of the free forecast are included via the vLAPS analyses.

Table 1. Experimental design. Groups are HS = hot start, WS = warm start, ON = observation nudging, and AN = analysis nudging (VLAPS3AO contains both analysis and observation nudging but is assigned to group AN).

Name	Initial Condition Source	Pre-Forecast Length (h)	Nudging		Group
			Analysis	Obs	
HRRR0	HRRR	0	N	N	WS
HRRR3	HRRR	3	N	N	WS
VLAPS0	vLAPS	0	N	N	HS
VLAPS3	vLAPS	3	N	N	WS
VLAPS3O	vLAPS	3	N	Y	ON
VLAPS3A	vLAPS	3	Y	N	AN
VLAPS3AO	vLAPS	3	Y	Y	AN

The experiments initialized with HRRR0 and HRRR3 served as the no-assimilation control experiments for the 0 h pre-forecast and 3 h pre-forecast experiments, respectively. HRRR0 and HRRR3 differ from one another in that, for a given cycle, the HRRR3 experiment starts integrating 3 h earlier to allow WRF to spin up. For example, the 20 UTC cycle of HRRR0 (referred to as $HRRR0_{20}$) starts integrating at 20 UTC using the 5 h forecast from the 15 UTC HRRR cycle, whereas $HRRR3_{20}$ starts integrating at 17 UTC using the 2 h forecast from the 15 UTC HRRR cycle, but the output between 17 and 20 UTC is not considered part of the forecast since it is during the spinup time. The experiments initialized with vLAPS that applied no additional nudging were used to determine the potential value of adding nudging. VLAPS0 represents the normal way in which vLAPS (and other 3DVAR analyses) is used, while VLAPS3 allows the effect of a 3 h pre-forecast without additional nudging assimilation to be ascertained. Both experiments used the vLAPS analysis as the initial conditions for the model integration. The other three experiments used vLAPS for the initial conditions, and during a 3 h pre-forecast, they analysis nudged toward the thirteen vLAPS analyses valid during this period (VLAPS3A), observation nudged toward observations (VLAPS3O), or both (VLAPS3AO). These experiments explore the value of applying a series of vLAPS 3D analyses rather than just one, and whether observation nudging might add value even when vLAPS analyses are used.

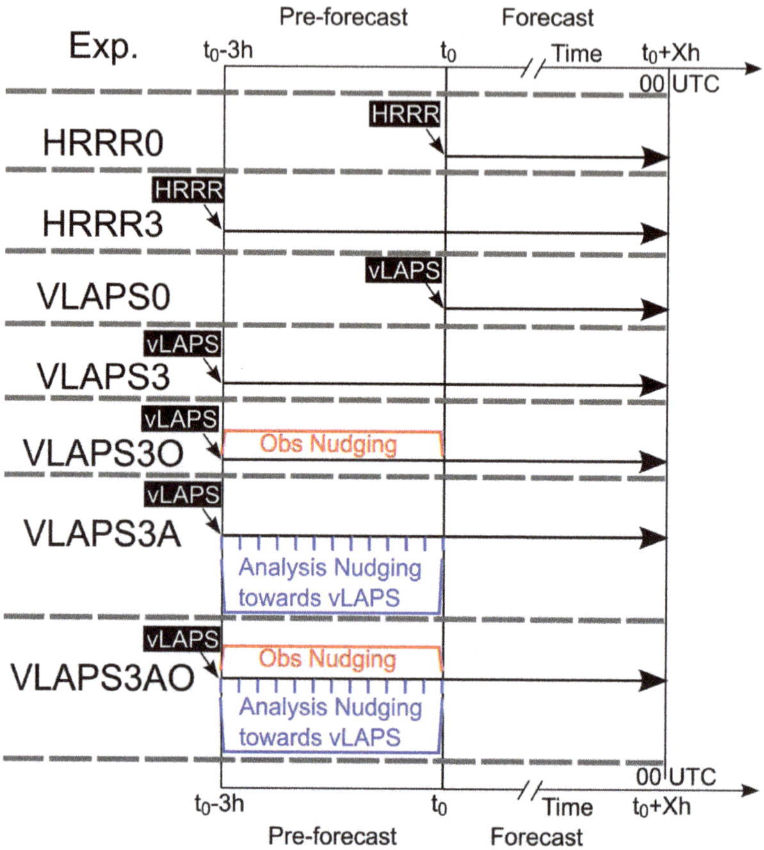

Figure 3. Experimental design. The initial conditions for the experiment are labeled in white letters. The short vertical blue lines represent the valid times of the variational version of the Local Analysis and Prediction System (vLAPS) analyses which were used in analysis nudging for a time period centered on the valid time. The rampdown period of nudging as described in Section 2.3 is omitted from this figure for clarity. Each experiment was run with t_0 values of 1800, 1900, 2000, 2100, 2200, and 2300 UTC. Adapted from Reen et al. [20].

For the purpose of comparing the experiments in the objective verification section, we broke them up into groups based upon the general method of data assimilation applied. These groups are referred to as hot start (HS; VLAPS0), analysis nudge (AN; VLAPS3A and VLAPS3AO), observation nudge (ON; VLAPS3O), and warm start (WS; HRRR0, HRRR3, and VLAPS3). The experiment that used both analysis and observation nudging was placed within the AN group, and the WS group included all experiments that "spun up" from either a vLAPS- or HRRR-generated background state without further data assimilation application. The motivation for including the non-vLAPS experiments in the WS group was that the HRRR 15 UTC forecast cycle fields provide sufficiently "spun up" mesoscale information (including hydrometeor fields), due to the model's own high-quality data assimilation methodology and 3 km native grid spacing, meaning it is essentially a warm start. In summary (Table 1), HS included VLAPS0, AN included VLAPS3A and VLAPS3AO, ON included VLAPS3O, and WS included HRRR0, HRRR3, and VLAPS3.

For each experiment, independent cycles were carried out hourly with the integration period of all cycles ending at 0000 UTC on 21 May 2013 (Figure 4). The 0 h forecast time is the end of the pre-forecast and the beginning of the forecast (t_0 in Figures 3 and 4), which was 1800 UTC for the first cycle and 2300 UTC for the final cycle (i.e., cycles were run with the following t_0 values: 1800, 1900, 2000, 2100, 2200, and 2300 UTC). Each cycle is referred to by its 0 h forecast time (t_0). Each WRF cycle includes the integration of WRF through the 3 h pre-forecast period up to t_0 (except for VLAPS0 and HRRR0, which do not have a pre-forecast period) and the integration of WRF from t_0 until the end of the forecast at 0000 UTC. For example, the 18 UTC cycle of VLAPS3AO (referred to as $VLAPS3AO_{18}$) starts integration at 1500 UTC, applies analysis and observation nudging data assimilation during the integration in the 3 h pre-forecast period ending at 1800 UTC (t_0), and then continues integrating until the end of the forecast period at 0000 UTC. Each cycle is independent in that it is not affected by any of the other cycles.

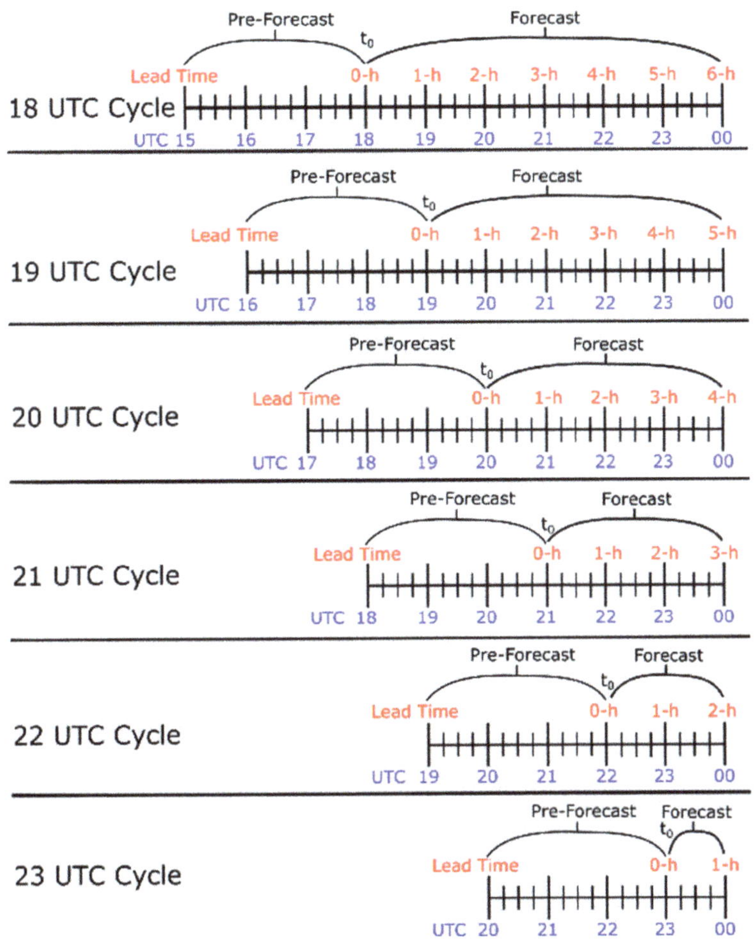

Figure 4. Illustration of the six WRF cycles used for each experiment and which are referred to by the 0 h forecast time (t_0). As shown in Figure 3, the 3 h pre-forecast period wherein any analysis or observation nudging data assimilation was applied is omitted in some experiments (HRRR0 and VLAPS0). The hourly forecast lead times for each cycle are labeled in red; the output was created at a 15 min temporal spacing.

All experiments used the 1500 UTC cycle of HRRR as the starting point to determine the initial conditions and boundary conditions, with the vLAPS analyses using this cycle as the first-guess field. This simplifies the experimental design and is more representative of the limited frequency of updated model data available to drive finer-scale model simulations in battlefield conditions.

In order to more concretely illustrate the experimental configuration, Figure 5 shows how one specific experiment (VLAPS3AO) was configured for one specific cycle (2100 UTC), i.e., $VLAPS3AO_{21}$. The lower portion of Figure 5 illustrates how analysis nudging was applied during the 3 h pre-forecast (1800–2100 UTC for this cycle) using 15 min vLAPS analyses by showing the analysis nudging details for the first hour of this period (i.e., 1800–1900 UTC). Each analysis was applied to the model over 30 min centered on the valid time of the analysis. The weight at which it was applied linearly increased in the 15 min prior to the valid time of the analysis and linearly decreased in the 15 min following the valid time. Observation nudging applied observations over a time window centered on each individual observation; as described in Section 2.3, a 2 h time window was used for surface observations, and a 3 h time window was used for above-surface observations.

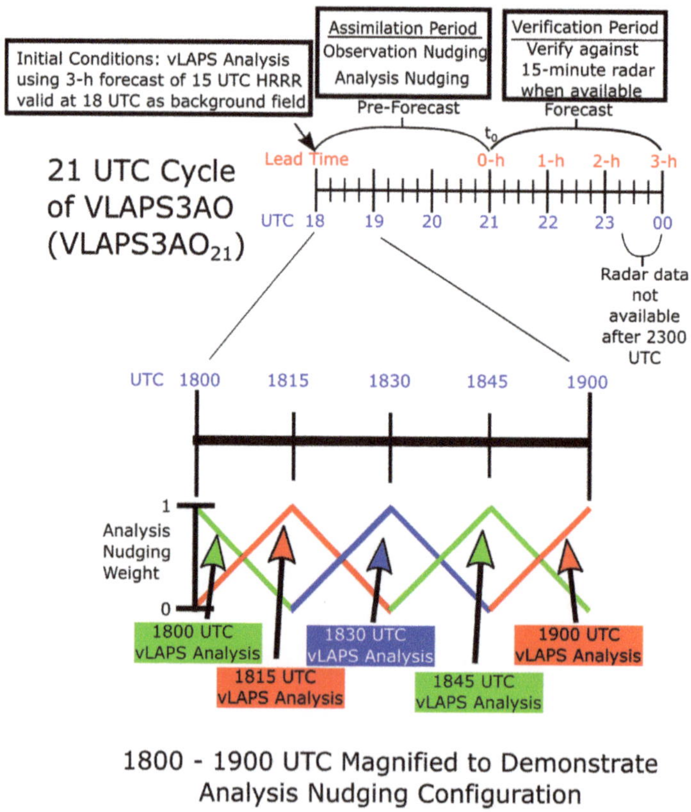

Figure 5. Schematic illustrating the details of a specific experiment (VLAPS3AO) for a specific cycle (21 UTC), i.e., $VLAPS3AO_{21}$. While the lower half of the figure shows the details of how vLAPS analyses were applied using analysis nudging between 1800 and 1900 UTC, analysis nudging towards vLAPS analyses took place throughout the 3 h pre-forecast period.

3. Results

Here, the evaluation of the combination of a 3DVAR-based technique (vLAPS) and nudging is limited to reflectivity. To demonstrate the variation among the experiments, a single forecast time for a single cycle is examined subjectively. However, given the number of experiments, cycles, and forecast times, the objective evaluation that follows allows for a more holistic understanding of the results.

3.1. Subjective Evaluation

The observed composite base reflectivity is compared to the WRF-ARW lowest-level reflectivity for the 18 UTC cycle of each experiment in Figure 6 for 2015 UTC 20 May 2013. This is near the time of the peak of the EF5 tornado and was chosen as a time where strong convection is present. The observed reflectivity (Figure 6c) shows convection along a southwest-to-northeast-oriented line from Texas through the southeast corner of Kansas. There are weaker echoes (<30 dBZ) in a short line in central Kansas, and scattered echoes in the northwest corner of the domain. All of the experiments forecast the main line of convection and some echoes in the northwest corner of the domain, but there are notable differences among the experiments.

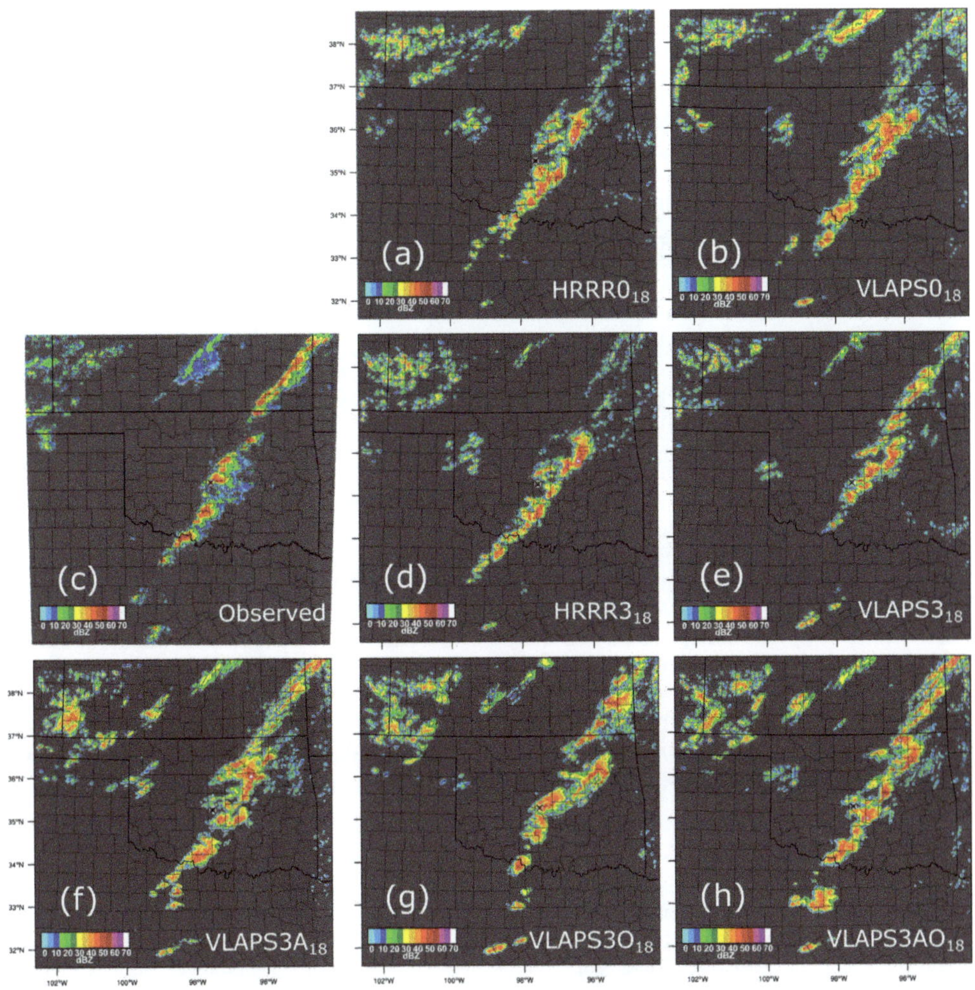

Figure 6. Reflectivity at 2015 UTC on 20 May 2013 from (**a**) HRRR0$_{18}$; (**b**) VLAPS0$_{18}$; (**c**) observations; (**d**) HRRR3$_{18}$; (**e**) VLAPS3$_{18}$; (**f**) VLAPS3A$_{18}$; (**g**) VLAPS3O$_{18}$; (**h**) VLAPS3AO$_{18}$. The black X within the white square shows the approximate location of the EF5 tornado. The observed reflectivity is the composite base reflectivity, and the WRF-ARW reflectivity is the lowest model-level radar reflectivity. The observed reflectivity is from the Iowa Environmental Mesonet (http://mesonet.agron.iastate.edu/GIS/apps/rview/warnings.phtml; accessed on 10 November 2021).

The northern portion of the main line of convection is not well simulated by the experiments initialized with HRRR (HRRR0_{18} in Figure 6a and HRRR3_{18} in Figure 6d). The use of vLAPS as the initial condition slightly strengthens convection in this area (VLAPSO$_{18}$ in Figure 6b), and the addition of the 3 h pre-forecast without nudging (VLAPS3_{18} in Figure 6e) brings the strength closer to the observations. However, it appears that the experiment analysis nudging to the vLAPS analyses and applying observation nudging (VLAPS3AO$_{18}$ in Figure 6h) best reproduces the continuous strength of the convection along this line in Kansas. The southern edge of the main line of convection does not extend far enough south in VLAPS3_{18} (Figure 6e), but most of the other experiments appear to extend it too far south. All experiments show echoes in the general region of the observed EF5 tornado, but only three experiments (VLAPSO$_{18}$ in Figure 6b, VLAPS3_{18} in Figure 6e, and VLAPS3O$_{18}$ in Figure 6g) show echoes at the location, with two of these experiments (VLAPSO$_{18}$ in Figure 6b and VLAPS3O$_{18}$ in Figure 6g) showing echoes $\geq$ 40 dBZ very close to the location.

The area of weak echoes in central Kansas is not well forecast by any of the experiments. The HRRR-initialized experiment without a pre-forecast (HRRR0_{18} in Figure 6a) shows the feature stronger and smaller than observed and, unlike the observations, does not show an area with no echoes between it and the echoes in the northwestern portion of the domain. The vLAPS-initialized experiment without a pre-forecast (VLAPS0_{18} in Figure 6b) shows a pair of lines of convection in this area that are both much stronger than observed. The 3 h pre-forecast (HRRR3_{18} in Figure 6d) allows the HRRR initialization to reproduce more of the separation between the feature and echoes in the northwestern portion of the domain observed but shows the feature with much less coverage than observed. For experiments with the 3 h pre-forecast, using vLAPS as the initial conditions broadens the area covered by this feature compared to the HRRR initialization (HRRR3_{18}) and removes the second line of convection seen in the vLAPS experiment without the 3 h pre-forecast (VLAPSO$_{18}$). The experiments with the area covered by this feature that most closely match the observations appear to be the experiments analysis nudging the vLAPS analyses (VLAPS3A$_{18}$ in Figure 6f and VLAPS3AO$_{18}$ in Figure 6h). However, all of the experiments that produce the feature show it stronger than observed. The echoes in the northwest quadrant of the domain, in general, appear to be stronger than observed in the experiments, perhaps most so for experiments analysis nudging towards vLAPS analyses or observation nudging (VLAPS3A$_{18}$ in Figure 6f, VLAPS3O$_{18}$ in Figure 6g, and VLAPS3AO$_{18}$ in Figure 6h).

All of the experiments produce model base reflectivity fields that are at least generally consistent with the observed composite radar reflectivity, but there is notable variation among the experiments. While it is difficult to subjectively determine which experiment performs best at the time shown here for this cycle, the results suggest that the experiment analysis nudging to the vLAPS analyses and applying observation nudging performed reasonably well (VLAPS3AO$_{18}$ in Figure 6h).

3.2. *Objective Evaluation*

The previous subsection presented subjective support that the model simulations realistically reproduced the general convective outbreak event on the afternoon of 20 May, including aspects associated with the timing, spatial location, alignment, and convective mode. Reen et al. [20] showed that the model had some ability in capturing realistic gross structures seen in supercellular storms. However, the evolution of the convective outbreak and structural details of individual convective elements differed among the experiments; this variation offers a spread of potential solutions, consistent with what might be expected based on the findings in Zhang et al. [49]. The following section provides a more objective evaluation using several forecast performance measures to compare all of the different experiments aggregated across the hourly cycles during the afternoon of May 20.

Although the results are only valid for this single unique case study event, they may provide valuable clues as to how the various techniques can perform in a more general sense. We are also interested in whether our combined 3DVAR analysis/observation nudging strategy might be of value in supporting stand-alone operational nowcasting systems run on a more modest computational hardware platform.

3.2.1. Metrics Used in Objective Evaluation

To quantitatively measure the performances of each cycle across the full spectrum of our experiments, a trio of well-established forecast evaluation measures was used to examine the short-range convective forecasts by comparing observed to model forecast radar reflectivity fields. The fractions skill score, or FSS [51], is a spatial verification metric often used for assessing the performance of precipitation forecasts from numerical weather prediction (NWP) models (frequently for convective precipitation). The critical success index, or CSI [52], is another common metric used to evaluate categorical forecasts, taking into account hits, misses, and false alarms. Finally, the frequency bias score (FBIAS) [53,54] is also used to assess the quality of model-derived radar reflectivity field forecasts against real radar observations [19,53]. The FBIAS is simply the ratio between the number of model grid points where the model forecasts reflectivity above a certain threshold and the number of model grid points where the observed radar reflectivity values exceed the same threshold.

The FSS was computed for different neighborhood sizes aggregated from the native grid resolution. The aggregation was also conducted across all of the relevant cycles for each lead forecast time in each experiment (not all cycles are relevant for each forecast lead time since some cycles are too short to include some forecast lead times). For example, the FSS for the 1.25 h forecast lead time of VLAPS3AO aggregates the verification of $VLAPS3AO_{18}$ at 1915 UTC, $VLAPS3AO_{19}$ at 2015 UTC, $VLAPS3AO_{20}$ at 2115 UTC, and $VLAPS3AO_{21}$ at 2215 UTC (Figure 4 shows the relationship between forecast lead time and valid time for each cycle). Since model forecast skill often varies based on the forecast lead time, aggregating statistics by forecast lead time reveals how the temporal evolution of error varies among methodologies. The CSI and FBIAS were not computed using neighborhoods but were both calculated on the highest available resolution grid (here, our native model grid spacing of 1 km) and, as with FSS, were aggregated across cycles for each lead forecast time in each experiment. The FBIAS of reflectivity measures how much the model either overforecasts (forecast is more than the actual) or underforecasts (forecast is less than the actual) coverage of reflectivity values exceeding a threshold. No verification takes place during the pre-forecast (where data assimilation occurs), but verification is instead limited to the forecast period.

As noted previously (and shown in Figure 4), a model integration with a t_0 of each hour from 1800 UTC to 2300 UTC was launched (i.e., an hourly refresh) for each experiment and produced 15 min forecasts from t_0 out to 0000 UTC of 21 May. The latest time verified is 2300 UTC because radar data provided by vLAPS were not available after this time. The output was produced in 15 min intervals, which thus produced forecasts for verification with lead times ranging from 15 min to 5 h (resulting in statistics at 20 different forecast lead times). Due to the 0000 UTC end time of all cycles, but radar data not being available after 2300 UTC, only the 1800 UTC cycle produced a 5 h forecast lead time. Because of various pre-forecast requirements across different experiments, some experiments start their integration before their 0 h lead time (e.g., a model cycle with a base time of 1800 UTC with a 3 h pre-forecast requirement was actually initialized at 1500 UTC with lead times of forecasts based on the 1800 UTC mark). Since all simulations terminated at 0000 UTC, the model integrations which started at 2300 UTC had only a maximum 1 h lead forecast available. Because radar data were not available after 2300 UTC, all verification forecast times of the 2300 UTC cycle (other than the 0 h forecast time) are after the end of radar data availability, and thus the 2300 UTC cycle is not included in the objective verification statistics. WRF-ARW "restart" forecasts were not used to initialize from a previous cycle's forecast—all first-guess initial condition (and lateral boundary tendency) fields for all cycles leveraged the 1500 UTC HRRR cycle forecast fields available at hourly increments. A single

HRRR cycle was used to more closely approximate the cycle frequency that is available outside of the United States for models such as the Global Forecast System that could be used as boundary conditions for fine-scale simulations.

Each experiment was aggregated across its simulations to create statistics, with each value representing that experiment's performance for the forecast lead time, reflectivity threshold, and (for FSS) neighborhood size being evaluated. Similar to other studies, the column maximum radar reflectivity from both the model and the National Weather Service WSR-88Ds (provided by the National Oceanic and Atmospheric Administration; NOAA) was used to evaluate the convective forecasts [55]. The ground truth column maximum radar reflectivity fields obtained from the NOAA are the same type as those used for previous studies involving the original LAPS product [33]. The model column maximum reflectivity was computed using the WRF-ARW diagnostic method [56] that uses prognostic hydrometeor fields from the model microphysics parameterization.

The FSS for two radar reflectivity thresholds (25 dBZ and 35 dBZ) using a 9 km neighborhood size is shown in Figure 7. The CSI and FBIAS metrics are shown in Figures 8 and 9. The neighborhood size of 9 km is here considered to be a reasonable estimate for the "effective resolution" of the WRF-ARW when run on a mesh with a native 1 km grid spacing [57]. The number of cycles included in the statistics varies by forecast lead time, with five cycles included at the 1 h forecast and one cycle included at the 5 h forecast.

3.2.2. Outcome of Objective Evaluation

For the purpose of comparing the experiments in this section, we utilized the groups described in Section 2.5 which are based upon the general method of data assimilation applied. The various FBIAS, FSS, and CSI curves throughout the lead forecast period of study (as seen in Figures 7–9) show similarity within these assimilation methodology groups, namely, hot start (HS), analysis nudge (AN), observation nudge (ON), and warm start (WS). Note that the time subscript is omitted when referring to the evaluation of the WRF forecasts in this section because each experiment is evaluated based on data that are combined across the cycles.

Although VLAPS0 is considered as a HS, it does differ from previous LAPS/vLAPS HS experiments [19,58] in that it did not directly insert the vertical velocity from the analysis into the WRF-ARW initial conditions. However, the hydrometeor fields produced in the vLAPS analyses are used in the appropriate initial WRF-ARW arrays. Additionally, since both reflectivity and radial wind information from the WSR-88D radars were incorporated into the vLAPS analyses, it is expected that the initial vertical velocity fields adjusted quickly through continuity considerations to the convective-scale information provided by the radar data. Any experiment that used a pre-forecast period and which started from an initial vLAPS analysis also filled the initial WRF-ARW arrays in the same fashion (Section 2 provides more details on the use of vLAPS analyses for model initialization). When the vLAPS analyses are used instead solely as a source for 15 min intermittent analysis nudging across a pre-forecast period by WRF-ARW, the analyses are only used to nudge the potential temperature, water vapor mixing ratio, and u- and v-wind components, as these are the variables for which analysis nudging is available in WRF-ARW.

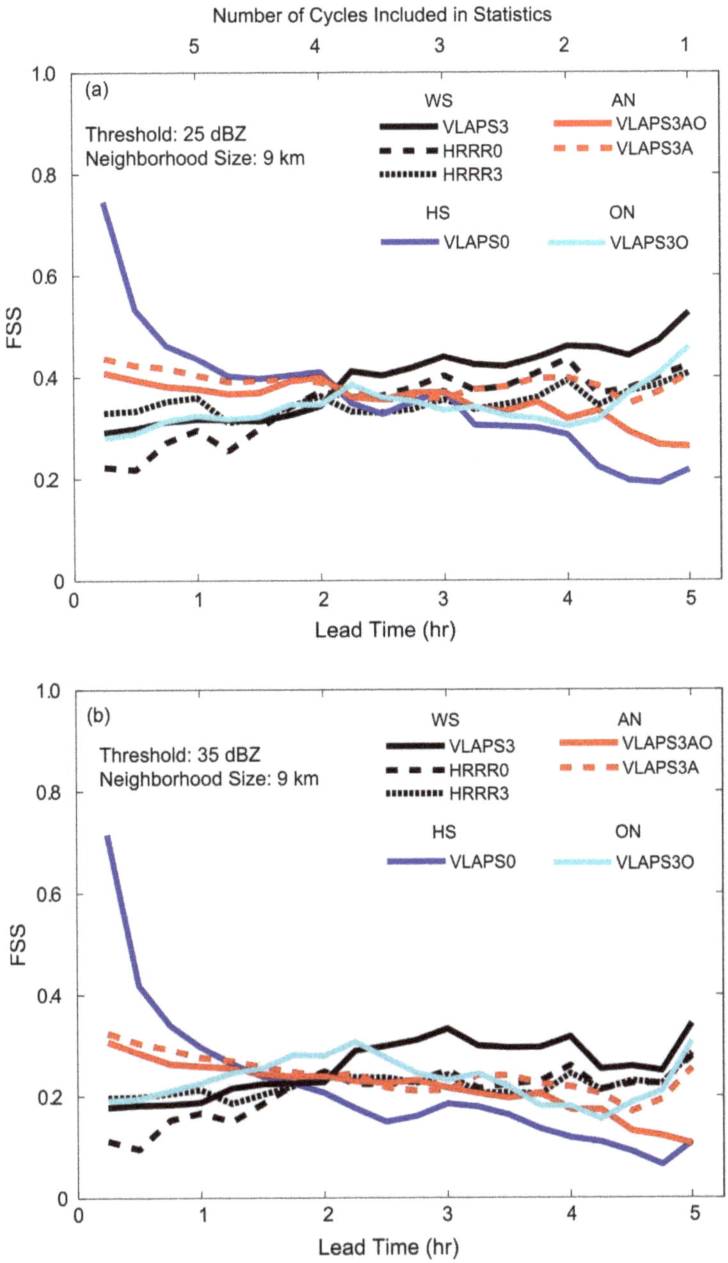

Figure 7. FSS by lead forecast time (by experiment) and for (**a**) 25 dBZ and (**b**) 35 dBZ reflectivity threshold levels (for a 9 km neighborhood size). Line colors indicate whether an experiment is in the warm start (WS), hot start (HS), analysis nudge (AN), or observation nudge (ON) group. The number of cycles included in the statistics at each integer lead time hour is shown along the top of the figure.

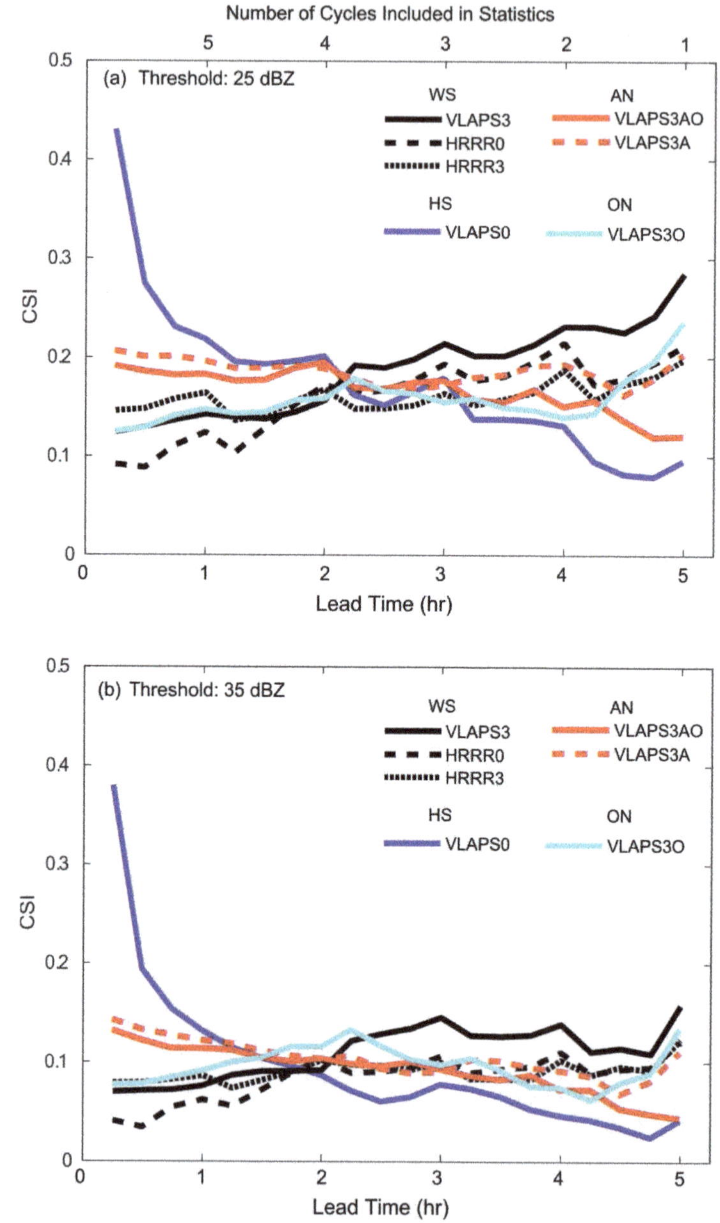

Figure 8. CSI by lead forecast time (by experiment) and for (**a**) 25 dBZ and (**b**) 35 dBZ reflectivity threshold levels. Line colors indicate whether an experiment is in the warm start (WS), hot start (HS), analysis nudge (AN), or observation nudge (ON) group. The number of cycles included in the statistics at each integer lead time hour is shown along the top of the figure.

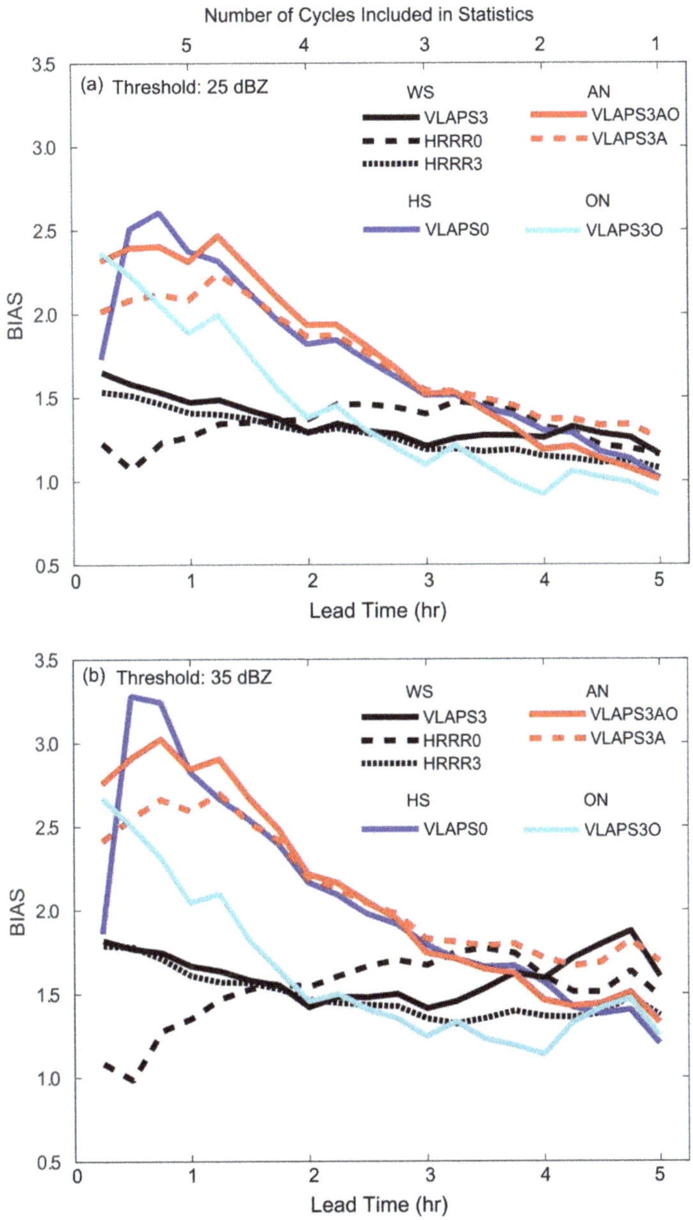

Figure 9. FBIAS by lead forecast time (by experiment) and for (**a**) 25 dBZ and (**b**) 35 dBZ reflectivity threshold levels. Line colors indicate whether an experiment is in the warm start (WS), hot start (HS), analysis nudge (AN), or observation nudge (ON) group. The number of cycles included in the statistics at each integer lead time hour is shown along the top of the figure.

Upon examining Figures 7–9, the first thing of immediate note is that for this case study event, the HS experiment stands out from the others, mostly within the initial lead forecast hour. The FSS and CSI scores for the HS experiment are clearly superior to all other groups during at least the first 30 min of lead forecast time (independent of whether the 25 dBZ or 35 dBZ reflectivity threshold value is used). This strong advantage for very short term nowcasting of vLAPS is most likely due to its use of a diabatic hot start methodology and has been noted previously [19,58]. The motivation with vLAPS has always been to compete with and exceed the skill of simpler nowcasting methods such as feature-based advection/extrapolation or basic persistence during the initial hour. This is a very challenging goal for convective-scale NWP modeling. In at least this study, the flip side to HS seems to be that the FSS and CSI scores quickly degrade back to those of the other methods by approximately a 2 h lead time and, by a 5 h lead time, even appear to result in somewhat worse scores. Perhaps this is some evidence of an overfitting to the stronger reflectivity echoes at time 0 h, and that some degree of imbalance could still remain between the full set of mass and momentum model fields produced by the hot start analysis itself.

The AN group of experiments exhibited the highest FSS and CSI scores during the first lead hour (outside of HS). The AN group shows that both FSS and CSI gradually declined to approximately 2 h and then remained mostly flat until approximately 4 h. During most of the period after approximately the 1 h lead forecast time, the AN experiments clump in with the other groups in terms of general FSS and CSI scores. An exception is the VLAPS3AO experiment in the AN group, which shows a gradual decrease in FSS and CSI scores after the 4 h lead forecast—all other experiments in all groups show a slight increase in FSS and CSI scores starting at approximately the 4.5 h lead forecast. This may be an artifact of the very limited number of cycles (and thus very small sample size) available to compute metrics at the 4 h and 5 h lead forecasts. It could also point to some need to fine-tune aspects of the combined analysis and observation nudging criteria used by VLAPS3AO, particularly when combining meso/synoptic-scale conventional observations via observation nudging with convective-scale observation data assimilated through analysis nudging to the 15 min updated vLAPS analyses (when the radar data are incorporated).

The ON group fell in the middle to lower half of the pack of all groups, both for the short and longer lead forecast times, in terms of FSS and CSI scores. The FSS and CSI scores for the VLAPS3O experiment generally remained fairly consistent at all lead forecast times.

The FSS and CSI scores for the WS group tended to track fairly closely to those of the ON group, although HRRR0 was a bit lower initially, while VLAPS3 beyond 2.5 h was a notable exception (higher FSS at further lead times). For the 25 dBZ threshold, the WS group (including HRRR0 and HRRR3) tended to outperform the ON group in FSS and CSI through much of the lead forecast period after approximately 3 h. VLAPS3 generally showed the highest FSS and CSI scores of all experiments from lead forecasts of 2.5 h and beyond, particularly for the 25 dBZ reflectivity threshold.

Tables 2 and 3 show the FSS scores at lead forecast times of 0.25 h, 1.75 h, 3.50 h, and 5.00 h for each experiment, and for the 25 dBZ and 35 dBZ thresholds, respectively (for a 9 km neighborhood size). The tables more clearly show the data values, which can also be estimated from Figure 7.

Table 2. FSS for 25 dBZ threshold and 9 km neighborhood size, at four different lead forecast times (h).

Experiment	FSS by Lead Forecast (h)			
	0.25	1.75	3.50	5.00
HRRR0	0.22	0.34	0.38	0.42
HRRR3	0.33	0.34	0.35	0.41
VLAPS0	0.75	0.40	0.30	0.22
VLAPS3	0.29	0.33	0.42	0.53
VLAPS3A	0.44	0.40	0.38	0.41
VLAPS3AO	0.41	0.39	0.33	0.26
VLAPS3O	0.28	0.35	0.32	0.46

Table 3. FSS for 35 dBZ threshold and 9 km neighborhood size, at four different lead forecast times (h).

Experiment	FSS by Lead Forecast (h)			
	0.25	1.75	3.50	5.00
HRRR0	0.11	0.22	0.22	0.28
HRRR3	0.20	0.22	0.21	0.28
VLAPS0	0.72	0.23	0.16	0.11
VLAPS3	0.18	0.23	0.30	0.34
VLAPS3A	0.33	0.25	0.24	0.26
VLAPS3AO	0.31	0.24	0.20	0.11
VLAPS3O	0.19	0.28	0.22	0.31

Examining the FBIAS among the groups (Figure 9), we can determine whether the model was forecasting too much (overforecasting; FBIAS > 1.0) or too little (underforecasting; FBIAS < 1.0) areal coverage of reflectivity at or above a given reflectivity threshold. Early in the lead forecast period, nearly all the groups of experiments showed a significant tendency to overforecast the stronger convection (higher reflectivity) areas, such as at the higher 35 dBZ threshold. HRRR0 from the WS group produced the least biased FBIAS results for both thresholds early in the lead forecast time, but with increasing FBIAS through approximately the 3 h lead forecast. The VLAPS3 and HRRR3 experiments, also from the WS group, tended to also be less high biased during the early lead forecast period. All of the other experiments from the different groups started in mostly the 2.0 to 3.0 FBIAS range for both the 25 dBZ and 35 dBZ thresholds but dropped steadily towards FBIAS values of 1.0 to 1.5 by the end of the 5 h lead forecast period. At the 25 dBZ threshold, the ON group began to underpredict areal coverage by the end of the forecast period (VLAPS3O).

At the 25 dBZ and 35 dBZ thresholds, overforecasting was the main forecast feature across all experiments, particularly early in the lead forecast period. The single HS experiment, VLAPS0, showed the most exaggerated overforecasting at approximately the 30 to 45 min lead forecast for the 35 dBZ threshold (as high as approximately 3.25) but dropped steadily towards ≈1.25 by the end of the 5 h forecast period. The experiments VLAPS3AO, VLAPS3A, and VLAPS3O followed a similar FBIAS evolution (in terms of rate of increase and decrease) to VLAPS0 for both the 25 dBZ and 35 dBZ thresholds after the 1 h lead time. During the 0 h to 1 h lead forecast period, the FBIAS decreased in the VLAPS3O forecasts, whereas the VLAPS3A and VLAPS3AO FBIAS remained fairly constant. In contrast to all of the other experiments, the VLAPS0 FBIAS sharply increased between the 0.25 h and 0.50 h forecasts, suggesting that the initial analysis solution may not be in complete balance (perhaps from overfitting to the radar observations) with the WRF-ARW model on the 1 km domain. Tables 4 and 5 show the specific FBIAS scores at lead forecast times of 0.25 h, 1.75 h, 3.50 h, and 5.00 h for each experiment, and for the 25 dBZ and 35 dBZ thresholds, respectively. The overforecasting at the 25 dBZ and 35 dBZ thresholds is generally more pronounced for forecasts recently influenced by observations (via vLAPS analyses directly inserted at t_0 in VLAPS0, via vLAPS analyses being analysis nudged towards in VLAPS3A and VLAPS3AO, or via observation nudging in VLAPS3O). It may be that the methods used to incorporate observations into vLAPS analyses and through observation nudging make assumptions that do not work optimally for this specific case and the specific observations available; future work should explore the generality of this bias and seek to determine how the methodologies could be improved to lower the overforecast bias.

The variation in performance by neighborhood size and reflectivity threshold (as measured by FSS) for this case study can be seen more clearly in Table 6. In Table 6, the FSS is shown for a few different lead forecast times near the start and end of the forecast period (0.50 h and 5.00 h), and for different neighborhood sizes (1 km, 9 km, 17 km). The lower 10 dBZ threshold is focused upon in Table 6, although for the 9 km neighborhood (closest to the effective resolution of the WRF 1 km grid spacing output), additional thresholds of 25 dBZ and 35 dBZ are also shown. The tendency of FSS to improve with lower reflectivity thresholds and larger neighborhood sizes, as seen in most previous NWP convection-allowing precipitation studies, is repeated here.

Table 4. FBIAS for 25 dBZ threshold at four different lead forecast times (h).

Experiment	FBIAS by Lead Forecast (h)			
	0.25	1.75	3.50	5.00
HRRR0	1.23	1.36	1.46	1.16
HRRR3	1.53	1.33	1.18	1.08
VLAPS0	1.73	1.97	1.44	1.02
VLAPS3	1.65	1.37	1.27	1.16
VLAPS3A	2.02	1.98	1.49	1.25
VLAPS3AO	2.32	2.10	1.42	1.01
VLAPS3O	2.37	1.55	1.10	0.91

Table 5. FBIAS for 35 dBZ threshold at four different lead forecast times (h).

Experiment	FBIAS by Lead Forecast (h)			
	0.25	1.75	3.50	5.00
HRRR0	1.08	1.56	1.77	1.48
HRRR3	1.79	1.53	1.53	1.37
VLAPS0	1.86	2.40	1.66	1.20
VLAPS3	1.82	1.55	1.53	1.60
VLAPS3A	2.41	2.42	1.79	1.69
VLAPS3AO	2.76	2.49	1.65	1.33
VLAPS3O	2.67	1.64	1.23	1.25

Table 6. FSS as it varies across three different neighborhood sizes for the 10 dBZ threshold, and for two different lead forecast times (near the start and end of the forecast period). Note that additional thresholds of 25 dBZ and 35 dBZ are shown only for the 9 km neighborhood size (since 9 km is of interest due to it closely representing the effective resolution of the WRF output produced from the 1 km grid spacing nest).

		FSS by Neighborhood Size (NS), Threshold (T), and Lead Time (LT)					
	NS	1 km		9 km		17 km	
	T	10 dBZ		10/25/35 dBZ		10 dBZ	
Experiment	LT	0.25 h	5.00 h	0.25 h	5.00 h	0.25 h	5.00 h
HRRR0		0.43	0.65	0.50/0.22/0.11	0.71/0.42/0.28	0.54	0.74
HRRR3		0.50	0.61	0.59/0.33/0.20	0.67/0.41/0.28	0.64	0.70
VLAPS0		0.67	0.56	0.77/0.75/0.72	0.62/0.22/0.11	0.80	0.65
VLAPS3		0.51	0.61	0.60/0.29/0.18	0.67/0.53/0.34	0.64	0.70
VLAPS3A		0.54	0.64	0.64/0.44/0.33	0.70/0.41/0.26	0.68	0.72
VLAPS3AO		0.51	0.55	0.60/0.41/0.31	0.61/0.26/0.11	0.64	0.64
VLAPS3O		0.45	0.52	0.51/0.28/0.19	0.57/0.46/0.31	0.55	0.59

4. Discussion

This study's main goal was to complete a preliminary investigation of how nudging and vLAPS 3DVAR compare, and how combining them could improve short-range nowcasting. This paper's focus was more upon the overall methods, and less upon the specific differences across the forecast results, because statistical evaluation of additional cases is needed to draw conclusive statements.

From a subjective standpoint, the experiments captured the main features/structures and general timing of the severe convective outbreak over Oklahoma on the afternoon of 20 May 2013 (some more realistically than others). This was a strongly forced convective event and was likely to be inherently more predictable [59–62]. The FSS and CSI metrics showed that the experiments performed similarly around the 2 h forecast, while differences among the experiments were somewhat larger at other forecast hours. VLAPS0 is an outlier with much higher scores over the initial lead forecast hour (but generally the lowest scores during the 3 to 5 h lead forecast period). The experiments that applied data assimilation across a short pre-forecast period, including the VLAPS3AO hybrid observation/analysis nudging approach introduced by the authors of this paper, appeared to improve FSS and CSI at the start of the nowcast cycles compared to not performing data assimilation during this pre-forecast period (i.e., VLAPS3AO, VLAPS3A, and VLAPS3O compared to VLAPS3). However, they could not match the FSS and CSI gained by inserting vLAPS analyses at the beginning of the simulation without a pre-forecast period (VLAPS0).

The VLAPS3 combination of starting the model integration from a 3D variational analysis (leveraging radar data), followed by a subsequent 3 h pre-forecast period, performed well relative to most of the other experiments for bias in the first portion of the forecast period (HRRR0 had better bias and HRRR3 was very similar to VLAPS3). It also performed better than all of the other experiments for CSI/FSS in the latter portion of the forecast period.

While analysis nudging toward high-resolution 3DVAR vLAPS analyses shows potential promise for improving short-term convection forecasts, the results suggest further work is needed to best leverage high-resolution analyses. The rapid increase in overforecast bias in VLAPS0 at the beginning of the forecast and CSI and FSS performing worse than any other experiment later in the forecast suggest that the VLAPS0 method of directly inserting the analysis at the beginning of the forecast may not be optimal. Analysis nudging towards a series of vLAPS analyses may be a way to improve the assimilation of vLAPS analyses. However, two factors suggest room for improvement in the methodology used to analysis nudge towards the vLAPS analyses (VLAPS3A, VLAPS3AO). One factor is that the analysis nudging experiments have a noticeably higher overforecast bias than an experiment started with the same initial conditions but no subsequent data assimilation (VLAPS3). The second factor is that during the first hour, directly inserting vLAPS analyses at the beginning of the forecast (VLAPS0) performs much better (in terms of CSI and FSS) than analysis nudging towards the vLAPS analyses. Analysis nudging is not normally used to assimilate analyses with the high temporal (15 min) and horizontal (1 km) spacing used here. Thus, informing the analysis nudging weights by previous research may have resulted in analysis nudging having been applied more weakly or more strongly than would best assimilate the analyses. Additionally, in order to fully assimilate the vLAPS analyses, nudging towards additional fields from the analyses may be needed to supplement those currently available for analysis nudging in WRF. Given the verification results, it is not clear that using observation nudging (VLAPS3A vs. VLAPS3AO) adds values in this case; this may be because the high-temporal and spatial resolution observations used in the vLAPS analysis limit the added value of assimilating individual observations at a time centered on their valid times.

Both HRRR0 and HRRR3 (also in the WS group) proved competitive in FSS and CSI scores after approximately the first 90 min of the lead forecast and out through the longer lead times (and for FBIAS throughout the forecast); this is consistent with both the high-resolution nature and the sophisticated data assimilation methodology built into the HRRR cycling model itself. Since the 15 UTC HRRR cycle forecasts were leveraged in every experiment across all cycle times, the oldest HRRR forecast used to provide a lateral boundary condition (at 00 UTC 21 May) had a 9 h lead time. Some recent studies [63–65] indicate that convection-resolving NWP models, using sophisticated data assimilation approaches (including radar data ingest), can show skill and predictability out to 6–12 h. This is especially true for strongly forced convection, which is the type of convection in this study. Weakly forced convection (such as the type common over the summer in the southeastern US) tends to be considerably less predictable at these same resolutions, due to less certainty in capturing the convective initiation mechanisms [66].

Two advantages our simulations had were that (1) a high-quality 3 km HRRR forecast produced at 15 UTC was available for creating lateral boundary tendencies (and either the initial conditions or the first guess for the analysis used as the initial conditions); and (2) multiple Doppler WSR-88D radars were available to the vLAPS analyses. If we consider a forward-deployed army unit running a similar modeling configuration on a laptop with restrictive communication bandwidth availability, conditions such as (1) and (2) above will likely not typically be present. However, this study shows that atmospheric data assimilation approaches not reliant on a full EnKF [67] or 4DVAR [68] approach, and not in need of high-performance computing cluster solutions, can still provide value-added short-range nowcast guidance to local human forecasters and automated forecast systems and weather decision support tools.

While this study used a single, large domain to simplify the experiment setup (made possible by the availability of the high-resolution HRRR output), in order to make this tractable for a forward-deployed system, nested grids with a much smaller innermost domain would be needed. In the case of a tactical NWP-based nowcasting system (a rapid update cycling model), the ability to collect and effectively assimilate special tactical weather observations (sensors on unmanned aerial systems, lidar, radar, etc.) along with non-traditional sources of weather observations (for example, the Air Force Global Synthetic Weather Radar product [69], which leverages satellite observations and other data sources) will be an important avenue to full success. Additionally, while this study used vLAPS analyses, the techniques investigated could be used with other high-resolution analyses.

In terms of convective forecasting, which was the focus of this paper, there is still ample work remaining towards improvement upon the bias and skill issues, especially at the higher reflectivity threshold levels. Future work should investigate additional cases to explore the generality of the results seen in this case, including exploring whether the high reflectivity bias seen in the current case is seen in other cases and determining the causes of this bias. Given that this was a strongly forced and large-scale convective outbreak, these issues are likely going to be even more challenging in weakly forced convective environments [66]. This case study describes methodologies that can be used to apply 3DVAR analyses in conjunction with nudging, demonstrates that applying these methodologies for vLAPS analyses for this case study does not show a clear overall improvement, and suggests areas for future research to explore how to improve the combination of 3DVAR analyses and nudging. In this case study, directly applying the vLAPS analyses at the 0 h forecast time without analysis nudging resulted in higher FSS and CSI than the other experiments throughout the 0–1 h forecast, but this was also the only experiment with bias increasing substantially during this time period. Similarly, Jiang et al. [19] found that in WRF forecasts initialized with vLAPS, the equitable threat scores were highest at the beginning of the forecasts

At a national level, convection-permitting models are already performing with good skill and with a high degree of sophistication in data assimilation strategies and NWP physics [70]. Running limited-area mesoscale modeling configurations on hardware technology as restricted as a desktop or laptop, while maintaining a flexibility to adjust quickly for operations over diverse and often data-restricted locations around the world, will require making intelligent adjustments to the national approaches at convection-permitting scales. The 1 km (and even finer) grid spacing is important for resolving many boundary layer flow phenomena that impact both military and civilian interests near the earth's surface, and convective forecasting is just one aspect of those (although a potentially highly impactful one).

Author Contributions: Conceptualization, B.P.R., H.C. and R.E.D.; methodology, B.P.R., H.C., R.E.D., Y.X. and S.A.; software, B.P.R., Y.X. and S.A.; validation, H.C. and J.W.R.; formal analysis, H.C. and J.W.R.; investigation, B.P.R., R.E.D., Y.X. and S.A.; writing—original draft preparation, B.P.R. and R.E.D.; writing—review and editing, B.P.R., H.C., R.E.D., Y.X., S.A. and J.W.R.; visualization, B.P.R. and H.C. All authors have read and agreed to the published version of the manuscript.

Funding: The US Army DEVCOM Army Research Laboratory (ARL) authors (Brian Reen, Huaqing Cai, Robert Dumais, and John Raby) were funded by ARL. Yuanfu Xie and Steve Albers were partially funded by ARL and partially funded by NOAA/GSD (and Steve Albers by CIRA).

Data Availability Statement: MADIS observations for assimilation were downloaded from https://madis-data.ncep.noaa.gov/ (accessed on 10 November 2021). Pilot observations, WSR-88D radar data, and GOES data were obtained from NOAA Global Systems Laboratory (GSL). The WRF verification against radar reflectivity that is the basis for the Results section is available at https://doi.org/10.6084/m9.figshare.14818497.v1 (accessed on 10 November 2021). The remaining code and datasets have not been approved by the authors' institutions for public release.

Acknowledgments: Hongli Jiang (formerly NOAA ESRL and CIRA) contributed in various ways to this study including providing guidance on using vLAPS analyses in WRF and running an earlier version of some of the experiments. The Real-Time Mesoscale Analysis use/reject lists provided by Steve Levine at the National Centers for Environmental Prediction's Environmental Modeling Center greatly facilitated making full use of the MADIS observational dataset. Cindy Bruyère (NCAR) provided the modified version of Ungrib used in this work to improve quality control of the observations for nudging. John Halley Gotway (NCAR) provided a script that facilitated ingestion of vLAPS-processed radar data into the Model Evaluation Tools (MET) for verification. Jeffrey Smith (ARL) assisted with file processing related to verification. This work was supported, in part, by high-performance computer time and resources from the DoD High Performance Computing Modernization Program. This study was made possible, in part, due to the data made available to NOAA by various data providers for inclusion in MADIS. MET was used in this study and was developed at NCAR through grants from the National Science Foundation, NOAA, the United States (US) Air Force, and the US Department of Energy. NCAR is sponsored by the US National Science Foundation.

Conflicts of Interest: The authors declare no conflict of interest.

References

1. Hu, M.; Xue, M.; Brewster, K. 3DVAR and Cloud Analysis with WSR-88D Level-II Data for the Prediction of the Fort Worth, Texas, Tornadic Thunderstorms. Part I: Cloud Analysis and Its Impact. *Mon. Weather Rev.* **2006**, *134*, 675–698. [CrossRef]
2. Xiao, Q.; Sun, J. Multiple-Radar Data Assimilation and Short-Range Quantitative Precipitation Forecasting of a Squall Line Observed during IHOP_2002. *Mon. Weather Rev.* **2007**, *135*, 3381–3404. [CrossRef]
3. Schroeder, A.J.; Stauffer, D.R.; Seaman, N.L.; Deng, A.; Gibbs, A.M.; Hunter, G.K.; Young, G.S. An Automated High-Resolution, Rapidly Relocatable Meteorological Nowcasting and Prediction System. *Mon. Weather Rev.* **2006**, *134*, 1237–1265. [CrossRef]
4. Reen, B.P.; Stauffer, D.R. Data assimilation strategies in the planetary boundary layer. *Bound.-Layer Meteorol.* **2010**, *137*, 237–269. [CrossRef]
5. Hu, M.; Benjamin, S.G.; Ladwig, T.T.; Dowell, D.C.; Weygandt, S.S.; Alexander, C.R.; Whitaker, J.S. GSI Three-Dimensional Ensemble–Variational Hybrid Data Assimilation Using a Global Ensemble for the Regional Rapid Refresh Model. *Mon. Weather Rev.* **2017**, *145*, 4205–4225. [CrossRef]
6. Lee, M.; Kuo, Y.; Barker, D.M.; Lim, E. Incremental Analysis Updates Initialization Technique Applied to 10-km MM5 and MM5 3DVAR. *Mon. Weather Rev.* **2006**, *134*, 1389–1404. [CrossRef]
7. Brewster, K.A.; Stratman, D.R. Tuning an analysis and incremental analysis updating assimilation for an efficient high-resolution forecast system. In Proceedings of the 20th Conference on Integrated Observing and Assimilation Systems for the Atmosphere, Oceans and Land Surface (IAOS-AOLS), New Orleans, LA, USA, 11–14 January 2016; Available online: https://ams.confex.com/ams/96Annual/webprogram/Paper289235.html (accessed on 10 November 2021).
8. Lorenc, A.C.; Rawlins, F. Why does 4D-Var beat 3D-Var? *Quart. J. Roy. Meteor. Soc.* **2005**, *131*, 3247–3257. [CrossRef]
9. Stauffer, D.R.; Seaman, N.L. Use of Four-Dimensional Data Assimilation in a Limited-Area Mesoscale Model. Part I: Experiments with Synoptic-Scale Data. *Mon. Weather Rev.* **1990**, *118*, 1250–1277. [CrossRef]
10. Stauffer, D.R.; Seaman, N.L.; Binkowski, F.S. Use of Four-Dimensional Data Assimilation in a Limited-Area Mesoscale Model Part II: Effects of Data Assimilation within the Planetary Boundary Layer. *Mon. Weather Rev.* **1991**, *119*, 734–754. [CrossRef]
11. Shaw, B.L.; Carpenter, R.L.; Spencer, P.L.; DuFran, Z. Commercial Implementation of WRF with Efficient Computing and Advanced Data Assimilation. In Proceedings of the 9th Annual WRF User's Workshop, Boulder, CO, USA, 23–27 June 2008; Available online: http://www2.mmm.ucar.edu/wrf/users/workshops/WS2008/abstracts/2-05.pdf (accessed on 10 November 2021).
12. Liu, Y.; Yu, W.; Warner, T.; Sun, J.; Xiao, Q.; Pace, J. Short-term explicit forecasting and nowcasting of convection systems with WRF using a hybrid radar data assimilation approach. In Proceedings of the 23rd Conference on Weather Analysis and Forecasting/19th Conference on Numerical Weather Prediction, Omaha, NE, USA, 1–5 June 2009; Available online: https://ams.confex.com/ams/23WAF19NWP/webprogram/Paper154264.html (accessed on 10 November 2021).

13. Lei, L.; Stauffer, D.R.; Haupt, S.E.; Young, G.S. A hybrid nudging- ensemble Kalman filter approach to data assimilation. Part I: Application in the Lorenz system. *Tellus* **2012**, *64A*, 18484. [CrossRef]
14. Lei, L.; Stauffer, D.R.; Deng, A. A hybrid nudging-ensemble Kalman filter approach to data assimilation. Part II: Application in a shallow-water model. *Tellus* **2012**, *64A*, 18485. [CrossRef]
15. Lei, L.; Stauffer, D.R.; Deng, A. A hybrid nudging-ensemble Kalman filter approach to data assimilation in WRF/DART. *Q. J. R. Meteorol. Soc.* **2012**, *138*, 2066–2078. [CrossRef]
16. Skamarock, W.C.; Coauthors. *A Description of the Advanced Research WRF Version 3*; NCAR Technical Note NCAR/TN-4751STR; National Center for Atmospheric Research: Boulder, CO, USA, 2008; p. 113. Available online: http://dx.doi.org/10.5065/D68S4MVH (accessed on 10 November 2021).
17. Liu, Y.; Wu, Y.; Pan, L.; Bourgeois, A.; Roux, G.; Knievel, J.C.; Hacker, J.; Pace, J.; Gallagher, W.; Halvorson, S.F. Recent developments of NCAR's 4D-Relaxation Ensemble Kalman Filter system. In Proceedings of the 19th Conference on Integrated Observing and Assimilation Systems for the Atmosphere, Oceans, and Land Surface (IOAS-AOLS), Phoenix, AZ, USA, 5–8 January 2015; Available online: https://ams.confex.com/ams/95Annual/webprogram/Paper269311.html (accessed on 10 November 2021).
18. Lei, L.; Hacker, J.P. Nudging, Ensemble, and Nudging Ensembles for Data Assimilation in the Presence of Model Error. *Mon. Weather Rev.* **2015**, *143*, 2600–2610. [CrossRef]
19. Jiang, H.; Albers, S.; Xie, Y.; Toth, Z.; Jankov, I.; Scotten, M.; Picca, J.; Stumpf, G.; Kingfield, D.; Birkenheuer, D.; et al. Real-Time Applications of the Variational Version of the Local Analysis and Prediction System (vLAPS). *Bull. Am. Meteorol. Soc.* **2015**, *96*, 2045–2057. [CrossRef]
20. Reen, B.P.; Xie, Y.; Cai, H.; Albers, S.; Dumais, R.E., Jr.; Jiang, H. *Incorporating Variational Local Analysis and Prediction System (vLAPS) Analyses with Nudging Data Assimilation: Methodology and Initial Results*; ARL-TR-8145; U.S. Army Research Laboratory: Adelphi, MD, USA, 2017; p. 76. Available online: https://apps.dtic.mil/sti/pdfs/AD1039383.pdf (accessed on 10 November 2021).
21. Rogers, R.E.; Deng, A.; Stauffer, D.R.; Gaudet, B.J.; Jia, Y.; Soong, S.; Tanrikulu, S. Application of the Weather Research and Forecasting Model for Air Quality Modeling in the San Francisco Bay Area. *J. Appl. Meteorol. Climatol.* **2013**, *52*, 1953–1973. [CrossRef]
22. Expósito, F.J.; González, A.; Pérez, J.C.; Díaz, J.P.; Taima, D. High-Resolution Future Projections of Temperature and Precipitation in the Canary Islands. *J. Clim.* **2015**, *28*, 7846–7856. [CrossRef]
23. Huo, Z.; Liu, Y.; Wei, M.; Shi, Y.; Fang, C.; Shu, Z.; Li, Y. Hydrometeor and Latent Heating Nudging for Radar Reflectivity Assimilation: Response to the Model States and Uncertainties. *Remote Sens.* **2021**, *13*, 3821. [CrossRef]
24. Janjić, Z.I. *Nonsingular Implementation of the Mellor–Yamada Level 2.5 Scheme in the NCEP Meso Model*; NCEP Office Note 437; National Centers for Environmental Prediction: Camp Springs, MD, USA, 2001; p. 61. Available online: http://www.lib.ncep.noaa.gov/ncepofficenotes/files/on437.pdf (accessed on 10 November 2021).
25. Lee, J.A.; Kolczyinski, W.C.; McCandless, T.C.; Haupt, S.E. An Objective Methodology for Configuring and Down-Selecting an NWP Ensemble for Low-Level Wind Prediction. *Mon. Weather Rev.* **2012**, *140*, 2270–2286. [CrossRef]
26. Reen, B.P.; Schmehl, K.J.; Young, G.S.; Lee, J.A.; Haupt, S.E.; Stauffer, D.R. Uncertainty in Contaminant Concentration Fields Resulting from Atmospheric Boundary Layer Depth Uncertainty. *J. Appl. Meteorol. Climatol.* **2014**, *53*, 2610–2626. [CrossRef]
27. Thompson, G.; Field, P.R.; Rasmussen, R.M.; Hall, W.D. Explicit Forecasts of Winter Precipitation Using an Improved Bulk Microphysics Scheme. Part II: Implementation of a New Snow Parameterization. *Mon. Weather Rev.* **2008**, *136*, 5095–5115. [CrossRef]
28. Mlawer, E.J.; Taubman, S.J.; Brown, P.D.; Iacono, M.J.; Clough, S.A. Radiative transfer for inhomogeneous atmospheres: RRTM, a validated correlated-k model for the longwave. *J. Geophys. Res. Atmos.* **1997**, *102*, 16663–16682. [CrossRef]
29. Dudhia, J. Numerical Study of Convection Observed during the Winter Monsoon Experiment Using a Mesoscale Two-Dimensional Model. *J. Atmos. Sci.* **1989**, *46*, 3077–3107. [CrossRef]
30. Chen, F.; Dudhia, J. Coupling an Advanced Land Surface–Hydrology Model with the Penn State–NCAR MM5 Modeling System. Part I: Model Implementation and Sensitivity. *Mon. Weather Rev.* **2001**, *129*, 569–585. [CrossRef]
31. McGinley, J.A.; Albers, S.C.; Stamus, P.A. Validation of a Composite Convective Index as Defined by a Real-Time Local Analysis System. *Weather Forecast.* **1991**, *6*, 337–356. [CrossRef]
32. Hiemstra, C.A.; Liston, G.E.; Pielke, R.A.; Birkenheuer, D.L.; Albers, S.C. Comparing Local Analysis and Prediction System (LAPS) Assimilations with Independent Observations. *Weather Forecast.* **2006**, *21*, 1024–1040. [CrossRef]
33. Albers, S.C.; McGinley, J.A.; Birkenheuer, D.L.; Smart, J.R. The Local Analysis and Prediction System (LAPS): Analyses of Clouds, Precipitation, and Temperature. *Weather Forecast.* **1996**, *11*, 273–287. [CrossRef]
34. Xie, Y.; Koch, S.; McGinley, J.; Albers, S.; Bieringer, P.E.; Wolfson, M.; Chan, M. A Space–Time Multiscale Analysis System: A Sequential Variational Analysis Approach. *Mon. Weather Rev.* **2011**, *139*, 1224–1240. [CrossRef]
35. Xie, Y.F.; MacDonald, A.E. Selection of momentum variables for a three-dimensional variational analysis. *Pure Appl. Geophys.* **2012**, *169*, 335–351. [CrossRef]
36. Xie, Y.; Lu, C.; Browning, G.L. Impact of Formulation of Cost Function and Constraints on Three-Dimensional Variational Data Assimilation. *Mon. Weather Rev.* **2002**, *130*, 2433–2447. [CrossRef]

37. Alexander, C.; Smirnova, T.; Hu, M.; Weygandt, S.; Dowell, D.; Olson, J.; Benjamin, S.; James, E.; Brown, J.; Hofmann, P.; et al. Recent advancements in convective-scale storm prediction with the coupled Rapid Refresh (RAP)/High-Resolution Rapid Refresh (HRRR) forecast system. In Proceedings of the 14th Annual WRF Users' Workshop, Boulder, CO, USA, 24–28 June 2013; Available online: http://www2.mmm.ucar.edu/wrf/users/workshops/WS2013/ppts/2.3.pdf (accessed on 10 November 2021).
38. Stauffer, D.R.; Seaman, N.L. Multiscale Four-Dimensional Data Assimilation. *J. Appl. Meteorol. Climatol.* **1994**, *33*, 416–434. [CrossRef]
39. Gilliam, R.C.; Godowitch, J.M.; Rao, S.T. Improving the horizontal transport in the lower troposphere with four dimensional data assimilation. *Atmos. Environ.* **2012**, *53*, 186–201. [CrossRef]
40. Reen, B.P. A Brief Guide to Observation Nudging in WRF. 2016. Available online: http://www2.mmm.ucar.edu/wrf/users/docs/ObsNudgingGuide.pdf (accessed on 10 November 2021).
41. De Pondeca, M.S.; Manikin, G.S.; DiMego, G.; Benjamin, S.G.; Parrish, D.F.; Purser, R.J.; Wu, W.; Horel, J.D.; Myrick, D.T.; Lin, Y.; et al. The Real-Time Mesoscale Analysis at NOAA's National Centers for Environmental Prediction: Current Status and Development. *Weather Forecast.* **2011**, *26*, 593–612. [CrossRef]
42. National Center for Atmospheric Research. User's Guide for the Advanced Research WRF (ARW) Modeling System Version 3.8. 2016. Available online: https://www2.mmm.ucar.edu/wrf/users/docs/user_guide_V3/user_guide_V3.8/contents.html (accessed on 10 November 2021).
43. Reen, B.P.; Dumais, R.E.; Passner, J.E. Mitigating Excessive Drying from the Use of Observations in Mesoscale Modeling. *J. Appl. Meteorol. Climatol.* **2016**, *55*, 365–388. [CrossRef]
44. Atkins, N.T.; Butler, K.M.; Flynn, K.R.; Wakimoto, R.M. An Integrated Damage, Visual, and Radar Analysis of the 2013 Moore, Oklahoma, EF5 Tornado. *Bull. Am. Meteorol. Soc.* **2014**, *95*, 1549–1561. [CrossRef]
45. Burgess, D.; Ortega, K.; Stumpf, G.; Garfield, G.; Karstens, C.; Meyer, T.; Smith, B.; Speheger, D.; Ladue, J.; Smith, R.; et al. 20 May 2013 Moore, Oklahoma, Tornado: Damage Survey and Analysis. *Weather Forecast.* **2014**, *29*, 1229–1237. [CrossRef]
46. Kurdzo, J.M.; Bodine, D.J.; Cheong, B.L.; Palmer, R.D. High-Temporal Resolution Polarimetric X-Band Doppler Radar Observations of the 20 May 2013 Moore, Oklahoma, Tornado. *Mon. Weather Rev.* **2015**, *143*, 2711–2735. [CrossRef]
47. Hanley, K.E.; Barrett, A.I.; Lean, H.W. Simulating the 20 May 2013 Moore, Oklahoma tornado with a 100-metre grid-length NWP model. *Atmos. Sci. Lett.* **2016**, *17*, 453–461. [CrossRef]
48. Zhang, Y.; Zhang, F.; Stensrud, D.J.; Meng, Z. Practical Predictability of the 20 May 2013 Tornadic Thunderstorm Event in Oklahoma: Sensitivity to Synoptic Timing and Topographical Influence. *Mon. Weather Rev.* **2015**, *143*, 2973–2997. [CrossRef]
49. Zhang, Y.; Zhang, F.; Stensrud, D.J.; Meng, Z. Intrinsic Predictability of the 20 May 2013 Tornadic Thunderstorm Event in Oklahoma at Storm Scales. *Mon. Weather Rev.* **2016**, *144*, 1273–1298. [CrossRef]
50. Snook, N.; Jung, Y.; Brotzge, J.; Putnam, B.; Xue, M. Prediction and Ensemble Forecast Verification of Hail in the Supercell Storms of 20 May 2013. *Weather Forecast.* **2016**, *31*, 811–825. [CrossRef]
51. Skok, G.; Roberts, N. Analysis of fractions skill score properties for random precipitation fields and ECMWF forecasts. *Q. J. R. Meteorol. Soc.* **2016**, *142*, 2599–2610. [CrossRef]
52. Schaefer, J.T. The critical success index as an indicator of warning skill. *Weather Forecast.* **1990**, *5*, 570–575. [CrossRef]
53. Wilks, D.S. *Statistical Methods in the Atmospheric Sciences*, 2nd ed.; Academic Press: San Diego, CA, USA, 2005; p. 264.
54. Hong, S.-Y.; Park, H.; Cheong, H.-B.; Kim, J.-E.E.; Koo, M.-S.; Jang, J.; Ham, S.; Hwang, S.-O.; Park, B.-K.; Chang, E.-C.; et al. The global/regional integrated model system (GRIMs). *Asia Pac. J. Atmos. Sci.* **2013**, *49*, 219–243. [CrossRef]
55. Sun, J.; Xue, M.; Wilson, J.W.; Zawadzki, I.; Ballard, S.P.; Onvlee-Hooimeyer, J.; Joe, P.; Barker, D.M.; Li, P.-W.; Golding, B.; et al. Use of NWP for nowcasting convective precipitation: Recent progress and challenges. *Bull. Am. Meteorol. Soc.* **2014**, *95*, 409–426. [CrossRef]
56. Koch, S.; Ferrie, B.; Stoelinga, M.; Szoke, E.; Weiss, J.; Kain, S.J. The use of simulated radar reflectivity fields in the diagnosis of mesoscale phenomena from high-resolution WRF model forecasts. In Proceedings of the 11th Conference on Mesoscale Processes/32nd Conference on Radar Meteorology, Albuquerque, NM, USA, 24–28 October 2005; Available online: https://ams.confex.com/ams/32Rad11Meso/techprogram/paper_97032.htm (accessed on 10 November 2021).
57. Skamarock, W.C.; Park, S.-H.; Klemp, J.B.; Snyder, C. Atmospheric kinetic energy spectra from global high-resolution non-hydrostatic simulations. *J. Atmos. Sci.* **2014**, *71*, 4369–4381. [CrossRef]
58. Shaw, B.L.; Albers, S.; Birkenheuer, D.; Brown, J.; McGinley, J.A.; Schultz, P.; Smart, J.R.; Szoke, E. Application of the Local Analysis and Prediction System (LAPS) diabatic initialization of mesoscale numerical weather prediction models for the IHOP—2002 field experiment. In Proceedings of the 20th Conference on Weather Analysis and Forecasting/16th Conference on Numerical Weather Prediction, Seattle, WA, USA, 12–15 January 2004; Available online: https://ams.confex.com/ams/84Annual/techprogram/paper_69700.htm (accessed on 10 November 2021).
59. Wapler, K.; Harnisch, F.; Pardowitz, T.; Senf, F. Characterisation and predictability of a strong and a weak forcing severe convective event a multi-data approach. *Meteorol. Z.* **2015**, *24*, 393–410. [CrossRef]
60. Keil, C.; Heinlein, F.; Craig, G.C. The convective adjustment time-scale as indicator of predictability of convective precipitation. *Q. J. R. Meteorol. Soc.* **2014**, *140*, 480–490. [CrossRef]
61. Stensrud, D.J.; Fritsch, J.M. Mesoscale convective systems in weakly forced large-scale environments. Part II: Generation of a mesoscale initial condition. *Mon. Weather Rev.* **1994**, *122*, 2068–2083. [CrossRef]

62. Duda, J.D.; Gallus, W.A. The impact of large-scale forcing on skill of simulated convective initiation and upscale evolution with convection-allowing grid spacings in the WRF. *Weather Forecast.* **2013**, *28*, 994–1018. [CrossRef]
63. Stratman, D.R.; Coniglio, M.C.; Koch, S.E.; Xue, M. Use of Multiple Verification Methods to Evaluate Forecasts of Convection from Hot- and Cold-Start Convection-Allowing Models. *Weather Forecast.* **2013**, *28*, 119–138. [CrossRef]
64. Carlberg, B.R.; Gallus, W.A.; Franz, K.J. A Preliminary Examination of WRF Ensemble Prediction of Convective Mode Evolution. *Weather Forecast.* **2018**, *33*, 783–798. [CrossRef]
65. Pan, L.; Liu, Y.; Lei, Y.; Xu, M.; Child, P.; Jacobs, N. Development of a CONUS radar data assimilation WRF-RTFDDA system for convection-resolvable analysis and prediction. In Proceedings of the 16th WRF Annual User's Workshop, Boulder, CO, USA, 15–19 June 2015; Available online: https://www2.mmm.ucar.edu/wrf/users/workshops/WS2015/WorkshopPapers.php (accessed on 10 November 2021).
66. Zeng, Y.; Janjić, T.; de Lozar, A.; Blahak, U.; Reich, H.; Keil, C.; Seifert, A. Representation of model error in convective-scale data assimilation: Additive noise, relaxation methods, and combinations. *J. Adv. Model. Earth Syst.* **2018**, *10*, 2889–2911. [CrossRef]
67. James, E.P.; Benjamin, S.G. Observation System Experiments with the Hourly Updating Rapid Refresh Model Using GSI Hybrid Ensemble-Variational Data Assimilation. *Mon. Weather Rev.* **2017**, *145*, 2897–2918. [CrossRef]
68. Ban, J.; Liu, Z.; Zhang, X.; Huang, X.-Y.; Wang, H. Precipitation data assimilation in WRFDA 4D-Var: Implementation and application to convection-permitting forecasts over United States. *Tellus* **2017**, *69*, 1368310. [CrossRef]
69. Veillette, M.S.; Iskenderian, H.; Lamey, P.M.; Mattioli, C.J.; Banerjee, A.; Worris, M.; Porschitsky, A.B.; Ferris, R.F.; Manwelyan, A.; Rajogoplalan, S.; et al. Global Synthetic Weather Radar in AWS GovCloud for the U.S. Air Force. In Proceedings of the 19th Conference on Artificial Intelligence for Environmental Science, Boston, MA, USA, 13–16 January 2020; Available online: https://ams.confex.com/ams/2020Annual/meetingapp.cgi/Paper/363150 (accessed on 10 November 2021).
70. Bytheway, J.L.; Kummerow, C.D. Toward an object-based assessment of high-resolution forecasts of long-lived convective precipitation in the central U.S. *J. Adv. Model. Earth Syst.* **2015**, *7*, 1248–1264. [CrossRef]

atmosphere

Article

WRF Rainfall Modeling Post-Processing by Adaptive Parameterization of Raindrop Size Distribution: A Case Study on the United Kingdom

Qiqi Yang [1], Shuliang Zhang [1,*], Qiang Dai [1,2] and Hanchen Zhuang [1]

1 Key Laboratory of VGE of Ministry of Education, Nanjing Normal University, Nanjing 210023, China; yangqiqi@nnu.edu.cn (Q.Y.); q.dai@njnu.edu.cn (Q.D.); zhuanghanchen@nnu.edu.cn (H.Z.)
2 Department of Civil Engineering, University of Bristol, Bristol BS8 1TR, UK
* Correspondence: zhangshuliang@njnu.edu.cn

Abstract: Raindrop size distribution (RSD) is a key parameter in the Weather Research and Forecasting (WRF) model for rainfall estimation, with gamma distribution models commonly used to describe RSD under WRF microphysical parameterizations. The RSD model sets the shape parameter (μ) as a constant of gamma distribution in WRF double-moment bulk microphysics schemes. Here, we propose to improve the gamma RSD model with an adaptive value of μ based on the rainfall intensity and season, designed using a genetic algorithm (GA) and the linear least-squares method. The model can be described as a piecewise post-processing function that is constant when rainfall intensity is <1.5 mm/h and linear otherwise. Our numerical simulation uses the WRF driven by an ERA-interim dataset with three distinct double-moment bulk microphysical parameterizations, namely, the Morrison, WDM6, and Thompson aerosol-aware schemes for the period of 2013–2017 over the United Kingdom at a 5 km resolution. Observations were made using a disdrometer and 241 rain gauges, which were used for calibration and validation. The results show that the adaptive-μ model of the gamma distribution was more accurate than the gamma RSD model with a constant shape parameter, with the root-mean-square error decreasing by averages of 23.62%, 11.33%, and 22.21% for the Morrison, WDM6, and Thompson aerosol-aware schemes, respectively. This model improves the accuracy of WRF rainfall simulation by applying adaptive RSD parameterization and can be integrated into the simulation of WRF double-moment microphysics schemes. The physical mechanism of the RSD model remains to be determined to improve its performance in WRF bulk microphysics schemes.

Keywords: Weather Research and Forecasting; raindrop size distribution; adaptive shape parameter model; gamma distribution; double-moment microphysics schemes

Citation: Yang, Q.; Zhang, S.; Dai, Q.; Zhuang, H. WRF Rainfall Modeling Post-Processing by Adaptive Parameterization of Raindrop Size Distribution: A Case Study on the United Kingdom. *Atmosphere* **2022**, *13*, 36. https://doi.org/10.3390/atmos13010036

Academic Editors: Zuohao Cao, Huaqing Cai and Xiaofan Li

Received: 18 November 2021
Accepted: 24 December 2021
Published: 27 December 2021

1. Introduction

Raindrop size distribution (RSD) provides fundamental information for characterizing the microphysical properties of precipitation and is an important factor in determining the accuracy of rainfall retrieval for radar-based quantitative precipitation estimation and rainfall simulation of numerical weather prediction systems [1–3]. Moreover, RSD also plays an important role in calculating rainfall kinetic energy [4], which is a dominant parameter affecting soil erosion [5,6].

Ground-based disdrometers are a precise tool for measuring RSD, studying rain microphysics, and verifying the rainfall retrievals obtained through remote sensing via radar and satellite or numerical weather forecast models [7]. However, disdrometer measurements can only represent point information on rainfall characteristics and have a limited ability to spatially represent larger areas. The numerical Weather Research and Forecasting (WRF) model can provide RSD spectra for each grid in the simulating results to cover a broader area than that obtained with the disdrometer [8].

The WRF model, with advanced dynamics, physics, and numerical schemes [9,10], has been increasingly used in water resource studies and hydrologic modeling [11]. However, predicting rainfall is extremely challenging for numerical models due to the lack of physical representation of rainfall generation and the limitation in computation that must simplify droplet generation as well as the limitation in grid resolution, among others. The WRF model exhibits relatively poor performance for estimating rainfall because there are many model biases. The cloud microphysical scheme with various RSD parameterizations is one of the main factors leading to the uncertainty of WRF rainfall simulation [2,12]. Bulk microphysics parameterization (BMP) schemes, which are standard approaches in representing cloud processes used in the WRF model [13–16], assume a hydrometeor particle size distribution function in the form of a gamma distribution with three parameters (intercept parameter N_0, shape parameter μ, and slope parameter λ). The BMP schemes can be classified into one-, two-, and three-moment schemes based on physical quantities. Several studies have shown that the two-moment BMP schemes outperformed one-moment schemes in a convection-scale simulation and supercell storm and performed similarly to the three-moment schemes with higher computational complexity [13,14,17,18]. Two-moment schemes apply a specific case of the gamma distribution in which the value of the shape parameter is set to 0, with a few exceptions (e.g., WDM6 scheme with $\mu = 1$) [14]. Thus, these schemes reduce the gamma RSD model from three parameters to two.

However, Ulbrich [19] found that the three-parameter gamma RSD model can characterize a broader range of raindrop size distributions than two-parameter (e.g., exponential) distributions while flexibly describing the relative number concentration of large raindrops. The shape parameter is directly related to the median volume diameter D_0 [19,20]. Milbrandt and Yau [21] suggested that it may be an improvement over using a fixed value of μ in two-moment schemes using a monotonically increasing function of the mean-mass drop diameter, and the shape parameter μ. Yang et al. [2] suggested that a gamma RSD model with an adaptive value of μ should be developed using the WRF model. Other studies [22,23] revealed that D_m increases with the rain rate, suggesting that the value of μ is also related to rainfall intensity. Nevertheless, studies aimed at improving the accuracy of WRF rainfall simulation from the perspective of investigating an adaptive or dynamic shape parameter framework of the RSD model in WRF double-moment microphysical schemes remain limited.

Therefore, we predicted that the gamma RSD model with an adaptive shape parameter could improve the accuracy of WRF rainfall simulation compared to the model with a fixed shape parameter of double-moment microphysics parameterizations (MPs). Additionally, we hypothesized that the shape parameter is related to rainfall intensity and can be used to construct an adaptive-μ model of gamma distribution for WRF double-moment bulk schemes [2].

To test this hypothesis, we conducted several long-term experiments and data assessments covering the period from 2013 to 2017. We downscaled five years of ERA-Interim [24,25] data to 5 km spatial and 1 h temporal resolutions using the WRF model based on three different double-moment microphysics schemes. Measurements from a disdrometer in southern England were used to constrain the value of the shape parameter of the gamma RSD model within a reasonable range, and measurements from 241 rain gauges with hourly resolution spread across the United Kingdom (UK) were used to evaluate the WRF model rainfall simulation results. Among the rain gauges, 211 were selected to construct the adaptive-μ model, whereas the others were used to validate the proposed model against three different error indices. This paper is organized as follows: Section 2 describes the study area and data source; Section 3 describes the WRF model configurations, RSD model, experimental designs, and evaluation indicators; Section 4 presents the results and validation of the proposed method; and Sections 5 and 6 provide a comprehensive discussion and conclude the main points of this study, respectively.

2. Study Area and Data

The study area is in the UK off the northwest coast of Europe within a range of 49°46′ N–60°43′ N and 8°25′ W–3°35′ E. This region covers approximately 248,532 km^2 and comprises mainly lowland terrain with a maximum elevation of 1345 m.

The ERA-Interim datasets from the third generation of the European Centre for Medium-Range Weather Forecasts (ECMWF) reanalysis [26] were used to drive the WRF model used in the study. ERA-Interim is a significant atmospheric data source covering a data-rich period from January 1979 to the (near-real-time) present with a spatial resolution of approximately 80 km and a 3 h time interval [26]. A 2006 release of the Integrated Forecasting System (IFS-Cy31r2) was used to support the data assimilation system to produce the ERA-Interim. The system includes a four-dimensional variational analysis with a 12 h analysis window. In this study, we selected the ERA-Interim reanalysis dataset since it has been used extensively in WRF downscaling modeling studies [27,28] and performs well in rainfall simulation [29–31]. More details on the ERA-Interim dataset can be found on the ECMWF website at http://apps.ecmwf.int/ (accessed on 20 September 2020). For our study, five years of ERA-Interim data covering the study area from January 2013 to December 2017 were downscaled using the WRF model at temporal and spatial resolutions of 1 h and 5 km, respectively.

The tipping bucket rain gauge data were sourced from the Met Office Integrated Data Archive System Land and Marine Surface Stations data set (1853–present) [32]. These data comprise daily, hourly, and sub-hourly rain measurements, including land and marine surface observations. The bucket size was 0.2 mm, and the tip time was up to 10 s time resolution [33]. Measurements from 241 rain gauges with 1 h rainfall accumulations collected from 2013 to 2017 by the UK Met Office weather station network were selected to evaluate the accuracy of WRF rainfall simulation. Thirty rain gauges were randomly selected to validate the experimental results. The locations of the rain gauges and validation points are shown in Figure 1.

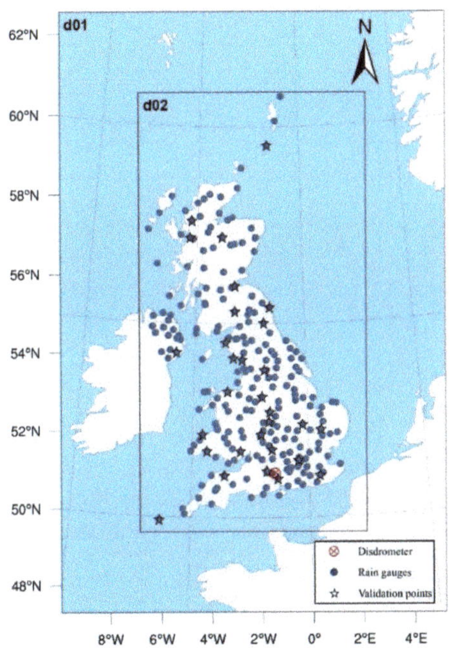

Figure 1. Distributions of the rain gauges and the domain configurations used in the WRF model over the United Kingdom.

An impact-type Joss–Waldvogel (JW) disdrometer was used to analyze the appropriate interval of the shape parameter. The disdrometer was sited in southern England at 51°08′ N, 1°26′ W (Figure 1) and measured drop sizes in 127 bins from 0.3 to 5.0 mm with a sampling period and collector area of 10 s and 50 cm^2, respectively. The disdrometer data provided by the British Atmospheric Data Centre cover an extended period from April 2003 to July 2018; in this study, data covering 2013 to 2017 were used to correspond with the rain gauge data. There are some uncertainties in the JW disdrometer data; for example, drops larger than approximately 5.0 mm diameter cannot be distinguished and the effect of vertical air motion on drop fall speed is neglected [34]. Therefore, raindrops larger than 5.0 mm and drizzle (rainfall intensity < 0.1 mm/h) recorded by a disdrometer were excluded from analysis in this study. In addition, to filter out the time variation previously reported [22,35,36], 10 s data measurements were averaged into 1 min periods.

3. Methodology

3.1. WRF Model Configurations

WRF model version 3.8, an Advanced Research WRF dynamical core, was used to downscale the ERA-Interim reanalysis data. The doubly nested domain configuration used in the WRF model was centered at 55°19′ N, 2°21′ W and applied at a downscaling ratio of 1:5, a finest grid of 5 km, and a temporal resolution of 1 h. In this study, we selected the 1:5 downscaling ratio rather than the 1:3 ratio, since it provides higher resolution results than the 1:3 ratio and also results in good performance of the WRF model [37]. Combining the domain center, domain size, and grid spacing, most areas in the UK can be covered as shown in Figure 1. A detailed list of the parameters used in this domain configuration is provided in Table 1. Both domains include 28 vertical pressure levels, with the top-level set at 50 hPa in each.

Table 1. Configurations of the two nested WRF model domains.

Domain	Domain Size (km)	Grid Spacing (km)	Grid Size	Downscaling Ratio
d01	1125 × 1675	25	45 × 67	-
d02	655 × 1230	5	131 × 46	1:5

The simulations were performed using three different bulk double-moment microphysical parameterizations, namely, the Morrison [38], WDM6 (WRF double-moment 6-class) [15,39], and aerosol-aware Thompson [40] schemes. The Morrison scheme can predict the number concentrations of ice, snow, rain, and graupel particles; the WDM6 scheme can predict the number concentrations of cloud droplets, rain, and cloud condensation nuclei [15,39]; and the Thompson aerosol-aware scheme can predict the number concentrations of cloud droplets and ice, rain, cloud condensation nuclei, and ice nuclei [40]. For the cumulus scheme, the Kain–Fritsch scheme [41] is used, whereas the cumulus scheme is not used in the inner domain because convective rainfall generation is definitely resolved when the model grid spacing is ≤5 km [37,42]. Other physical parameterizations include the Mellor–Yamada–Janjić planetary boundary layer scheme [43], RRTM longwave radiation scheme [44], Dudhia shortwave radiation scheme [45], and Noah land-surface model [46].

3.2. Gamma RSD Model

The normalized gamma distribution is widely used to model RSD spectra because it facilitates straightforward comparisons of the microphysical characteristics of rainfall, such as raindrop diameter and number concentration [36,47], which are in the form of:

$$(D) = N_w f(\mu)\left(\frac{D}{D_m}\right)^{\mu} exp\left[-(4+\mu)\frac{D}{D_m}\right] \quad (1)$$

$$f(\mu) = \frac{6(4+\mu)^{\mu+4}}{4^4\Gamma(\mu+4)} \quad (2)$$

where N_w (mm^{-1} mm^{-3}), μ, and D_m (mm) are the generalized intercept, shape, and mass-weighted mean diameter parameters defining the RSD spectra $N(D)$; D is the equivalent volume diameter in mm; and $\Gamma(n)$ is the Euler gamma function. However, a three-parameter gamma distribution [19,47,48] is used to model RSD spectra in the WRF model, which is expressed as:

$$N(D) = N_0 D^{\mu} e^{-\lambda D}, \tag{3}$$

where N_0 and λ are the intercept and slope parameters. In two-moment microphysical schemes of the WRF model, the intercept and slope parameters are obtained from the predicted number concentration N, mixing ratio q, and fixed μ as follows [38]:

$$N_0 = \frac{N\lambda^{\mu+1}}{\Gamma(\mu+1)}, \tag{4}$$

$$\lambda = \left[\frac{cN\Gamma(\mu+d+1)}{q\Gamma(\mu+1)}\right]^{\frac{1}{d}}, \tag{5}$$

where c and d are the coefficients of an assumed power law between mass and diameter $m = cD^d$ [38]. Based on a comparison of Equations (1) and (3), μ in the two formulas is the same, and the N_0 and λ parameters of the three-parameter gamma distribution can be converted to D_m and N_w parameters of normalized gamma distribution using the following equations:

$$D_m = \frac{4+\mu}{\lambda}, \tag{6}$$

$$N_w = \frac{N_0\left(\frac{4+\mu}{\lambda}\right)^{\mu}}{f(\mu)}. \tag{7}$$

The RSD measured by the JW disdrometer at time instant t (s) can be obtained as follows [22,36]:

$$N_m(D_i, t) = \frac{n_i(t)}{A \cdot \Delta t \cdot V_i \cdot \Delta D_i}, \tag{8}$$

where m is a measured quantity, $N_m(D_i, t)$ is the number of raindrops per unit volume in channel i at time t $(\mathrm{mm}^{-1} m^{-3})$, $n_i(t)$ is the number of raindrops counted in channel i at t, A is the sensor area (m^2), ΔD_i is the width of channel i (mm), and V_i is the terminal speed of the raindrops [49] (m/s), which is expressed as:

$$V_i = 3.78 \cdot D_i^{0.67}. \tag{9}$$

The distribution of $N_m(D_i, t)$ was adopted from the normalized gamma distribution fit $N(D)$ [19,50,51]. μ, D_m, and N_w, can be estimated based on $N(D_i)$. The n-order matrix m_n of $N(D_i)$ can be calculated by:

$$m_n = \sum_{j=1}^{n_c} D_i^n N(D_i) \Delta D_i. \tag{10}$$

μ can be estimated as follows [52]:

$$\mu = \frac{(7-11\gamma) - \sqrt{(7-11\gamma)^2 - 4(\gamma-1)(30\gamma-12)}}{2(\gamma-1)}. \tag{11}$$

where $\gamma = \frac{m_4^2}{m_2 m_6}$. D_m is calculated by:

$$D_m = \frac{m_4}{m_3}, \tag{12}$$

and N_w can be estimated from:

$$N_w = \frac{256}{6}\frac{m_3^5}{m_4^4}. \tag{13}$$

The rainfall intensity (mm/h) R in each rainfall step of the WRF or JW disdrometer is obtained from the RSD as:

$$R = 6\pi \cdot 10^{-4} \int_0^{\infty} D^3 V(D) N(D) dD. \quad (14)$$

3.3. Experimental Designs

We determined whether a gamma RSD model with an adaptive shape parameter of WRF double-moment microphysical parameterization could improve the accuracy of WRF rainfall retrieval. An adaptive-μ model of the gamma distribution based on rainfall intensity was tested over six scenarios comprising three double-moment schemes and two seasons. To focus on liquid precipitation analyses, hail and snow were excluded from this study according to temperature, drop size, and ground weather reports.

Figure 2 presents the experimental flowchart of this study. As an initial step, five years of ERA-Interim data were downscaled by applying the WRF model under different double-moment schemes, i.e., Morrison, WDM6, and Thompson aerosol-aware MPs, to obtain long-term WRF simulation results. To simulate the continuous WRF results for each year, we ran the model monthly with a spin-up time of 12 h [53] for the three MPs using the configurations described in Section 3.1. This study only used the lowest-level and inner-domain of the WRF results. The RSDs of different double-moment schemes were calculated based on the predicted number concentration and mixing ratio variables of the WRF outputs using Equations (4) and (5).

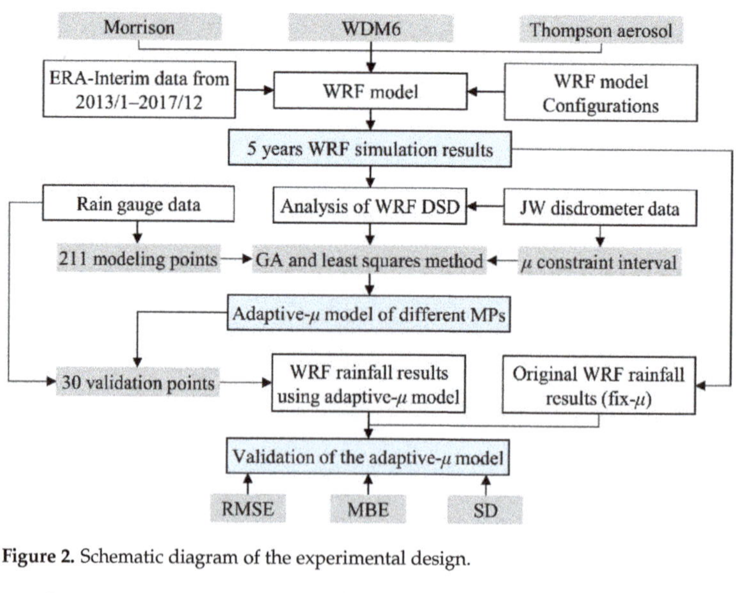

Figure 2. Schematic diagram of the experimental design.

Second, we analyzed the characteristics of the WRF RSD simulation results of three schemes. The $D_m - log_{10}N_w$ relationship and $D_m - R$ relationship of the three schemes in the WRF model were investigated by comparison of these relationships with the JW disdrometer in this study. The rainfall characteristics of different seasons are very different, which is worth studying. Concerning the climatic seasonal characteristics of the UK, we separated the four seasons into cold (23 September–19 March) and warm seasons (20 March–22 September) [22]. Based on the five years of disdrometer-observed RSD, the constraint interval of the shape parameter for different seasons was calculated.

Third, we used the three-parameter RSD model by applying the adaptive-μ method to improve the performance of WRF rainfall simulation. We modified the shape parameter μ

in the RSD model to introduce one extra degree of freedom to it to improve its fitness. We searched for the optimal μ within the constraint interval of rainfall intensities following the principle of minimizing the WRF rainfall simulation error by comparing observed rainfall data for different seasons with results produced by three double-moment microphysics schemes. According to the three-parameter gamma RSD model and Equations (4) and (5), the N_0 and λ parameters are computed based on the shape parameter μ; thus, the N_0 and λ parameters will also be changed by μ. The root-mean-square error (RMSE) in mm was selected as the objective function of the optimization algorithm, as shown in Equation (15), and can range from zero to infinity, with a lower value corresponding to a better fit to the observed data:

$$\text{RMSE} = \sqrt{\frac{1}{N}\sum_{i=1}^{N}(R_i - G_i)^2}, \tag{15}$$

where G_i and R_i are the rainfall intensity of each rainfall step at time i derived from the rain gauge rainfall data and the corresponding location of the WRF RSD data, respectively, and N is the total number of rainfall steps. The genetic algorithm (GA), which is inspired by natural selection processes involved in biological evolution [54], can solve both constrained and unconstrained optimization problems and can avoid being trapped at locally optimal solutions [55]. The main characteristics of GA include coding of the parameter set rather than the parameters themselves, initiating its search from multiple points rather than a single point, applying payoff information rather than derivatives, and using probabilistic transition rules rather than deterministic ones [56]. There are three main genetic operators of GA, namely, reproduction, crossover, and mutation [56]. The optimization search procedure in GA is based on the principle of natural selection and natural genetics. We implemented GA optimization to minimize the value of the RMSE when searching for the optimal μ at different rainfall intensities. Depending on the relationship between the optimal μ and rainfall intensity, a linear adaptive-μ model was proposed for each combination of the season and double-moment scheme.

Finally, we applied the adaptive-μ models to the 30 validation points to test their reliability and applicability relative to a fixed-μ model. As the adaptive-μ models were established to improve the precision of WRF rainfall simulation, two other commonly used model performance evaluation indices, mean bias error (MBE) and standard deviation (SD) (Equations (16) and (17), respectively), were used in addition to RMSE to validate the model [37,57]:

$$\text{MBE} = \frac{1}{N}\sum_{i=1}^{N}(R_i - G_i), \tag{16}$$

$$\text{SD} = \sqrt{\frac{1}{N-1}\sum_{j=1}^{N}(R_i - G_i - \text{MBE})^2}, \tag{17}$$

A perfect MBE score of 0 indicates a low overall magnitude of bias in the simulated data, which is essential in hydrological applications. The SD can be used to measure random error; a low value indicates a slight variation from the MBE.

4. Results

4.1. WRF RSD Simulation Results of Different Double-Moment MPs

To build a model that produces an optimal gamma distribution shape parameter reflecting changes in rainfall intensity by season, we ran the WRF model under different double-moment schemes using long-term ERA-Interim data. The three-dimensional values of N_0 and λ parameters in the gamma RSD model were calculated from the N and q variables of the lowest level in the WRF simulation outputs based on Equations (4) and (5). Using Equations (6) and (7), the values of D_m and N_w were then calculated. Figure 3 shows the map of the average values of D_m and $log_{10}N_w$ and the annual rainfall from 2013 to 2017 under the Morrison scheme. This figure indicates that the spatial distribution trend of the average D_m and $log_{10}N_w$ as well as that of annual rainfall in different years was consistent, although the number varied by year. The value of average D_m presented a trend of gradual increase from north to south and west to east, whereas the values of $log_{10}N_w$ and annual rainfall showed a trend of gradual decrease from north to south and from west to east. These trends indicate that rainfall information can be reflected through the values of the RSD parameters in the WRF model, such that a smaller average value of D_m or larger average value of $log_{10}N_w$ indicates higher annual rainfall. The annual average value of RSD and annual rainfall results of WDM6 and Thompson aerosol-aware schemes displayed similar patterns; therefore, the analysis does not present these schemes separately.

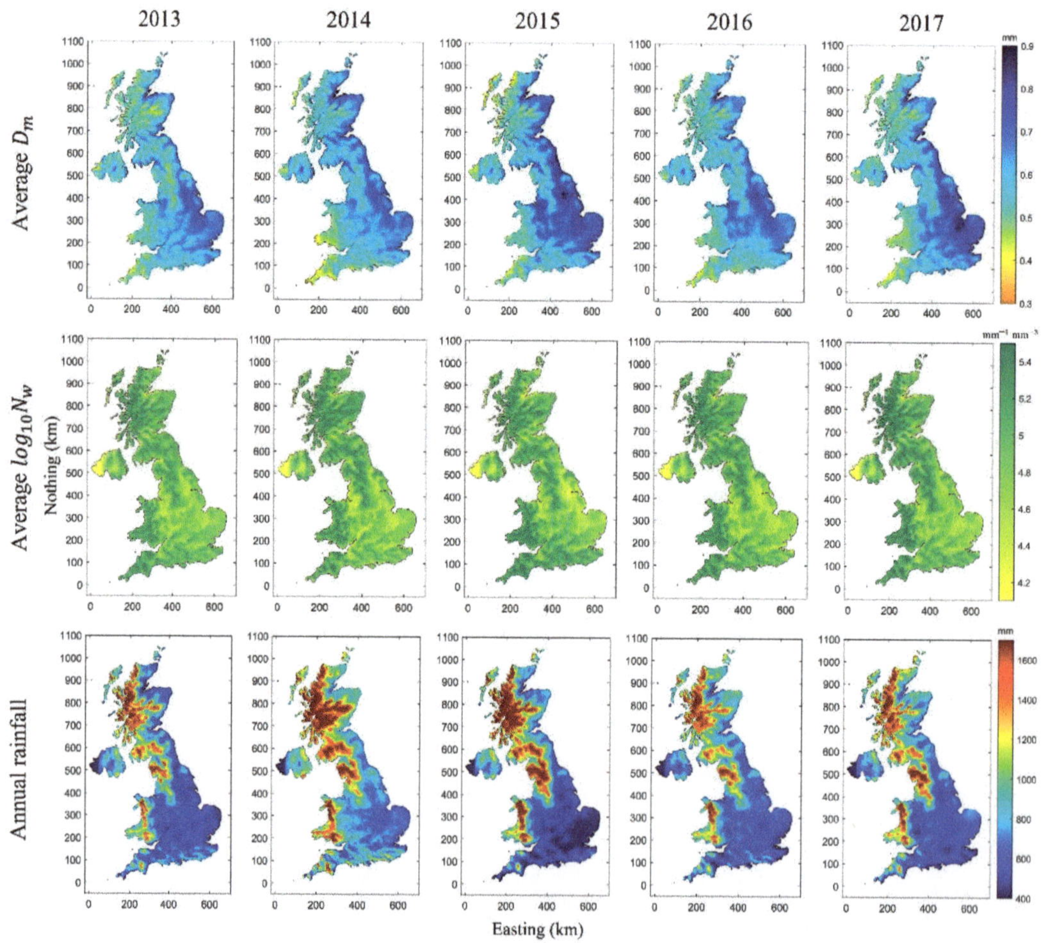

Figure 3. Spatial distribution of the average value of D_m and $log_{10}N_w$ and annual rainfall across the UK from 2013 to 2017 under the Morrison scheme.

The D_m and N_w in the observation were calculated from the raindrop data recorded by the JW disdrometer. To accurately compare the observed and simulated D_m and N_w, we extracted the grid (coordinates are 51°08′ N and 1°26′ W with a grid size of 5 km × 5 km) in the WRF simulation that was closest to the JW disdrometer with the same time span. Figure 4 shows the occurrences of the relationship between D_m and $log_{10}N_w$ of the Morrison, WDM6, and Thompson aerosol-aware schemes, and the JW disdrometer. The $D_m - log_{10}N_w$ density scatter plots show a fan shape centered at the D_m between approximately 0.5 mm and 1.5 mm. The $D_m - log_{10}N_w$ relation showed a negative correlation, which can be fitted by a quadratic polynomial function in the form of $log_{10}N_w = aD_m^2 + bD_m + c$ [58]. The fitting curves between D_m and $log_{10}N_w$ and corresponding R-squared (R^2) statistic of different schemes and the JW disdrometer are also presented in Figure 4. R^2 is a metric that measures the degree of dependence between variables in a regression model with a range of 0 to 1, where a higher value of R^2 generally reflects a better fit between the model and data. Compared to the fitting curve of the JW disdrometer, the fitting coefficients of the WRF model were quite close to those of the JW disdrometer, with the values of R^2 larger than 0.70. Among the three schemes, the fitting curve of Morrison showed the highest degree of similarity with that of the JW disdrometer. These results demonstrate that the WRF model can identify relationships between D_m and $log_{10}N_w$, whereas there are some uncertainties between different schemes.

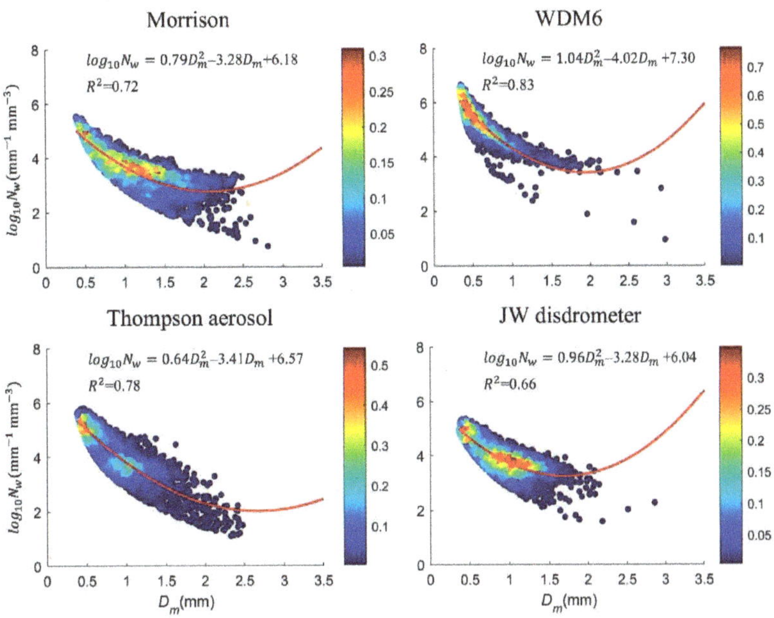

Figure 4. Density and fitting curves between D_m and $log_{10}N_w$ of different double-moment schemes in the WRF model and JW disdrometer, and the corresponding R^2 values.

Figure 5 shows the occurrences of the relationship between R and D_m of the Morrison, WDM6, and Thompson aerosol-aware schemes, and the JW disdrometer. R and D_m showed a positive correlation, that is, the size of raindrops increased gradually with increasing rainfall intensity. The R-D_m relationship can be fitted by a power-law function in the form of $D_m = aR^b$ [59,60]. The fitting curves between R and D_m and the corresponding R^2 of different schemes and the JW disdrometer are also presented in Figure 5. All double-moment schemes identified a power-law relationship between R and D_m, whereas the values of fitting coefficients and R^2 of different schemes showed a certain gap. Among the three schemes, the fitting curve of the Morrison scheme was closest to that of the JW disdrometer, whereas the R^2 value of the Thompson aerosol-aware scheme was relatively small.

4.2. Shape Parameter Constraint Interval

Figure 6 shows the reasonable ranges of the gamma RSD model shape parameter for warm and cold seasons, respectively, based on the five years of disdrometer data. Given these wide ranges, the middle 95% values were adopted as constraint intervals (gray-highlighted regions in the figure) to improve the efficiency of the optimization algorithm.

The warm season had a wider constraint interval, ranging from 0 to 48, than the cold season, which ranged from 0 to 39.

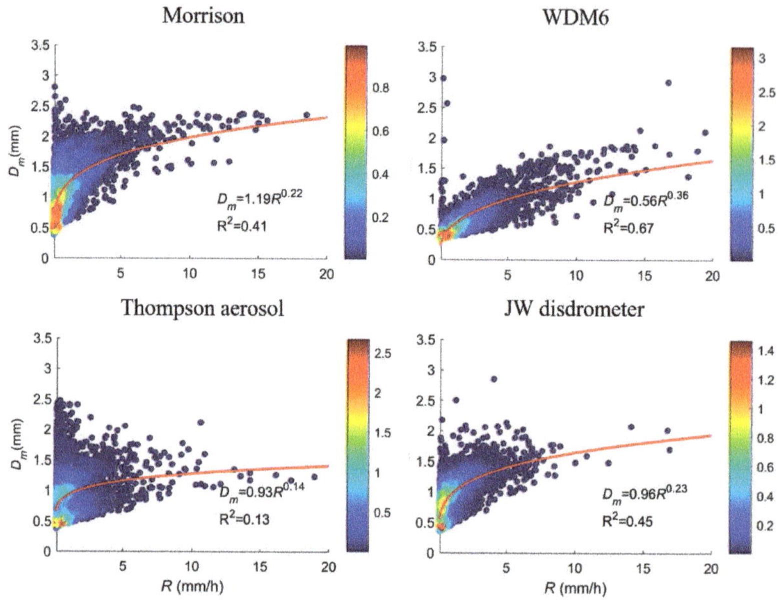

Figure 5. Density and fitting curves between D_m and R of different double-moment schemes in the WRF model and JW disdrometer, and the corresponding R^2 values.

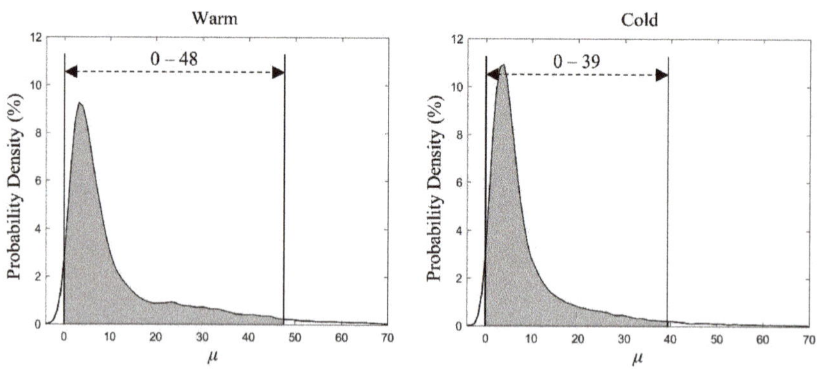

Figure 6. Shape parameter constraint intervals for different seasons derived from the disdrometer data throughout 2013–2017.

4.3. *Empirical Formula of Adaptive Shape Parameter in the WRF RSD Model*

We modified the shape parameter μ in the RSD model to bring one extra degree of freedom to it to improve its fitness. The shape parameter μ is computed based on the μ–R relationship by adopting the GA method to search for the optimal μ for different seasons and microphysical schemes. The search goal of GA is to minimize the value of RMSE between the observed and WRF-simulated rainfall. There are five phases in the GA search procedure, including initial population, fitness function, selection, crossover, and mutation. Since extreme rainfall data were insufficient within the study area, the optimal μ value under a rainfall intensity >9.5 mm/h will not be representative and can lead to uncertainty for the adaptive-μ model. Therefore, only rainfall with intensities of

≤9.5 mm/h was considered in this study. Figure 7 shows the 211 rain gauges observed and the corresponding WRF-simulated rainfalls produced at fixed and optimal values of μ for the two seasons and three double-moment schemes. For all three schemes, adopting the optimal μ resulted in a distinct decrease in RMSE relative to the fixed-μ case, although in each case, this difference was difficult to distinguish when the rainfall intensity was <1.5 mm/h, indicating that at rainfall intensities <1.5 mm/h, the adaptive-μ model cannot improve the performance of WRF rainfall simulation. Therefore, the fixed-μ gamma RSD model is suitable for double-moment microphysics in this low-rainfall-intensity regime. In other words, μ equaled 0 for the Morrison and Thompson aerosol-aware schemes and μ equaled 1 for the WDM6 scheme when rainfall intensities were <1.5 mm/h. We restricted our further investigation of the relationship between optimal μ and rainfall intensity to intensities > 1.5 mm/h.

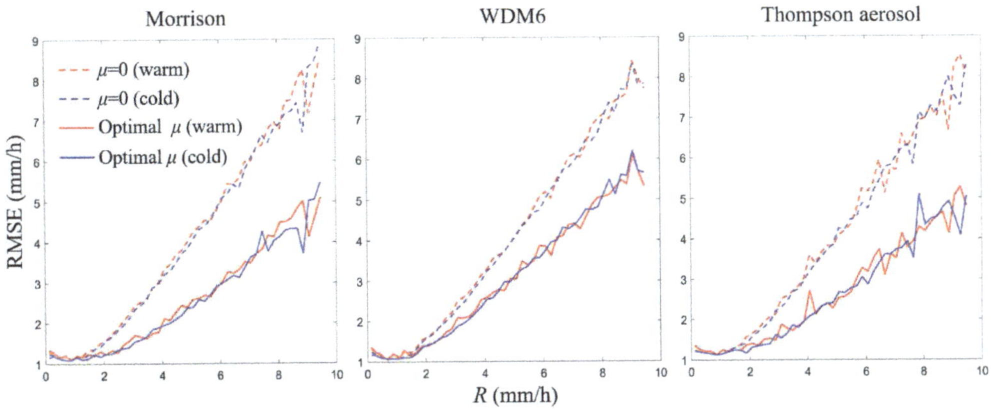

Figure 7. Root-mean-square errors (RMSEs, mm/h) of fixed- and optimal-μ gamma RSD models in WRF rainfall retrieval for different seasons and double-moment schemes.

Figure 8 shows the optimal μ of rainfall intensity by scenarios. The optimal value of μ tended to increase as rainfall intensity increased, with a clear linear relationship between optimal μ and rainfall intensity for all three schemes. We applied the ordinary least-squares regression method to select a linear function to relate the rainfall intensity and optimal μ data points at intensities ≥1.5 mm/h. The R^2 statistic for each scheme is also shown in Figure 8. Overall, the R^2 values between rainfall intensity and optimal μ indicate that the selected linear functions represented reasonable models for fitting the data. Notably, the values were higher in the cold season than in the warm season for all three double-moment schemes, suggesting that the linear relationship between rainfall intensity and optimal μ is more prominent during cold seasons.

Despite good fit, there remain differences among the respective linear empirical formulas. The constant term of the empirical Morrison scheme formula was higher than that in the Thompson aerosol-aware scheme, which was higher than that in the WDM6 scheme. In contrast, the independent variable coefficient under the Thompson scheme was lower than in the others.

The differing results by seasons suggest that the respective adaptive-μ models of the gamma RSD distribution under WRF double-moment bulk schemes based on rainfall intensity (mm/h) can be expressed as piecewise functions (equations listed in Table 2). In addition, as this study excluded rainfall data with R > 9.5 mm/h, we could not confirm whether the adaptive-μ model proposed in this study is suitable for extreme rainfall over the study area. This study suggests that the fixed-μ gamma RSD model can be adopted, i.e., it is not necessary to adjust the value of WRF-simulated R for extreme rainfall.

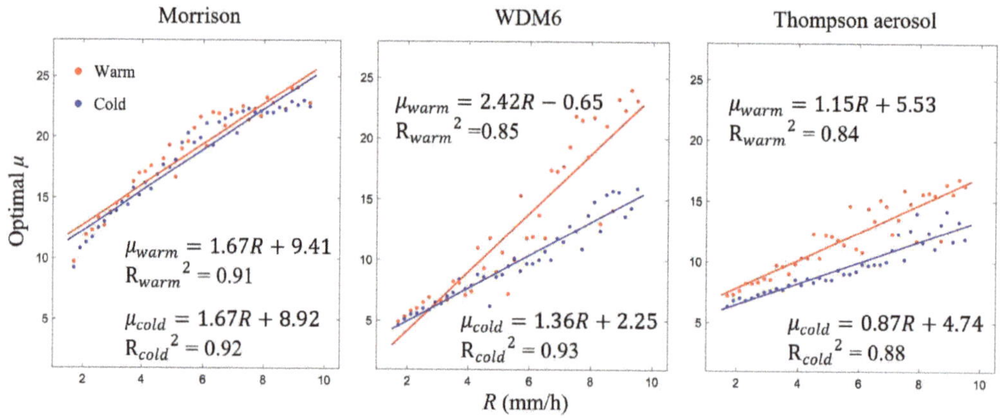

Figure 8. Optimal μ of different rain rates and fittings curves by season and double-moment schemes.

Table 2. Adaptive-μ models of the gamma RSD distribution in the WRF double-moment bulk microphysical parameterizations of different scenarios.

Schemes	Seasons	Adaptive-μ Models
Morrison	Warm	$\mu = \begin{cases} 0, & R < 1.5 \\ 1.67R + 9.41, & 1.5 \le R \le 9.5 \end{cases}$
	Cold	$\mu = \begin{cases} 0, & R < 1.5 \\ 1.67R + 8.92, & 1.5 \le R \le 9.5 \end{cases}$
WDM6	Warm	$\mu = \begin{cases} 1, & R < 1.5 \\ 2.42R - 0.65, & 1.5 \le R \le 9.5 \end{cases}$
	Cold	$\mu = \begin{cases} 1, & R < 1.5 \\ 1.36R - 2.25, & 1.5 \le R \le 9.5 \end{cases}$
Thompson aerosol	Warm	$\mu = \begin{cases} 0, & R < 1.5 \\ 1.15R - 5.53, & 1.5 \le R \le 9.5 \end{cases}$
	Cold	$\mu = \begin{cases} 0, & R < 1.5 \\ 0.87R - 4.74, & 1.5 \le R \le 9.5 \end{cases}$

4.4. WRF Rainfall Results of Different Scenarios

We computed the WRF rainfall using the adaptive-μ models of the gamma RSD distribution. Using the rain gauge rainfall as a benchmark, we calculated the WRF and rain gauge rainfall difference using the fixed-μ and adaptive-μ RSD models. Figures 9 and 10 show the spatial distribution of the absolute value of the annual rainfall amount difference between the WRF simulation and rain gauge of different scenarios in 2013 and 2016 (the results in 2014, 2015, and 2017 are shown in the Figures S1–S3). Values other than the position of the rain gauge were interpolated via the kriging method [61,62]. Kriging is an interpolation method stemming from regionalized variable theory for geographical information systems [61]. Due to the nature of the spatial interpolation process, the extreme values are typically associated with the rain gauge locations. From these two figures, among three schemes, the annual rainfall simulated by the Thompson aerosol-aware scheme was closest to the actual value, whereas the WDM6 results were the least concordant. Compared with the annual rainfall difference of the fixed-μ RSD model, the map's color had a larger area of blue for the adaptive-μ RSD model, particularly for the WDM6 model. This illustrates that for most sites, the annual rainfall differences in the adaptive-μ RSD model were smaller than those of the fixed-μ RSD model for different double-moment schemes. From the perspective of the spatial distribution of the annual rainfall difference, the adaptive-μ RSD models of Morrison and Thompson aerosol-aware schemes showed good performance in the eastern and southern regions of the UK, whereas some sites exhibited relatively poor performance in the northern region of the UK.

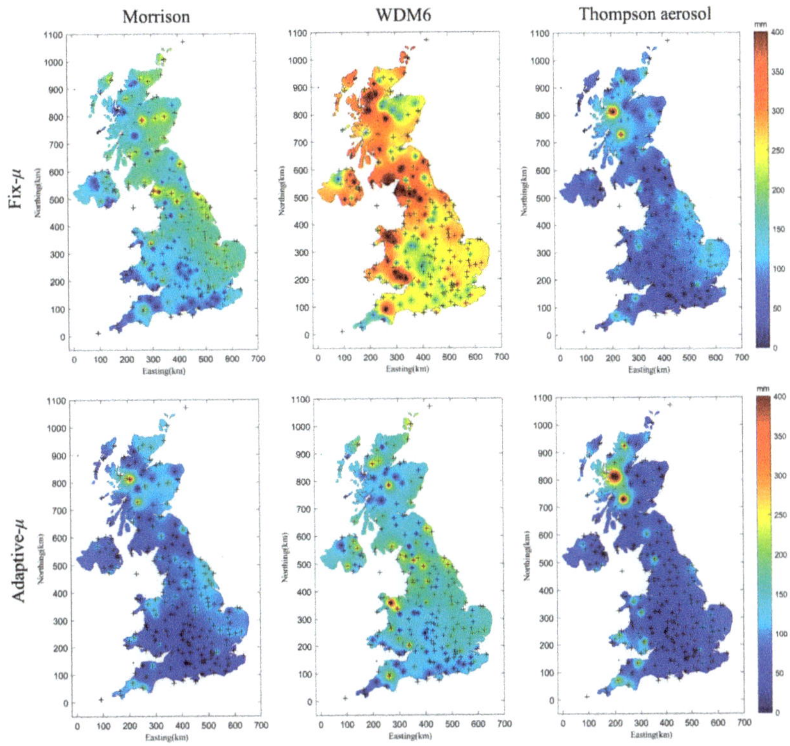

Figure 9. The absolute value of annual rainfall amount difference between the WRF simulation and rain gauge using fixed-μ and adaptive-μ RSD models of three double-moment schemes in 2013, plus symbols indicate the rain gauge positions.

4.5. Validation of the Adaptive-μ Model

Five years of WRF simulation and rain gauge data from 30 validation points were used to examine the applicability of the adaptive-μ model proposed in Section 3.2 to the study region. The RMSE, MBE, and SD error indices were used to investigate whether the adaptive-μ model for the WRF double-moment microphysics schemes could enhance the accuracy of WRF rainfall retrieval and to quantify any improvements obtained. Figure 11 shows the results of the fixed- and adaptive-μ indices at rainfall intensities higher than 1.5 mm/h for 30 validation points and different schemes. Compared with the results obtained using the fixed-μ model, all validation points of the Morrison and Thompson aerosol-aware schemes and 29 validation points of the WDM6 showed smaller RMSE values, and all rain gauges of the three schemes showed smaller MBE values. In contrast, only one point of the Morrison and Thompson aerosol-aware schemes and seven rain gauges of the WDM6 had larger SD values using the adaptive-μ model. These results demonstrate that the adaptive-μ model had better performance than the fix-μ model in almost all validation sites.

Tables 3–5 list the values of the fixed- and adaptive-μ indices at rainfall intensities higher than 1.5 mm/h for each of the three models along with the improvements (IMPROV% columns) obtained using adaptive-μ parameters as measured using each index. For each combination of the warm and cold seasons and application of the Morrison, WDM6, and Thompson aerosol-aware double-moment schemes, the RMSEs were decreased by 18.46%, 28.77%, 7.27%, 15.38%, 18.18%, and 26.24%, respectively. Similarly, the MBEs were decreased by 54.12%, 65.67%, 27.68%, 34.51%, 50.88%, and 58.27% for the respective combinations. Finally, the SDs were improved by 7.17%, 20.61%, 0.98%, 7.30%, 7.66%, and 19.15% under the respective season and scheme combinations.

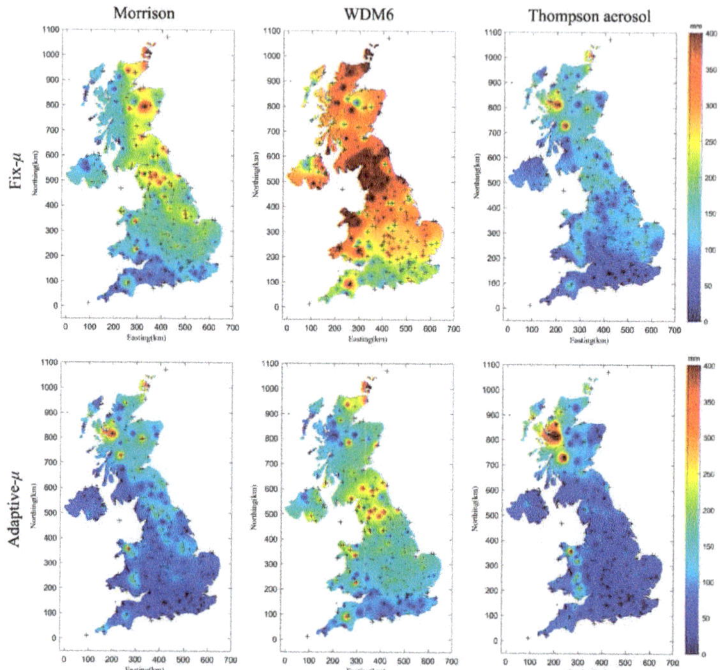

Figure 10. The absolute value of annual rainfall amount difference between the WRF simulation and rain gauge using fixed-μ and adaptive-μ RSD models of three double-moment schemes in 2016.

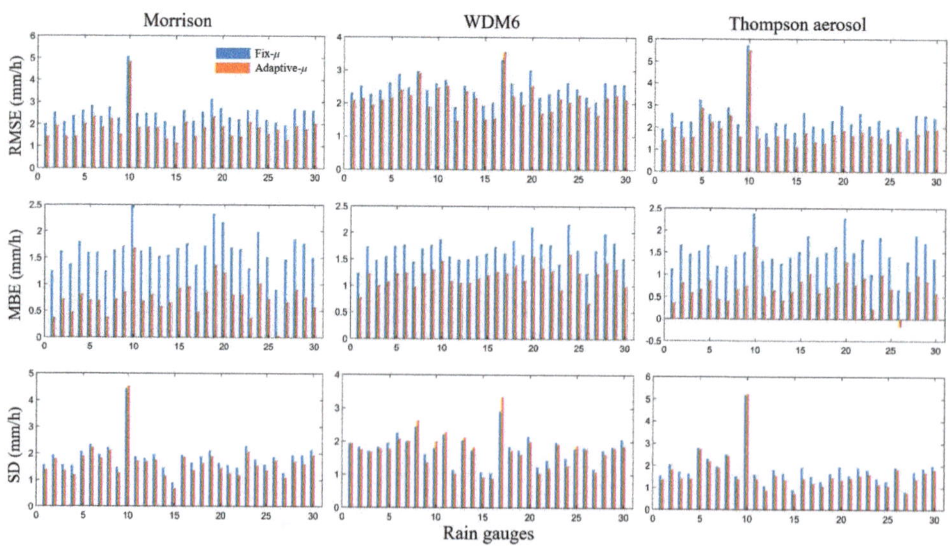

Figure 11. Comparison of the fixed- and adaptive-μ model results in terms of RMSE, MBE, and SD indices for 30 validation points and three double-moment schemes.

Table 3. Comparison of the fixed- and adaptive-μ model results in terms of RMSE, MBE, and SD indices for different seasons under the Morrison scheme.

Indices	Season	Fix-μ	Adaptive-μ	IMPROV (%)
RMSE (mm/h)	Warm	2.98	2.43	18.46%
	Cold	2.12	1.51	28.77%
MBE (mm/h)	Warm	1.70	0.78	54.12%
	Cold	1.34	0.46	65.67%
SD (mm/h)	Warm	2.23	2.07	7.17%
	Cold	1.31	1.04	20.61%

Table 4. Comparison of the fixed- and adaptive-μ model results in terms of RMSE, MBE, and SD indices for different seasons under the WDM6 scheme.

Indices	Season	Fix-μ	Adaptive-μ	IMPROV (%)
RMSE (mm/h)	Warm	2.89	2.68	7.27%
	Cold	2.21	1.87	15.38%
MBE (mm/h)	Warm	1.77	1.28	27.68%
	Cold	1.42	0.93	34.51%
SD (mm/h)	Warm	2.05	2.03	0.98%
	Cold	1.37	1.27	7.30%

Table 5. Comparison of the fixed- and adaptive-μ model results in terms of RMSE, MBE, and SD indices for different seasons under the Thompson aerosol-aware scheme.

Indices	Season	Fix-μ	Adaptive-μ	IMPROV (%)
RMSE (mm/h)	Warm	2.97	2.43	18.18%
	Cold	2.21	1.63	26.24%
MBE (mm/h)	Warm	1.71	0.84	50.88%
	Cold	1.39	0.58	58.27%
SD (mm/h)	Warm	2.22	2.05	7.66%
	Cold	1.41	1.14	19.15%

These validation results demonstrate that the linear adaptive-μ model can more accurately capture rainfall intensity than the fixed-μ model. Applying the adaptive-μ model during the cold season results in a more remarkable improvement in the accuracy of the WRF rainfall simulation results than is achieved by applying this model during the warm season. The relatively low level of improvement obtained for the WDM6 scheme suggests that the sensitivity to the adaptive-μ model sensitivity differed for each scheme.

5. Discussion

This study was conducted based on the assumption that double-moment bulk microphysical parameterizations of the WRF model using a fixed value of the gamma RSD model shape parameter can be improved. To improve WRF rainfall retrieval, we used an adaptive-μ model to study the rainfall records covering most of the UK landmass over five years (2013 to 2017). Under the assumption that the μ-value might vary by rainfall intensity, the GA method was used to search for optimal values of μ by rainfall intensity during different seasons and under varying double-moment bulk microphysics. Empirical formulas relating rainfall intensity to optimal μ were then built using the ordinary least-squares linear regression method and, based on these, adaptive-μ models of the gamma distribution were proposed for the respective scenarios. Our experimental results were used to address the three following primary issues.

The first issue related to whether using an adaptive value of the shape parameter of the gamma RSD model could improve the performance of WRF rainfall simulation. The results revealed a significant decline in the RMSE using the optimal μ of different rainfall intensities rather than the constant value of μ in the gamma RSD model, particularly for high rainfall intensity. However, the optimal μ of low rainfall intensity (R < 1.5 mm/h) did not clearly reduce the WRF-simulated rainfall error, indicating that the fixed-μ gamma RSD model is suitable when rainfall intensity is low (<1.5 mm/h) for the studied microphysics schemes. Given that the calculation efficiency could be reduced using the adaptive-μ model and the performance of WRF rainfall retrieval could not be improved, for low rainfall intensity, there is no need to determine the relationship between rainfall intensity and the optimal μ, and fixing the value of the shape parameter to 0 or 1 of the gamma RSD model is feasible for WRF double-moment microphysics schemes.

The second issue was whether a suitable model can accurately describe the relationship between rainfall intensity and the optimal μ under multiple scenarios. Our results demonstrate that, at rainfall intensities of $\geq$1.5 mm/h, a linear function can fit rainfall intensity to the optimal μ exceedingly well for different seasons and under various double-moment schemes. However, although there was a robust linear relationship between rainfall intensity and the optimized μ, the empirical formulas differed by scenario, suggesting that no unified adaptive-μ parameter formulation fits all scenarios (i.e., different seasons and microphysics schemes) over the specific study area. The empirical formulas for the Morrison scheme were relatively similar by seasons, whereas those for the other two schemes differed significantly. Thus, the adaptive-μ empirical formulas proposed in this study are universally applicable to the microphysics schemes and regions studied.

The third issue was validating the empirical formulas and examining how well the adaptive-μ model could improve WRF-simulated rainfall accuracy. Our validation results indicate that the adaptive-μ gamma RSD model outperformed a fixed-μ WRF double-moment microphysics model in terms of at least three error indices. The results also indicate that the improvements in WRF rainfall retrieval obtained by applying the adaptive-μ gamma RSD model differed by scenario.

The proposed model performed better during cold seasons and the various double-moment schemes assessed had distinct sensitivities to the model, which are reflected in the fact that the R^2 values of the fitted lines produced by the adaptive-μ model were higher in the warm season than in the cold season. These results may also relate to differences in the probability distribution of rainfall intensity between seasons. As shown in Figure 12, in the region studied, the cold season had a higher probability density for low rainfall intensity (i.e., R < 2.8 mm/h) and a lower probability density for high rainfall intensity than the warm season. This differentiation by season may enhance the WRF rainfall simulation accuracy during the cold season when applying the adaptive-μ model.

The RMSEs obtained under the Morrison and Thompson aerosol-aware schemes were decreased by more than 15% and 25%, respectively, during the cold season, whereas the respective MBEs were decreased by more than 50% and 55%, and the SDs were decreased by approximately 7% and close to 20%, respectively. The degree of error reduction under the WDM6 scheme, however, was much lower. The significant reduction in the error of WRF rainfall retrieval under the former two schemes using the adaptive-μ model suggests that the adaptive-μ model can successfully be combined with these schemes.

Because the data from 241 rain gauges used to validate the WRF rainfall simulations produced limited information on rainfall microphysics and RSD, the adaptive-μ gamma RSD models assessed in this study could only be examined through statistical error analysis. In contrast, the physical mechanisms underlying the model could not be explored. Two important data sources for validating numerically simulated WRF precipitation microphysics—disdrometer observations and dual-polarization radar reflectivity data—could be applied to the mechanistic modification of the gamma RSD model of WRF double-moment microphysics. As disdrometer samplings were already imitated within the study area, in future studies, dual-polarization radar data will be used to further examine the adaptive-μ gamma RSD model in terms of additional variables (e.g., median drop diameter D_0, differential reflectivity Z_{DR}, and specific differential phase K_{DP}) related to the microphysics of RSD.

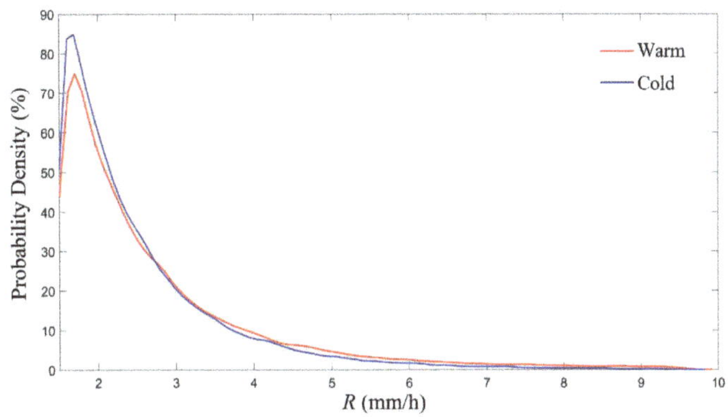

Figure 12. Probability density (%) of rainfall intensity ($1.5 \leq R \leq 9.5$, mm/h) for warm and cold seasons, measured from 241 data points.

6. Conclusions

To construct a model that would produce an optimal gamma distribution shape parameter reflecting changes in rainfall intensity by season, five years of ERA-Interim data were downscaled by applying the WRF model under different double-moment schemes; the results were compared with observations taken at 241 rain gauges and one disdrometer during the same period. We concluded the following from the analysis results.

1. The spatial characteristics of rainfall can be reflected by the WRF-simulated RSD, with a smaller average value of D_m or larger average value of $log_{10}N_w$, associated with higher annual rainfall. Similar to the RSD results of the JW disdrometer, the three tested WRF double-moment schemes showed a strong quadratic relationship between D_m and $log_{10}N_w$ and a clear power-law relationship between R and D_m, although there were some uncertainties between different schemes.
2. Although the fixed-μ gamma RSD model was suitable when rainfall intensity was <1.5 mm/h, linear empirical formulas relating rainfall intensity to optimal μ were successfully built to reflect different scenarios for which the rainfall intensity was $\geq$1.5 mm/h. Adaptive-μ models of the gamma distribution based on rainfall can be constructed using a piecewise function, as shown in the following equation:

$$\mu = \begin{cases} constant, & R < 1.5 \\ aR + b, & 1.5 \leq R \leq 9.5 \end{cases}, \quad (18)$$

   where R is the rainfall intensity in mm/h, a is the coefficient of the independent variable, and b is the constant term of the linear function. Adaptive-μ models of the gamma distribution were constructed to apply the Morrison, WDM6, and Thompson aerosol-aware double-moment schemes to two seasons (Table 2).
3. The consistency and usability of the adaptive-μ model were also demonstrated by using three error indices by applying the model to 30 validation points. A higher degree of error reduction was observed during the cold season, whereas the Morrison and Thompson aerosol-aware schemes achieved higher degrees of error reduction overall compared to the WDM6 scheme. The adaptive-μ model showed improved predictability for the three tested double-moment schemes compared to the fixed-μ model, indicated by the decreases in RMSE by 23.62%, 11.33%, and 22.21%; decreases in MBE by 59.90%, 31.10%, and 54.58%; and decreases in SD by 13.89%, 4.14%, and 13.41% for the Morrison, WDM6, and Thompson aerosol-aware schemes, respectively.

Based on the above results, the adaptive-μ model of gamma distribution can be successfully integrated into WRF double-moment microphysics scheme simulations to improve the accuracy of WRF rainfall retrieval, particularly during cold seasons and when using the Morrison and Thompson aerosol-aware schemes. This method and the relevant optimal μ-values are appropriate for the study area in the UK and can potentially be incorporated into the WRF model as part of its simulation process. The adaptive-μ model was designed to reduce simulated rainfall error in the WRF model by applying the observed rain gauge data. As we could not explore the physical mechanisms underpinning the adaptive-μ results in this study, further application of additional data sources (e.g., dual-polarization radar data) should be performed to study the RSD model in WRF bulk microphysics schemes based on additional variables related to precipitation microphysics.

Supplementary Materials: The following supporting information can be downloaded at: https://www.mdpi.com/article/10.3390/atmos13010036/s1. Figure S1. The absolute value of the annual rainfall amount difference between the WRF simulation and rain gauge using fixed-μ and adaptive-μ RSD models of three double-moment schemes in 2014. Figure S2. The absolute value of the annual rainfall amount difference between the WRF simulation and rain gauge using fixed-μ and adaptive-μ RSD models of three double-moment schemes in 2015. Figure S3. The absolute value of the annual rainfall amount difference between the WRF simulation and rain gauge using fixed-μ and adaptive-μ RSD models of three double-moment schemes in 2017.

Author Contributions: Conceptualization, S.Z. and Q.D.; methodology, Q.Y. and S.Z.; software, Q.Y.; validation, Q.Y. and H.Z.; formal analysis, Q.D.; investigation, Q.Y.; resources, Q.Y.; data curation, H.Z.; writing—original draft preparation, Q.Y.; writing—review and editing, S.Z.; visualization, Q.D.; supervision, S.Z.; project administration, S.Z. and Q.Y.; funding acquisition, S.Z. All authors have read and agreed to the published version of the manuscript.

Funding: This research was funded by the National Natural Science Foundation of China (Nos: 42071364, 42101416, and 41871299) and the China Postdoctoral Science Foundation Funded Project (No. 2021M691628).

Institutional Review Board Statement: Not applicable.

Informed Consent Statement: Not applicable.

Data Availability Statement: Not applicable.

Acknowledgments: We thank the Advanced Computing Research Centre at the University of Bristol for providing the access to the high-performance computing (HPC) system BlueCrystal. The ERA-Interim data driving the WRF model can be downloaded from the ECMWF Public Datasets web interface (https://www.ecmwf.int/ (accessed on 18 May 2021)). The gauge data sets and disdrometer were sourced from the UK Met Office and British Atmospheric Data Centre at http://archive.ceda.ac.uk/ (accessed on 18 May 2021).

Conflicts of Interest: The authors declare no conflict of interest.

References

1. Cao, Q.; Zhang, G. Errors in Estimating Raindrop Size Distribution Parameters Employing Disdrometer and Simulated Raindrop Spectra. *J. Appl. Meteorol. Climatol.* **2009**, *48*, 406–425. [CrossRef]
2. Yang, Q.; Dai, Q.; Han, D.; Chen, Y.; Zhang, S. Sensitivity analysis of raindrop size distribution parameterizations in WRF rainfall simulation. *Atmos. Res.* **2019**, *228*, 1–13. [CrossRef]
3. Ma, Y.; Ni, G.; Chandra, C.V.; Tian, F.; Chen, H. Statistical characteristics of raindrop size distribution during rainy seasons in Beijing urban area and implications for radar rainfall estimation. *Hydrol. Earth Syst. Sci.* **2019**, *23*, 4153–4170. [CrossRef]
4. Zhu, J.; Zhang, S.; Yang, Q.; Shen, Q.; Zhuo, L.; Dai, Q. Comparison of rainfall microphysics characteristics derived by numerical weather prediction modelling and dual-frequency precipitation radar. *Meteorol. Appl.* **2021**, *28*, e2000. [CrossRef]
5. Caracciolo, C.; Napoli, M.; Porcù, F.; Prodi, F.; Dietrich, S.; Zanchi, C.; Orlandini, S. Raindrop size distribution and soil erosion. *J. Irrig. Drain. Eng.* **2012**, *138*, 461–469. [CrossRef]
6. Janapati, J.; Seela, B.K.; Lin, P.-L.; Wang, P.K.; Kumar, U. An assessment of tropical cyclones rainfall erosivity for Taiwan. *Sci. Rep.* **2019**, *9*, 1–14.
7. Michaelides, S.; Levizzani, V.; Anagnostou, E.; Bauer, P.; Kasparis, T.; Lane, J.E. Precipitation: Measurement, remote sensing, climatology and modeling. *Atmos. Res.* **2009**, *94*, 512–533. [CrossRef]
8. Dai, Q.; Zhu, J.; Zhang, S.; Zhu, S.; Han, D.; Lv, G. Estimation of rainfall erosivity based on WRF-derived raindrop size distributions. *Hydrol. Earth Syst. Sci.* **2020**, *24*, 5407–5422. [CrossRef]

9. Skamarock, W.C.; Klemp, J.B.; Dudhia, J.; Gill, D.O.; Barker, D.M.; Wang, W.; Powers, J.G. A Description of the Advanced Research WRF Version 2. National Center for Atmospheric Research Boulder Co Mesoscale and Microscale Meteorology Div 2005. Available online: https://apps.dtic.mil/sti/citations/ADA487419 (accessed on 25 December 2021).
10. Skamarock, W.C.; Klemp, J.B.; Dudhia, J.; Gill, D.O.; Barker, D.M.; Wang, W.; Powers, J.G. A Description of the Advanced Research WRF Version 3. NCAR Technical Note-475+STR 2008. Available online: http://citeseerx.ist.psu.edu/viewdoc/summary?doi=10.1.1.484.3656 (accessed on 25 December 2021).
11. Dai, Q.; Han, D. Exploration of discrepancy between radar and gauge rainfall estimates driven by wind fields. *Water Resour. Res.* **2014**, *50*, 8571–8588. [CrossRef]
12. Crétat, J.; Pohl, B.; Richard, Y.; Drobinski, P. Uncertainties in simulating regional climate of Southern Africa: Sensitivity to physical parameterizations using WRF. *Clim. Dyn.* **2012**, *38*, 613–634. [CrossRef]
13. Dawson, D.T.; Xue, M.; Milbrandt, J.A.; Yau, M. Comparison of evaporation and cold pool development between single-moment and multimoment bulk microphysics schemes in idealized simulations of tornadic thunderstorms. *Mon. Weather. Rev.* **2010**, *138*, 1152–1171. [CrossRef]
14. Johnson, M.; Jung, Y.; Dawson, D.T.; Xue, M. Comparison of simulated polarimetric signatures in idealized supercell storms using two-moment bulk microphysics schemes in WRF. *Mon. Weather. Rev.* **2016**, *144*, 971–996. [CrossRef]
15. Lim, K.-S.S.; Hong, S.-Y. Development of an effective double-moment cloud microphysics scheme with prognostic cloud condensation nuclei (CCN) for weather and climate models. *Mon. Weather Rev.* **2010**, *138*, 1587–1612. [CrossRef]
16. Jung, Y.; Xue, M.; Tong, M. Ensemble Kalman filter analyses of the 29–30 May 2004 Oklahoma tornadic thunderstorm using one-and two-moment bulk microphysics schemes, with verification against polarimetric radar data. *Mon. Weather Rev.* **2012**, *140*, 1457–1475. [CrossRef]
17. Putnam, B.J.; Xue, M.; Jung, Y.; Snook, N.; Zhang, G. The analysis and prediction of microphysical states and polarimetric radar variables in a mesoscale convective system using double-moment microphysics, multinetwork radar data, and the ensemble Kalman filter. *Mon. Weather Rev.* **2014**, *142*, 141–162. [CrossRef]
18. Brown, B.R.; Bell, M.M.; Frambach, A.J. Validation of simulated hurricane drop size distributions using polarimetric radar. *Geophys. Res. Lett.* **2016**, *43*, 910–917. [CrossRef]
19. Ulbrich, C.W. Natural variations in the analytical form of the raindrop size distribution. *J. Clim. Appl. Meteorol.* **1983**, *22*, 1764–1775. [CrossRef]
20. Zhang, G.; Vivekanandan, J.; Brandes, E.A.; Meneghini, R.; Kozu, T. The shape–slope relation in observed gamma raindrop size distributions: Statistical error or useful information? *J. Atmos. Ocean. Technol.* **2003**, *20*, 1106–1119. [CrossRef]
21. Milbrandt, J.; Yau, M. A multimoment bulk microphysics parameterization. Part I: Analysis of the role of the spectral shape parameter. *J. Atmos. Sci.* **2005**, *62*, 3051–3064. [CrossRef]
22. Islam, T.; Rico-Ramirez, M.A.; Thurai, M.; Han, D. Characteristics of raindrop spectra as normalized gamma distribution from a Joss–Waldvogel disdrometer. *Atmos. Res.* **2012**, *108*, 57–73. [CrossRef]
23. Harikumar, R.; Sampath, S.; Kumar, V.S. Variation of rain drop size distribution with rain rate at a few coastal and high altitude stations in southern peninsular India. *Adv. Space Res.* **2010**, *45*, 576–586. [CrossRef]
24. Simmons, A. ERA-Interim: New ECMWF reanalysis products from 1989 onwards. *ECMWF Newsl.* **2007**, *110*, 29–35.
25. Uppala, S.; Dee, D.; Kobayashi, S.; Berrisford, P.; Simmons, A. Towards a climate data assimilation system: Status update of ERA-Interim. *ECMWF Newsl.* **2008**, *115*, 12–18.
26. Dee, D.P.; Uppala, S.M.; Simmons, A.; Berrisford, P.; Poli, P.; Kobayashi, S.; Andrae, U.; Balmaseda, M.; Balsamo, G.; Bauer, d.P. The ERA-Interim reanalysis: Configuration and performance of the data assimilation system. *Q. J. R. Meteorol. Soc.* **2011**, *137*, 553–597. [CrossRef]
27. Cardoso, R.M.; Soares, P.M.; Lima, D.C.; Miranda, P.M. Mean and extreme temperatures in a warming climate: EURO CORDEX and WRF regional climate high-resolution projections for Portugal. *Clim. Dyn.* **2019**, *52*, 129–157. [CrossRef]
28. Politi, N.; Vlachogiannis, D.; Sfetsos, A.; Nastos, P.T. High-resolution dynamical downscaling of ERA-Interim temperature and precipitation using WRF model for Greece. *Clim. Dyn.* **2021**, *57*, 799–825. [CrossRef]
29. Bieniek, P.A.; Bhatt, U.S.; Walsh, J.E.; Rupp, T.S.; Zhang, J.; Krieger, J.R.; Lader, R. Dynamical downscaling of ERA-Interim temperature and precipitation for Alaska. *J. Appl. Meteorol. Climatol.* **2016**, *55*, 635–654. [CrossRef]
30. Fu, G.; Charles, S.P.; Timbal, B.; Jovanovic, B.; Ouyang, F. Comparison of NCEP-NCAR and ERA-Interim over Australia. *Int. J. Climatol.* **2016**, *36*, 2345–2367. [CrossRef]
31. Liu, Z.; Liu, Y.; Wang, S.; Yang, X.; Wang, L.; Baig, M.H.A.; Chi, W.; Wang, Z. Evaluation of spatial and temporal performances of ERA-Interim precipitation and temperature in mainland China. *J. Clim.* **2018**, *31*, 4347–4365. [CrossRef]
32. Met-Office. Met Office Integrated Data Archive System (MIDAS) Land and Marine Surface Stations Data (1853-Current). NCAS British Atmospheric Data Centre: 2012. Available online: http://catalogue.ceda.ac.uk/uuid/220a65615218d5c9cc9e4785a3234bd0 (accessed on 20 September 2020).
33. Dai, Q.; Bray, M.; Zhuo, L.; Islam, T.; Han, D. A Scheme for Rain Gauge Network Design Based on Remotely Sensed Rainfall Measurements. *J. Hydrometeorol.* **2017**, *18*, 363–379. [CrossRef]
34. Tokay, A.; Wolff, K.; Bashor, P.; Dursun, O. *On the Measurement Errors of the Joss-Waldvogel Disdrometer*; ResearchGate: Berlin, Germany, 2003.

35. Song, Y.; Han, D.; Rico-Ramirez, M.A. High temporal resolution rainfall rate estimation from rain gauge measurements. *J. Hydroinformatics* **2017**, *19*, 930–941. [CrossRef]
36. Montopoli, M.; Marzano, F.S.; Vulpiani, G. Analysis and synthesis of raindrop size distribution time series from disdrometer data. *IEEE Trans. Geosci. Remote Sens.* **2008**, *46*, 466–478. [CrossRef]
37. Liu, J.; Bray, M.; Han, D. Sensitivity of the Weather Research and Forecasting (WRF) model to downscaling ratios and storm types in rainfall simulation. *Hydrol. Processes* **2012**, *26*, 3012–3031. [CrossRef]
38. Morrison, H.; Thompson, G.; Tatarskii, V. Impact of cloud microphysics on the development of trailing stratiform precipitation in a simulated squall line: Comparison of one-and two-moment schemes. *Mon. Weather Rev.* **2009**, *137*, 991–1007. [CrossRef]
39. Hong, S.-Y.; Lim, K.-S.S.; Lee, Y.-H.; Ha, J.-C.; Kim, H.-W.; Ham, S.-J.; Dudhia, J. Evaluation of the WRF double-moment 6-class microphysics scheme for precipitating convection. *Adv. Meteorol.* **2010**, *2010*, 707253. [CrossRef]
40. Thompson, G.; Eidhammer, T. A study of aerosol impacts on clouds and precipitation development in a large winter cyclone. *J. Atmos. Sci.* **2014**, *71*, 3636–3658. [CrossRef]
41. Kain, J.S. The Kain–Fritsch Convective Parameterization: An Update. *J. Appl. Meteorol.* **2004**, *43*, 170–181. [CrossRef]
42. Cai, J.; Zhu, J.; Dai, Q.; Yang, Q.; Zhang, S. Sensitivity of a weather research and forecasting model to downscaling schemes in ensemble rainfall estimation. *Meteorol. Appl.* **2020**, *27*, e1806. [CrossRef]
43. Janjić, Z.I. The step-mountain eta coordinate model: Further developments of the convection, viscous sublayer, and turbulence closure schemes. *Mon. Weather Rev.* **1994**, *122*, 927–945. [CrossRef]
44. Mlawer, E.J.; Taubman, S.J.; Brown, P.D.; Iacono, M.J.; Clough, S.A. Radiative transfer for inhomogeneous atmospheres: RRTM, a validated correlated-k model for the longwave. *J. Geophys. Res. Atmos.* **1997**, *102*, 16663–16682. [CrossRef]
45. Dudhia, J. Numerical study of convection observed during the winter monsoon experiment using a mesoscale two-dimensional model. *J. Atmos. Sci.* **1989**, *46*, 3077–3107. [CrossRef]
46. Ek, M.; Mitchell, K.; Lin, Y.; Rogers, E.; Grunmann, P.; Koren, V.; Gayno, G.; Tarpley, J. Implementation of Noah land surface model advances in the National Centers for Environmental Prediction operational mesoscale Eta model. *J. Geophys. Res. Atmos.* **2003**, *108*. [CrossRef]
47. Bringi, V.; Chandrasekar, V.; Hubbert, J.; Gorgucci, E.; Randeu, W.; Schoenhuber, M. Raindrop size distribution in different climatic regimes from disdrometer and dual-polarized radar analysis. *J. Atmos. Sci.* **2003**, *60*, 354–365. [CrossRef]
48. Bringi, V.; Huang, G.-J.; Chandrasekar, V.; Gorgucci, E. A methodology for estimating the parameters of a gamma raindrop size distribution model from polarimetric radar data: Application to a squall-line event from the TRMM/Brazil campaign. *J. Atmos. Ocean. Technol.* **2002**, *19*, 633–645. [CrossRef]
49. Montero-Martínez, G.; Kostinski, A.B.; Shaw, R.A.; García-García, F. Do all raindrops fall at terminal speed? *Geophys. Res. Lett.* **2009**, *36*. [CrossRef]
50. Testud, J.; Oury, S.p.; Black, R.A.; Amayenc, P.; Dou, X. The Concept of "Normalized" Distribution to Describe Raindrop Spectra: A Tool for Cloud Physics and Cloud Remote Sensing. *J. Appl. Meteorol.* **2001**, *40*, 1118–1140. [CrossRef]
51. Willis, P.T. Functional fits to some observed drop size distributions and parameterization of rain. *J. Atmos. Sci.* **1984**, *41*, 1648–1661. [CrossRef]
52. Ulbrich, C.W.; Atlas, D. Rainfall microphysics and radar properties: Analysis methods for drop size spectra. *J. Appl. Meteorol.* **1998**, *37*, 912–923. [CrossRef]
53. Kleczek, M.A.; Steeneveld, G.-J.; Holtslag, A.A. Evaluation of the weather research and forecasting mesoscale model for GABLS3: Impact of boundary-layer schemes, boundary conditions and spin-up. *Bound.-Layer Meteorol.* **2014**, *152*, 213–243. [CrossRef]
54. Mitchell, M. *An Introduction to Genetic Algorithms*; MIT Press: Cambridge, MA, USA, 1998.
55. Whitley, D. A genetic algorithm tutorial. *Stat. Comput.* **1994**, *4*, 65–85. [CrossRef]
56. Osman, M.S.; Abo-Sinna, M.A.; Mousa, A.A. A solution to the optimal power flow using genetic algorithm. *Appl. Math. Comput.* **2004**, *155*, 391–405. [CrossRef]
57. Jolliffe, I.T.; Stephenson, D.B. *Forecast Verification: A Practitioner's Guide in Atmospheric Science*; John Wiley & Sons: Hoboken, NJ, USA, 2012.
58. Wu, Z.; Zhang, Y.; Zhang, L.; Lei, H.; Xie, Y.; Wen, L.; Yang, J. Characteristics of summer season raindrop size distribution in three typical regions of western Pacific. *J. Geophys. Res. Atmos.* **2019**, *124*, 4054–4073. [CrossRef]
59. Chen, B.; Yang, J.; Pu, J. Statistical characteristics of raindrop size distribution in the Meiyu season observed in eastern China. *J. Meteorol. Soc. Japan. Ser. II* **2013**, *91*, 215–227. [CrossRef]
60. Wen, L.; Zhao, K.; Zhang, G.; Xue, M.; Zhou, B.; Liu, S.; Chen, X. Statistical characteristics of raindrop size distributions observed in East China during the Asian summer monsoon season using 2-D video disdrometer and Micro Rain Radar data. *J. Geophys. Res. Atmos.* **2016**, *121*, 2265–2282. [CrossRef]
61. Oliver, M.A.; Webster, R. Kriging: A method of interpolation for geographical information systems. *Int. J. Geogr. Inf. Syst.* **1990**, *4*, 313–332. [CrossRef]
62. Stein, M.L. *Interpolation of Spatial Data: Some Theory for Kriging*; Springer Science & Business Media: Berlin/Heidelberg, Germany, 2012.

atmosphere

MDPI

Article

Optimizing Analog Ensembles for Sub-Daily Precipitation Forecasts

Julia Jeworrek [1,*], Gregory West [1,2] and Roland Stull [1]

1 Department of Earth, Ocean and Atmospheric Sciences, University of British Columbia, Vancouver, BC V6T 1Z4, Canada
2 BC Hydro, Vancouver, BC V3N 4X8, Canada
* Correspondence: jjeworrek@eoas.ubc.ca

Abstract: This study systematically explores existing and new optimization techniques for analog ensemble (AnEn) post-processing of hourly to daily precipitation forecasts over the complex terrain of southwest British Columbia, Canada. An AnEn bias-corrects a target model forecast by searching for past dates with similar model forecasts (i.e., analogs), and using the verifying observations as ensemble members. The weather variables (i.e., predictors) that select the best past analogs vary among stations and seasons. First, different predictor selection techniques are evaluated and we propose an adjustment in the forward selection procedure that considerably improves computational efficiency while preserving optimization skill. Second, temporal trends of predictors are used to further enhance predictive skill, especially at shorter accumulation windows and longer forecast horizons. Finally, this study introduces a modification in the analog search that allows for selection of analogs within a time window surrounding the target lead time. These supplemental lead times effectively expand the training sample size, which significantly improves all performance metrics—even more than the predictor weighting and temporal-trend optimization steps combined. This study optimizes AnEns for moderate precipitation intensities but also shows good performance for the ensemble median and heavier precipitation rates. Precipitation is most challenging to predict at finer temporal resolutions and longer lead times, yet those forecasts see the largest enhancement in predictive skill from AnEn post-processing. This study shows that optimization of AnEn post-processing, including new techniques developed herein, can significantly improve computational efficiency and forecast performance.

Keywords: analog ensembles; precipitation; WRF; statistical post-processing; Pacific north west; complex terrain

Citation: Jeworrek, J.; West, G.; Stull, R. Optimizing Analog Ensembles for Sub-Daily Precipitation Forecasts. *Atmosphere* **2022**, *13*, 1662. https://doi.org/10.3390/atmos13101662

Academic Editors: Xiaofan Li, Huaqing Cai and Zuohao Cao

Received: 26 August 2022
Accepted: 9 October 2022
Published: 12 October 2022

1. Introduction

Numerical weather prediction (NWP) models are impaired by imperfect initial conditions and simplified approximations of physical concepts. Statistical post-processing can improve forecast quality by reducing systematic model errors [1]. Common bias-correction procedures include regression methods [2–4], model output statistics [1,5–8], Kalman-filtering [9], moving- and weighted-average techniques [10], Bayesian model averaging [11,12], machine learning [13–15], and analog ensembles (AnEns; [16,17]); many of these categories have overlapping techniques. Previous research has demonstrated successful applications of AnEns to temperature [17,18], wind speed [17,19–23], wind power [24–26], solar radiation [24,27], air quality [28–30], precipitation [16,31–33], and streamflow predictions [34].

Originally, analog methods were based on the assumption that atmospheric conditions tend to follow recurring patterns and hence could estimate future weather from similar past developments. More successful recent versions of the technique operate on the assumption that past similar conditions will have similar model errors. In an operational framework, the AnEn technique post-processes a target model forecast by searching for similar (i.e., "analog") model forecasts in the past and using their corresponding observations to compose an ensemble [9,16–18]. Since the AnEn samples the observed distribution without assuming a target distribution, this nonparametric method directly corrects systematic model error. The analog procedure constructs probabilistic ensemble forecasts from a history of raw deterministic model forecasts from only one NWP model run. Hence, they require a fraction of the computational expense of a traditional multi-model or multi-run NWP ensemble.

The success of AnEns has two main requirements: (a) the availability of a consistent high-quality meteorological archive of model forecasts and concurrent observations—the longer the archive the better; and (b) a definition of "similarity", which measures the degree of analogy of past forecasts with a target forecast to identify a set of best analogs that construct the AnEn.

Similarity is assessed primarily using predictors, for instance via a multivariate Euclidean-distance measure. Usually predictor variables have varying degrees of importance (for example, related to location and season [35]) and can be weighted accordingly. Some studies neglect these dependencies and utilize equal [9,17,18,36,37] or arbitrary weights [16,22,38] for a subjectively reasonable selection of predictors. Other studies use domain knowledge to design "levels of analogy" [32,39–43], where sequential selection levels sort successive subsets of analog candidates from the meteorological archive. More recent studies derive predictor weights objectively from correlation coefficients with the predictand [30], or obtain them directly from a predictor selection procedure, such as brute force (BF; [35]), forward selection (FS; [44]), principal component analysis (PCA; [24,26,45,46]), or genetic algorithms [42,47].

BF is a popular optimization approach as it identifies optimal weights by testing all possible combinations. However, the computational expense of running and evaluating AnEns in numerous variations becomes infeasible for a large number of predictors. Therefore many studies [14,21,33,48] pre-select a reduced number of predictor candidates before BF optimization. The pre-selection is sometimes based on domain knowledge [19,21] or an efficient filter method (such as correlation analysis; [33]), however it may limit the BF predictor optimization by unintentionally disregarding useful predictors to begin with. References [44,49] developed a more efficient BF approach, the stepwise FS (also used in [33,50]), which iteratively tests the AnEn performance by adding any of the predictor candidates to the previously selected predictors one step at a time. This approach resembles BF results closely [33], while the computational savings enable the exploration of more predictor variables. However, compared to filter methods, the FS approach still requires significant time and resources to train and analyse the predictor optimization. Despite the known advantage of optimized predictor weights [35], the computational effort continues to represent an obstacle.

The similarity measure for the analog search can also account for the temporal trend of the predictors across sequential forecast lead times [17]. This consideration has been shown to improve AnEns [9]. A few studies [9,17,25] investigate the impact of the time window length, while others use an arbitrary window length of $+/-1$ forecast steps [21,29,35,44,50], which can result in different total window widths depending on the forecast interval. Moreover, different physical variables can have different autocorrelation characteristics and hence the ideal time window to match the temporal predictor trends may vary accordingly.

Real-time AnEns require operational NWP with a long and consistent data history. Reforecast datasets therefore have a great potential for data-driven methods like AnEns [38,51], especially if the same model configuration is tuned for local characteristics and is still used operationally in real time. However, many studies [31–33,43,52] train the AnEn on long reanalysis datasets targeting the development of statistically downscaled data products, or serve as proof of concept for an operational framework (i.e., perfect prognosis). References [43,53] showed that the quality of reanalyses affects the AnEn skill significantly—sometimes even more than the choice of predictors. Since reanalyses includes more data assimilation of atmospheric measurements, operational forecasts have faster (and possibly different) error growth and NWP quality is likely to have a similar or larger impact on AnEns than the reanalyses.

Compared to other meteorological variables, precipitation has high spatial and temporal variability, making post-processing particularly challenging. Precipitation distributions are also skewed towards zero and can be represented with a discrete distribution regarding event thresholds (e.g., rain vs. no-rain), or a continuous distribution when considering quantitative amounts.

The decreasing number of precipitation events with increasing intensity represents a statistical disadvantage for post-processing high-impact events with data-driven methods, such as AnEns. A long data history is required to ensure sufficient sampling when searching for analogs of relatively rare events. For the purpose of expanding the analog search data pool, ref. [38] introduced the concept of supplemental locations, which uses additional grid points or stations with similar climatology and terrain characteristics. This technique significantly improves heavy precipitation forecast calibration [38] but requires a relatively large domain with numerous stations. Instead of expanding the sample size using spatial supplements, ref. [54] suggested a moving-time-window approach to inflate the meteorological archive using temporal supplements. When searching for analogs for a daily-precipitation target, ref. [54] considered not only the same lead times, but also 24-h totals that result from sub-daily offsets to the target lead time.

Most precipitation AnEn studies investigate daily accumulation totals [16,32,40,42,43,52,55–57], which exhibit better predictability than sub-daily amounts [58]. However, resolving the sub-daily variability of precipitation has value to many weather forecast applications, such as flood management, infrastructure maintenance (e.g., transportation and construction), agriculture (e.g., soil erosion and crop damage), and smaller hydroelectric operations where flows are managed at sub-daily time steps. Reference [59] temporally disaggregated daily precipitation analog forecasts to hourly time steps. Other AnEn studies directly derive 12-hourly [38,60] or 6-hourly precipitation amounts [33,36,41,61,62], whereas NWP forecasts are commonly considered at hourly intervals.

Our study domain is southwest British Columbia (BC), Canada. BC's complex terrain amplifies a variety of forecasting challenges, such as imperfect numerical and physical approximations in NWP and the high spatial variability of the surface conditions. For example, the prediction of orographic precipitation enhancement on windward slopes and lee-side rain-shadow requires adequate representation of topography (e.g., slope steepness), initial state (e.g., upstream conditions over the Pacific Ocean), dynamics (e.g., flow and stability), and physics (e.g., mixed-phase microphysical processes). Inaccuracies in any NWP components cumulatively contribute to model errors and make post-processing a crucial factor in improving precipitation forecast quality over regions of complex terrain such as BC. Southwest BC sees copious precipitation during the cool season, which, among other impacts, has led to catastrophic flooding events [63–65]. Additionally, skillful precipitation forecasts are crucial because hydropower contributes the bulk of the total electricity production in BC, and precipitation forecasts are used to plan generation system operations and mitigate flood risk.

This study demonstrates the optimization of AnEn forecasts for sub-daily precipitation in southwest BC by post-processing one of the regionally best performing configurations following [58]. As described in Section 2, we apply the AnEn method as a station-based post-processing tool to statistically downscale the deterministic model forecasts—an approach that is suitable for real-time operational model post-processing. This paper builds on existing methods to optimize AnEn parameters and explores new variations in the AnEn methodology that either improve forecast performance or computational efficiency.

In Section 3, we first compare predictor selection techniques and suggest a more efficient FS approach. Next, we investigate the impact of the temporal trend similarity, while assessing accumulations from daily to hourly intervals. As such, to our knowledge, this is the first paper demonstrating successful AnEns for hourly precipitation forecasts in this form. Finally, we redesign [54]'s moving-time-window approach for shorter accumulation windows to make the best use of a limited meteorological archive in finding the best available analogs. Our verification shows the improvement in each optimization step and reveals the trade-off between temporal resolution and precipitation predictive skill following AnEn post-processing. Section 4 gives the summary and conclusions.

2. Materials and Methods

2.1. Data

The NWP data for this study are from the Weather Research and Forecasting (WRF) model [66] version 3.8.1 with the Advanced Research WRF (ARW) dynamical core, initialized with the Global Deterministic Prediction System (GDPS) model [67,68] from Environment and Climate Change Canada (ECCC). The model setup, including initialization, domains, and physics, was chosen based on [58]—a study that evaluated precipitation forecasts from over 100 systematically varied model configurations. We use the WRF configuration from that study that performed above average across verification scores and performed best for 75th-percentile (75p) equitable threat score (ETS) and probability of detection (POD)—namely, the WRF single-moment 5-class microphysics scheme (WSM5; [69]), the Kain-Fritsch cumulus scheme (KF; [70]), the Yonsei University turbulence scheme (YSU; [71]), and the multiphysics Noah land surface model (Noah-MP; [72,73]).

Since [58] found smaller raw forecast errors for coarser grid spacings, but finer grid spacings are assumed to better resolve the spatial variability over complex terrain, this study focuses on the mid-size domain with $\Delta x = 9$-km. WRF runs are initialized daily at 00 UTC and provide 3 days (72 h) of hourly forecast data after 9 h of spinup. Hence, each forecast day starts at 0900 UTC (0100 Local Standard Time). For further details on other WRF settings and the verification results, see [58].

Utilizing this regionally optimized WRF configuration, we generated a 5.75-year reforecast dataset from January 2016 through September 2021. Table 1 lists 22 physical variables that were extracted or derived from the WRF output, some at different vertical levels, resulting in 41 variables total. The list includes general atmospheric parameters that characterize moisture, thermal, stability, and wind conditions. Some variables like MI1 and MI2 were inspired by other precipitation AnEn studies [43,52,54] that used these as predictors.

The model variable "PCP" includes precipitation in all forms (i.e., rainfall, snow, sleet, etc.) and is represented as liquid equivalent. However, BC's South Coast has a mild climate year-around and freezing temperatures and snowfall are rarely observed at lower elevations.

The original temporal resolution of WRF output is hourly. When assessing longer accumulation windows in this study (i.e., 3-, 6-, 12-hourly, and daily intervals), the sum of hourly PCP is calculated, whereas all other model variables are estimated by their time average. This study considers discrete (non-overlapping) and rolling (overlapping) windows for the accumulated datasets. Discrete-window results are useful for comparison with other studies that use such windows. However, they can split precipitation events unfavorably, making them seem longer and weaker.

Rolling windows, on the other hand, sample the same precipitation events in hourly offsets and ensure the capture of maximum rates for any event and accumulation interval. Therefore they provide a more complete picture in assessing the impact of temporal resolution on predictability. The hourly time step of rolling windows can further benefit temporal trend similarities and supplemental lead times (described below), for which longer-accumulation discrete windows often have too large of a time step. However, it is important to remember that rolling windows possess overlap from one time step to another, which makes them temporally correlated.

Table 1. Physical variables extracted from the WRF model output and considered as predictors.

Variable	Abbreviation	Levels
Total Precipitation	PCP	Surface
Integrated Water Vapor *	IWV	Column
Integrated Vapor Transport *	IVT	Column
Water Vapor Mixing Ratio	r	2 m, 70 kPa, 50 kPa
Specific Humidity *	SH	70 kPa, 50 kPa
Relative Humidity *	RH	70 kPa, 50 kPa
Moisture Index 1 * ($RH_{70kPa} \times IWV$)	MI1	
Moisture Index 2 * ($RH_{70kPa} \times W_{70kPa}$)	MI2	
Temperature	T	2 m, 70 kPa, 50 kPa
Potential Temperature	Th	2 m, 70 kPa, 50 kPa
Dewpoint Temperature	Td	70 kPa, 50 kPa
Total Totals Index *	TT	
K-Index *	KI	
U-component Wind	U	10 m, 70 kPa, 50 kPa
V-component Wind	V	10 m, 70 kPa, 50 kPa
W-component Wind	W	70 kPa, 50 kPa
Wind Direction *	WD	10 m, 70 kPa, 50 kPa
Wind Speed *	WS	10 m, 70 kPa, 50 kPa
Sea Level Pressure	SLP	Sea Level
Surface Pressure	SfcP	Surface
Geopotential Height	GPH	70 kPa, 50 kPa
Boundary Layer Height	PBLH	

* Derived from WRF output variables.

46 stations from two networks (ECCC and BC Hydro) provide hourly precipitation observations within the domain of interest, shown in Figure 1. BC Hydro station observations are manually quality controlled at BC Hydro [74]. Additional quality control checks on both observational networks ensure that

- each station has at least 90% of data available, and missing data is not systematically distributed (e.g., in the same season, at the same time of day, or over a large consecutive period); and
- outliers are reasonable considering the synoptic situation (e.g., convection), nearby stations, and the available station climatology.

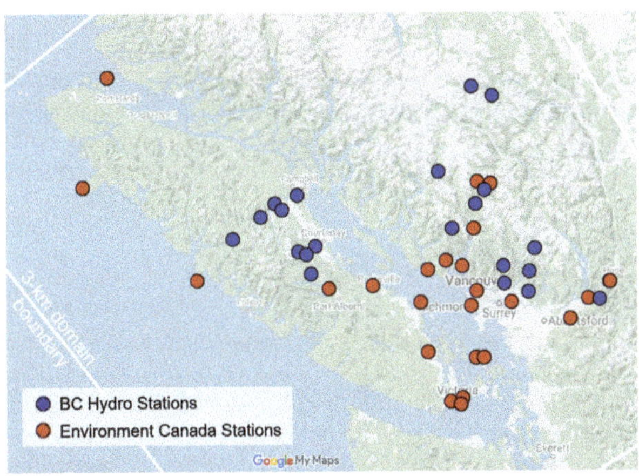

Figure 1. Domain of interest in southwest British Columbia with locations of 46 stations that provide hourly precipitation observations from two networks.

The gridded model data is spatially interpolated to the station locations using the nearest-neighbor approach. The matched station and model dataset is split into 4.75 years (January 2016 through September 2020) for training/optimization using a leave-one-out approach, and one year (the 2021 water year: October 2020 through September 2021) for independent testing/verification. The optimization and verification process both search for analogs from the same 4.75-year training dataset. However, to ensure data independence during optimization, the leave-one-out approach excludes a buffer of ±15 days surrounding the targeted initialization.

Since higher precipitation rates have larger impact and are more challenging to forecast, this study optimizes for performance on "moderate" or heavier precipitation intensities—specifically, 75th percentiles (75p). Section 3.4 provides additional verification results for 90th-percentile (90p) events, i.e., "heavy" precipitation rates. The thresholds are calculated at each station based on observations after excluding values <0.25 mm (which for many rain gauges is the smallest measurable amount). Percentile values vary with accumulation window; examples of frequency distributions are shown in Appendix A.

2.2. *Analog Ensemble Methodology*

Analog model forecasts (AnFcsts) are a set of past model forecasts (PaFcsts) that, in regard to selected variables and similarity metrics, are most similar to the target model forecast (TaFcst) at a given lead time and location. The past verifying station measurements that correspond to the AnFcsts—the analog observations (AnObs)—are used as ensemble members to compose the AnEn (see Figure 2). The AnEn is considered to be the post-processed version of the deterministic raw TaFcst, and hence should provide a better forecast for the verifying observation (VerifObs) at the target time. Thus, the analog selection is determined solely from the model space, whereas the AnEn is composed solely of samples from the observation space.

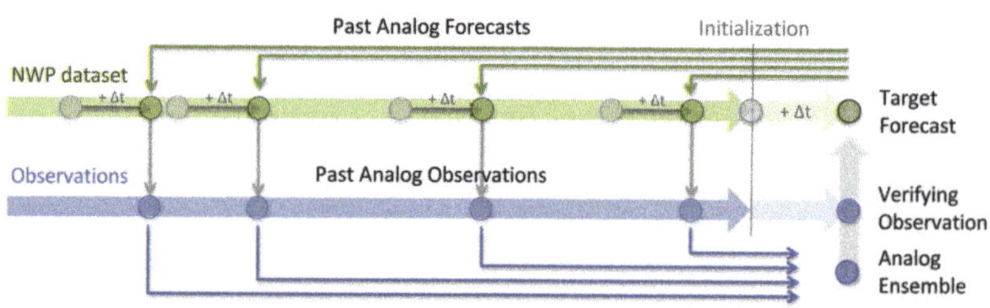

Figure 2. Illustration of the analog ensemble (AnEn) methodology.

Although this study investigates univariate AnEn forecasts for precipitation (i.e., the predictand), a distance measure assesses the multivariate similarity between the TaFcst and all PaFcsts. The best ranking PaFcsts are chosen to be the AnFcsts and are determined independently for each station location and forecast lead time. We use a popular similarity metric developed by [9] that calculates similarity scores

$$\|\mathrm{TaFcst}_t, \mathrm{PaFcsts}_t\| = \sum_{v=1}^{N_v} \frac{w_v}{\sigma_{v,t}} \sqrt{\sum_{j=-\tau}^{\tau} (\mathrm{TaFcst}_{v,t+j} - \mathrm{PaFcsts}_{v,t+j})^2}, \quad (1)$$

for each PaFcst with the TaFcst at lead-time t relative to model initialization. N_v is the number of physical variables v (i.e., predictors) for which closeness between AnFcsts and the TaFcst is desired. The variable weights w_v can assign larger importance to variables with stronger predictor relationships. The division by each variable's standard deviation $\sigma_{v,t}$ in the model training dataset at lead time t standardizes the variables and makes the similarity scores dimensionless. τ defines a lead-time window that centers the target lead time t and includes additional lead times over which the similarity in the temporal trends of the predictors is computed.

The physical variables that are used for the analog search should exhibit good predictor relationships with the predictand. Precipitation AnEn studies often use model PCP and IWV as predictors [14,16,38] with weights 0.7 and 0.3, respectively, [16,38]. We use these variables as reference when evaluating the predictors that we obtain from other predictor selection techniques (presented in Section 2.2.1). The control run further uses $\tau = 0$, thus only matching the predictor values of TaFcst and PaFcsts at lead time t. The AnEn sensitivity to τ is investigated in Section 3.2. With these reference settings, the optimal AnEn size was found using approximately 30 AnFcst members on average (not shown), which agrees with [33], and is therefore used throughout this study.

To prioritize forecast performance for significant events that are more critical for decision makers, in this study the AnEn parameters are optimized using the 75p threshold-weighted continuous ranked probability score (twCRPS; [7], see Appendix B.1).

2.2.1. Predictor Selection Procedures

The predictor selection procedure objectively assesses which of the 41 model variables in Table 1 are the best predictors for precipitation. Initial investigation of predictor relationships between all variables and observed precipitation using filter methods (e.g., correlation analysis) reveals variability in predictor importance across stations and months (see Appendix C). This is expected since locations and seasons in southwest BC can exhibit different characteristics due to topography and climate. Therefore, we investigate optimal predictors and their weights independently at each station location and meteorological season.

A brute force (BF) approach [35] is a method that runs the AnEn optimization on all combinations of predictor weights (to a defined precision), and determines the best predictor combination according to an evaluation score. Since this method is computationally very expensive, [44] suggested a step-wise forward selection (FS) method, which sequentially selects one predictor at a time by BF testing all weighting options only among the step-wise selected predictors. The first step tests all N_v variables as single predictors with $w_v = 1$ (see Equation (1)), and the variable resulting in the best evaluation score is chosen as the first predictor. The next step selects the second predictor by testing the remaining $N_v - 1$ variables in combination with the already selected first predictor. The weights applied to each variable are tested by BF, i.e., all possible combinations. For instance, using weight increments of 0.1 in the interval [0, 1] with the constraint that the sum of weights is always 1, results in 9 options for two variables. Selecting the third predictor has $(N_v - 2) * 84$, and the fourth predictor has $(N_v - 3) * 126$ options, etc. Predictors are selected if they improve the evaluation score by a chosen increment compared to the score in the previous FS step (e.g., mean absolute error of 3% in [44]). This way, different stations can receive different numbers of predictors. Although this FS approach is considerably faster than BF, the computational cost is still significant for large N_v and high weight precision (i.e., smaller increments).

As a further computational reduction of [44]'s FS, we propose an "efficient FS" (EFS) as follows. Assuming that those variables first selected as predictors have larger importance, we constrain the weighting options to $w_{Predictor1} \geq w_{Predictor2} \geq \dots \geq w_{PredictorN_v}$. This way the second predictor has $(N_v - 1) * 5$, the third predictor has $(N_v - 2) * 8$, the third predictor has $(N_v - 3) * 9$, and the fourth predictor has only $(N_v - 4) * 7$ options. We further define the first predictor to be PCP without testing. We proceed to optimize predictors for twCRPS until the improvement drops below 1%.

We investigate four variants of FS to determine predictor weights:

- All-EFS: Using the EFS to test all 40 variables in addition to PCP as predictors.
- DC-FS: Using [44]'s FS to test a subset of 10 variables as predictors. The 10 predictor candidates are pre-selected based on the highest distance correlation coefficient

(DCorr; a measure that identifies both linear and non-linear relationships [75]) with observed precipitation.

- DC-EFS: Using the EFS to test the same subset of 10 predictor candidates as in DC-FS.
- DCV-EFS: Using the EFS to test a subset of 10 variables as predictors, except here, the predictor candidates are based on the best DCorr, as well as the variance inflation factor (VIF, a measure of multicollinearity among variables). Specifically, we grow a set of 10 predictor candidates by sequentially adding one variable at a time, starting from the best ranking DCorr, provided the VIF among the growing set of predictor candidates stays below a threshold value of 10. If this threshold is exceeded it means that the variable exhibits strong correlation with other variables that were already selected and we assume that this variable contributes no additional value as a predictor for the AnEn. Since some of our 41 variables are related (e.g., the same variables at different vertical levels), the VIF check limits the use of correlated and presumably redundant variables in the FS.

These experiments aim to investigate (a) whether the EFS is competitive with [44]'s FS, and (b) whether DCorr or DCorr and VIF can effectively pre-filter meaningful predictor candidates, reducing computational expense compared to testing all variables in EFS. For the feasibility of testing all these methods, we conducted this optimization of predictor weights only for 3-hourly discrete accumulation windows and day-1 forecasts.

We further conducted principal component analysis (PCA) on the standardized datasets of all variables and the 10-variable subsets resulting from DCorr and DCorr-VIF analyses. Different experiments with the principal components (PC) as predictors and their weights either using the eigenvalues or from EFS on the PCs, were not competitive with the (E)FS methods described above and are therefore not shown in this paper. Ref. [35] obtained similar results comparing BF and PCA predictor-selection methods.

2.2.2. The Supplemental-Lead-Time (SLT) Approach

The original AnEn approach, as described above, searches AnFcsts only across those lead times in the training period that match the target lead time of the TaFcst. The aim of matching lead times is to detect AnFcst candidates with error characteristics similar to the TaFcst. It further ensures that the AnFcsts are sampled from the same time of day, which is particularly important for predictands like temperature and wind that exhibit diurnal cycles. However, precipitation in southwest BC has no significant diurnal cycle in the cool-season when most precipitation occurs; and it is plausible that better AnFcst candidates are available at other lead times surrounding the target lead time, for which model errors are still similar.

Thus, we explore the use of "supplemental lead times" (SLTs) in Section 3.3. As exemplified in Figure 3, AnFcst candidates are considered over a range of offsets from the target lead time. Since AnFcsts from the same past model initialization would be temporally correlated, we only select the best single PaFcst among SLTs from each initialization. Hence, this method selects the AnFcst from a SLT (different from the target lead time) only when the score resulting from Equation (1) indicates closer similarity (i.e., a better analogy).

Accumulation-window treatment must be considered when using SLTs. For instance, ± 2 SLTs applied to hourly forecasts selects the best out of five PaFcst options over a lead-time window width of five hours; ± 2 SLTs applied to 3-hourly rolling forecasts considers five PaFcst options over a window width of eight hours; and ± 2 SLTs applied to 3-hourly discrete forecasts considers five PaFcst options over a total window width of 15 h. Since SLT consideration for longer discrete windows quickly inflates the effective lead-time window width over which model error growth can be significant, we examine SLTs only for short and rolling accumulation windows.

This method differs from [54], who inspected 12-, 6-, and 3-hourly offsets of 24-h accumulations for daily precipitation targets by including all offsets within the 24-h period. For example, [54]'s best-performing 3-hourly offsets result in 8 times as many AnFcst candidates as the original AnEn approach, hereby artificially inflating the meteorological archive with temporally correlated PaFcsts. Our SLT approach, on the other hand, increases the number of AnFcst candidates indirectly by choosing only the single best candidate in the SLT window centering the target lead time, without risking the selection of multiple correlated PaFcsts.

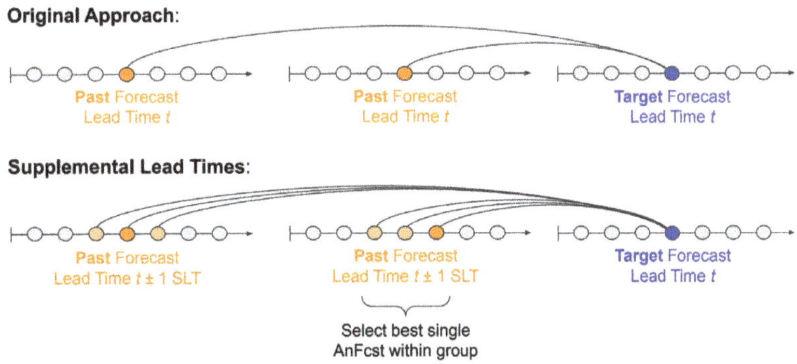

Figure 3. Graphic of the supplemental-lead-time approach (SLT; bottom), compared to the original approach (top). The circles along the arrows represent lead times of an initialization. This example illustrates the analog search at lead time t. For the first past-forecast (PaFcst) initialization, the SLT approach using ± 1 SLTs selects the analog forecast (AnFcst) at lead time t as in the original approach. For the second PaFcst initialization, the SLT approach finds a better AnFcst at lead time $t + 1$.

3. Results and Discussion

3.1. Predictor Selection Optimization

The four iterative FS and EFS approaches described in Section 2.2.1 determine predictor weights from the training dataset only. Following optimization for 75p twCRPS, all methods require that the resulting predictors yield better or equal twCRPS compared to the control predictors over the training period. Because the set of 10 predictor candidates used in DC-EFS or DCV-EFS are each a subset of the variables used in All-EFS, it is also required that All-EFS yields better or equal twCRPS during training. It is further required that DC-FS is better than DC-EFS during training, since both methods are preconditioned by the same predictor candidates and the FS approach has more weighting options than EFS. However, running the AnEn over the testing period with the predictor weights determined from the training period reveals whether the selection procedure actually led to the required improvement or whether it overfitted the training dataset.

Figure 4 compares twCRPSS among the four methods, segregated by training and testing. Recall that predictor weights are optimized independently for each station and meteorological season. Therefore, Figure 4 summarizes the twCRPSS of all 46 stations in boxplots and panels for each season.

For all seasons the majority of stations see improvement in twCRPS following predictor optimization with any method. This is true for both training and testing, which indicates that all methods are beneficial. However, the consistent shift between training and testing scores suggests some overfitting; namely, the best predictors during training are still good—but potentially not the best—predictors during testing.

Since summer in BC is by far the driest season, differences in small values of twCRPS between control and optimized predictors are amplified in twCRPSS, hence the larger improvement on average and the large spread. Warm-season dry periods can extend into spring and fall months. The winter season contains the most consistent precipitation pattern.

All methods significantly (see Appendix B.2 for significance testing) improve twCRPS compared to the control during training and testing. However, the differences in twCRPSS among methods are relatively small. Compared to the training, the testing score distributions across stations are less often significantly different among methods. However, winter All-EFS scores are still significantly better than DC-FS and DCV-EFS during testing. Although not always significant, All-EFS is consistently best during training and continues to be best on average (not shown) during testing in summer, fall, and winter. Therefore, we use the predictors resulting from the All-EFS method (see Appendix D) hereinafter. DC-EFS and DCV-EFS are competitive with All-EFS but require considerably less optimization time by assessing only a quarter of the variables. Therefore, they would be viable alternatives for predictor optimization in other studies.

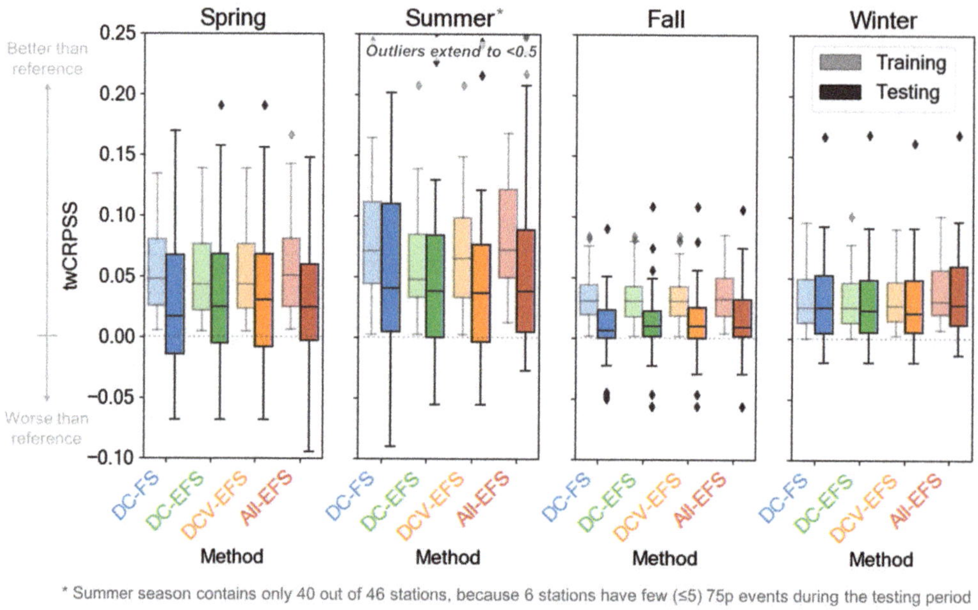

Figure 4. Box-and-whisker plots of 75p twCRPSS distributions across stations (46 stations in each boxplot, except in summer) after predictor optimization with four methods. The dotted zero line separates values that indicate improvement (positive values) vs. deterioration (negative values) compared to the reference twCRPS using control predictors. Performance differences between training (lighter colors) and testing (darker colors) informs about the degree of overfitting.

These results show that the EFS approach is capable of effectively reducing the computational effort of predictor tuning compared to [44]'s FS method, while maintaining similar (and sometimes even better) improvements compared to static control predictors.

3.2. *Temporal Trend Similarity*

In addition to predictor-variable choice and their relative weights, the temporal trends of predictors can also help to better identify AnFcsts. Temporal trend similarity (TTS) is assessed over a time window centered on the target lead time, and window width results from the definition of τ ranging over a number of time steps. The total window width depends on the accumulation window and steps. For instance, a TTS window using ± 2 time steps (i.e., $\tau = 2$ in Equation (1)) covers a 5-h period for hourly precipitation, or a 7-h period for 3-hourly rolling windows, whereas it covers an effective 15-h period for 3-hourly discrete windows.

We investigate the impact of TTS with τ ranging from 1 to 5 time steps on twCRPS relative to using no TTS ($\tau = 0$). Since autocorrelation between the predictors and the predictand is unlikely to exceed half a day, for long discrete accumulation windows we discard TTS calculations that would result in effective window widths >36 h. Therefore, 24-hourly discrete windows are not considered for TTS, and 12-hourly discrete windows are assessed only for a TTS with $\tau = 1$. Note also that there are no forecast values available preceding (following) the first (last) lead time in a forecast series from one initialization; therefore, a few lead times at the beginning and end of a forecast series do not experience the full effect of TTS.

Figures 5 and 6 show that hourly and rolling windows, all of which have hourly time steps, benefit most from TTS. Discrete windows with longer accumulations, and hence larger time steps, cover longer total lead-time widths over which TTS is often less effective—here only on days 2 and 3 is twCRPSS sometimes positive (better) for $\tau > 0$.

Longer forecast horizons generally obtain better twCRPSS for longer time windows. Due to error growth with lead time, day-1 forecasts exhibit better predictability than day-3 forecasts and thus, day-3 forecasts have greater total potential for improvement. At longer forecast horizons it appears that the added temporal dimension in the similarity consideration somewhat balances the abating quality of the predictor variables, whereas at shorter forecast horizons the instant predictor values preserve higher predictability.

The magnitude of improvement is slightly larger in spring and summer seasons, likely because warm-season precipitation is often convective and has more of a diurnal pattern. However, despite the differences in predictor variables among seasons, the best value of τ is relatively similar across seasons.

Since we determined the optimal values for τ as a function of season, forecast day, and accumulation window, we apply the best significant τ value according to Figures 5 and 6 in the following parts of this study.

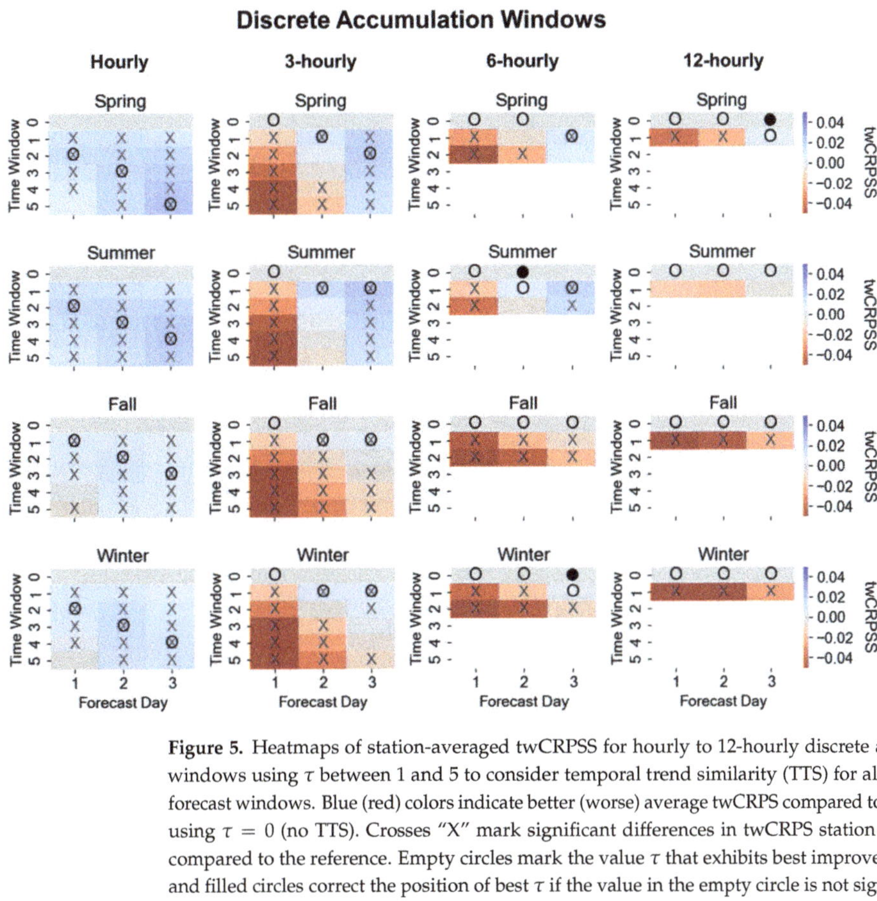

Figure 5. Heatmaps of station-averaged twCRPSS for hourly to 12-hourly discrete accumulation windows using τ between 1 and 5 to consider temporal trend similarity (TTS) for all seasons and forecast windows. Blue (red) colors indicate better (worse) average twCRPS compared to the reference using $\tau = 0$ (no TTS). Crosses "X" mark significant differences in twCRPS station distributions compared to the reference. Empty circles mark the value τ that exhibits best improvement overall, and filled circles correct the position of best τ if the value in the empty circle is not significant.

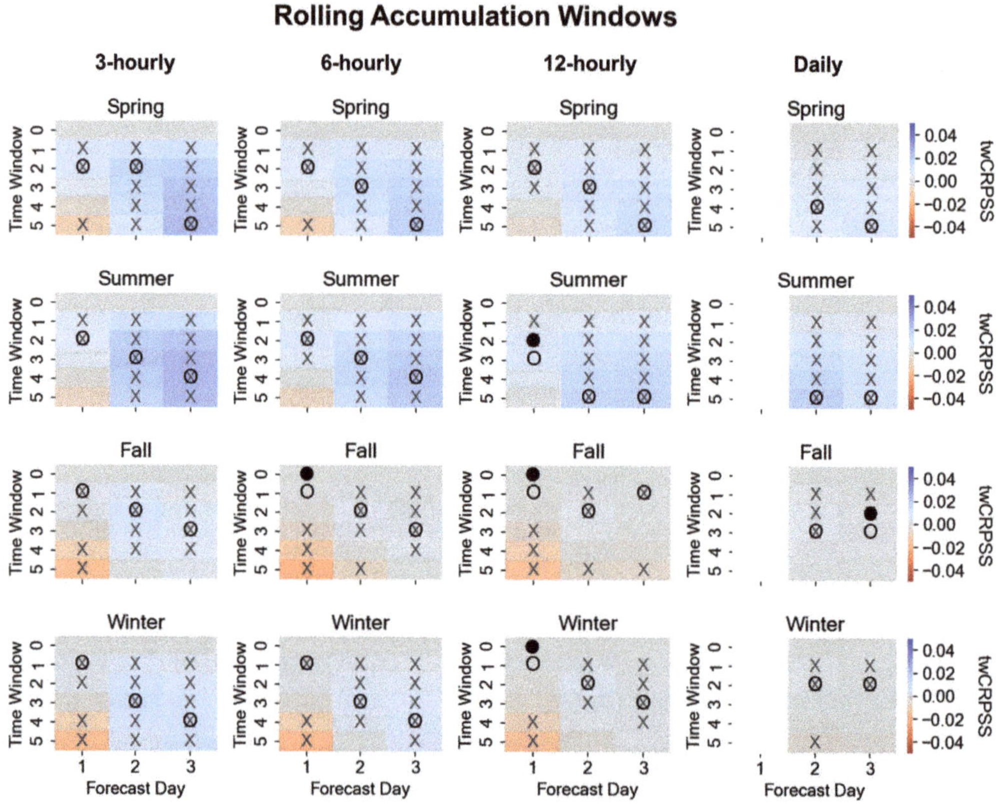

Figure 6. As in Figure 5 but for 3-hourly to daily rolling windows. Again, empty circles mark the value τ that exhibits best improvement overall, and filled circles correct the position of best τ if the value in the empty circle is not significant. Forecast day 1 for daily accumulations is removed, since it contains only 1 instead of 24 lead-time samples as on the other forecast days.

3.3. *Supplemental Lead Times (SLTs)*

The SLT approach (described in Section 2.2.2) is tested using windows with up to ±10 SLTs. At this window width the best out of 21 AnFcst candidates is considered, rather than only the one PaFcst at the target lead time as in the original approach. Since we wish to retain error characteristics from similar lead times, we do not assess the effect of SLTs on discrete windows, nor beyond a lead-time offset of 10 h.

Across stations and forecast days, there is a clear tendency that twCRPSS values improve with increasing SLT window (Figure 7). The steepest improvement occurs within ±3 SLTs and levels out at ±6 SLTs on forecast day 3, but day-1 and day-2 average scores keep improving at a decreasing rate until our maximum of ±10 SLTs. Both, twCRPS and twCRPSS changes are significant in every step of growing SLT windows.

While TTS considerations yield best improvements for longer forecast horizons, the SLT approach benefits shorter forecast horizons most. A reason for this could be that on forecast day 1, the limiting factor in AnEn performance is the availability of good analogs from the provided sample size, whereas on day 3, the limiting factor is the quality of the forecast itself.

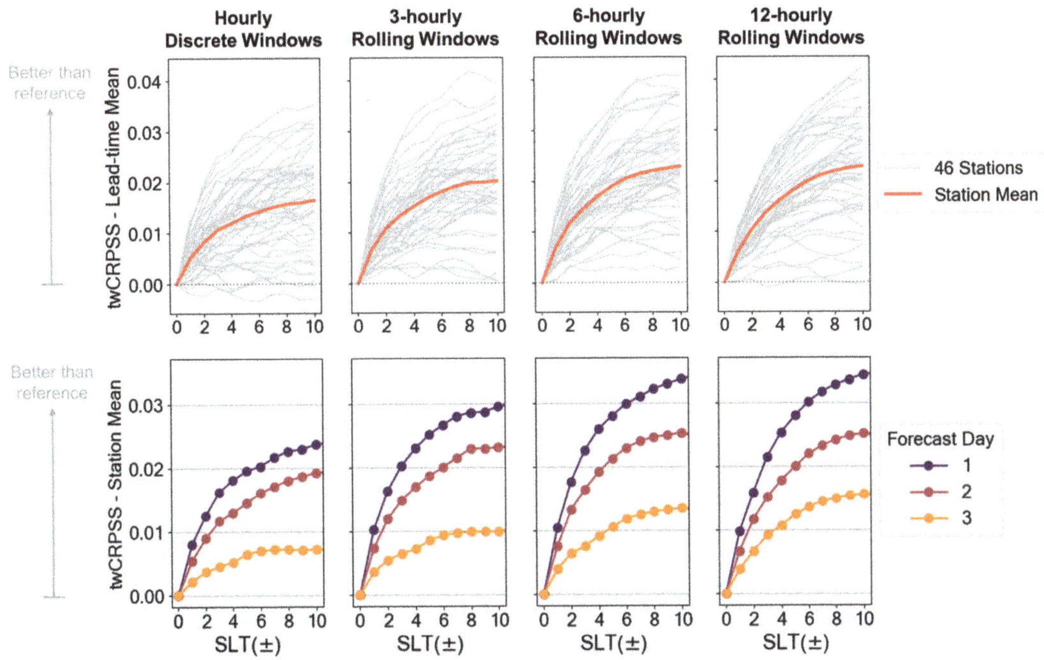

Figure 7. Lead-time aggregated (**top**) and station-aggregated (**bottom**) twCRPSS from the supplemental-lead-time (SLT) experiments. The reference for twCRPSS is the original approach with optimized predictors but without SLT. Positive twCRPSS indicate improvement compared to the reference.

3.4. Verification

Not all methods could be performed on the discrete accumulations due to their larger time steps, and the predictor selection was in part already evaluated in Section 3.1; therefore, this verification section focuses on hourly and rolling windows only. Verification is conducted by running the AnEn over the independent 1-year testing period with the best tuning parameters determined in Sections 3.1–3.3. We show the stepwise improvement by comparing:

- the control AnEn using reference predictors, no TTS, and no SLTs (Control),
- the AnEn with optimized predictors, but no TTS, and no SLTs (Step 1),
- the AnEn with optimized predictors and optimized TTS consideration, but no SLT (Step 2), and
- the AnEn with optimized predictors and optimized TTS consideration, and using SLTs in a window of ± 6 (Step 3).

First, the performance improvement over the raw WRF forecasts is assessed. For comparison we transform the probabilistic AnEn into a deterministic forecast by taking the ensemble median and calculate the mean absolute error (MAE) with the verifying observations. Analogous to Equation (A2) for twCRPSS, the MAE skill score (MAESS) is computed, where positive values represent improvement over the MAE of the raw WRF forecasts.

As seen in Figure 8, all of the AnEn post-processing methods generally have higher rates of improvement for shorter accumulation windows and longer lead times. Even the Control AnEn significantly improves the raw WRF forecasts, and the additional steps further enhance the AnEn performance. For example, the Control AnEn improves WRF MAEs by about 13.3% for hourly forecasts (averaged over forecast days) and about 5.5% for daily forecasts, while the Step 3 AnEn reduces WRF MAEs by an additional 2.9% and 2.0%, respectively. Compared to the Control, our Step 3 AnEn improves hourly-precipitation

MAESS by 30.6% on day 1, 26.3% on day 2, and 9.6% on day 3; whereas 12-hourly MAESS are improved by 83.8%, 74.0%, and 41.6%, respectively.

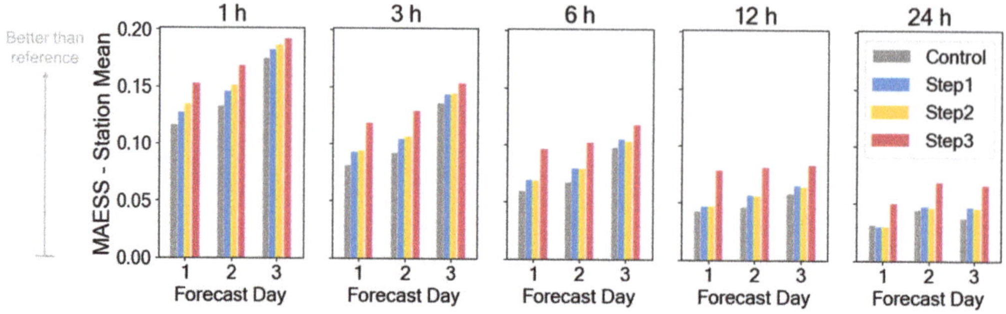

Figure 8. Station mean of the mean absolute error skill score (MAESS) by forecast day and for hourly to daily rolling windows, relative to the raw NWP forecast.

The optimization steps show step-wise improvement, except Step 2 shows a slight drop in MAESS compared to Step 1 for accumulations larger than 3 h. This indicates that the TTS does not always yield the expected improvement as seen during training. The largest improvement is consistently seen in Step 3 from using SLTs.

Recall that the predictor selection was optimized on only forecast day 1 using 3-hourly windows. Yet across accumulation windows and forecast days the same predictors improve the Step 1 MAE, except for daily precipitation on forecast day 1. This study shows results up to daily intervals only for reference, however, users who desire daily-precipitation AnEn forecasts are advised to re-trained the predictor selection to assess whether different variables and weights would make better predictors.

To date, in this paper, we have focused on AnEn improvements for 75p events. Figures 9 and 10 show 90p results to assess if the AnEn forecasts remain skillful for the heavier precipitation events.

The reliability diagrams in Figure 9 show that AnEn probabilities compare well to observed relative frequency across forecast days and accumulation windows; i.e., the AnEns are calibrated and reliable. Relative to the dashed line that represents perfect reliability, most points in the calibration functions have a small deviation towards the left side. This means that the AnEns have a small dry bias for 90p events. This is a common property of AnEns, especially for high-impact events for which only a smaller number of good AnFcsts are available [22,38,48].

Compared to the Control, Step 1 and Step 2 slightly worsen this bias, however, Step 3 moves the calibration function back closer to the line of perfect reliability. This is particularly meaningful because the SLT approach further improves sharpness. Sharper AnEn forecasts were expected as a result of SLTs, because the larger number of considered PaFcsts provides a better chance for the selection of closer AnFcst. These results agree with [54], and they also agree with [38]'s supplemental-locations approach, which uses spatial rather than temporal supplements to inflate the PaFcst sample size.

The receiver operating characteristic (ROC) diagram in Figure 10 shows the discrimination between 90p events and non-events. Larger values of the area under the curve (AUC) score are better, corresponding to a higher true-positive rate (i.e., POD or hit rate) and a lower false-positive rate (i.e., false-alarm rate).

The AnEn AUC scores increase with each optimization step, however, the improvement is larger for shorter accumulations and longer forecast horizons, both of which exhibit worse discrimination to begin with. Although in Figures 8 and 9 Step 2 (TTS) showed smaller improvements or sometimes even worse performance in comparison, TTS contributes considerable improvement with regard to AUC, in particular on forecast day 3 for shorter accumulation windows.

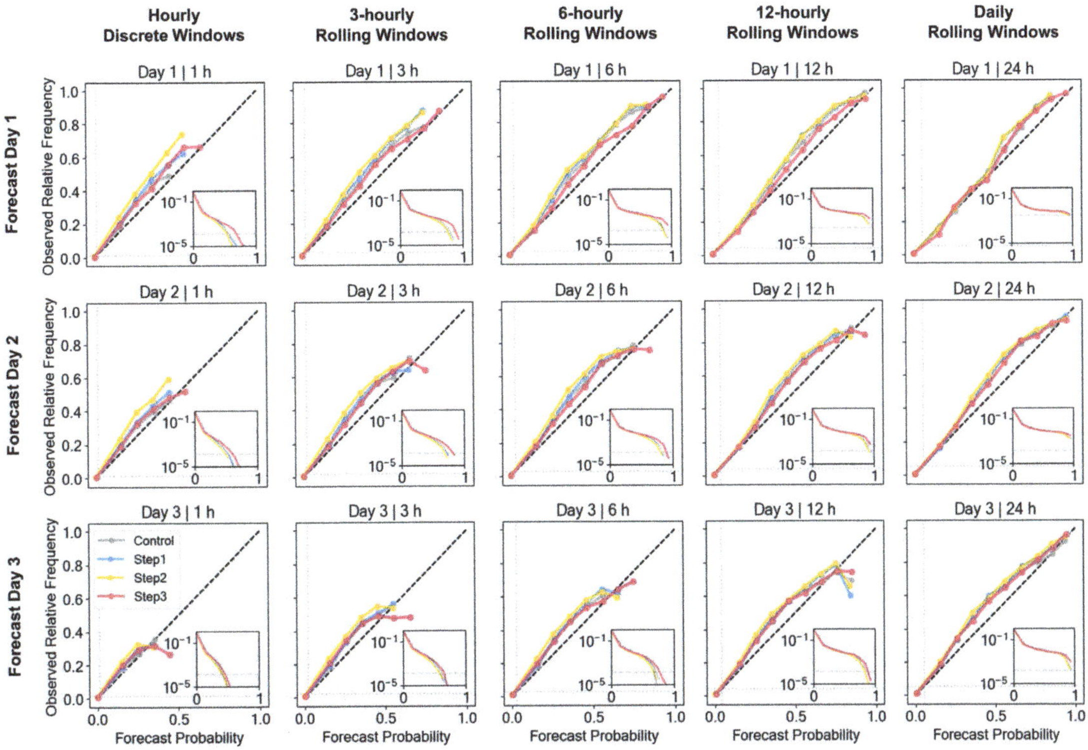

Figure 9. Station-aggregated 90p reliability diagrams on all forecast days for hourly discrete to daily rolling accumulation windows. The dashed black line is the reference for perfect reliability, the grey dotted lines show climatological probability. The inset in the lower right corner displays the corresponding sharpness diagram, which shows the relative frequency of forecasts that fall into each bin. Due to the skewed nature of precipitation distributions, the y-axis in the sharpness diagram is plotted on a logarithmic scale. The reliability diagram displays only bins that include at least 50 samples in total (i.e., only those points above the dashed grey line in the sharpness diagram).

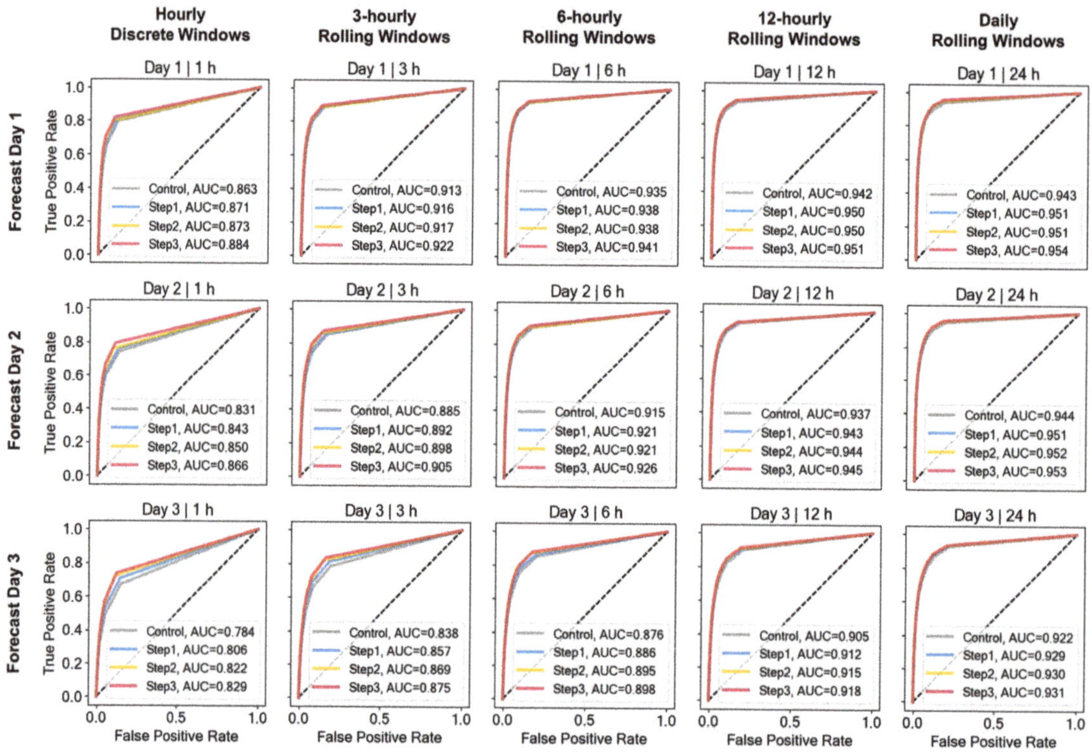

Figure 10. Station-aggregated 90p receiver operating characteristic (ROC) diagrams for all forecast days and hourly discrete to daily rolling accumulation windows. The area under the curve (AUC) is given in each legend and has a perfect score of 1. The dashed black line represents the line of no skill with AUC = 0.5 corresponding to climatology.

4. Summary and Conclusions

This study demonstrates the benefits of existing and new optimization techniques for AnEn post-processing on sub-daily precipitation forecasts over southwest BC. Lower precipitation rates are easier to predict and have less impact, but since they are far more common they are likely to dominate optimization procedures. Therefore, this study tuned the AnEn parameters for moderate and heavier events based on the 75p twCRPS, instead of the full CRPS as in most other studies.

First, we objectively optimized the choice of predictor variables and their weights by evaluating four variants of forward selection (FS). Since common predictor optimization techniques come at significant computational expense, we suggested the efficient FS (EFS)–an adaptation of [44]'s FS. Limiting the weighting options in sequence with the selected predictors significantly reduces computational cost while maintaining similar optimization performance. Predictor tuning is beneficial even if trained on a portion of the dataset (i.e., only one forecast day instead of the full forecast horizon) and even if the initial set of variables on which the (E)FS is conducted is pre-selected by filter methods such as DCorr. However, EFS on a larger number of meteorological variables can result in minor additional improvements and could be considered in other studies if the computational capacity exists.

Next, we explored the impact of the time-window width over which the temporal predictor trends are matched. This investigation revealed that longer time windows are most beneficial for longer forecast horizons and shorter accumulation windows—a relationship that is often neglected in other studies. Although TTS was shown in the verification Section 3.4 to increase the unconditional dry bias, it improves discrimination of high-impact events. Perhaps the method described in Section 3.2 to assess optimal time windows could be generalized across seasons, such as a staggered implementation for forecast days 1, 2, and 3 using τ equal to 1 or 2, 2 or 3, and 3 or 4, respectively. Caution is advised when using TTS for discrete windows of accumulations that are longer than hourly, because we obtained mixed results dependent on lead time.

Finally, we implemented a new methodology that uses the concept of supplemental lead times (SLTs). It enhances the chance of finding better AnFcsts by allowing the algorithm to choose from forecast lead times within a time window around the target lead time, that should maintain similar error characteristics. This approach is similar to the idea in [54], but it is suitable for shorter accumulations and prevents the selection of temporally dependent AnFcsts. SLTs could be used in addition to [38]'s supplemental locations, or as an alternative if the domain or station sample size is not sufficient (as in this study).

The use of SLTs had the largest impact on AnEn performance, often exceeding the effects of predictor and TTS optimization, especially for verification statistics including the ensemble-median MAE and 90p reliability and sharpness. The time window width for which SLTs showed performance increase is relatively wide in this study, likely because precipitation in BC has no pronounced diurnal cycle in its cool/wet season. Other predictands with more pronounced diurnal cycle would likely require shorter SLT windows and may experience smaller relative improvements. It is conceivable that a longer dataset history would result in similar improvements, dampening the impact of SLTs. However, when relatively short training periods are available (<5 years in this study), the SLT approach somewhat compensates for the small sample size. This opens up opportunities for AnEn applications on shorter but locally optimized and operational data products.

One NWP model forecast produces three-dimensional multivariate deterministic predictions, whereas the AnEn method in this study creates a univariate probabilistic point forecast by post-processing NWP. Compared to NWP ensembles, AnEns are extremely efficient in creating reliable probabilistic point forecasts—that is, if a sufficiently long reforecast dataset is available. Although our algorithm was not optimized for efficiency, a single 3-day forecast at one point location takes on average only 0.5 s to create the Control or Step 1 AnEn on a macOS computer (with 3.2 GHz Intel Core i5 processor), 1 s to create the Step 2 AnEn (with TTS), and <10 s to create the Step 3 AnEn (with TTS and $\pm$6 SLTs). This computational time applies to the AnFcst search and AnEn composition only (i.e., excluding time for running the NWP TaFcst and interpolation to station locations) and would have to be multiplied by the number of point locations at which forecasts are desired (if not run simultaneously in parallel). In comparison, the three-domain WRF runs used in this study took on average 80 min run time using 48 cores (Intel-compiled WRF code run on an HPC cluster using Open MPI and no hyperthreading on Intel Xeon Processor E5–2683 v4 compute nodes with 2.10 GHz). While at least one NWP run is required to make an AnEn forecast, running an equivalent-sized 30-member WRF ensemble would require 10 runs, which would take approximately 27 core days of run time in serial mode using configurations (e.g., domain setup) as in [58].

In this study, we improved the computational efficiency of AnEn optimization, while also significantly improving AnEn forecast performance by up to 83.8% compared to a reference AnEn. This increase in AnEn performance can be attributed mainly to a new SLT technique. Temporal variability and forecast-error growth cause finer-temporal-resolution and longer-lead-time forecasts to have inherently worse performance, especially over the complex terrain of southwest BC—yet those forecasts benefit the most from the optimized AnEn post-processing. This is an important result in a world where end users desire evermore accurate predictions at finer resolutions and longer outlooks.

Author Contributions: Conceptualization, J.J., G.W. and R.S.; methodology, J.J.; software, J.J.; validation, J.J., G.W. and R.S.; formal analysis, J.J.; investigation, J.J.; resources, R.S. and J.J.; data curation, J.J.; writing—original draft preparation, J.J.; writing—review and editing, G.W. and R.S.; visualization, J.J.; supervision, G.W. and R.S.; project administration, J.J., G.W. and R.S.; funding acquisition, R.S. and G.W. All authors have read and agreed to the published version of the manuscript.

Funding: Computational and storage resources to create the re-forecast dataset and to optimize the AnEns were provided by WestGrid (westgrid.ca) and the Digital Research Alliance of Canada (alliancecan.ca) through the Resource Allocation Competition (RAC) awards 2019–2022. The research was enabled by funding support provided by Mitacs (Grants IT07224 and IT28208), BC Hydro (Contracts 00089063 and 00091424), the Natural Science and Engineering Research Council (NSERC; Discovery Grant RGPIN-2017-03849), and the University of British Columbia (UBC). We thank William Wei Hsieh for supporting this research through the Chih-Chuang and Yien-Ying Wang Hsieh Memorial Scholarship (#5357).

Institutional Review Board Statement: Not applicable.

Informed Consent Statement: Not applicable.

Data Availability Statement: WRF model point forecasts are not publicly archived, but can be reproduced following [58] or made available upon request [contact Roland Stull (rstull@eoas.ubc.ca)]. ECCC station data used for verification are available at https://climate.weather.gc.ca/historical_data/search_historic_data_e.html (accessed on 2 November 2021), whereas BC Hydro station data may be obtained by contacting Gregory West (greg.west@bchydro.com).

Acknowledgments: We thank Timothy Chun-Yiu Chui, Yingkai Sha, Henryk Modzelewski, and Roland Schigas for their technical support with this study.

Conflicts of Interest: The authors declare no conflict of interest.

Abbreviations

The following abbreviations are used in this manuscript:

AnEn(s)	Analog Ensemble(s)
AnFcst(s)	Analog Forecast(s)
AnObs	Analog Observation(s)
TaFcst(s)	Target Forecast(s)
VerifObs	Verifying Observation(s)
TTS	Temporal Trend Similarity
SLT(s)	Supplemental Lead Time(s)
FS	Forward Selection
EFS	Efficient Forward Selection

Appendix A. Percentiles

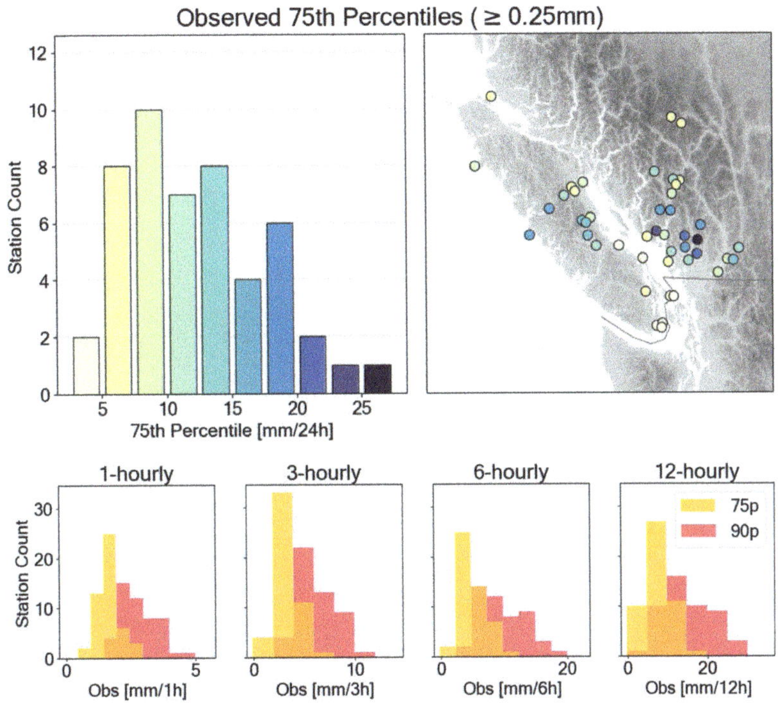

Figure A1. Top row: Histogram of binned daily observed 75th percentiles (75p) at all stations (top left), and corresponding geographic distribution of the 75p relative to topography (top right); Bottom row: Histograms of 75p and 90p for other accumulations. Although not plotted here, these frequency distributions show similar geographic variations to the top right panel.

Appendix B. Evaluation

Appendix B.1. Threshold-Weighted Continuous Ranked Probability Score

The 75th-percentile (75p) threshold-weighted continuous ranked probability score [7] across forecasts j is defined as

$$\mathrm{twCRPS}(\mathrm{AnEn}_j, \mathrm{VerifObs}_j) = \int_{-\infty}^{\infty} \mathbb{1}_{\{x \geq 75\mathrm{p}\}} (\mathrm{AnEn}_j(x) - \mathbb{1}_{\{\mathrm{VerifObs}_j \leq x\}})^2 \, dx, \quad \text{(A1)}$$

where $\mathbb{1}$ denotes an indicator function, which is 1 under the sub-scripted condition, and 0 otherwise, while the conventional CRPS can be interpreted as the integral of the Brier scores over the range of possible thresholds [50,76], the twCRPS with the additional $\mathbb{1}_{\{x \geq 75\mathrm{p}\}}$ can be interpreted as the integral of the Brier scores over the thresholds larger than a desired value–75p in our study. In other words, the twCRPS is the fraction of the conventional CRPS that assesses events above the given threshold (the right tail of the distribution).

The relative performance between methods and optimization steps is compared using the threshold-weighted continuous ranked probability skill score

$$\mathrm{twCRPSS} = 1 - (\overline{\mathrm{twCRPS}} / \overline{\mathrm{twCRPS}_{\mathrm{ref}}}), \quad \text{(A2)}$$

which yields positive values when skill is improved compared to the reference.

Appendix B.2. Statistical Tests

If the Shapiro–Wilk test for normality [77] is rejected over the distributions of results across stations, we use the non-parametric two-sided Wilcoxon signed-rank test [78] to assess whether the paired station-result samples from different methods originate from the same distribution at the $\alpha = 0.05$ level. Otherwise, we use the paired t-test.

Appendix C. Correlation Analysis

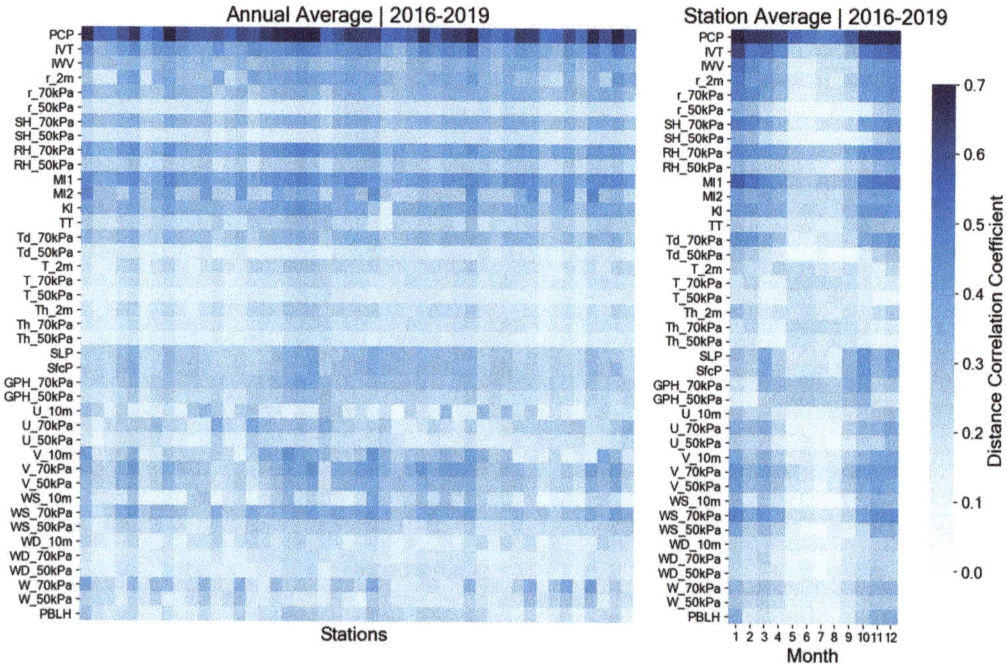

Figure A2. Distance Correlation coefficients (DCorr) of all model variables (see Table 1) with observed precipitation. Variability across 46 stations aggregated over time (left), and variability across months aggregated over stations (right). The time period covers four complete years from the optimization period (rather than the full 4.75-year period) to ensure similar sample size across months.

Appendix D. Predictor Weights

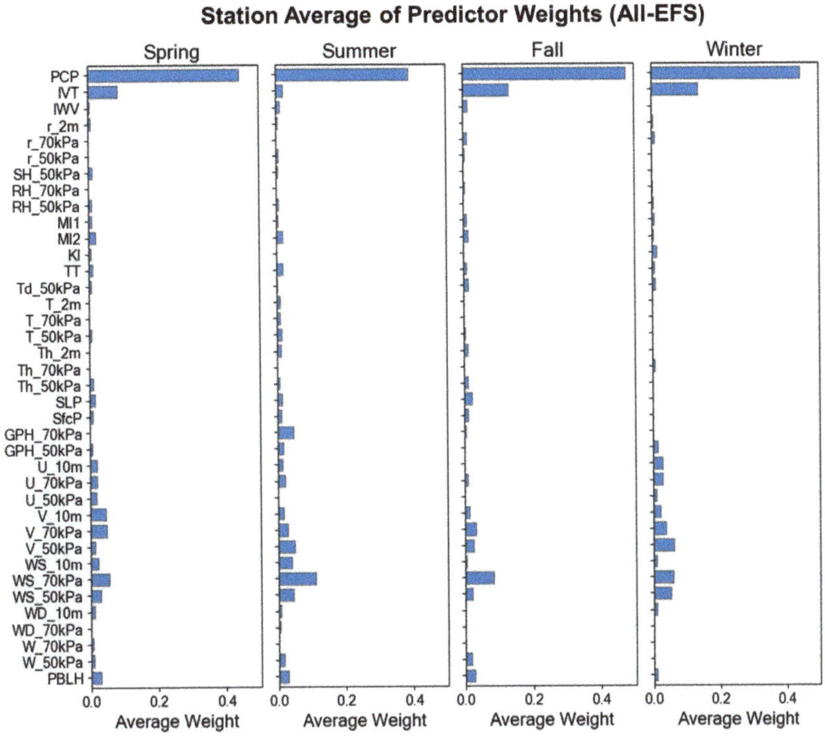

Figure A3. Station average of the final predictor weights resulting from the All-EFS method (see Section 2.2.1) for each season. Variables that are in Table 1 but not in the x-axis were never selected by any station at any season.

References

1. Wilks, D.S. *Statistical Methods in the Atmospheric Sciences*; Elsevier/Academic Press: Cambridge, MA, USA, 2011; p. 676.
2. Wilks, D.S. Extending logistic regression to provide full-probability-distribution MOS forecasts. *Meteorol. Appl.* **2009**, *16*, 361–368. [CrossRef]
3. Roulin, E.; Vannitsem, S. Postprocessing of Ensemble Precipitation Predictions with Extended Logistic Regression Based on Hindcasts. *Mon. Weather Rev.* **2012**, *140*, 874–888. [CrossRef]
4. Bakker, K.; Whan, K.; Knap, W.; Schmeits, M. Comparison of statistical post-processing methods for probabilistic NWP forecasts of solar radiation. *Sol. Energy* **2019**, *191*, 138–150. [CrossRef]
5. Carter, G.M.; Dallavalle, J.P.; Glahn, H.R. Statistical Forecasts Based on the National Meteorological Center's Numerical Weather Prediction System. *Weather Forecast.* **1989**, *4*, 401–412. [CrossRef]
6. Stensrud, D.J.; Yussouf, N. Short-range ensemble predictions of 2-m temperature and dewpoint temperature over New England. *Mon. Weather Rev.* **2003**, *131*, 2510–2524. [CrossRef]
7. Gneiting, T.; Ranjan, R. Comparing density forecasts using threshold and quantile-weighted scoring rules. *J. Bus. Econ. Stat.* **2011**, *29*, 411–422. [CrossRef]
8. Scheuerer, M. Probabilistic quantitative precipitation forecasting using Ensemble Model Output Statistics. *Q. J. R. Meteorol. Soc.* **2014**, *140*, 1086–1096. . [CrossRef]
9. Delle Monache, L.; Nipen, T.; Liu, Y.; Roux, G.; Stull, R. Kalman Filter and Analog Schemes to Postprocess Numerical Weather Predictions. *Mon. Weather Rev.* **2011**, *139*, 3554–3570. [CrossRef]
10. McCollor, D.; Stull, R. Hydrometeorological accuracy enhancement via postprocessing of numerical weather forecasts in complex terrain. *Weather Forecast.* **2008**, *23*, 131–144. [CrossRef]
11. Raftery, A.E.; Gneiting, T.; Balabdaoui, F.; Polakowski, M. Using Bayesian Model Averaging to Calibrate Forecast Ensembles. *Mon. Weather Rev.* **2005**, *133*, 1155–1174. [CrossRef]

12. Faidah, D.Y.; Kuswanto, H.; Suhartono. The comparison of Bayesian model averaging with gaussian and gamma components for probabilistic precipitation forecasting. *AIP Conf. Proc.* **2019**, *2192*, 090003. [CrossRef]
13. Yuan, H.; Gao, X.; Mullen, S.L.; Sorooshian, S.; Du, J.; Juang, H.M.H. Calibration of Probabilistic Quantitative Precipitation Forecasts with an Artificial Neural Network. *Weather Forecast.* **2007**, *22*, 1287–1303. [CrossRef]
14. Sha, Y.; Gagne, D.J., II; West, G.; Stull, R. A hybrid analog-ensemble, convolutional-neural-network method for post-processing precipitation forecasts. *Mon. Weather. Rev.* **2022**, *150*, 1495–1515. [CrossRef]
15. Cho, D.; Yoo, C.; Son, B.; Im, J.; Yoon, D.; Cha, D.H. A novel ensemble learning for post-processing of NWP Model's next-day maximum air temperature forecast in summer using deep learning and statistical approaches. *Weather Clim. Extrem.* **2022**, *35*, 100410. [CrossRef]
16. Hamill, T.M.; Whitaker, J.S. Probabilistic Quantitative Precipitation Forecasts Based on Reforecast Analogs: Theory and Application. *Mon. Weather Rev.* **2006**, *134*, 3209–3229. [CrossRef]
17. Delle Monache, L.; Eckel, F.A.; Rife, D.L.; Nagarajan, B.; Searight, K. Probabilistic Weather Prediction with an Analog Ensemble. *Mon. Weather Rev.* **2013**, *141*, 3498–3516. [CrossRef]
18. Eckel, F.A.; Delle Monache, L. A Hybrid NWP–Analog Ensemble. *Mon. Weather Rev.* **2016**, *144*, 897–911. [CrossRef]
19. Junk, C.; Monache, L.D.; Alessandrini, S. Analog-Based Ensemble Model Output Statistics. *Mon. Weather Rev.* **2015**, *143*, 2909–2917. [CrossRef]
20. Frediani, M.E.B.; Hopson, T.M.; Hacker, J.P.; Anagnostou, E.N.; Delle Monache, L.; Vandenberghe, F. Object-Based Analog Forecasts for Surface Wind Speed. *Mon. Weather Rev.* **2017**, *145*, 5083–5102. [CrossRef]
21. Sperati, S.; Alessandrini, S.; Delle Monache, L. Gridded probabilistic weather forecasts with an analog ensemble. *Q. J. R. Meteorol. Soc.* **2017**, *143*, 2874–2885. [CrossRef]
22. Odak Plenković, I.; Delle Monache, L.; Horvath, K.; Hrastinski, M. Deterministic Wind Speed Predictions with Analog-Based Methods over Complex Topography. *J. Appl. Meteorol. Climatol.* **2018**, *57*, 2047–2070. [CrossRef]
23. Yang, J.; Astitha, M.; Monache, L.D.; Alessandrini, S. An Analog Technique to Improve Storm Wind Speed Prediction Using a Dual NWP Model Approach. *Mon. Weather Rev.* **2018**, *146*, 4057–4077. [CrossRef]
24. Davò, F.; Alessandrini, S.; Sperati, S.; Delle Monache, L.; Airoldi, D.; Vespucci, M.T. Post-processing techniques and principal component analysis for regional wind power and solar irradiance forecasting. *Sol. Energy* **2016**, *134*, 327–338. [CrossRef]
25. Alessandrini, S.; Delle Monache, L.; Sperati, S.; Nissen, J. A novel application of an analog ensemble for short-term wind power forecasting. *Renew. Energy* **2015**, *76*, 768–781. [CrossRef]
26. Martín, M.; Valero, F.; Pascual, A.; Sanz, J.; Frias, L. Analysis of wind power productions by means of an analog model. *Atmos. Res.* **2014**, *143*, 238–249. [CrossRef]
27. Alessandrini, S.; Delle Monache, L.; Sperati, S.; Cervone, G. An analog ensemble for short-term probabilistic solar power forecast. *Appl. Energy* **2015**, *157*, 95–110. [CrossRef]
28. Djalalova, I.; Delle Monache, L.; Wilczak, J. PM2.5 analog forecast and Kalman filter post-processing for the Community Multiscale Air Quality (CMAQ) model. *Atmos. Environ.* **2015**, *108*, 76–87. [CrossRef]
29. Delle Monache, L.; Alessandrini, S.; Djalalova, I.; Wilczak, J.; Knievel, J.C.; Kumar, R. Improving Air Quality Predictions over the United States with an Analog Ensemble. *Weather Forecast.* **2020**, *35*, 2145–2162. [CrossRef]
30. Raman, A.; Arellano, A.F.; Delle Monache, L.; Alessandrini, S.; Kumar, R. Exploring analog-based schemes for aerosol optical depth forecasting with WRF-Chem. *Atmos. Environ.* **2021**, *246*, 118134. [CrossRef]
31. Horton, P.; Jaboyedoff, M.; Metzger, R.; Obled, C.; Marty, R. Spatial relationship between the atmospheric circulation and the precipitation measured in the western Swiss Alps by means of the analogue method. *Nat. Hazards Earth Syst. Sci.* **2012**, *12*, 777–784. [CrossRef]
32. Ben Daoud, A.; Sauquet, E.; Bontron, G.; Obled, C.; Lang, M. Daily quantitative precipitation forecasts based on the analogue method: Improvements and application to a French large river basin. *Atmos. Res.* **2016**, *169*, 147–159. [CrossRef]
33. Keller, J.D.; Monache, L.D.; Alessandrini, S. Statistical Downscaling of a High-Resolution Precipitation Reanalysis Using the Analog Ensemble Method. *J. Appl. Meteorol. Climatol.* **2017**, *56*, 2081–2095. [CrossRef]
34. Yang, C.; Yuan, H.; Su, X. Bias correction of ensemble precipitation forecasts in the improvement of summer streamflow prediction skill. *J. Hydrol.* **2020**, *588*, 124955. [CrossRef]
35. Junk, C.; Delle Monache, L.; Alessandrini, S.; Cervone, G.; von Bremen, L. Predictor-weighting strategies for probabilistic wind power forecasting with an analog ensemble. *Meteorol. Z.* **2015**, *24*, 361–379. [CrossRef]
36. Li, N.; Ran, L.; Jiao, B. An analogy-based method for strong convection forecasts in China using GFS forecast data. *Atmos. Ocean. Sci. Lett.* **2020**, *13*, 97–106. [CrossRef]
37. Liu, Y.Y.; Li, L.; Liu, Y.S.; Chan, P.W.; Zhang, W.H.; Zhang, L. Estimation of precipitation induced by tropical cyclones based on machine-learning-enhanced analogue identification of numerical prediction. *Meteorol. Appl.* **2021**, *28*, e1978. [CrossRef]
38. Hamill, T.M.; Scheuerer, M.; Bates, G.T. Analog Probabilistic Precipitation Forecasts Using GEFS Reforecasts and Climatology-Calibrated Precipitation Analyses. *Mon. Weather Rev.* **2015**, *143*, 3300–3309. [CrossRef]
39. Obled, C.; Bontron, G.; Garçon, R. Quantitative precipitation forecasts: A statistical adaptation of model outputs through an analogues sorting approach. *Atmos. Res.* **2002**, *63*, 303–324. [CrossRef]
40. Marty, R.; Zin, I.; Obled, C.; Bontron, G.; Djerboua, A. Toward Real-Time Daily PQPF by an Analog Sorting Approach: Application to Flash-Flood Catchments. *J. Appl. Meteorol. Climatol.* **2012**, *51*, 505–520. [CrossRef]

41. Bellier, J.; Zin, I.; Siblot, S.; Bontron, G. Probabilistic flood forecasting on the Rhone River: Evaluation with ensemble and analogue-based precipitation forecasts. *E3S Web Conf.* **2016**, *7*, 18011. [CrossRef]
42. Horton, P.; Jaboyedoff, M.; Obled, C. Global Optimization of an Analog Method by Means of Genetic Algorithms. *Mon. Weather Rev.* **2017**, *145*, 1275–1294. [CrossRef]
43. Horton, P.; Brönnimann, S. Impact of global atmospheric reanalyses on statistical precipitation downscaling. *Clim. Dyn.* **2019**, *52*, 5189–5211. [CrossRef]
44. Alessandrini, S.; Delle Monache, L.; Rozoff, C.M.; Lewis, W.E. Probabilistic Prediction of Tropical Cyclone Intensity with an Analog Ensemble. *Mon. Weather Rev.* **2018**, *146*, 1723–1744. [CrossRef]
45. Fernández, J.; Sáenz, J. Improved field reconstruction with the analog method: Searching the CCA space. *Clim. Res.* **2003**, *24*, 199–213. [CrossRef]
46. Cannon, A.J. Nonlinear analog predictor analysis: A coupled neural network/analog model for climate downscaling. *Neural Netw.* **2007**, *20*, 444–453. [CrossRef] [PubMed]
47. Horton, P.; Jaboyedoff, M.; Obled, C. Using genetic algorithms to optimize the analogue method for precipitation prediction in the Swiss Alps. *J. Hydrol.* **2018**, *556*, 1220–1231. [CrossRef]
48. Alessandrini, S.; Sperati, S.; Delle Monache, L. Improving the Analog Ensemble Wind Speed Forecasts for Rare Events. *Mon. Weather Rev.* **2019**, *147*, 2677–2692. [CrossRef]
49. Alessandrini, S.; Delle Monache, L.; Rozoff, C.; Lewis, W. Probabilistic Prediction of Hurricane Intensity with an Analog Ensemble. In Proceedings of the 96th American Meteorological Society Annual Meeting, New Orleans, LA, USA, 10–14 January 2016.
50. Odak Plenković, I.; Schicker, I.; Dabernig, M.; Horvath, K.; Keresturi, E. Analog-based post-processing of the ALADIN-LAEF ensemble predictions in complex terrain. *Q. J. R. Meteorol. Soc.* **2020**, *146*, 1842–1860. [CrossRef]
51. Hamill, T.M.; Whitaker, J.S.; Mullen, S.L. Reforecasts: An Important Dataset for Improving Weather Predictions. *Bull. Am. Meteorol. Soc.* **2006**, *87*, 33–46. [CrossRef]
52. Meech, S.; Alessandrini, S.; Chapman, W.; Delle Monache, L. Post-processing rainfall in a high-resolution simulation of the 1994 Piedmont flood. *Bull. Atmos. Sci. Technol.* **2020**, *1*, 373–385. [CrossRef]
53. Dayon, G.; Boé, J.; Martin, E. Transferability in the future climate of a statistical downscaling method for precipitation in France. *J. Geophys. Res. Atmos.* **2015**, *120*, 1023–1043. [CrossRef]
54. Horton, P.; Obled, C.; Jaboyedoff, M. The analogue method for precipitation prediction: Finding better analogue situations at a sub-daily time step. *Hydrol. Earth Syst. Sci.* **2017**, *21*, 3307–3323. [CrossRef]
55. Gibergans-Báguena, J.; Llasat, M. Improvement of the analog forecasting method by using local thermodynamic data. Application to autumn precipitation in Catalonia. *Atmos. Res.* **2007**, *86*, 173–193. [CrossRef]
56. Ren, F.; Ding, C.; Zhang, D.L.; Chen, D.; li Ren, H.; Qiu, W. A Dynamical-Statistical-Analog Ensemble Forecast Model: Theory and an Application to Heavy Rainfall Forecasts of Landfalling Tropical Cyclones. *Mon. Weather Rev.* **2020**, *148*, 1503–1517. [CrossRef]
57. Saminathan, S.; Medina, H.; Mitra, S.; Tian, D. Improving short to medium range GEFS precipitation forecast in India. *J. Hydrol.* **2021**, *598*, 126431. [CrossRef]
58. Jeworrek, J.; West, G.; Stull, R. WRF Precipitation Performance and Predictability for Systematically Varied Parameterizations over Complex Terrain. *Weather Forecast.* **2021**, *36*, 893–913. [CrossRef]
59. Marty, R.; Zin, I.; Obled, C. Sensitivity of hydrological ensemble forecasts to different sources and temporal resolutions of probabilistic quantitative precipitation forecasts: Flash flood case studies in the Cévennes-Vivarais region (Southern France). *Hydrol. Process.* **2013**, *27*, 33–44. [CrossRef]
60. Hamill, T.M.; Hagedorn, R.; Whitaker, J.S. Probabilistic forecast calibration using ECMWF and GFS ensemble reforecasts. Part II: Precipitation. *Mon. Weather Rev.* **2008**, *136*, 2620–2632. [CrossRef]
61. Fernández-Ferrero, A.; Sáenz, J.; Ibarra-Berastegi, G. Comparison of the performance of different analog-based bayesian probabilistic precipitation forecasts over Bilbao, Spain. *Mon. Weather Rev.* **2010**, *138*, 3107–3119. [CrossRef]
62. Chapman, W.E.; Delle Monache, L.; Alessandrini, S.; Subramanian, A.C.; Ralph, F.M.; Xie, S.P.; Lerch, S.; Hayatbini, N. Probabilistic Predictions from Deterministic Atmospheric River Forecasts with Deep Learning. *Mon. Weather Rev.* **2021**, *150*, 215–234. [CrossRef]
63. PCIC. *Atmospheric Rivers State of Knowledge Report*; Technical Report; Pacific Climate Impacts Consortium: Victoria BC, Canada, 2013. https://www.pacificclimate.org/sites/default/files/publications/Atmospheric%20Report%20Final%20Revised.pdf
64. Gillett, N.P.; Cannon, A.J.; Malinina, E.; Schnorbus, M.; Anslow, F.; Sun, Q.; Kirchmeier-Young, M.; Zwiers, F.; Seiler, C.; Zhang, X.; et al. Human influence on the 2021 British Columbia floods. *Weather Clim. Extrem.* **2022**, *36*, 100441. [CrossRef]
65. Vasquez, T. How an Atmopsheric River Flooded British Columbia. *Weatherwise* **2022**, *75*, 19–23. [CrossRef]
66. Skamarock, W.; Klemp, J.; Dudhi, J.; Gill, D.; Barker, D.; Duda, M.; Huang, X.Y.; Wang, W.; Powers, J. *A Description of the Advanced Research WRF Version 3*; Technical Report; University Corporation for Atmospheric Research: Boulder, CO, USA, 2008. [CrossRef]
67. Côté, J.; Gravel, S.; Méthot, A.; Patoine, A.; Roch, M.; Staniforth, A. The Operational CMC–MRB Global Environmental Multiscale (GEM) Model. Part I: Design Considerations and Formulation. *Mon. Weather Rev.* **1998**, *126*, 1373–1395. [CrossRef]
68. Girard, C.; Plante, A.; Desgagné, M.; McTaggart-Cowan, R.; Côté, J.; Charron, M.; Gravel, S.; Lee, V.; Patoine, A.; Qaddouri, A.; et al. Staggered Vertical Discretization of the Canadian Environmental Multiscale (GEM) Model Using a Coordinate of the Log-Hydrostatic-Pressure Type. *Mon. Weather Rev.* **2014**, *142*, 1183–1196. [CrossRef]
69. Hong, S.Y.; Dudhia, J.; Chen, S.H. A Revised Approach to Ice Microphysical Processes for the Bulk Parameterization of Clouds and Precipitation. *Mon. Weather Rev.* **2004**, *132*, 103–120. [CrossRef]

70. Kain, J.S. The Kain–Fritsch Convective Parameterization: An Update. *J. Appl. Meteorol.* **2004**, *43*, 170–181. [CrossRef]
71. Hong, S.Y.; Noh, Y.; Dudhia, J. A New Vertical Diffusion Package with an Explicit Treatment of Entrainment Processes. *Mon. Weather Rev.* **2006**, *134*, 2318–2341. [CrossRef]
72. Niu, G.Y.; Yang, Z.L.; Mitchell, K.E.; Chen, F.; Ek, M.B.; Barlage, M.; Kumar, A.; Manning, K.; Niyogi, D.; Rosero, E.; et al. The community Noah land surface model with multiparameterization options (Noah-MP): 1. Model description and evaluation with local-scale measurements. *J. Geophys. Res. Atmos.* **2011**, *116*, 1–19. [CrossRef]
73. Yang, Z.L.; Niu, G.Y.; Mitchell, K.E.; Chen, F.; Ek, M.B.; Barlage, M.; Longuevergne, L.; Manning, K.; Niyogi, D.; Tewari, M.; et al. The community Noah land surface model with multiparameterization options (Noah-MP): 2. Evaluation over global river basins. *J. Geophys. Res. Atmos.* **2011**, *116*. [CrossRef]
74. Sha, Y.; Gagne, D.J., II; West, G.; Stull, R. Deep-Learning-Based Precipitation Observation Quality Control. *J. Atmos. Ocean. Technol.* **2021**, *38*, 1075–1091. [CrossRef]
75. Székely, G.J.; Rizzo, M.L.; Bakirov, N.K. Measuring and testing dependence by correlation of distances. *Ann. Stat.* **2007**, *35*, 2769–2794. [CrossRef]
76. Smith, L.A.; Suckling, E.B.; Thompson, E.L.; Maynard, T.; Du, H. Towards improving the framework for probabilistic forecast evaluation. *Clim. Chang.* **2015**, *132*, 31–45. [CrossRef]
77. Shapiro, S.S.; Wilk, M.B. An Analysis of Variance Test for Normality (Complete Samples). *Biometrika* **1965**, *52*, 591–611. [CrossRef]
78. Wilcoxon, F. Individual Comparisons by Ranking Methods. *Biom. Bull.* **1945**, *1*, 80–83. [CrossRef]

atmosphere

MDPI

Article

Short-Term Intensive Rainfall Forecasting Model Based on a Hierarchical Dynamic Graph Network

Huosheng Xie [1,*], Rongyao Zheng [1] and Qing Lin [2,3]

1 College of Computer and Data Science, Fuzhou University, Fuzhou 350108, China; N190327102@fzu.edu.cn
2 Fujian Key Laboratory of Severe Weather, Fuzhou 350008, China; qlin07@hotmail.com
3 Fujian Meteorological Observatory, Fuzhou 350008, China
* Correspondence: xiehs@fzu.edu.cn

Abstract: Accurate short-term forecasting of intensive rainfall has high practical value but remains difficult to achieve. Based on deep learning and spatial–temporal sequence predictions, this paper proposes a hierarchical dynamic graph network. To fully model the correlations among data, the model uses a dynamically constructed graph convolution operator to model the spatial correlation, a recurrent structure to model the time correlation, and a hierarchical architecture built with graph pooling to extract and fuse multi-level feature spaces. Experiments on two datasets, based on the measured cumulative rainfall data at a ground station in Fujian Province, China, and the corresponding numerical weather grid product, show that this method can model various correlations among data more effectively than the baseline methods, achieving further improvements owing to reversed sequence enhancement and low-rainfall sequence removal.

Keywords: short-term intensive rainfall forecast; spatial–temporal sequence prediction; hierarchical dynamic graph network; graph convolutional network; numerical weather prediction

Citation: Xie, H.; Zheng, R.; Lin, Q. Short-Term Intensive Rainfall Forecasting Model Based on a Hierarchical Dynamic Graph Network. *Atmosphere* **2022**, *13*, 703. https://doi.org/10.3390/atmos13050703

Academic Editors: Zuohao Cao, Huaqing Cai and Xiaofan Li

Received: 3 March 2022
Accepted: 26 April 2022
Published: 28 April 2022

1. Introduction

Short-term intensive rainfall is generally defined as a cloudburst event in which the accumulated rainfall reaches or exceeds 30 mm within 3 h (Fujian Provincial Meteorological Observatory) [1]. It is usually caused by strong convective weather and is characterized by extreme suddenness, high destructiveness, and a short duration. It can easily cause natural disasters such as mountain torrents, mudslides, and urban floods. The forecasting accuracy of short-term intensive rainfall is usually lower than that of ordinary rainfall events in China [2]. Inaccurate forecasts may lead to a serious loss of life and property. Therefore, improving the accuracy of short-term forecasting is important.

We focus on two types of short-term intensive rainfall forecasting methods [3]. The radar extrapolation method uses historical radar echo maps with a high spatial–temporal resolution, as drawn by meteorological radars, to forecast the rain and cloud movement. It uses the optical flow method [4], precipitation cloud extrapolation [5], or other methods [6] to predict the movement. In addition, it subsequently uses rainfall rate–reflectivity relationships or other means to invert the results into rainfall data. The Numerical Weather Prediction (NWP) method, based on historically accumulated observation data, uses numerical calculations to solve the fluid mechanics and thermodynamic equations that represent the weather evolution under certain conditions. The result is a computer-simulated NWP product. Based on this, forecasters combine various monitoring products and their own experience to conduct comprehensive analyses and corrections, thus finally obtaining forecasting results.

With the increased scale of deployment of metering equipment and meteorological data expansion, previous studies have integrated deep learning with meteorological forecasting methods. Recent extrapolation methods transform the problem into a video-like prediction, such that it is easier to model the long-range spatial–temporal relationship when compared with that in traditional methods [7]. Some attempts have focused on using the NWP products as input data for deep learning methods to perform prediction tasks [8,9].

However, forecasting short-term intensive rainfall using deep learning still faces challenges: (1) the distribution of meteorological data is complex and involves multi-modal dynamics, which are difficult to model; (2) statistics show that samples have a robust scale-free structure in the atmospheric rainfall field [10], which indicates a data imbalance problem; (3) when using data with a temporal resolution of 3 h, modeling is more difficult than radar extrapolation at high spatial–temporal resolutions and small neighborhood variations; and (4) NWP products are not the actual measured results, as their accuracy is limited by the characteristics of the models they use.

To manage these challenges, based on our previous study [1], we combined rainfall data from ground stations and related data from an NWP product to propose the Hierarchical Dynamic Graph Network (HDGN), a new model based on spatial–temporal sequence prediction and a Graph Convolutional Network (GCN). By designing the corresponding structure, we comprehensively captured the correlations among time, space, and features, which facilitated the prediction of short-term intensive rainfall.

The remainder of this paper is organized as follows. Section 2 introduces the background related to this study, including the NWP, spatial–temporal sequence prediction, study area, and data sources. Section 3 presents the methods applied for short-term intensive rainfall prediction, which involve data preprocessing and the HDGN model. Section 4 presents the configuration of the experiments and interpretations of their results. Section 5 provides the conclusions of the study.

2. Background

2.1. Numerical Weather Prediction

Multi-scale forecasts are provided by operational NWP centers, which involve small to planetary-scale emulations at time resolutions from the minute to seasonal scale. The numerical calculation model in the NWP is frequently updated with the aid of new observation data and forecasting technologies, thereby improving the physical simulation performance and uncertainty quantification of the model. This also improves the effect of model forecasting and data assimilation.

We used fine-grid numerical forecasting products from the European Centre for Medium-Range Weather Forecasts (ECMWF) and the Weather Research and Forecasting (WRF) method, a unified mesoscale weather-forecasting model.

However, NWP has certain limitations related to the cumulative error resulting from the high complexity of the simulation process. The ECMWF is disadvantageously characterized by a weak intensity forecast [11]. The WRF model performs relatively poorly when estimating the rainfall value [12]; its rainfall forecasting results may not be as optimal as those of the ECMWF [13]. Generally, the performance of the NWP in the convective period of a precipitation forecast is relatively poor, despite the occurrence of short-term intensive rainfall during the convective period. The lifetime of convective storm cells is generally <30 min [14], such that it is difficult to accurately predict short-term intensive rainfall events using a single NWP simulation. In this study, we combined the measured cumulative rainfall data from ground stations and related data from the NWP product on the input side to overcome the limitations associated with a single set of NWP data. Additionally, based on these data, the concept of spatial–temporal sequence prediction was employed to predict future rainfall conditions.

2.2. Spatial–Temporal Sequence Prediction

As a sub-field of deep learning, spatial–temporal sequence prediction is suitable for uncovering the spatial–temporal correlations among data, such as rainfall-related information for forecasting tasks based on time-sequence prediction. Classical models for time-sequence prediction include Long Short-Term Memory (LSTM) [15], which is a recurrent neural network with long- and short-term memory cells, and the Deep Belief Network (DBN) [16], which is a multi-layered probabilistic generative neural network. They have a simple structure with a low cost; however, they are poor in integrating spatial information from our data. To enhance the performances of these methods, we can classify the spatial–temporal sequence prediction problems into grid and non-grid scenarios [17].

Grid spatial–temporal sequence prediction uses fixed space coordinates to characterize the spatial–temporal relationship among data [18–20]. Convolutional LSTM (ConvLSTM) [21] combines LSTM and a three-dimensional (3-D) convolutional neural network [22], which is a deep learning network for extracting information from spatial data. It is portable and can be the building block of a predictive network, but it lacks bidirectional information flow between the different layers in the temporal direction. Sequence-to-Sequence (Seq2Seq) [23] is a basic recurrent architecture used in our model to perform frame-by-frame predictions. PhyDNet [24] uses an encoder–predictor–decoder architecture, which includes the mutual conversion of data and physical feature spaces. Furthermore, a previous study developed a video prediction model based on a multi-level feature space [25]; our study extends this idea to the graph domain. Multi-level feature spaces increase the model complexity but facilitate the extraction of feature correlations among data. Finally, a model for serially generating two-dimensional (2-D) convolutional kernels, with a sliding window [26], inspired this study regarding the hierarchical generation of graph convolution operators.

Non-grid spatial–temporal sequence prediction uses additional structures, such as graphs (a structure composed of nodes and edges), to characterize the spatial–temporal relationships among data [27–29]. The Attention-based Spatial–Temporal Graph Convolutional Network (ASTGCN) [30] alternately calculates the temporal and spatial attention within the data, which act as antecedent auxiliary transformations to the graph convolution operator. However, its high cost of spatial attention prevents it from being used in large graphs, where ours can be used. The Spatial–Temporal Graph Ordinary Differential Equation (STGODE) network [31] models the semantic adjacency matrix of a graph via the dynamic time-warping algorithm. Graph Convolution embedded LSTM (GC-LSTM) [32] uses the Inverse Distance Weight (IDW) to calculate the weights of a graph; the graph convolution operator selects one- to K-hop neighbors. In the HDGN, we also use semantic distances, which are more indicative of the correlations between node pairs, as weights instead of fixed geographical distance, and this method can increase the model dynamics at a low cost. Our graph convolution operator has the same capabilities as those of the GC-LSTM. The Dynamic Graph Convolutional Recurrent Network (DGCRN) [33] uses a highly dynamic graph construction method, whereas we proposed a hierarchical graph generation process; compared to the previous method, our approach trades a small reduction in flexibility for a faster graph construction speed.

2.3. Study Area and Data Sources

Fujian Province is located in Southeastern China, with a total land area of 12.4 million km^2. It has a subtropical maritime monsoon climate characterized by an average annual temperature of 15.0 to 21.7 °C, with hot summers and warm winters; its annual precipitation ranges from 1132 to 2059 mm, where March to September accounts for 81.4% of the annual precipitation. The topography of Fujian is high in the northwest and low in the southeast. It has two mountain belts that trend from the northeast to the southwest: the Wuyi Mountains in northwestern Fujian and the Jiufeng and Daiyun mountains in northeastern to central Fujian. Owing to the influence of the terrain, these areas are the centers of heavy rainfall in Fujian [34]. Figure 1 depicts the occurrence of short-term intensive rainfall in Fujian from February 2015 to December 2018. Fujian is one of the areas of high-frequency intensive rainfall in China, which often leads to severe flooding and geological hazards.

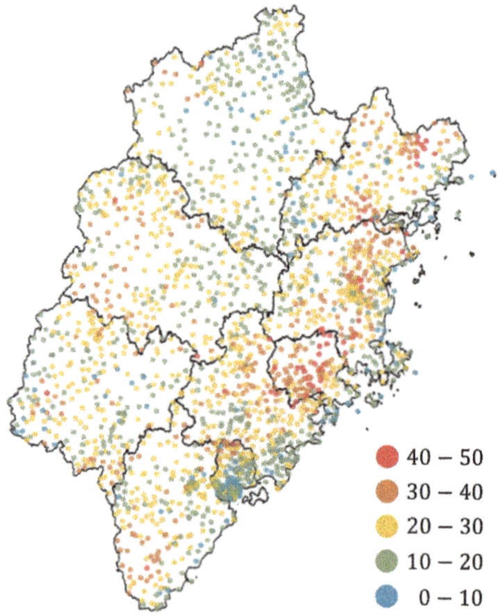

Figure 1. Distribution of each ground station in Fujian Province. The colors indicate the total number of observed short-term intensive rainfall events in February 2015 to December 2018 for each station [1].

The four original datasets issued by the Fujian Meteorological Observatory were used in this study (see Table 1 for details); the grid points refer to a series of nodes arranged in rows (latitude) and columns (longitude). These datasets can be divided into three categories.

Table 1. Details of the original datasets.

Dataset	Stations	ECMWF250	ECMWF125	WRF
Horizontal range	23.52°–28.37° N, 115.84°–120.67° E	23.5°–28.5° N, 115.5°–121.0° E		23.5°–28.45° N, 115.5°–120.99° E
Time range	February 2015 to December 2018			January 2017 to December 2018
Time resolution	3 h			1 h
Number of stations/grid points	2170	23 × 21	45 × 41	62 × 56
Grid spacing	—	0.25° × 0.25°	0.125° × 0.125°	0.09° × 0.09°
Number of available features	3	113	26	24
Starting time of the forecast	—	08:00 and 20:00 (UTC + 8)		

Stations: A dataset of the observed rainfall, comprising data collected by 2170 available ground stations in Fujian. It contains three features, i.e., the longitude, latitude, and measured 3 h accumulated rainfall.

ECMWF: It comprises the ECMWF250 and ECMWF125 datasets for Fujian. With the exception of a few features, such as the dew point temperature, their feature sets do not overlap with each other.

WRF: A dataset containing the Fujian WRF grid data, which is divided into 3-h interval groups; only the forecasting results from the third hour were used for alignment with the other datasets.

3. Methodology

3.1. Problem Description

Let the data obtained from the samplings performed at equal time intervals belong to one frame. We set sequence prediction as the task of outputting the predicted sequence data as close as possible to the ground truth based on the historical data, which can be expressed as follows:

$$\{X_0, \ldots, X_{T_{in}-1}\} \rightarrow \{X_{T_{in}}, \ldots, X_{T_{in}+T_{out}-1}\}, \quad (1)$$

where X represents a frame, and T_{in} and T_{out} represent the historical and predicted sequence lengths, respectively.

The structure of X differs for different types of prediction problems. $X \in \mathbb{R}^F$ represents the time-sequence prediction, $X \in \mathbb{R}^{N \times F}$ represents the non-grid spatial–temporal sequence prediction, and $X \in \mathbb{R}^{M \times F}$ represents grid prediction, where F, N, and M denote the number of features, nodes, and measurement dimensions of the grid, respectively. When the grid is 2-D, $M = H \times W$, where H and W represent the height and width of the grid, respectively.

We set our object as a non-grid spatial–temporal sequence prediction problem, where $X_{t,0} = X_t$ in the HDGN. The grid spatial–temporal sequence prediction was applied after separating the latitude and longitude coordinates from the data points. The time-sequence prediction methods can individually predict each node and then combine them.

3.2. Data Preprocessing

For the stations dataset, we used the IDW interpolation method [35] on each frame to interpolate the rainfall values to the ECMWF250 and WRF grid points. Thus, the measured cumulative rainfall and corresponding NWP features shared identical spatial coordinates, which avoided forecasting difficulties caused by a lack of measured rainfall data. The station dataset is important because the model will perform poorly if the percentage of missing values is high. In this context, the observed rainfall data must usually be obtained from multiple sources to prevent potential problems caused by missing data when the data are obtained from a single source.

For the NWP datasets, we used the forecasting period between 12 and 33 h owing to numerical instability in the first 12 h of the NWPs. We then selected the forecasting data closest to the start time to reduce the influence of long-term forecasting errors. Because each feature has a different impact on network training and prediction performance, the Box Difference Index (*BDI*) was used for feature selection [36] to reduce the volume of data and avoid feature overlap. The higher the index, the stronger the feature's ability to distinguish whether the data point was a short-term intensive rainfall event. The *BDI* of each feature was calculated as follows:

$$BDI = \frac{|m_{30} - m_0|}{\sigma_{30} + \sigma_0}, \quad (2)$$

where m_0 and m_{30} represent the characteristic mean values of the rainfall for data points between 0 and 30 and above 30 mm, respectively, and σ_0 and σ_{30} represent the standard deviations of the rainfall values for the data points between 0 and 30 mm and above 30 mm, respectively. After calculation, a list of each feature in descending order of *BDI* value was obtained, and the features with the highest *BDI* were selected in turn. Note that features with a high percentage of missing data owing to equipment failure, etc., were not used because they degrade the performance of the model; therefore, we manually skipped these features and replaced them with features with lower *BDIs*. Table 2 lists the features that we selected following the above process and used in the subsequent steps.

Table 2. Description of the selected features.

Data Source	Feature Name	Meaning
ECMWF125	Q_850	Specific humidity 850 kPa
	Ki	k index
	Td_850	Dew point temperature 850 kPa
	GH_1000	Geopotential height 1000 kPa
	Tt_850	Temperature 850 kPa
ECMWF250	TCWV	Atmospheric water vapor content
	MSL	Sea-level pressure
	2D	Dew point temperature 2 m
WRF	TCDC	Total cloud cover
	RH_850	Relative humidity 850 kPa
	LCDC	Low cloud cover
	CR	Combined reflectance
	GUST	Wind gust
	CAPE	Convective avail pot energy
	REFD_1000	Radar-derived reflectivity 1000 kPa
	MSLP	Mean sea-level pressure

After feature selection, we constructed the sequential datasets adapted to the HDGN and other sequence prediction models, i.e., S-ECMWF and S-WRF, where S denotes the sequence. Their construction methods are shown below: (1) As the ECMWF dataset comprises two groups of data with different grid spacings, they must be merged. The selected ECMWF125 retained only the features of the 23 × 21 grid points that overlapped with ECMWF250. The features of both were then concatenated according to the grid points. (2) The interpolated rainfall data were spliced into the two datasets using the operation described in (1). (3) Linear interpolation was used to supplement the missing values. The data were standardized with the z-score. (4) Sequence samples were generated using a sliding window with a step size of one frame. (5) The graph, G_0, was constructed according to the grid of the data, where each grid point was treated as a node and each node formed an edge with the nearest node in eight directions (N, E, S, W, NE, SE, SW, and NW); a direction with no nodes in it was skipped. Graphs in the HDGN were stored as a compressed sparse matrix structured as $\mathbb{R}^{2\times E}$, where E represents the number of edges in the graph. The edges of the graphs in the HDGN were all undirected edges, unless otherwise specified. (6) S-ECMWF and S-WRF contained the sequence samples and G_0, respectively.

As an optional step, we performed data augmentation on the S-ECMWF and S-WRF datasets before training to improve the prediction performance. This involved two methods: (1) Reversed sequence enhancement: the reverse form, $\{X_{T_{in}-1}, \ldots, X_0\}$, of the historical sequence, $\{X_0, \ldots, X_{T_{in}-1}\}$, was generated and added to the training data; the related sequence prediction task is shown in Equation (3). (2) Low-rainfall sequence removal: 10% of the training samples with the highest number of data points characterized by zero rainfall in the historical sequence were removed. This configuration was used unless otherwise specified.

$$\{X_{T_{in}-1}, \ldots, X_0\} \rightarrow \{X_{-1}, \ldots, X_{-T_{out}}\} \tag{3}$$

3.3. Hierarchical Dynamic Graph Network

We proposed an HDGN model for short-term intensive rainfall forecasting, which is shown in Figure 2. We note that the structure of the Hierarchical Graph Convolutional Network (HGCN) should correspond to the hierarchical graph generation process. The components of this model were implemented based on Multi-Layer Perceptron (MLP), a trivial forward-structured artificial neural network, unless otherwise specified.

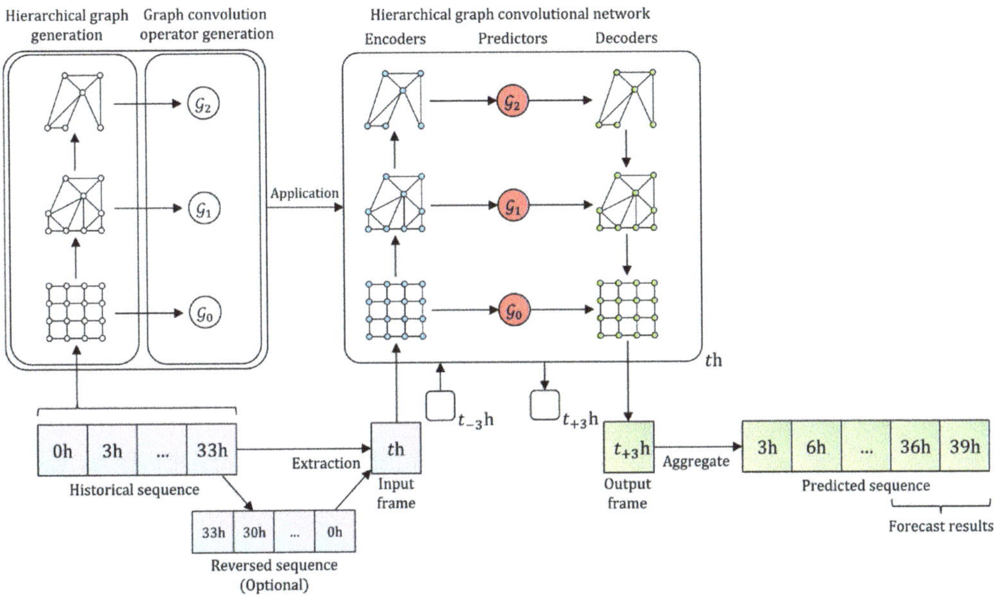

Figure 2. The overall architecture of the proposed Hierarchical Dynamic Graph Network (HDGN) model. It consists of three main modules: hierarchical graph generation, graph convolution operator generation, and Hierarchical Graph Convolutional Network (HGCN). The model is dynamic in nature because it uses different graphs for different sequences.

The steps of the sequence prediction were as follows. (1) The model read a historical sequence and generated multi-level graphs based on it. It then generated the corresponding graph convolution operators based on these graphs, finally using these results to initialize the HGCN. (2) The HGCN read each frame on the historical sequence in chronological order and output the corresponding predicted frames while updating its own state. When the HGCN reached the end of the historical sequence, the last predicted frame was re-input into the HGCN as the historical frame such that continuous prediction could be achieved. (3) The model output the forecasting results for this historical sequence.

3.3.1. Hierarchical Graph Generation

Graph pooling was used to dynamically construct multi-level graphs, which serve as the basis for the subsequent steps. Figure 3 shows the process of hierarchical graph generation.

We used a set of MLP encoders, $\mathcal{C}$, $\mathcal{A}$, and $\mathcal{W}$, to extract the auxiliary information according to the historical sequence. Each node cluster feature was $c_i, i \in \{0, \ldots, N-1\}$, and the edge weight adjustment feature was $a_i, i \in \{0, \ldots, E-1\}$. These were calculated using $\mathcal{C}$ and $\mathcal{A}$.

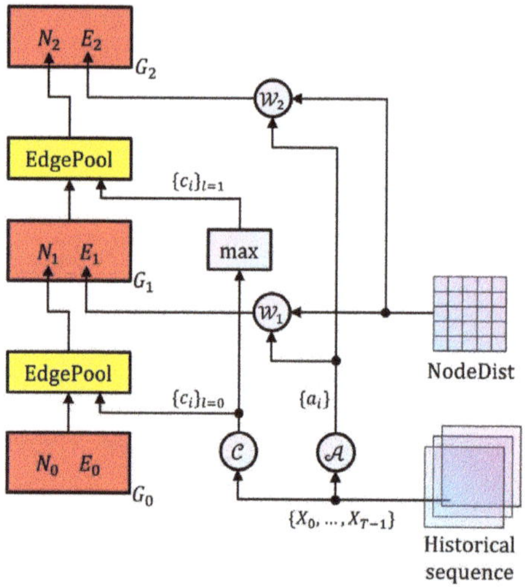

Figure 3. Hierarchical graph generation process generates high-level graphs G_l ($l > 0$) from initial graph G_0, historical sequence, and geographic distance between nodes (NodeDist).

EdgePool [37] was adopted to generate a new non-weighted graph based on edge shrinking. The related process was as follows. (1) The correlation score of each edge was calculated using Equations (4) and (5), where concat is the concatenate operation and softmax_i is a normalization function on all adjacent edges of node i. (2) The edge with the highest score was shrunk, followed by merging of its two adjacent nodes into a new node with a clustering feature of $\max(c_i, c_j)$. The adjacent edges of this new node no longer participated in the shrinking of this layer. (3) Step (2) was repeated until all edges were processed. We note that at least 50% of the nodes were always reserved for each pooling. For a graph with numerous nodes, multiple EdgePools were arranged instantaneously to reduce the number of layers.

$$s_{i,j} = \text{MLP}(\text{concat}(c_i, c_j, e_{i,j})) \tag{4}$$

$$s_{ij} = \max(0.5 + \text{softmax}_i(s_{i,j}),\ 0.5 + \text{softmax}_j(s_{i,j})) \tag{5}$$

We generated the weights for this new graph using Equation (6), where $d_{i,\,j}$ is the geographic distance between i and j, multiplied by a multiplier for correction. The multiplier allowed learnable weights and aided in the modeling of the semantic distance between the nodes based on $d_{i,\,j}$. We constrained the multiplier within (0.5, 1.5) by the sigmoid function σ, to obtain more stable weights.

$$e_{i,j} = d_{i,\,j}(0.5 + \sigma(\mathcal{W}(a_i, a_j))) \tag{6}$$

3.3.2. Graph Convolution Operator Generation

Based on a given graph, G, we generated a graph convolution operator, θ, using graph Fourier transform theory [38].

First, we calculated the symmetric normalized Laplace matrix of the graph via Equation (7), where A and D are the adjacency and degree matrices of the graph, respectively, and I is the identity matrix corresponding to A.

$$L_{sym} = ID^{-\frac{1}{2}}AD^{-\frac{1}{2}} \tag{7}$$

Next, eigenvalue decomposition was performed on L_{sym}; this step is complex, especially when there are many nodes on the graph. Therefore, the Chebyshev polynomial approximation was used to accelerate the solution process [39]. The graph convolution operator, θ, based on L_{sym}, was approximated as a superposition of K parts, with the k-th part extracting relevant information from k-hop neighbors around the target node, as shown in Equation (8), where λ_{max} is the maximum eigenvalue of L_{sym} and T_k represents the k-th term of the first type of recursive Chebyshev polynomial; i.e., Equation (9), where $T_0(X) = I$, $T_1(X) = X$, and θ_k is the k-th learnable graph convolution kernel [40]. The value of K is important; if it is small, the graph convolution operator does not have a good mapping ability; if it is large, it causes over-smoothing, i.e., the data on the graph converge rapidly, which severely affects the subsequent process.

$$\theta(L_{sym}) = \sum_{k=0}^{K-1} \theta_k T_k\left(\frac{2L_{sym}}{\lambda_{max} - I}\right) \tag{8}$$

$$T_k(X) = 2XT_{k-1}(X) - T_{k-2}(X) \tag{9}$$

In summary, we calculated $\{T_0, \ldots, T_{K-1}\}$ through G, followed by implementation of the graph convolution process.

3.3.3. Hierarchical Graph Convolutional Network

We proposed the HGCN, as shown in Figure 4, which is a multi-layered encoder–predictor–decoder network. HGCN extracts the high-level features from the data to produce a multi-level description of the data, which is useful for prediction. A feature space consisted of a set of features used to describe the data. Layer 0 feature space, i.e., the meteorological features within the dataset, and other feature spaces were latent spaces with learnable anonymous features. The encoders were responsible for mapping the low-level feature space to the higher space, whereas the decoders were responsible for performing the opposite process.

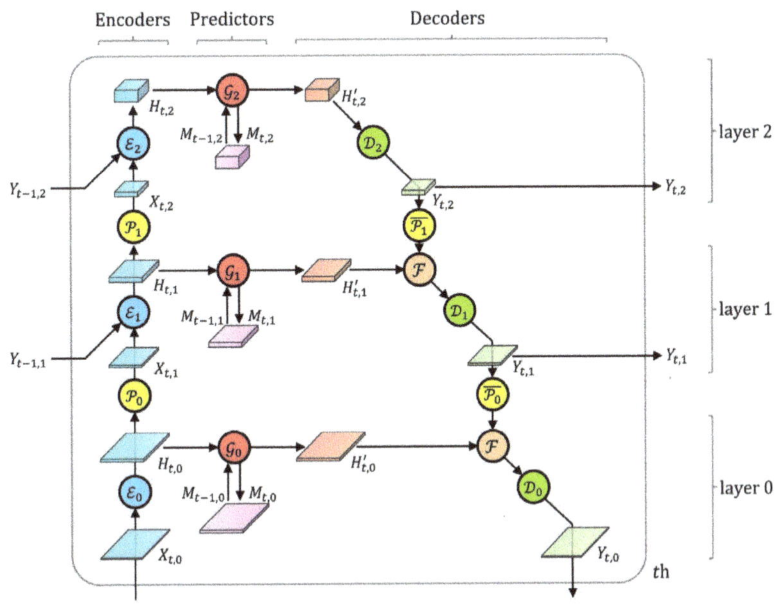

Figure 4. Proposed hierarchical graph convolutional network as the sequence prediction part of HDGN. Schematic shows the detailed structure of a three-layer network, where circles are components and rectangles are data. A rectangle's bottom area and height characterize the size of graph and feature space, respectively.

The components in the HGCN network are as follows. (1) The encoder, $\mathcal{E}$, and decoder, $\mathcal{D}$, map feature spaces to higher- or lower-level hidden spaces through MLP. Based on the residual connection [41], we proposed a cross-frame connection. When $t > 0$ and $l > 0$, the encoder uses the form shown in Equation (10), where $Y_{t-1,l}$ represents the output feature space of the same layer in the previous frame. The cross-frame connection aids in stabilizing inter-frame and inter-layer information transmission, shortens the transmission path of the high-level information, and alleviates the gradient explosion problem. (2) The predictor, $\mathcal{G}$, communicates information between the nodes in the graph and produces data for the next moment, which are expressed by Equations (11) and (12). $M_{t,l}$ was used to adaptively adjust the update magnitude of $H_{t,l}$, and $M_{-1,l}$ is an empty matrix. $T_{l,k}$ was calculated from G_l, is the element-wise product, and U_l, W_l, $\theta_{l,\,k}$, a_l, and b_l are learnable matrices. (3) Graph data pooling, $\mathcal{P}$, and graph data de-pooling, $\overline{\mathcal{P}}$, convert the data between the graphs in adjacent layers. $\mathcal{P}$ copies the corresponding lower-level node with the largest rainfall as the new node, while $\overline{\mathcal{P}}$ copies the node to every corresponding lower-level node. (4) The fusion operation, $\mathcal{F}$, was used to integrate the data, as shown in Equation (13). To avoid learning of constant transformations by the model, we used the maximum function to achieve $\mathcal{F}$.

$$H_{t,l} = \mathcal{E}_l(\text{concat}(Y_{t-1,l}, X_{t,l})) \tag{10}$$

$$M_{t,l} = \sigma(U_l M_{t-1,l} + W_l H_{t,l} + a_l) \tag{11}$$

$$H'_{t,l} = (1 - M_{t,l})H_{t,l} + M_{t,l}\tanh\left(\sum_{k=0}^{K-1} \theta_{l,\,k} T_{l,\,k} H_{t,l} + b_l\right) \tag{12}$$

$$Y'_{t,l} = \mathcal{F}\left(H'_{t,l}, Y_{t,l+1}\right). \tag{13}$$

According tohe number of nodes in the S-ECMWF and S-WRF datasets, we designed the corresponding HGCN structures as shown in Figure 5.

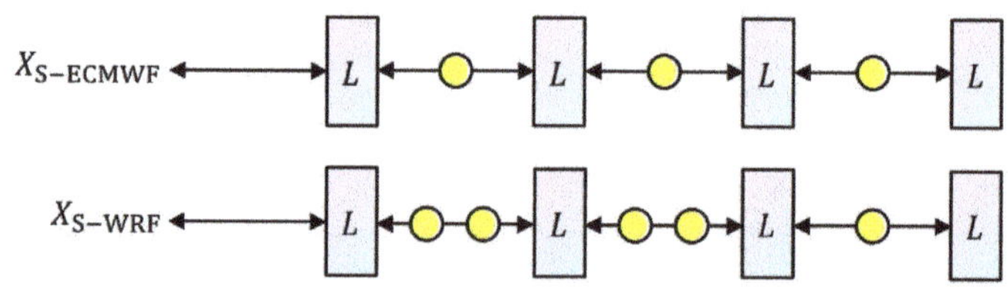

Figure 5. Hierarchical graph convolutional networks adapted to S-ECMWF or S-WRF datasets, where boxes represent one layer, circles represent graph data pooling and de-pooling operations.

3.3.4. Loss Function

After obtaining the prediction results of the model, a loss function was used to evaluate the degree of difference between the predicted values, x, and reference values, y, to guide model training. Owing to the imbalance in the rainfall values, we implemented a set of rainfall thresholds, $\phi = \{0.1, 1, 5, 10, 20, 30\}$, and grouped all data points into seven categories. The weighted mean absolute error was used as the loss function.

$$loss = \frac{1}{N}\sum_{i\in\Omega} \frac{\Omega}{\Omega_k}|x_i - y_i|, \tag{14}$$

where Ω represents the total number of data points involved in the evaluation and Ω_k represents the number of the k-th category, i.e., within the i-th data point. Categories with smaller sample sizes had larger proportions of the prediction error.

4. Experimental Settings and Results

4.1. Experimental Settings

4.1.1. Model Configuration

We implemented the HDGN model on PyTorch [42] 1.6.0 using an NVIDIA Tesla P100-PCIE-16 GB GPU for experiments on a Windows workstation. After shuffling the order of the sequence of samples in the S-ECMWF and S-WRF datasets, they were grouped into three subsets: training, validation, and test sets at a ratio of 6:2:2. For all prediction tasks, the length of the historical input sequence was 12 frames (spanning 36 h) and the length of the forecasting results was 2 frames (spanning 6 h). In the training phase, the model used an Adam optimizer: the initial learning rate was set to 0.0005; early stopping was configured, which could adaptively adjust the learning rate during the training process and stop training when the loss could not be reduced further. The batch size was set to 80 or 12, the number of layers was set to 4, and the term of the Chebyshev polynomial, K, was set to 3. We fixed the parameters of the model for the validation and test phases; the validation phase fine-tuned the parameters and the test phase output the evaluation indicators for the forecasting results.

The selection of these hyperparameters was influenced by various aspects. If the length of the historical sequence was short, it was difficult for the model to obtain sufficient information, whereas longer sequences did not significantly improve the prediction effect but increased the time and space cost of the model. The effect of the initial learning rate was reduced after configuring early stopping. A larger batch size could accelerate model training; owing to the large scale of the S-WRF dataset, the upper limit of the GPU load, i.e., 12, was selected. Model performance degraded when K was 2 or 4. The number of layers could significantly affect the prediction accuracy (see Section 4.2.3 for details).

4.1.2. Evaluation Index

We mainly focused on the classification performance of short-term intensive rainfall events. Rainfall evaluation indicators were based on the following three categories of statistical scoring methods: (1) Critical Success Index (CSI), which is a commonly used indicator to measure the rainfall forecasting results. Its values range from $[0, 1]$; the higher the value, the better the result. (2) Equitable Threat Score (ETS), which is used to measure the degree of improvement in the rainfall forecasting results relative to random forecasting results under the same configuration. Its values range from $[-1/3, 1]$; the higher the value, the better the result. An ETS of 0 indicates that the model's prediction results are comparable to random results, whereas ETS ≤ 0 is not acceptable. (3) False Alarm Ratio (FAR), which is the proportion of misclassified data included in the prediction results. Its values range from of $[0, 1]$; the lower the value, the better the result.

Table 3 presents the rainfall classification with 1 and 30 mm as the threshold.

Table 3. Rainfall classification table.

Actual Class	Predicted Class		
	0–1 mm	1–30 mm	>30 mm
0–1 mm	D	B_1	B_2
1–30 mm	C_1	A_1	B_2
>30 mm	C_2	C_2	A_2

With 1–30 mm as the first category and >30 mm as the second category, the indicators for each category were calculated as follows:

$$CSI_k = \frac{A_k}{A_k + \sum B_k + \sum C_k}, \quad (15)$$

$$ETS_k = \frac{A_k - R_k}{A_k + \sum B_k + \sum C_k - R_k}, \tag{16}$$

$$FAR_k = \frac{\sum B_k}{A_k + \sum B_k}, \tag{17}$$

where A, B, C, and D represent the number of event hits, empty reports, missed reports, and number of successful predictions of non-events, respectively, and R represents the result of the random forecasting model evaluated as follows:

$$R_k = \frac{(A_k + \sum B_k)(A_k + \sum C_k)}{A_k + \sum B_k + \sum C_k + D}. \tag{18}$$

We referred to each indicator with a k of 1 as a type 1 indicator and that with a k of 2 as a type 2 indicator. The CSI_2 and ETS_2 indicators generally had values <0.1 in Fujian; values >0.1 were considered major breakthroughs. Repeated comparison experiments revealed that for each deep learning method considered herein, the first four decimal places of the type 2 indictors remained unchanged, while the subsequent decimal places showed fluctuations; thus, we retained only the initial four decimal digits to make the results reasonable. Because of the same reasons, we retained three decimal digits for the type 1 indicators. We consider that a model with only one or two stable decimal digits for type 1 indicators may be unstable or ineffective, and a better one can be trained using our data because our data corresponding to 1–30 mm of cumulative rainfall are adequate in terms of scale and diversity.

4.2. Results

4.2.1. Comparison

We implemented several baselines and our proposed method on the S-ECMWF and S-WRF datasets; Tables 4 and 5 show the results. We adjusted the hyperparameters in the data for all non-NWP baselines; other configurations were set at the default settings.

The simulation results of the ECMWF and WRF were obtained from the original data; their related indicators were directly calculated as experimental results for predicting the first frame. The History Average (HA) uses the average frame in the historical sequence as the prediction result. The LSTM and DBN are time-sequence prediction methods. ConvLSTM belongs to the grid spatial–temporal prediction method, whereas ASTGCN and our HDGN model are non-grid methods. As there is no artificially definable period of short-term intensive rainfall, we only used the proximity sub-module in the ASTGCN network to fit the data.

Table 4. Short-term intensive rainfall prediction performance of the baseline and proposed methods on the S-ECMWF dataset in the future first and second frames.

Methods	**Predict of First Frame**						**Predict of Second Frame**					
	CSI_1	**ETS_1**	**FAR_1**	**CSI_2**	**ETS_2**	**FAR_2**	**CSI_1**	**ETS_1**	**FAR_1**	**CSI_2**	**ETS_2**	**FAR_2**
ECMWF	0.205	0.159	0.722	0.0020	0.0018	0.9937	—	—	—	—	—	—
HA	0.102	0.086	0.578	0.0000	0.0000	1.0000	0.088	0.067	0.862	0.0000	0.0000	1.0000
LSTM	0.130	0.095	0.830	0.0043	0.0019	0.9963	0.110	0.073	0.867	0.0019	0.0008	0.9968
ConvLSTM	0.173	0.139	0.815	0.0059	0.0031	0.9940	0.163	0.131	0.814	0.0031	0.0016	0.9972
DBN	0.180	0.145	0.791	0.0069	0.0047	0.9923	0.165	0.131	0.805	0.0035	0.0022	0.9968
ASTGCN	0.157	0.074	0.841	0.0114	0.0105	0.9787	0.138	0.052	0.860	0.0051	0.0034	0.9913
HDGN	0.115	0.026	0.885	0.0211	0.0202	0.9576	0.107	0.018	0.892	0.0086	0.0068	0.9790

Table 5. Short-term intensive rainfall prediction performance of the baseline and proposed methods on the S-WRF dataset in the future first and second frames.

Methods	Predict of First Frame						Predict of Second Frame					
	CSI_1	ETS_1	FAR_1	CSI_2	ETS_2	FAR_2	CSI_1	ETS_1	FAR_1	CSI_2	ETS_2	FAR_2
WRF	0.143	0.076	0.839	0.0047	0.0037	0.9953	—	—	—	—	—	—
HA	0.174	0.143	0.513	0.0000	0.0000	1.0000	0.088	0.067	0.561	0.0000	0.0000	1.0000
LSTM	0.151	0.135	0.823	0.0057	0.0019	0.9944	0.124	0.110	0.853	0.0025	0.0008	0.9980
ConvLSTM	0.233	0.201	0.691	0.0088	0.0050	0.9851	0.217	0.188	0.722	0.0049	0.0027	0.9959
DBN	0.214	0.179	0.741	0.0090	0.0061	0.9805	0.195	0.164	0.763	0.0042	0.0029	0.9954
ASTGCN	0.146	0.052	0.851	0.0004	0.0004	0.9374	0.128	0.031	0.869	0.0002	0.0001	0.9762
HDGN	0.134	0.037	0.865	0.0343	0.0336	0.9219	0.123	0.025	0.876	0.0140	0.0003	0.9604

Our model achieved better results for short-term intensive rainfall prediction than the other models, thus reflecting the advantages of the proposed method. The CSI_2 and ETS_2 of HA are equal to 0, indicating that it could not forecast the short-term intensive rainfall events. This indicates that they were rare and short in duration; hence, it was necessary to comprehensively consider the adjacent spatial–temporal information. The prediction effect of ConvLSTM was better than that of LSTM, indicating that adjacent spatial information is valuable. DBN had a higher density of network connections than the previous models; hence, its learning ability was stronger. However, the stronger the modeling capability, the higher the training time cost of the model.

ASTGCN uses a structure, with a space complexity of $O\left(N^2\right)$, to directly model the relationship between pairs of nodes, thus achieving a spatial attention mechanism; therefore, running ASTGCN on a large dataset, such as the S-WRF, was difficult. We employed the dynamically designed graph convolution operator implemented using the compressed sparse matrix to model the spatial correlation, which significantly reduced the number of parameters to $O\left(E\right)$. ASTGCN and HDGN showed better prediction for type 2 indicators owing to the use of graph representation and spatial–temporal modeling methods with a greater complexity. However, their performance in terms of the first category decreased with improvements in the second category, indicating that the performance of the model was limited by the data after partial improvement. In other words, there was a trade-off in the forecasting accuracy between the different categories. These methods also had more training time than the other sequence prediction models. The training time for HDGN was slightly higher than that of ASTGCN because the latter was static in nature, whereas the former was dynamic. In the testing phase, the forecasting time of each sequence prediction model was lower than their training time because their parameters were fixed.

We then analyzed the overall results. (1) The S-WRF dataset had a higher spatial resolution and generally provided more information than S-ECMWF; therefore, HDGN had a better prediction effect on it. (2) Over time, the performance of all sequence prediction models decayed. As frame-by-frame prediction models reached the end of the historical sequence, the last predicted frame was re-inputted, following which the forecasting errors accumulated over time. Additionally, the decay for type 2 indicators was generally larger than that for type 1 indictors, implying that predicting short-term intensive rainfall events was more difficult. (3) The FAR values of all results were unsatisfactory. This was because short-term intensive rainfall prediction is difficult and reducing FAR_2 is complex. However, the methods adopted in this study were biased to enhance short-term intensive rainfall forecasting. For example, we selected 30 mm as the threshold of the *BDI* in the feature selection, resulting in a corresponding increase in FAR_1. An alert analysis method can be used to reduce FAR_2; specifically, all short-term intensive rainfall prediction results can be input into a downstream module, which will analyze these data and reject misreported predictions. This module can be implemented using specially designed meteorological or deep learning models or by manual analysis. (4) Further inspection revealed that the classification errors were concentrated at the marginal area within our data. Our observed rainfall data originated from ground stations in Fujian Province, such that the interpolation of rainfall for nodes outside Fujian Province was relatively inaccurate. Better results may be obtained by combining data around Fujian Province.

4.2.2. Reversed Sequence Enhancement

Traditionally, data enhancement increases the amount of data to improve the model performance. For oversampling, learning rules from a sparse number of >30 mm data points and generated data similar to actual situations were not easier than the prediction task owing to the complexity of our data. Moreover, there was a greater probability of data overfitting. For undersampling, separating the 0–30 mm data was difficult.

Assuming that meteorological thermodynamics is a reversible process, the display of the reverse process aids in model learning [43]. This is especially the case for the increase in and attenuation of rainfall, as they are important characteristics that affect short-term intensive events. The experiments revealed that, after reverse sequence enhancement, the proportion of each classification was almost invariable; but, the results improved, as shown in Table 6. This provides another means of improving the forecasting accuracy: data should be available to input more valuable information into the model, thereby reducing the difficulty associated with model learning.

Table 6. Data processing to demonstrate the effectiveness of reversed sequence enhancement. The improvement with the S-ECMWF dataset was more significant than that with the S-WRF dataset. The symbol '—' indicates that the model did not function owing to excessive number of parameters.

Methods	S-ECMWF Predict of First Frame						S-WRF Predict of First Frame					
	CSI_1	ETS_1	FAR_1	CSI_2	ETS_2	FAR_2	CSI_1	ETS_1	FAR_1	CSI_2	ETS_2	FAR_2
ASTGCN	0.157	0.074	0.841	0.0114	0.0105	0.9787	0.146	0.052	0.851	0.0004	0.0004	0.9374
ASTGCN + reverse	0.166	0.085	0.830	0.0153	0.0147	0.9714	—	—	—	—	—	—
HDGN	0.115	0.026	0.885	0.0211	0.0202	0.9576	0.134	0.037	0.865	0.0343	0.0336	0.9219
HDGN + reverse	0.123	0.035	0.877	0.0254	0.0252	0.9442	0.152	0.060	0.845	0.0401	0.0395	0.9457

However, the reversed sequence enhancement method doubles the amount of data, which almost doubles the training time and increases the space cost of the model. Therefore, there is a trade-off between effectiveness and cost.

4.2.3. Low-Rainfall Sequence Removal

Before training, we removed a portion of the training samples with the least number of data points characterized by non-zero rainfall in the historical sequence. The results in Table 7 indicate that, owing to the complexity of the rainfall data, this type of elimination could not fundamentally change the imbalance in the data; however, it still improved the prediction performance for short-term intensive rainfall forecasting.

For type 2 indicators, the HDGN model achieved the highest performance when 10% of the data were removed; moreover, the proportion of the >30 mm data points was the highest. Above 10%, the negative effect of the simultaneous decrease in the proportion and volume of the >30 mm data points was observed, which led to underfitting of the model and a reduced model learning ability. Significant performance degradation occurred when the removal ratio exceeded 20%.

Table 7. Ablation experiment investigating the effect of low-rainfall sequence removal on the performance of the HDGN model.

Settings	Data Proportion		S-ECMWF Predict of First Frame			
	1–30 mm	>30 mm	CSI_1	ETS_1	CSI_2	ETS_2
del 0%	0.1240	0.001399	0.113	0.024	0.0204	0.0196
del 5%	0.1239	0.001398	0.118	0.030	0.0202	0.0195
del 10%	0.1281	0.001544	0.115	0.026	0.0211	0.0202
del 15%	0.1246	0.001511	0.115	0.026	0.0202	0.0194
del 20%	0.1241	0.001495	0.115	0.027	0.0199	0.0191
del 50%	0.1196	0.001423	0.117	0.029	0.0179	0.0173

For type 1 indicators, the correlation between the removal ratio and indicators was low. Both the proportion and number of the 1–30 mm data points were significantly higher than those of the >30 mm data points; hence, the effect of removing 10% of the data points was relatively smaller. However, it exceeded the effect of not removing data points.

4.2.4. Ablation Study of HDGN

We conducted ablation experiments to analyze the optimal means of designing the HDGN architecture. From Table 8, the use of a greater number of layers yielded enhanced performance benefits, which aided in the extraction of the correlations between the data. However, this benefit was marginal and restricted by the complexity of the model. Although the spatial size of the high-level feature space was smaller, it corresponded to a larger number of hidden features. The use of a greater number of layers increased the model's time and space costs. Therefore, we selected a suitable value that yielded satisfactory prediction effects at a low cost.

Table 8. Ablation experiments conducted on two datasets to examine the relationship between the number of layers and prediction effect of the HDGN model.

Settings	S-ECMWF Predict of First Frame			S-WRF Predict of First Frame		
	CSI_2	ETS_2	FAR_2	CSI_2	ETS_2	FAR_2
4 layers	0.0211	0.0209	0.9576	0.0343	0.0341	0.9219
3 layers	0.0186	0.0181	0.9631	0.0237	0.0229	0.9390
2 layers	0.0007	0.0000	0.9976	0.0016	0.0011	0.9809
1 layer	0.0000	0.0000	1.0000	0.0000	0.0000	1.0000

Another issue was the degree of influence of each dynamic building block on the final result of the HDGN. We compared several schemes under the same premise used for the other configurations, whose results are shown in Table 9, where the w/o multiplier denotes the case where the actual distances between the node pairs were used as the weights of the graphs, while the w/o dynamic graphs denote the case where the dynamic graph generation process was replaced with the supplied fixed graphs. The experimental results showed that the dynamic graph construction scheme was effective. The semantic weights slightly improved the results, whereas removing the entire dynamic graph pooling, including the semantic weights, resulted in serious performance anomalies. Similar to the HDGN model with one layer, their CSI_2 and ETS_2 are smaller than 5×10^{-5}. Significant overfitting of the HDGN model was observed because of the higher decrease in modeling ability than in model complexity in the case where modules were removed. We argue that, in this case, the model can be considered as one that does not have a relevant predictive capability.

Table 9. Influence of the configuration of each dynamic building block on the forecasting results of the HDGN model.

Settings	S-ECMWF Predict of First Frame			S-WRF Predict of First Frame		
	CSI_2	ETS_2	FAR_2	CSI_2	ETS_2	FAR_2
HDGN	0.0211	0.0209	0.9576	0.0343	0.0341	0.9219
w/o multiplier	0.0205	0.0203	0.9573	0.0088	0.0082	0.9647
w/o dynamic graphs	0.0000	0.0000	1.0000	0.0000	0.0000	1.0000

5. Conclusions

In this study, we aimed to improve the prediction performance of a short-term intensive rainfall prediction model. To achieve this goal, we described the short-term intensive rainfall prediction task as a spatial–temporal sequence prediction problem and then proposed a non-grid spatial–temporal sequence prediction model, HDGN, which can optimally extract the potential correlations between meteorological data and obtain more accurate prediction results. It consists of three modules: hierarchical graph generation, which is responsible for dynamically generating graphs for multi-level representation of the data from the historical sequences; graph convolution operator generation, which generates the graph convolution operators corresponding to these graphs; and hierarchical graph convolution network, which performs hierarchical feature space extraction and fusion based on the results of the first two modules, followed by frame-by-frame short-term intensive rainfall prediction. The design of the HDGN draws on relevant experience in the field of sequence prediction. To further improve the prediction performance, we also proposed two data enhancement methods for spatial–temporal sequences, namely, reversed sequence enhancement and low-rainfall sequence removal. They are relatively simple to implement, and the training effect is optimized by adding or removing training samples strategically.

The proposed method involves interpolation of rainfall, feature selection for NWP, construction of sequence datasets, data augmentation (optional), training of the HDGN model, and, finally, prediction using the trained model. Compared with the baselines, which included several sequence prediction methods based on deep learning, the experimental results obtained with real-world data from Fujian Province showed that our proposed method significantly improves the short-term intensive rainfall forecasting performance beyond that achieved with pure NWP simulations. On the first prediction frame of the S-ECMWF and S-WRF datasets, CSI2 improved by 9.55 and 6.30 times, and ETS2 improved by 10.22 and 8.08 times, respectively, compared with those of ECMWF and WRF. This method also outperforms the graph-based spatial–temporal sequence prediction model ASTGCN, with improvements of 85.09% and 92.38% in CSI2 and ETS2, respectively, on the first prediction frame of S-ECMWF. On S-WRF, ASTGCN cannot make predictions because of the large size of the graph, whereas HDGN can. Additionally, the proposed reversed sequence enhancement and low-rainfall sequence removal further improved the performance of the HDGN.

The HDGN has the following advantages: (1) It treats different features equally across time and space; thus, the data do not require additional processing. (2) The model's structure can be adjusted to adapt to different dataset sizes. (3) The prediction speed of the model is high after training.

However, our proposed method has some disadvantages: (1) The HDGN has difficulties when modeling the meteorological evolution at sub-grid and inter-frame scales, characterized by poor predictions for margin regions. (2) Additional measures are needed to further reduce the relatively high FAR of our model. (3) The cost-effectiveness of the reversed sequence enhancement is low.

Owing to this issue, there is much work needed before achieving an ideal short-term intensive rainfall prediction model. Future research should focus on the following aspects: (1) achieve learnable data fusion based on the nature of graphs to avoid errors introduced by the interpolation process; and (2) enhance the modeling capability and response to special regions without losing the generalization ability by combining, for example, the meteorological physical rules.

Author Contributions: Conceptualization, H.X. and R.Z.; methodology, H.X., R.Z. and Q.L.; software, R.Z.; validation, H.X.; formal analysis, R.Z.; writing—original draft preparation, R.Z.; writing—review and editing, H.X.; supervision, H.X.; funding acquisition, H.X. and Q.L. All authors have read and agreed to the published version of the manuscript.

Funding: This study was funded by the National Key Research and Development Program of China (2018YFC1506905) and the Guided Key Program of Social Development of Fujian Province of China (2017Y-008).

Institutional Review Board Statement: Not applicable.

Informed Consent Statement: Not applicable.

Data Availability Statement: Restrictions apply to the availability of these data. Data were obtained from the Fujian Meteorological Observatory and are available with their permission.

Acknowledgments: We thank the Fujian Meteorological Observatory for data support, as well as the reviewers for their critical comments and suggestions.

Conflicts of Interest: The authors declare no conflict of interest.

Abbreviations

The following abbreviations are used in this manuscript:

HDGN	Hierarchical Dynamic Graph Network
GCN	Graph Convolutional Network
LSTM	Long Short-Term Memory
DBN	Deep Belief Network
ConvLSTM	Convolutional LSTM
Seq2Seq	Sequence-to-Sequence
ASTGCN	Attention-based Spatial–Temporal Graph Convolutional Network
STGODE	Spatial–Temporal Graph Ordinary Differential Equation
GC-LSTM	Graph Convolution embedded LSTM
IDW	Inverse Distance Weight
DGCRN	Dynamic Graph Convolutional Recurrent Network
BDI	Box Difference Index
HGCN	Hierarchical Graph Convolutional Network
MLP	Multi-Layer Perceptron
HA	History Average

References

1. Xie, H.; Wu, L.; Xie, W.; Lin, Q.; Liu, M.; Lin, Y. Improving ECMWF short-term intensive rainfall forecasts using generative adversarial nets and deep belief networks. *Atmos. Res.* **2021**, *249*, 105281. [CrossRef]
2. Liu, C.; Sun, J.; Yang, X.; Jin, S.; Fu, S. Evaluation of ECMWF precipitation predictions in China during 2015–2018. *Weather Forecast.* **2021**, *36*, 1043–1060. [CrossRef]
3. Wang, T.; Liu, Y.; Dong, C.; Li, J. A summary of research on short-term precipitation forecast method and its application. *Electron. World* **2019**, *41*, 11–13. [CrossRef]
4. Woo, W.; Wong, W. Operational Application of Optical Flow Techniques to Radar-Based Rainfall Nowcasting. *Atmosphere* **2017**, *8*, 48. [CrossRef]
5. Huang, X.; He, L.; Zhao, H.; Wu, Y.; Huang, Y. Long-Term intelligent calculation and prediction model for heavy precipitation satellite cloud Images. In Proceedings of the 2018 4th International Conference on Advances in Energy Resources and Environment Engineering, ICAESEE 2018, Chengdu, China, 7–9 December 2018.
6. Shi, E.; Li, Q.; Gu, D.; Zhao, Z. A Method of Weather Radar Echo Extrapolation Based on Convolutional Neural Networks. In Proceedings of the 24th International Conference on MultiMedia Modeling, MMM 2018, Bangkok, Thailand, 5–7 February 2018; pp. 16–28.
7. Bouget, V.; Béréziat, D.; Brajard, J.; Charantonis, A.; Filoche, A. Fusion of Rain Radar Images and Wind Forecasts in a Deep Learning Model Applied to Rain Nowcasting. *Remote Sens.* **2021**, *13*, 246. [CrossRef]

8. Boukabara, S.-A.; Krasnopolsky, V.; Stewart, J.Q.; Maddy, E.S.; Shahroudi, N.; Hoffman, R.N. Leveraging modern artificial intelligence for remote sensing and NWP: Benefits and challenges. *Bull. Am. Meteorol. Soc.* **2019**, *100*, ES473–ES491. [CrossRef]
9. Zhou, K.; Zheng, Y.; Li, B.; Dong, W.; Zhang, X. Forecasting different types of convective weather: A deep learning approach. *J. Meteorol. Res.* **2019**, *33*, 797–809. [CrossRef]
10. Qian, Z.; Zhou, Q.; Liu, L.; Feng, G. A preliminary study on the characteristics of quiet time and intrinsic dynamic mechanism of precipitation events in the rainy season in eastern China. *Acta Meteorol. Sin.* **2020**, *78*, 914. [CrossRef]
11. Yin, S.; Ren, H. Performance verification of medium-range forecasting by T639, ECMWF and Japan models from September to November 2017. *Meteorogical Mon.* **2018**, *44*, 326–333.
12. Yang, Q.; Dai, Q.; Han, D.; Chen, Y.; Zhang, S. Sensitivity analysis of raindrop size distribution parameterizations in WRF rainfall simulation. *Atmos. Res.* **2019**, *228*, 1–13. [CrossRef]
13. Zhuo, S.; Zhang, J.; Yang, X.; Zi, L.; Qiu, H. Study on Comparison and Evaluation Index of Quantitative Rainfall Forecast Accuracy. *J. Water Resour. Res.* **2017**, *6*, 557–567. [CrossRef]
14. Chen, M.; Yu, X.; Tan, X.; Wang, Y. A brief review on the development of nowcasting for convective storms. *J. Appl. Meteorol. Sci.* **2004**, *15*, 754–766.
15. Hochreiter, S.; Schmidhuber, J. Long short-term memory. *Neural Comput.* **1997**, *9*, 1735–1780. [CrossRef] [PubMed]
16. Narejo, S.; Jawaid, M.M.; Talpur, S.; Baloch, R.; Pasero, E.G.A. Multi-step rainfall forecasting using deep learning approach. *PeerJ Comput. Sci.* **2021**, *7*, 1–23. [CrossRef]
17. Shi, X.; Yeung, D.-Y. Machine learning for spatiotemporal sequence forecasting: A survey. *arXiv* **2018**, arXiv:1808.06865.
18. Liu, Y.; Racah, E.; Correa, J.; Khosrowshahi, A.; Lavers, D.; Kunkel, K.; Wehner, M.; Collins, W. Application of deep convolutional neural networks for detecting extreme weather in climate datasets. *arXiv* **2016**, arXiv:1605.01156.
19. Agrawal, S.; Barrington, L.; Bromberg, C.; Burge, J.; Gazen, C.; Hickey, J. Machine learning for precipitation nowcasting from radar images. *arXiv* **2019**, arXiv:1912.12132.
20. Zhang, C.; Wang, H.; Zeng, J.; Ma, L.; Guan, L. Tiny-RainNet: A deep convolutional neural network with bi-directional long short-term memory model for short-term rainfall prediction. *Meteorol. Appl.* **2020**, *27*, e1956. [CrossRef]
21. Shi, X.; Chen, Z.; Wang, H.; Yeung, D.-Y.; Wong, W.-K.; Woo, W.-C. Convolutional LSTM network: A machine learning approach for precipitation nowcasting. In Proceedings of the 29th Annual Conference on Neural Information Processing Systems, NIPS 2015, Montreal, QC, Canada, 7–12 December 2015; pp. 802–810.
22. Ji, S.; Xu, W.; Yang, M.; Yu, K. 3D Convolutional neural networks for human action recognition. *IEEE Trans. Pattern Anal. Mach. Intell.* **2013**, *35*, 221–231. [CrossRef]
23. Sutskever, I.; Vinyals, O.; Le, Q.V. Sequence to sequence learning with neural networks. In Proceedings of the Advances in Neural Information Processing Systems, Montreal, QC, Canada, 8–13 December 2014; pp. 3104–3112.
24. Guen, V.L.; Thome, N. Disentangling physical dynamics from unknown factors for unsupervised video prediction. In Proceedings of the IEEE/CVF Conference on Computer Vision and Pattern Recognition, Seattle, WA, USA, 13–19 June 2020; pp. 11474–11484.
25. Lan, M.; Ning, S.; Li, Y.; Chen, Q.; Chen, X.; Han, X.; Cui, S. From single to multiple: Leveraging multi-level prediction spaces for video forecasting. *arXiv* **2021**, arXiv:2107.10068.
26. Su, J.; Byeon, W.; Kossaifi, J.; Huang, F.; Kautz, J.; Anandkumar, A. Convolutional tensor-train LSTM for spatio-temporal learning. In Proceedings of the 34th Conference on Neural Information Processing Systems, NeurIPS 2020, Virtual, Online, 6–12 December 2020.
27. Seo, S.; Liu, Y. Differentiable physics-informed graph networks. *arXiv* arXiv:1902.02950, 2019.
28. Wang, X.; Ma, Y.; Wang, Y.; Jin, W.; Wang, X.; Tang, J.; Jia, C.; Yu, J. Traffic flow prediction via spatial temporal graph neural network. In Proceedings of the Web Conference 2020, Taipei, Taiwan, 20–24 April 2020; pp. 1082–1092.
29. Zheng, C.; Fan, X.; Wang, C.; Qi, J. Gman: A graph multi-attention network for traffic prediction. In Proceedings of the AAAI Conference on Artificial Intelligence, New York, NY, USA, 7–12 February 2020; pp. 1234–1241.
30. Guo, S.; Lin, Y.; Feng, N.; Song, C.; Wan, H. Attention based spatial-temporal graph convolutional networks for traffic flow forecasting. In Proceedings of the AAAI Conference on Artificial Intelligence, Honolulu, HI, USA, 27 January–1 February 2019; pp. 922–929.
31. Fang, Z.; Long, Q.; Song, G.; Xie, K. Spatial-temporal graph ODE networks for traffic flow forecasting. In Proceedings of the 27th ACM SIGKDD Conference on Knowledge Discovery & Data Mining, Singapore, 14–18 August 2021; pp. 364–373.
32. Qi, Y.; Li, Q.; Karimian, H.; Liu, D. A hybrid model for spatiotemporal forecasting of PM2.5 based on graph convolutional neural network and long short-term memory. *Sci. Total Environ.* **2019**, *664*, 1–10. [CrossRef] [PubMed]
33. Li, F.; Feng, J.; Yan, H.; Jin, G.; Jin, D.; Li, Y. Dynamic graph convolutional recurrent network for traffic prediction: Benchmark and solution. *arXiv* **2021**, arXiv:2104.14917.
34. Lin, X.; Liu, A.; Lin, Y.; Xu, J. *Technical Manual of Weather Forecast of Fujian Province*; China Meteorological Press: Beijing, China, 2013.
35. Xie, H.; Zhang, S.; Hou, S.; Zheng, X. Comparison research on rainfall interpolation methods for small sample areas. *Res. Soil Water Conserv.* **2018**, *25*, 117–121.
36. Fu, B.; Peng, M.S.; Li, T.; Stevens, D.E. Developing versus nondeveloping disturbances for tropical cyclone formation. Part II: Western North Pacific. *Mon. Weather Rev.* **2012**, *140*, 1067–1080. [CrossRef]
37. Diehl, F. Edge contraction pooling for graph neural networks. *arXiv* **2019**, arXiv:1905.10990.

38. Henaff, M.; Bruna, J.; LeCun, Y. Deep convolutional networks on graph-structured data. *arXiv* **2015**, arXiv:1506.05163.
39. Simonovsky, M.; Komodakis, N. Dynamic edge-conditioned filters in convolutional neural networks on graphs. In Proceedings of the IEEE Conference on Computer Vision and Pattern Recognition, Honolulu, HI, USA, 21–26 July 2017; pp. 3693–3702.
40. Kipf, T.N.; Welling, M. Semi-supervised classification with graph convolutional networks. In Proceedings of the 5th International Conference on Learning Representations, ICLR 2017, Toulon, France, 24–26 April 2017.
41. He, K.; Zhang, X.; Ren, S.; Sun, J. Deep residual learning for image recognition. In Proceedings of the 29th IEEE Conference on Computer Vision and Pattern Recognition, CVPR 2016, Las Vegas, NV, USA, 26 June–1 July 2016; pp. 770–778.
42. Paszke, A.; Gross, S.; Massa, F.; Lerer, A.; Bradbury, J.; Chanan, G.; Killeen, T.; Lin, Z.; Gimelshein, N.; Antiga, L. Pytorch: An imperative style, high-performance deep learning library. *Adv. Neural Inf. Processing Syst.* **2019**, *32*, 8024–8035.
43. Foresti, L.; Sideris, I.V.; Nerini, D.; Beusch, L.; Germann, U. Using a 10-year radar archive for nowcasting precipitation growth and decay: A probabilistic machine learning approach. *Weather Forecast.* **2019**, *34*, 1547–1569. [CrossRef]

atmosphere

MDPI

Article

Application of AIRS Soundings to Afternoon Convection Forecasting and Nowcasting at Airports

Nan-Ching Yeh [1], Yao-Chung Chuang [2,*], Hsin-Shuo Peng [3] and Chih-Ying Chen [4]

1 Department of Military Meteorology, Air Force Institute of Technology, Kaohsiung 82047, Taiwan; Trea024@afats.khc.edu.tw
2 Department of Aviation & Communication Electronics, Air Force Institute of Technology, Kaohsiung 82047, Taiwan
3 Naval Meteorological and Oceanographic Office, Kaohsiung 81300, Taiwan; qqmoaso@gmail.com
4 Research Center for Environmental Changes, Academia Sinica, Taipei 11529, Taiwan; cycch6914@gate.sinica.edu.tw
* Correspondence: physicsmoon1213@gmail.com or Tree063@afats.khc.edu.tw

Abstract: In Taiwan, the frequency of afternoon convection increases in summer (July and August), and the peak hour of afternoon convection occurs at 1500–1600 local solar time (LST). Afternoon convection events are forecasted based on the atmospheric stability index, as computed from the 0800 LST radiosonde data. However, the temporal and spatial resolution and forecast precision are not satisfactory. This study used the observation data of Aqua satellite overpass near Taiwan around 1–3 h before the occurrence of afternoon convection. Its advantages are that it improves the prediction accuracy and increases the data coverage area, which means that more airports can use results of this research, especially those without radiosondes. In order to determine the availability of Atmospheric Infrared Sounder (AIRS) in Taiwan, 2010–2016 AIRS and radiosonde-sounding data were used to determine the accuracy of AIRS. This study also used 2017–2018 AIRS data to establish K index (KI) and total precipitable water (TPW) thresholds for the occurrence of afternoon convection of four airports in Taiwan. Finally, the KI and TPW were calculated using the independent AIRS atmospheric sounding (2019–2020) to forecast the occurrence of afternoon convection at each airport. The average predictive accuracy rate of the four airports is 84%. Case studies at Hualien Airport show the average predictive accuracy rate of this study is 81.8%, which is 9.1% higher than that of the traditional sounding forecast (72.7%) during the same period. Research results show that using AIRS data to predict afternoon convection in this study could not only increase data coverage area but also improve the accuracy of the prediction effectively.

Keywords: afternoon convection; atmospheric stability index; radiosonde; AIRS; K index; total precipitable water

Citation: Yeh, N.-C.; Chuang, Y.-C.; Peng, H.-S.; Chen, C.-Y. Application of AIRS Soundings to Afternoon Convection Forecasting and Nowcasting at Airports. *Atmosphere* **2022**, *13*, 61. https://doi.org/10.3390/atmos13010061

Academic Editors: Zuohao Cao, Huaqing Cai and Xiaofan Li

Received: 19 November 2021
Accepted: 28 December 2021
Published: 30 December 2021

1. Introduction

Heavy convection storms are relevant to flight safety during takeoff and landing. The World Meteorological Organization (WMO) defined nowcasting as forecasting with local detail, by any method, over a period from the present to six hours ahead, including a detailed description of the present weather. It is thus of great concern to aeronautical meteorological forecasting and nowcasting. When deep convection occurs in Taiwan, the temperature and humidity of each vertical layer of the atmospheric environment increase [1,2]. Relative to typhoons and the Meiyu front, the afternoon convection system exhibits a smaller spatial scale and shorter lifetime, so it is very difficult to predict the start time, initial location, and duration of afternoon convection [1,3,4]. Even if the rainfall pattern has great variability, it is very important to estimate the rainfall characteristics of different scales, seasons, and environments [5–8].

Radiosonde measurements of the atmospheric stability index can be used to predict severe weather development; such measurements are considered to be a representative of the synoptic scale environment [9]. Researchers have compared various atmospheric stability indices calculated from radiosonde observations to analyze the correlation between the stability index and cumulative precipitable water and have found the K index (KI) to be the most suitable index for forecasting heavy rain events [10,11]. Using numerical models to simulate rainfall studies, the KI can provide useful forecast guidance for rainfall events [12].

When a weather balloon is launched from the ground to the stratosphere, the balloon is horizontally displaced by tens of kilometers due to the wind field of the height [13]. Weather balloons are launched daily to record local vertical atmospheric parameters in the vicinity of where the balloon was launched. Taking the main island of Taiwan as an example, only two of the radiosonde observations of the Central Weather Bureau (CWB) can be obtained online. Thus, in Taiwan, the spatial distribution of weather balloons is insufficiently broad. The use of satellites, given their wide observational swath, compensates for these spatial distribution–related shortcomings in sounding.

Atmospheric stability index on pre-convective atmospheric stability and changes in boundary-layer structure are crucial [14]. Forecasters obtain data on atmospheric stability from the weather balloons launched at 00 UTC (i.e., 08am LST). They use such data as a basis for forecasting afternoon convection because atmospheric stability relates to the development of afternoon convection. The problem is the launch time of the weather balloons differs by 7–8 h from the afternoon convection's extremum (1500–1600 LST), and atmospheric-environmental changes during this period can result in forecasting error. Therefore, the use of weather balloons is inadequate because of limitations in data volume, coverage area, and immediacy.

As a remedy, Aqua satellite can be used. This satellite, which has an Atmospheric Infrared Sounder (AIRS) mounted on it, passes Taiwan 1–3 h before the extremum of afternoon convection. Numerous studies have shown that AIRS can provide a three-dimensional field with respect to variation characteristics of temperature, specific humidity, etc. [15–19]. Specifically, the AIRS observes the atmospheric environment at a time that is closer to the time at which afternoon convection occurs. Thus, relative to the use of radiosonde observation data, the use of Aqua satellite yields more accurate forecasts. However, there is a limitation in the use of polar-orbiting satellite Aqua. The satellites orbits and swaths will be shifted a bit, and there is no data to use outside of the swath. Total precipitable water (TPW) is the total amount of precipitable water in an atmospheric column between the Earth's surface and space. Regardless of the phases, its value, variability, and trends have a great influence on rainfall events [20–24].

The main objective of this study is to use satellite data to establish an atmospheric stability index and TPW threshold. These indicators can be used by forecasters to better predict the summer occurrence of afternoon convection in various airports in Taiwan, thus allowing them to anticipate possible weather changes.

2. Materials and Methods

2.1. Data

The data used in this study include Aqua satellite data, ground observation data, and radiosonde data. The data period is July and August from 2010 to 2020. Its purpose includes testing the reliability of satellite data, systematic error analysis and correction, the establishment of rainfall thresholds, and case verification. The details of the above data are as shown in Section 2.2.1, Section 2.2.2, Section 2.2.3.

2.1.1. AIRS

NASA's Aqua Satellite is part of the A-Train constellation of orbiting satellites. The satellite is equipped with six different earth observation systems, being able to obtain data on various parameters relating to the land, ocean, atmosphere, and biosphere [25]. AIRS has a scanning width of 2330 km, a nadir point horizontal resolution of 13.5 km, and can gather data on the entire planet in 2 days. AIRS comprises a hyperspectral sounder with 2378 infrared channels as well as four visible-light and near-infrared channels, allowing it to measure infrared radiation from the earth's surface and atmosphere [26,27]. Moreover, AIRS can measure various parameters pertaining to the physical properties of clouds and the thermodynamics of the atmosphere. In addition to weather monitoring, AIRS can also be applied to data model assimilation and the study of the climate [28–30].

AIRS can undertake high-precision atmospheric sounding due to its multiple channels under clear and partly cloudy conditions [15], but lower-tropospheric measurements are susceptible to sea conditions [31]. These differences significantly affect measurements of the stable structure of the low troposphere. In response, numerous studies have devised various methods for modifying AIRS-measured data [32–34]. This study used the AIRS Level 2 dataset (AIRX2RET) in July and August 2010–2020. The dataset provided daily global temperature and moisture profiles with an accuracy of 1 K per 1-km-thick and of 15% per 2-km-thick in the troposphere [35].

2.1.2. Atmospheric Sounding

Since 1958, the radiosonde has been the only instrument used for the long-term observation of temperature distributions in the troposphere to low stratosphere [36]. The radiosonde is often used as reference data in gauging the validity of water vapor obtained from other techniques [37–39]. Globally distributed sounding stations provide in situ radiosonde observations for assessing the state of the vertical atmosphere [31]. The weather balloon is launched from the ground to the stratosphere with an average horizontal displacement of about 50 km [13]. Therefore, the radiosonde observation represents the atmospheric conditions within a radius of 50 km from the balloon launch location. However, the temporal and spatial resolution of radiosondes is inadequate for use in forecasting.

This study used radiosonde data on Hualien and Banqiao from the CWB of Taiwan; the radiosonde locations are marked by a black star in Figure 1. In general, radiosonde data have errors from encoding, data transmission, and decoding. In this study, the method proposed by Chen (1994) [40] was used to verify the radiosonde data's accuracy.

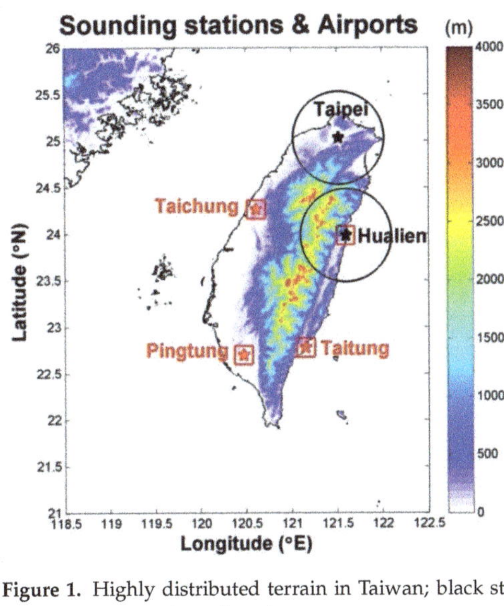

Figure 1. Highly distributed terrain in Taiwan; black stars represent sounding stations and red squares represent airport locations.

The uncertainty of temperature and relative humidity at upper-air network data observed by Vaisala RS-92 radiosonde was below 1 °C and 6% respectively [41]. Moreover, different brands of radiosondes have different observation errors in different seasons, regions, and even day and night [36,42–44]. Therefore, only 00-UTC radiosonde data required for practical applications were used for the conduct of its comparison of atmospheric sounding data. The models for both sets of data were Vaisala radiosonde models, which feature more consistent uncertainties. This study used atmospheric sounding data for the same time period as AIRS. Table 1 presents the radiosonde information used in this study.

Table 1. Information related to the radiosonde observations used to evaluate the AIRS thermodynamic profiles.

Site Name	Location	Launch Frequency (Per Day)	No. of Profiles (Day Time Only)	Time Period for Evaluation	Radiosonde Manufacturer/Model
Taipei	25.03° N 121.52° E	2	229	July–August from 2010–2016	Vaisala RS-92/RS-41
Hualien	23.98° N 121.6° E	2	219		

2.1.3. Surface Observations

The Meteorological Terminal Aviation Routine Weather Report (METAR) is a format for reporting aeronautical meteorological observations. A special (SPECI) report is drafted if weather conditions change significantly during the two METAR observation intervals. The METAR/SPECI observations include cloud coverage, phenomenon (e.g., rainfall), and cumulative rainfall. The above parameters are used in this study.

This study predicted the occurrence of convection in the weak synoptic scale for the airports in July and August. The determination of "weak weather scale" and "occurring convection" is based on the observation data of METAR/SPECI.

2.2. Methodology

This study used July and August 2010–2016 data from the Banqiao and Hualien radiosonde stations. Problematic data were deleted [40] and compared with data from within the AIRS' swath and from the radiosonde stations. The systematic errors in the AIRS data for Taiwan were analyzed and corrected according to the linear regression equations for each altitude, which were in terms of the temperature and dew point. Subsequently, the corresponding atmospheric stability index was calculated based on corrected and reliable temperature and dew point.

Finally, in conjunction with AIRS-retrieved TPW, the thresholds of afternoon convection at each airport were established and verified with independent data. More specifically, July and August 2017–2018 data were used for obtaining the threshold for afternoon convective rainfall; therefore, the accuracy of predicting afternoon convection used independent data from 2019–2020.

2.2.1. Validation of AIRS Temperature and Dew Point Profiles

Although the temperature and water vapor accuracy of AIRS in the troposphere are 1 K and 15%, respectively [35]. However, the uncertainty of sounding measurements varies with region [36,44], and the difference in temperature and humidity between AIRS and radiosondes in the low troposphere varies with seasons [31]. Therefore, soundings must be compared before use to ensure the correctness of the AIRS' measurements of temperature and humidity. In addition, this study focused only on afternoon convection in Taiwan and thus only analyzed the months (July and August) when afternoon convection most frequently occurs.

In this study, radiosonde data were used to validate AIRS' atmospheric sounding measurements. July and August 2010–2016 data were compared, and July and August 2017–2018 data were used for verifying the credibility of temperature and dew point of AIRS. Research has revealed differences between the deviations of day and night radiosonde observations [36,43]. Therefore, only 00 UTC data is used for comparison, correction, and verification between AIRS and radiosonde observations.

For reasons such as AIRS's swath not providing coverage up to the sounding station and sounding data failing the quality inspection procedure [40], the number of comparison samples was fewer than the number of soundings administered. In this study, because the temperature and humidity of 850, 700, and 500 hPa were used to calculate KI, the temperatures and humidity of these three levels were compared. The observations can be regarded as representative of atmospheric conditions within a 50-km radius [13]. Therefore, a radiosonde station was used as the center, where the average of all AIRS field of View (FOV) within a 50-km radius was used as the AIRS-retrieved measurements (as marked by the black circles in Figure 1).

2.2.2. Confirmation and Correction of Systemic Errors

Figure 2 presents the temperature and dew point scatter plot for AIRS and radiosonde measurements. The abscissa represents AIRS measurements, and the ordinate represents radiosonde measurements. Figure 2a illustrates the temperature distribution for 850, 700, and 500 hPa, and Figure 2b illustrates the dew point distribution for 850, 700, and 500 hPa. The blue dotted line represents the fitted straight-line equation, and the green solid line represents the reference equation x = y. Tables 2 and 3 present the correlation coefficients, temperature, and dew point fitting equations for each level.

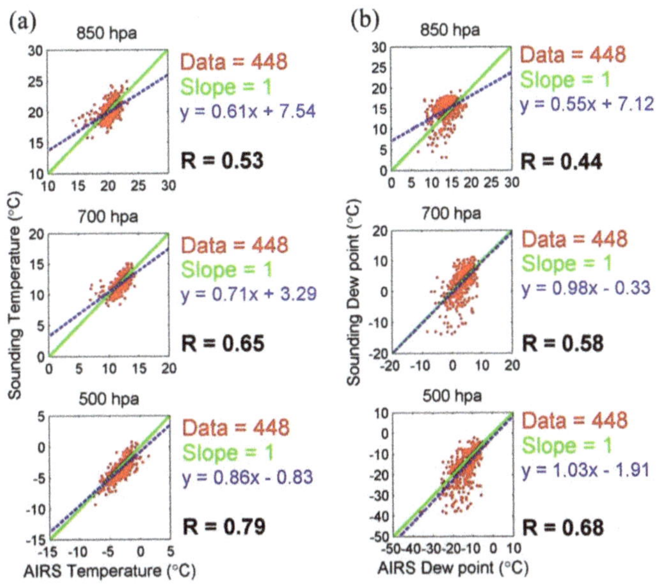

Figure 2. AIRS and sounding scatter plots of 850, 700, 500 hPa from top to bottom for (**a**) temperature and (**b**) dew point.

Table 2. Temperature correlation coefficients and linear regression equations for each level of 00Z for July and August of 2010–2016.

High-Altitude Level	Number of Data	Correlation Coefficient	Regression Equation
500 hPa	448	0.79	y = 0.86x − 0.83
700 hPa	448	0.65	y = 0.71x + 3.29
850 hPa	448	0.53	y = 0.61x + 7.54

Table 3. Dew point correlation coefficients and linear regression equations for each level of 00Z for July and August of 2010–2016.

High-Altitude Level	Number of Data	Correlation Coefficient	Regression Equation
500 hPa	448	0.68	y = 1.03x − 1.91
700 hPa	448	0.58	y = 0.98x − 0.33
850 hPa	448	0.44	y = 0.55x + 7.12

As evident in Figure 2, the temperature distribution is more concentrated than the dew point distribution. In other words, the change in temperature was less obvious than the change in dew point, consistent with the results of Ingleby (2017) [44], where dew point uncertainty was greater than that for temperature. Data spanning 7 years were used to establish the modified equations, and 2 years of independent data for each level were included in the corresponding modified equation. In doing these, the temperature and dew point data more accurately represented the actual atmospheric environment.

2.2.3. Forecast Rules and Probability Using K Index and Total Precipitable Water

This study identified KI to be the most suitable index from different stability indices for forecasting afternoon convection in Taiwan. There are similar results in several different areas [10–12]. KI is considered to have the static stability of the 850–500-mb layer, and the mathematical formula for KI is as follows [45].

$$KI = (T_{850} - T_{500}) + Td_{850} - (T_{700} - Td_{700}) \tag{1}$$

T_{850}, T_{700}, and T_{500} are the temperatures at 850, 700, and 500 hPa, respectively, and Td_{850} and Td_{700} are the dew points at 850 and 700 hPa, respectively. KI includes the factors of a lapse rate of temperature of 850–500 hPa, a dew point of 850 hPa, and a saturation level of 700 hPa; the sum of all three factors represents the potential of a thunderstorm and rainfall. KI is higher, and the chance of thunderstorms/rainfall is higher.

The middle-troposphere humidity is a vital factor explaining the occurrence and development of convection [46]. KI includes the 850–500 hPa lapse rate of temperature and the water vapor content in the middle and low troposphere [12]. Therefore, KI can be used as a reference for predicting afternoon convection. In addition to KI, TPW is an essential indicator [21,23,24]. When TPW is low, convection will not occur even if the atmospheric environment is unstable.

The afternoon convection threshold must be evaluated separately for each airport, because the threshold of thunderstorm occurrence changes depending on location [47]. The evaluation of predictive accuracy is illustrated in Figure 3; the abscissa is TPW, the ordinate is KI, and the red dotted lines are the thresholds of KI and TPW (hereafter abbreviated as K_h and T_h, respectively). Both dotted lines divide the atmospheric-environmental parameters into four quadrants, named quadrants 1 to 4 (Q1–Q4).

Q1 represents when TPW $> T_h$ and when the atmosphere is unstable (i.e., KI $> K_h$), entailing a forecast that convection will occur in the afternoon. By contrast, Q3 represents when TPW $< T_h$ and when the atmosphere is stable (i.e., KI $< K_h$), entailing a forecast that convection will not occur in the afternoon. Furthermore, Q2 (KI $> K_h$ and TPW $< T_h$) and Q4 (KI $< K_h$ and TPW $> T_h$) entail a forecast that precipitation will not occur.

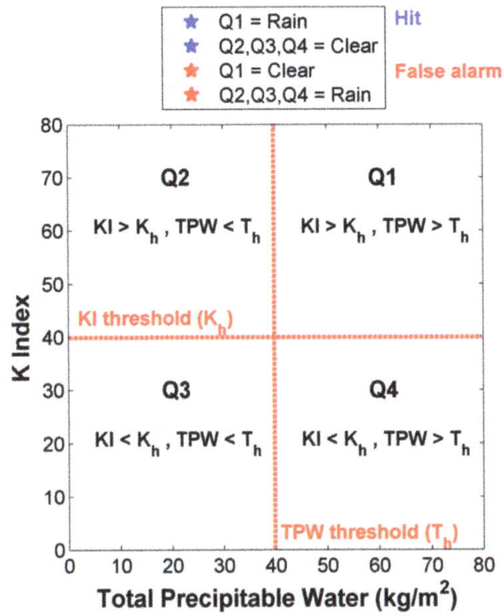

Figure 3. Evaluation of accuracy for TPW and KI of the afternoon convection of each airport.

3. Results and Discussion

3.1. AIRS Comparison Results

The comparison of temperature and dew point used 2010–2016 data, and 2017–2018 data were used to verify its accuracy. The 2019–2020 data is used to estimate how much the forecast results of this study have improved. The comparison between measurements obtained from AIRS atmospheric sounding and that obtained from radiosonde observations is illustrated in Figure 4. Figure 4a,b illustrate, for each level, the comparison between temperature and dew point, respectively. The blue line represents the AIRS retrieval value, and the red line represents the radiosonde observations (sample size: 448). For each level, the correlation coefficients of the temperature and humidity were 0.66 and 0.57, respectively. Regardless of temperature and humidity, the correlation of each level was optimum at 500 hPa, followed by 700 and 850 hPa. Moreover, with respect to correlation, that for the temperature was more satisfactory than that for the dew point.

These results are attributable to the following reasons. The first is topography: mountains within 50 km from sounding stations were covered, and the highest-altitude mountain range was approximately 4 km tall. This meant that only the 500-hPa level was unaffected. The second is the configuration of the measuring equipment. Specifically, the sounding station furnished only single-point observations, and the weather balloon shifted horizontally with the wind, whereas AIRS furnished plane observations. Therefore, the horizontal resolutions of both methods were different. The third is differences in observation time. Specifically, most atmospheric conditions change gradually. However, an approaching weather system can cause a large and rapid change in the atmospheric temperature and dew point, which results in errors. Nonetheless, measurements of the average AIRS temperature and dew point around the stations were still representative of vertical atmospheric conditions around the station.

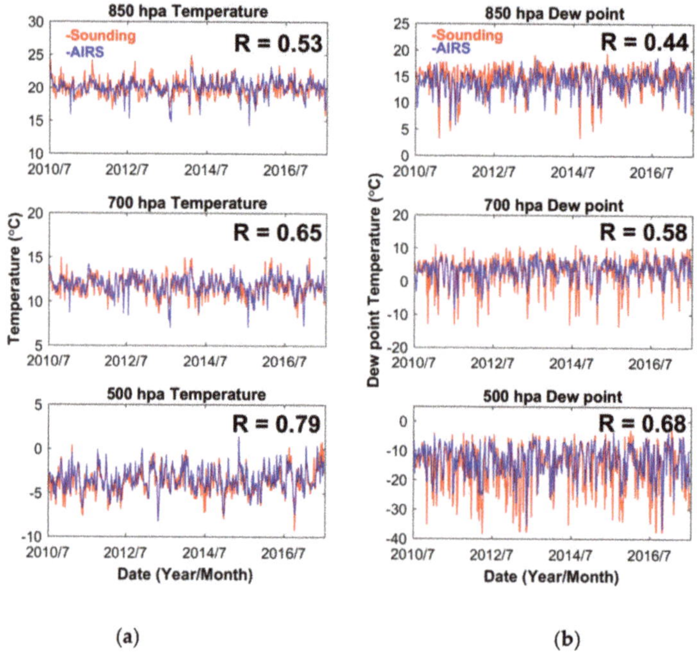

Figure 4. Comparison of AIRS (blue line) and radiosonde (red line) for (**a**) temperature, and (**b**) dew point for 850, 700, and 500 hPa (from top to bottom).

For radiosonde observations, the root-mean-square error (RMSE) and Standard deviation (SD) of the temperature and humidity, both before and after AIRS correction, are presented in Tables 4 and 5. In Table 4, temperature had the largest RMSE and SD at 850 hPa, followed by 700 hPa and 500 hPa. Subsequent to corrections through the equations in Table 2, the RMSE and SD of each level decreased, with 850 hPa having the highest correction margin, followed by 700 hPa and 500 hPa. This is attributable to the small error of the original level (500 hPa), resulting in a low correction margin. By contrast, the correction margin was larger at 850 hPa. Crucially, because of the lower RMSE and SD, the corrected AIRS temperature had a reduced dispersion that was closer to the radiosonde observation value, indicating that this study's modified equation effectively made the AIRS temperature measurements closer to their radiosonde counterparts.

Table 4. RMSE and SD for 2017–2018 temperature measurements before and after correction.

Levels	500 hPa		700 hPa		850 hPa	
	Before	After	Before	After	Before	After
root-mean-square error	1.08	1.05	1.22	1.12	1.48	1.27
Standard deviation	1.11	0.94	1.18	0.83	1.31	0.79

Table 5. RMSE and SD for 2017–2018 dew point measurements before and after correction.

Levels	500 hPa		700 hPa		850 hPa	
	Before	After	Before	After	Before	After
root-mean-square error	4.93	4.92	4.16	4.13	3.79	2.73
Standard deviation	5.82	5.80	3.08	3.01	2.69	1.48

For dew point correction, the pre- and post-conditions were similar to those for temperature, and the RMSE and SD of each level were reduced. These results indicate that this study's modified equations effectively reduced errors for the AIRS measurements of temperature and humidity in Taiwan, thus making the AIRS measurements closer to their sounding-observation counterparts. In addition, as the altitude becomes higher, the RMSE and SD of the temperature and dew point observed by AIRS will increase. This result is similar to the previous studies [41,42].

3.2. Threshold for Afternoon Convection and Probability of Precipitation

This study investigated thermodynamically induced afternoon convection, which necessitated the use of the weak synoptic scale in the case selection. The METAR/SPECI showed cloud coverage at 0800–1200 LST was less than four oktas, which is defined as a weak synoptic scale. There must also be no significant weather systems, such as a weather front and typhoon, approaching the vicinity of the airports before and after convection.

The definition of rainfall in this study is METAR/SPECI afternoon precipitation data at 1200–1800 LST were checked to detect convection. The coverage area for detecting convection includes all surface stations within 20 km, and the airports were used as the center points. An indication of precipitation by at least one station was interpreted, as the occurrence of rain at the airport. By contrast, an absence of indication of precipitation by all stations was interpreted as the absence of rainfall at the airport.

The Taichung, Pingtung, Hualien, and Taitung airports marked by a red square in Figure 1 were selected as the research areas. In short, a forecast is correct if KI and TPW located at Q1 of Figure 3 and it rained in the afternoon. A forecast is also correct if KI and TPW located at Q2, Q3, or Q4 of Figure 3 and it did not rain in the afternoon. By contrast, a forecast is wrong if KI and TPW located at Q1 of Figure 3 and it did not rain in the afternoon. A forecast is also wrong if KI and TPW located at Q2, Q3, or Q4 of Figure 3 and it rained in the afternoon.

This study analyzed the afternoon convection thresholds of four different airports, because the threshold of convection occurrence changes depending on location [48]. Figure 5a,b map the distribution of KI and TPW for the Taitung and Taichung airports, respectively. The abscissa represents TPW, and the ordinate represents KI. The blue dots indicate those cases of afternoon convection, and the black dots indicate those cases where afternoon convection did not occur. Two red dotted lines represent the threshold values for the establishment of KI and TPW at the airports, and the thresholds were obtained from the highest forecast accuracy of the July and August of 2017–2018 data. The verification data in Figure 5 were for the July and August of 2019 and 2020.

As recorded in Figure 5a, the KI and TPW thresholds of Taitung airport's afternoon convection were 32.1 and 47.1, respectively. In 19 days, the predictions were located at Q1, and convection was predicted because $KI > K_h$ and $TPW > T_h$ (situated in Q1). However, 4 cases had no indications of precipitation at all stations within 20 km of Taitung airport, whereas 15 cases had afternoon convection. Therefore, the forecast accuracy of Q1 was 78.9% (15 hits over 19 cases). By contrast, 18 cases located in the Q2, Q3, and Q4 were forecasted to have no afternoon convection. Nonetheless, 4 of them had precipitation records at the stations within 20 km of the airport, whereas 14 cases had no precipitation records. Therefore, the forecast accuracy of Q2 to Q4 was 77.8% (14 hits over 18 cases). Overall, 29 hits and 8 false alarms were identified among the 37 cases of Taitung airport. Therefore, the total forecast accuracy of Taitung airport was 29/37, or 78.4%.

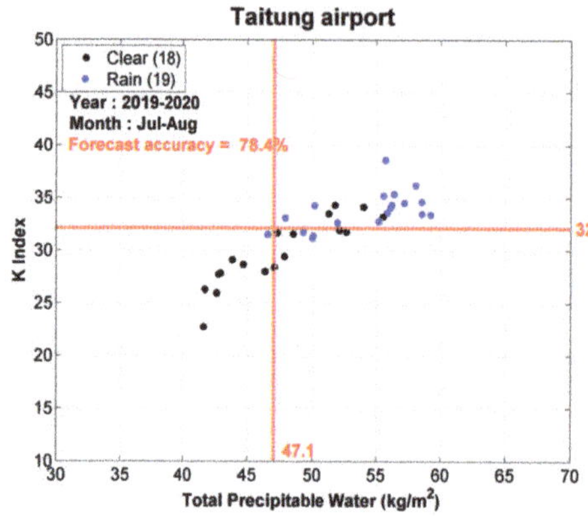

(a)

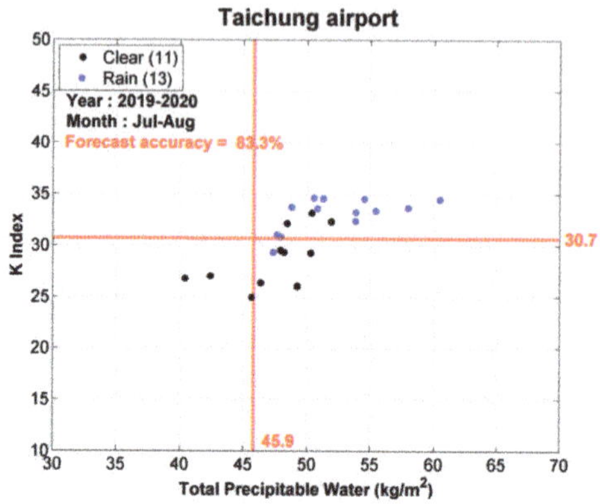

(b)

Figure 5. KI and TPW distribution map for (**a**) Taitung airport and (**b**) Taichung Airport. The blue and black dots are those cases with and without convection, respectively, and the red dotted lines indicate the thresholds for KI and TPW.

The same method for evaluating predictive accuracy was used for predictions for Taichung airport (Figure 5b). The forecast accuracy for afternoon convection in Q1 was 80% (12 hits over 15 cases), and the forecast accuracy for Q2, Q3, and Q4 was 88.8% (8 hits over 9 cases), for a total forecast accuracy of approximately 83.3% (20 hits over 24 cases).

All information such as the location of the four airports, the rainfall thresholds of KI and TPW for each airport, the accuracy of rainfall events (Q1 area), the accuracy of non-rainfall events (Q2–Q4 area), and the total accuracy are shown in Table 6. According to Table 6, the KI and TPW thresholds differ by region. This result indicates the stability index threshold of rainfall in different regions, which needs to be revised according to different locations.

Table 6. KI and TPW thresholds and forecast accuracy at the four airports.

Site Name	Location	Threshold of KI	Threshold of TPW	Forecast Accuracy of Q1	Forecast Accuracy of Q2–Q4	Total Accuracy
Hualien	23.98° N 121.6° E	27.7	44.3	82.1% (23/28)	80% (4/5)	81.8% (27/33)
Taitung	22.78° N 121.16° E	32.1	47.1	78.9% (15/19)	77.8% (14/18)	78.4% (29/37)
Pingtung	22.7° N 120.48° E	25.5	42.4	95.2% (20/21)	100% (0/0)	95.2% (20/21)
Taichung	24.26° N 120.62° E	30.7	45.9	80% (12/15)	88.8% (8/9)	83.3% (20/24)
All cases	-	-	-	84.3% (70/83)	81.3% (26/32)	83.5% (96/115)

The forecast accuracy and frequency of occurrence of afternoon convection in the four airports were further analyzed. The Taitung and Hualien airports (hereafter referred to as the eastern airports) are situated in more mountainous areas, as marked by the red square in Figure 1. By contrast, the Pingtung and Taichung airports (hereafter referred to as the western airports) were situated on almost flat terrain. As for the ratios of without rainfall to rainfall with afternoon convection, those for the eastern and western airports were approximately 1:1.6 and 1:2.8, respectively. The eastern airports also had fewer rainy days than did the western airports. Moreover, the precipitation forecast accuracy and the total forecast accuracy for the western (~89%) airports were more favorable than those of the eastern airport (80%).

This difference in performance is attributable to the following reasons. First, because the eastern airports were situated in more mountainous areas, convection also depended on dynamic factors (and, to some extent, thermodynamic factors). Therefore, forecasting performance was worse for eastern airports because this study only considered thermodynamic factors. Second, atmospheric sounding was more accurate for western than eastern airports because western airports were not situated in mountainous areas. This result suggests the rainfall is related to terrain, elevation, slope, shape, and wind structure and so on [48–50].

This result suggests that precipitation products as well as the forecast precipitation indices that were derived from satellite data were susceptible to the influence of terrain. This is particularly true in Taiwan where mountainous areas account for 70% of the terrain. Therefore, such errors must be corrected prior to the use of satellite precipitation–related products. These results are similar to those of Yeh et al. (2019) [51].

The total forecast accuracy was lowest for Taitung airport (at 78.4%) because a mountain range dominated its landscape (within a 20-km radius) at a degree greater than that for other airports. There are a total of 115 cases for verification in this study. A total of 96 cases were forecasted correctly, and 19 cases were forecasted incorrectly. The total forecast accuracy rate was 84%. Furthermore, the accuracy rate of rainfall events is 84.3%, and the accuracy rate of non-rainfall events is 81.3%.

3.3. Case Studies

In addition to upgrading point information to area information for better applicability to airport personnel, the use of satellite data for the forecast of afternoon convection improves forecasting accuracy. The improvement percentage will be discussed in the next section. In this section, Hualien airport was used as an example. Two cases were analyzed to illustrate how AIRS measurements yield more accurate predictions than their radiosonde counterparts.

The first case occurred on 6 July 2018. The KI value was 31, as calculated from the radiosonde observation of the atmospheric sounding on the morning of that day; this KI value exceeded the threshold for afternoon convection (Table 6). An afternoon convection is forecasted to occur at the airport if the forecaster judges it to be so through the use of this set of radiosonde data. Figure 6a,b map the distribution of AIRS-derived KI and TPW. As illustrated in the figures, KI did not reach 20 and TPW did not reach 40 kg/m^2; neither reached the threshold. Based on these data, the forecaster predicted no convection in the afternoon. No precipitation was recorded that afternoon at the observation stations within 20 km of the airport.

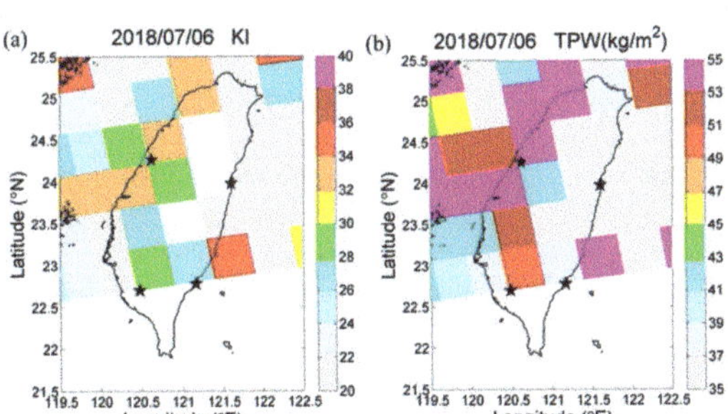

Figure 6. AIRS retrieval on 6 July 2018. (**a**) KI and (**b**) TPW distribution maps.

The second case occurred on August 9 of the same year. The radiosonde-derived KI was approximately 15, which did not reach the threshold. Figure 7a,b map the distribution of AIRS-derived KI and TPW. The KI near the airport was greater than 30, and the TPW was greater than 50 kg/m^2, both of which exceed the rainfall thresholds of Hualien airport. Thus, based on these radiosonde data, the forecasters predicted no convection at the airport. However, based on this study's satellite data, convection was forecasted at the airport. A rainfall event was determined to have occurred at the airport that afternoon.

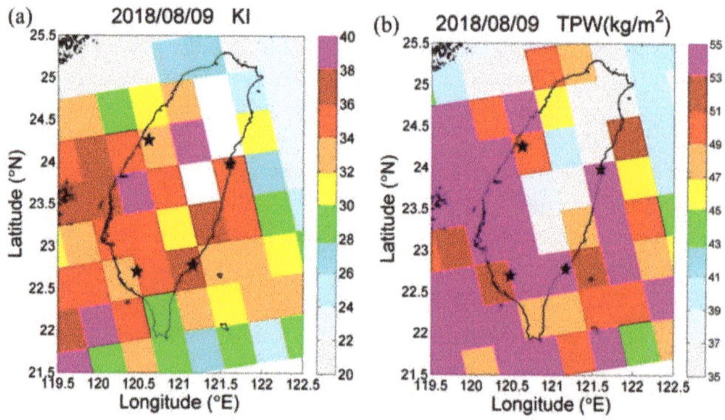

Figure 7. AIRS retrieval on 9 August 2018. (**a**) KI and (**b**) TPW distribution maps.

The aforementioned cases indicated that the atmosphere can change from stable to unstable (and vice versa) from the morning to the afternoon. Forecasters can reduce false weather forecasts if they use this study's satellite data instead of solely using radiosonde data. Because the atmospheric environment potentially changes every few hours, false forecasts are likely if changes in atmospheric stability are not accounted for. Moreover, because the time of the satellite scanning Taiwan is closer to the time when convection occurs in the afternoon, AIRS results in more effective forecasts of afternoon convection.

3.4. Improvement Percentage

In order to further determine the forecast accuracy of the method proposed in this study and see whether it is better than the traditional method, the results of this study were compared with those of radiosonde data, which are conventionally used by aeronautical meteorological forecasters, to compare the accuracy and practicability of both methods. In this study, only the radiosonde observations were used to forecast the afternoon convection at the airports in July and August of 2019 and 2020. The comparison was conducted only for Hualien airport because the available open-source radiosonde data covered only that airport.

Among the 33 cases that used radiosonde observations to forecast afternoon convection in Hualien airport, 24 cases were accurate and 9 cases were inaccurate, resulting in total forecast accuracy of 72.7%. By contrast, this study's total forecast accuracy of afternoon convection in Hualien airport using satellite data was, at 81.8%, higher, which was an improvement of 9.1%. Even if the forecast accuracy rate is improved compared with the previous method, the forecast accuracy rate still cannot reach 100%; this is the disadvantage of this research. In other words, there is still a possibility of incorrect forecasts using the forecasting method of this research, so it is still necessary to use manual assistance to observe weather changes to ensure flight safety. Another contribution of this study was its transformation of the original single-point sounding data into area data. This allows airports (such as the Taichung and Taitung airports) that have not launched weather balloons to use this study's method to forecast afternoon convection. Therefore, relative to the radiosonde method, this study's method is applicable to a wider range of airports.

4. Conclusions

This study uses satellite data to do airports nowcasting research. In other words, this study uses satellite data to predict the summer afternoon convection in the weak synoptic scale at Taiwan airports in Taichung, Pingtung, Hualien, and Taitung. Modified equations were established using 2010–2016 AIRS and radiosonde observation data, and 2017–2018 data were used to verify the accuracy of temperature and dew point of AIRS. The independent data (2019–2020) is used to verify the practicality and accuracy of this study's forecasting method. This study aimed to increase the number of airports that can be covered by the forecast by using the satellite's large swath. In addition, it also improves the accuracy of the forecast and the validity of the data, for the satellite scanning time is close to the time when convection occurs most frequently in Taiwan. The novel aspect of this research is to use satellite data closer to the time of convective rainfall than the radiosonde while considering two rainfall-related parameters to forecast afternoon convection. This study also takes into account that different airports have different environments, so different rainfall thresholds are established to improve the accuracy of rainfall forecasts.

AIRS atmospheric sounding products have good accuracy in the troposphere [35]. However, because of Taiwan's mountainous terrain as well as the difference between AIRS and radiosonde with respect to measurements of temperature and humidity, the deviation in low-troposphere measurements will differ depending on the season [31]. This resulted in an unsatisfactory correlation between temperature and humidity when AIRS measurements were directly compared with their radiosonde counterparts. Relevant factors must be considered, such as the FOV covering mountainous areas and various atmospheric conditions within 50 km of the radiosonde. After the numerical average, the correlation coefficients of temperature and humidity were increased by approximately 0.2 to 0.1.

For the temperature and dew point of the vertical altitude layer observed by AIRS and radiosonde, the correlation coefficient of 500 hPa is the best, because it is not affected by mountains. In addition, the temperature observed by AIRS is more accurate than the dew point. The reason is that the uncertainty of the dew point is greater than that of the temperature. Moreover, upon applying the modified equations established in this study, the RMSE and SD of temperature and humidity of each level were improved, thus demonstrating that the modified equations effectively reduce errors for AIRS measurements of temperature and humidity in Taiwan.

This study used AIRS data from 2017 and 2018 to obtain the airport afternoon convection threshold at Taichung, Pingtung, Hualien, and Taitung airports. Because the terrain around the western airports was relatively flat, the use of these thresholds to forecast the accuracy of afternoon convection was more favorable than that for the eastern airports. Using data from 2019–2020, the total forecast accuracy for the Taichung, Pingtung, Hualien, and Taitung airports was 83.3%, 95.2%, 81.8%, and 78.4%, respectively, with a total forecast accuracy of 84%. The main contribution of this research is to use scanning area to increase the available airports and improve the accuracy by 9.1% compared with traditional radiosonde forecasting methods. The improvement of forecast accuracy can reduce problems caused by inaccurate weather forecasts. These problems include flight safety issues, especially when aircraft take off and land; aircraft not being able to land at the scheduled airport, which affects the subsequent flight schedule; and the waste of fuel and increased costs entailed.

Author Contributions: Conceptualization, N.-C.Y.; methodology, N.-C.Y. and Y.-C.C.; data curation, H.-S.P. and C.-Y.C.; validation, N.-C.Y. and H.-S.P.; formal analysis, Y.-C.C. and C.-Y.C.; writing—original draft preparation, N.-C.Y.; writing—review and editing, Y.-C.C. and H.-S.P.; funding acquisition, Y.-C.C. and N.-C.Y. All authors have read and agreed to the published version of the manuscript.

Funding: This research was funded by Ministy of Science and Technology of Taiwan, grant number MOST 110-2111-M-344-001.

Institutional Review Board Statement: Not applicable.

Informed Consent Statement: Not applicable.

Data Availability Statement: The datasets of Aqua/AIRS and radiosonde used in this study are publicly available in the archives: https://disc.gsfc.nasa.gov/datasets/AIRS2RET_7.0/summary and https://dbar.pccu.edu.tw/, respectively (accessed on 1 October 2021).

Conflicts of Interest: The authors declare no conflict of interest.

References

1. Johnson, R.H.; Bresch, J.F. Diagnosed characteristics of Mei-Yu precipitation systems over Taiwan during the May–June 1987 TAMEX. *Mon. Weather Rev.* **1991**, *120*, 2540–2557. [CrossRef]
2. Lin, P.F.; Chang, P.L.; Jou, B.J.D.; Wilson, J.W.; Roberts, R.D. Warm season afternoon thunderstorm characteristics under weak synoptic-scale forcing over Taiwan Island. *Weather Forecast.* **2011**, *26*, 44–60. [CrossRef]
3. Wilson, J.W.; Mueller, C.K. Nowcasts of thunderstorm initiation and evolution. *Weather Forecast.* **1993**, *8*, 113–131. [CrossRef]
4. Chang, H.L.; Brown, B.G.; Chu, P.S.; Liou, Y.C.; Wang, W.H. Nowcast guidance of afternoon convection initiation for Taiwan. *Weather Forecast.* **2017**, *32*, 1801–1817. [CrossRef]
5. Wilson, J.W.; Schreiber, W.E. Initiation of convective storms by radar-observed boundary layer convergent lines. *Mon. Weather Rev.* **1986**, *114*, 2516–2536. [CrossRef]
6. Bohner, J. General climatic controls and topoclimatic variations in Central and High Asia. *Boreas* **2006**, *35*, 279–295. [CrossRef]
7. Ichiyanagi, K.; Yamanaka, M.D.; Muraji, Y.; Vaidya, B.K. Precipitation in Nepal between 1987 and 1996. *Int. J. Climatol.* **2007**, *27*, 1753–1762. [CrossRef]
8. Shrestha, D.; Sharma, S.; Hamal, K.; Jadoon, U.K.; Dawadi, B. Spatial Distribution of Extreme Precipitation Events and Its Trend in Nepal. *Environ. Sci.* **2021**, *9*, 58–66. [CrossRef]
9. Davis, R.E.; Stanmeyer, T.M.; Jones, G.V. A synoptic climatology of tornadoes in Virginia. *Phys. Geogr.* **1997**, *18*, 383–407. [CrossRef]
10. Kodama, K.; Barnes, G.M. Heavy rain events over the south-facing slopes of Hawaii: Attendant conditions. *Weather Forecast.* **1997**, *12*, 347–367. [CrossRef]
11. Derubertis, D. Recent trends in four common stability indices derived from U.S. radiosonde observations. *J. Clim.* **2006**, *19*, 309–323. [CrossRef]
12. Tu, C.C.; Chen, Y.L. Favorable conditions for the development of a heavy rainfall event over Oahu during the 2006 wet period. *Weather Forecast.* **2011**, *26*, 280–300. [CrossRef]
13. Seidel, D.J.; Sun, B.; Pettey, M.; Reale, A. Global radiosonde balloon drift statistics. *J. Geophys. Res.* **2011**, *116*, D07102. [CrossRef]
14. Li, J.; Liu, C.Y.; Zhang, P.; Schmit, T.J. Applications of full spatial resolution space-based advanced infrared soundings in the preconvection environment. *Weather Forecast.* **2012**, *27*, 515–524. [CrossRef]
15. Tobin, D.C.; Revercomb, H.E.; Knuteson, R.O.; Lesht, B.M.; Strow, L.L.; Hannon, S.E.; Feltz, W.F.; Moy, L.A.; Fetzer, E.J.; Cress, T.S. Atmospheric Radiation Measurement site atmospheric state best estimates for Atmospheric Infrared Sounder temperature and water vapor retrieval validation. *J. Geophys. Res.* **2006**, *111*, D09S14. [CrossRef]
16. Smith, W.L.; Revercomb, H.; Bingham, G.; Larar, A.; Huang, H.; Zhou, D.; Li, J.; Liu, X.; Kireev, S. Evolution, current capabilities, and future advance in satellite nadir viewing ultra-spectral IR sounding of the lower atmosphere. *Atmos. Chem. Phys.* **2009**, *9*, 5563–5574. [CrossRef]
17. Isioye, O.A.; Combrinck, L.; Botai, J.O. Retrieval and analysis of precipitable water vapour based on GNSS, AIRS, and reanalysis models over Nigeria. *Int. J. Remote Sens.* **2017**, *38*, 5710–5735. [CrossRef]
18. Jiang, J.; Zhou, T.; Zhang, W. Evaluation of Satellite and Reanalysis Precipitable Water Vapor Data Sets Against Radiosonde Observations in Central Asia. *Earth Space Sci.* **2019**, *6*, 1129–1148. [CrossRef]
19. Wang, X.; Jiang, H. A 13-year global climatology of tropical cyclone warm-core structures from AIRS data. *Mon. Weather Rev.* **2019**, *147*, 773–790. [CrossRef]
20. Tuller, S.E. The relationship between precipitable water vapor and surface humidity in New Zealand. *Mon. Weather Rev.* **1977**, *26*, 197–212. [CrossRef]
21. Liang, H.; Zhang, Y.; Cao, L.; Cao, Y. Temporal relations between precipitable water vapour and precipitation during wet seasons based on nearly two decades of data from the Lhasa River valley, Tibetan Plateau. *Int. J. Clim.* **2020**, *40*, 1656–1668. [CrossRef]
22. Makama, E.K.; Lim, H.S. Variability and Trend in Integrated Water Vapour from ERA-Interim and IGRA2 Observations over Peninsular Malaysia. *Atmosphere* **2020**, *11*, 1012. [CrossRef]
23. Kim, S.; Matyas, C.J.; Yan, G. Rainfall Symmetry Related to Moisture, Storm Intensity, and Vertical Wind Shear for Tropical Cyclones Landfalling over the U.S. Gulf Coastline. *Atmosphere* **2020**, *11*, 895. [CrossRef]
24. Taniguchi, K.; Minobe, Y. Impact of Global Warming on Extreme Heavy Rainfall in the Present Climate: Case Study of Heavy Rainfall in Kinugawa, Japan (2015). *Atmosphere* **2020**, *11*, 220. [CrossRef]
25. Parkinson, C.L. Aqua: An Earth-observing satellite mission to examine water and other climate variables. *IEEE Trans. Geosci. Remote Sens.* **2003**, *41*, 173–183. [CrossRef]

26. Aumann, H.H.; Chahine, M.T.; Gautier, C.; Goldberg, M.D.; Kalnay, E.; McMillin, L.M.; Revercomb, H.; Rosenkranz, P.W.; Smith, W.L.; Staelin, D.H.; et al. AIRS/AMSU/ HSB on the Aqua mission: Design, science objectives, data products, and processing systems. *IEEE Trans. Geosci. Remote Sens.* **2003**, *41*, 253–264. [CrossRef]
27. Chahine, M.T.; Pagano, T.S.; Aumann, H.H.; Atlas, R.; Barnet, C.; Blaisdell, J.; Chen, L.; Divakarla, M.G.; Fetzer, E.J.; Goldberg, M.; et al. AIRS: Improving weather forecasting and providing new data on greenhouse gases. *Bull. Am. Meteor. Soc.* **2006**, *87*, 911–926. [CrossRef]
28. Li, J.; Huang, H.L.; Liu, C.Y.; Yang, P.; Schmit, T.J.; Wei, H.; Weisz, E.; Guan, L.; Menzel, W.P. Retrieval of cloud microphyiscal properties from MODIS and AIRS. *J. Appl. Meteorol.* **2005**, *44*, 1526–1543. [CrossRef]
29. Weisz, E.; Li, J.; Menzel, W.P.; Heidinger, A.; Kahn, B.H.; Liu, C.Y. Comparison of AIRS, MODIS, CloudSat and CALIPSO cloud top height retrievals. *Geophys. Res. Lett.* **2007**, *34*, L17811. [CrossRef]
30. Li, J.; Li, J.; Otkin, J.; Schmit, T.J.; Liu, C.Y. Warning information in a preconvention environment from the geostationary advanced infrared sounding system—A simulation study using the IHOP case. *J. Appl. Meteorol. Climatol.* **2011**, *50*, 776–783. [CrossRef]
31. Sedlar, J.; Tjernström, M. 2019: A process-based climatological evaluation of AIRS level 3 tropospheric thermodynamics over the high-latitude arctic. *J. Appl. Meteor. Climatol.* **2019**, *58*, 1867–1886. [CrossRef]
32. Liu, C.Y.; Li, J.; Weisz, E.; Schmit, T.J.; Huang, H.L. Synergic use of AIRS and MODIS radiance measurement for atmospheric profiling. *Geophys. Res. Lett.* **2008**, *35*, L21805. [CrossRef]
33. Liu, C.Y.; Liu, G.R.; Lin, T.H.; Liu, C.C.; Ren, H.; Young, C.C. Using surface stations to improve sounding retrievals from hyperspectral infrared instruments. *IEEE Trans. Geosci. Remote Sens.* **2014**, *52*, 6957–6963. [CrossRef]
34. Liu, C.Y.; Li, J.; Ho, S.P.; Liu, G.R.; Lin, T.H.; Young, C.C. Retrieval of atmospheric thermodynamic state from synergistic use of radio occultation and hyperspectral infrared radiances observations. *IEEE J. Sel. Top. Appl. Earth Obs. Remote Sens.* **2015**, *9*, 744–756. [CrossRef]
35. Divakarla, M.G.; Barnet, C.D.; Goldberg, M.D.; McMillin, L.M.; Maddy, E.; Wolf, W.; Zhou, L.; Liu, X. Validation of Atmospheric Infrared Sounder temperature and water vapor retrievals with matched radiosonde measurements and forecasts. *J. Geophys. Res.* **2006**, *111*, D09S15.
36. Ho, S.P.; Peng, L.; Vömel, H. Characterization of the long-term radiosonde temperature biases in the upper troposphere and lower stratosphere using COSMIC and Metop-A/GRAS data from 2006 to 2014. *Atmos. Chem. Phys.* **2017**, *17*, 4493–4511. [CrossRef]
37. Ross, R.J.; Elliott, W.P. Tropospheric Water Vapor Climatology and Trends over North America: 1973–93. *J. Clim.* **1996**, *9*, 3561–3574. [CrossRef]
38. Chen, B.; Liu, Z. Global water vapor variability and trend from the latest 36 year (1979 to 2014) data of ECMWF and NCEP reanalyses, radiosonde, GPS, and microwave satellite. *J. Geophys. Res. Atmos.* **2016**, *121*, 11442–11462. [CrossRef]
39. Zhao, J.; Li, T.; Shi, K.; Qiao, Z.; Xia, Z. Evaluation of ERA-5 Precipitable Water Vapor Data in Plateau Areas: A Case Study of the Northern Qinghai-Tibet Plateau. *Atmosphere* **2021**, *12*, 1367. [CrossRef]
40. Chen, W.J. The use of AVHRR data in TOVS retrievals. Ph.D. Thesis, National Central University, Zhongli, Taiwan, 1994.
41. Dirksen, R.J.; Sommer, M.; Immler, F.J.; Hurst, D.F.; Kivi, R.; Vömel, H. Reference quality upper-air measurements: GRUAN data processing for the Vaisala RS92 radiosonde. *Atmos. Meas. Tech.* **2014**, *7*, 4463–4490. [CrossRef]
42. Sun, B.; Reale, A.; Seidel, D.J.; Hunt, D.C. Comparing radiosonde and COSMIC atmospheric profile data to quantify differences among radiosonde types and the effects of imperfect collocation on comparison statistics. *J. Geophys. Res.* **2010**, *115*, D23104. [CrossRef]
43. Sun, B.; Reale, A.; Schroeder, S.; Seidel, D.J.; Ballish, B. Toward improved corrections for radiation-induced biases in radiosonde temperature observations. *J. Geophys. Res. Atmos.* **2013**, *118*, 4231–4243. [CrossRef]
44. Ingleby, B. *An Assessment of Different Radiosonde Types 2015/2016*; ECMWF Technical Memoranda: Reading, UK, 2017; p. 807.
45. George, J.J. *Weather Forecasting for Aeronautics*; Academic Press: Waltham, MA, USA, 1960.
46. Burpee, R.W. Peninsula-scale convergence in the south Florida sea breeze. *Mon. Weather Rev.* **1979**, *107*, 852–860. [CrossRef]
47. Li, J.E.; Liu, C.C.; Chao, C.C.; Yeh, N.C. A preliminary analysis in potential environment of convections in the summer by using satellite data. *Atmos. Sci.* **2005**, *33*, 189–214.
48. Shrestha, D.; Singh, P.; Nakamura, K. Spatiotemporal variation of rainfall over the central Himalayan region revealed by TRMM Precipitation Radar. *J. Geophys. Res. Atmos.* **2012**, *117*, D22106. [CrossRef]
49. Chen, Y.; Sharma, S.; Zhou, X.; Yang, K.; Li, X.; Niu, X.; Hu, X.; Khadka, N. Spatial performance of multiple reanalysis precipitation datasets on the southern slope of central Himalaya. *Atmos. Res.* **2021**, *250*, 105365. [CrossRef]
50. Sharma, S.; Khadka, N.; Nepal, B.; Ghimire, S.K.; Luintel, N.; Hamal, K. Elevation Dependency of Precipitation over Southern Slope of Central Himalaya. *Jalawaayu* **2021**, *1*, 1–14. [CrossRef]
51. Yeh, N.C.; Chuang, Y.C.; Peng, H.S.; Hsu, K.L. Bias adjustment of satellite precipitation estimation using ground-based observation: Mei-Yu front case studies in Taiwan. *Asia Pac. J. Atmos. Sci.* **2020**, *56*, 485–492. [CrossRef]

atmosphere MDPI

Article

Assessment of Quarterly, Semiannual and Annual Models to Forecast Monthly Rainfall Anomalies: The Case of a Tropical Andean Basin

Angel Vázquez-Patiño [1,2,3], Mario Peña [4] and Alex Avilés [5,6,*]

1 Departamento de Ingeniería Civil, Universidad de Cuenca, Cuenca 010207, Ecuador; angel.vazquezp@ucuenca.edu.ec
2 Facultad de Ingeniería, Universidad de Cuenca, Cuenca 010207, Ecuador
3 Facultad de Arquitectura, Universidad de Cuenca, Cuenca 010203, Ecuador
4 Departamento de Química Aplicada y Sistemas de Producción, Facultad de Ciencias Químicas, Universidad de Cuenca, Cuenca 010203, Ecuador; mario.penao@ucuenca.edu.ec
5 Grupo de Evaluación de Riesgos Ambientales en Sistemas de Producción y Servicios (RISKEN), Eco Campus Balzay, Universidad de Cuenca, Cuenca 0101168, Ecuador
6 Carrera de Ingeniería Ambiental, Facultad de Ciencias Químicas, Eco Campus Balzay, Universidad de Cuenca, Cuenca 0101168, Ecuador
* Correspondence: alex.aviles@ucuenca.edu.ec

Abstract: Rainfall forecasting is essential to manage water resources and make timely decisions to mitigate adverse effects related to unexpected events. Considering that rainfall drivers can change throughout the year, one approach to implementing forecasting models is to generate a model for each period in which the mechanisms are nearly constant, e.g., each season. The chosen predictors can be more robust, and the resulting models perform better. However, it has not been assessed whether the approach mentioned above offers better performance in forecasting models from a practical perspective in the tropical Andean region. This study evaluated quarterly, semiannual and annual models for forecasting monthly rainfall anomalies in an Andean basin to show if models implemented for fewer months outperform accuracy; all the models forecast rainfall on a monthly scale. Lagged rainfall and climate indices were used as predictors. Support vector regression (SVR) was used to select the most relevant predictors and train the models. The results showed a better performance of the annual models mainly due to the greater amount of data that SVR can take advantage of in training. If the training of the annual models had less data, the quarterly models would be the best. In conclusion, the annual models show greater accuracy in the rainfall forecast.

Keywords: forecasting; SVR; SVM; rainfall; anomalies; large-scale climate indices; Andean river basin

Citation: Vázquez-Patiño, A.; Peña, M.; Avilés, A. Assessment of Quarterly, Semiannual and Annual Models to Forecast Monthly Rainfall Anomalies: The Case of a Tropical Andean Basin. *Atmosphere* **2022**, *13*, 895. https://doi.org/10.3390/atmos13060895

Academic Editors: Zuohao Cao, Huaqing Cai and Xiaofan Li

Received: 26 April 2022
Accepted: 26 May 2022
Published: 31 May 2022

Publisher's Note: MDPI stays neutral with regard to jurisdictional claims in published maps and institutional affiliations.

1. Introduction

Rain is a phenomenon that significantly conditions human activity. Knowing its dynamics and forecasting its behavior is essential to optimize water use, for example, in human consumption, hydroelectric generation, agriculture [1], and industry. On the other hand, anticipating extreme rainfall events helps to take measures to mitigate possible adverse effects (e.g., landslides, floods, and droughts). Examples of such extreme events are the droughts in the Southwest US [2], the São Francisco river basin (Brazil) [3], the northeast region of Brazil [4,5], and over Brazil [6]. Additionally, rain is an essential atmospheric variable to characterize the climate [7]. Therefore, unveiling the drivers related to this hydrologic process is essential to understanding possible changes in its dynamics under low-frequency natural climate variability [8] or climatic change [9].

Different models allow us to anticipate rain behavior. There are different methods that can be used to make predictions. Dynamic models are physically consistent [10], but these have a tremendous computational burden. Instead, statistical methods are widely used to identify the main modes of climate variability at different spatial and temporal scales [11]. In addition, some models use a combination of the two approaches [12]. Weather and seasonal forecasting models or decadal prediction models could be developed depending on the forecast horizon. From an operational point of view, and in terms of relevance for short- and medium-term decision-makers, intraseasonal and seasonal forecasting models are the most important. They consider forecasts from two months to a little over a year [13,14] and help in tasks such as those listed above. Moreover, different data-based approaches are used when constructing forecasting models such as those based on autoregressive models (e.g., [15]), empirical models (e.g., [14]) or others that are more robust, such as those based on machine learning (ML) techniques (e.g., [16,17]) or network science (e.g., [18,19]). Likewise, models can be constructed based on different candidate predictors such as exogenous variables (e.g., [17,20]) or climate indices (e.g., [21,22]).

Seasonality is a remarkable feature in nature, and different precursors defining rainfall also have temporal variability [23] (change in mechanisms [24]). Moreover, heterogeneous rainfall magnitudes through the year could negatively affect the performance of a single model in forecasting the rainfall of each month of a year. Depending on the algorithm/model used to train the forecasting model, scaling and/or standardization are frequently used preprocessing methods [25]. The rainfall anomalies are commonly computed, which helps eliminate seasonality and allows unexpected values beyond trivial behavior (mainly affected by seasonal solar irradiance) to be forecasted. However, the changing influence of predictors over the target variable (rainfall) is an implicit feature that is still present. So, an alternative is to construct models for each semester, season or even month of the year [24] (e.g., [26]). Following the premise of changing mechanisms throughout the year, such an alternative aims to learn the relationships between predictors and rainfall in different seasons or periods of wet or dry behavior to have better-performing models. This is because the predictors may provide information that makes the models more robust. However, as far as the authors know, an assessment to determine if subannual models perform better has not been carried out, and less is known about mountain zones such as the Andes where complex processes dominate the rainfall behavior throughout the year [27–29].

This study aimed to determine if the performance of anomaly rainfall forecasting models improved as the models were developed for each quarter, each semester or the whole year with a horizon of one year. These models always aim to forecast monthly rainfall and differ in the months for which they can make the forecast. For example, one of the quarterly models allows one to forecast the rainfall of December, January, and February (there are four quarterly models in total), one of the semiannual models forecasts the rainfall from November to April (there are two semiannual models), and the annual model forecasts the rainfall of any month of the year. In addition, the influence of predictors based on large-scale climatic factors on these improvements was analyzed. For this, three sets of anomaly rainfall forecasting models were trained through the support vector regression algorithm. The first set constituted the four quarterly models to forecast rainfall each season. The second set contained the two semiannual models, and the third referred to the annual model. Each model used an independent subset of lagged climate indices and anomaly rainfall signals as predictors. Such subsets were chosen by employing the sequential feature selection algorithm. Finally, the models were assessed by utilizing seven evaluation metrics.

2. Materials and Methods

2.1. Study Zone

The Machángara river basin is located northeast of Cuenca, the capital city of Azuay, in the Andes mountain range of southern Ecuador (Figure 1). The basin area is approximately 325 km^2 and has a high altitudinal gradient extending from 2440 to 4420 m a.s.l. Natural areas form the upper part of this basin, agricultural activities mainly occupy the middle part, with small-urbanized patches, and urbanized sectors characterize the lower part [30].

The rainfall varies from 856 to 1309 mm, while the temperature ranges from 8.1 to 14 °C [31]. The Pacific Ocean, the Andes range and the Amazon basin mainly influence the climate of this region [29,32,33].

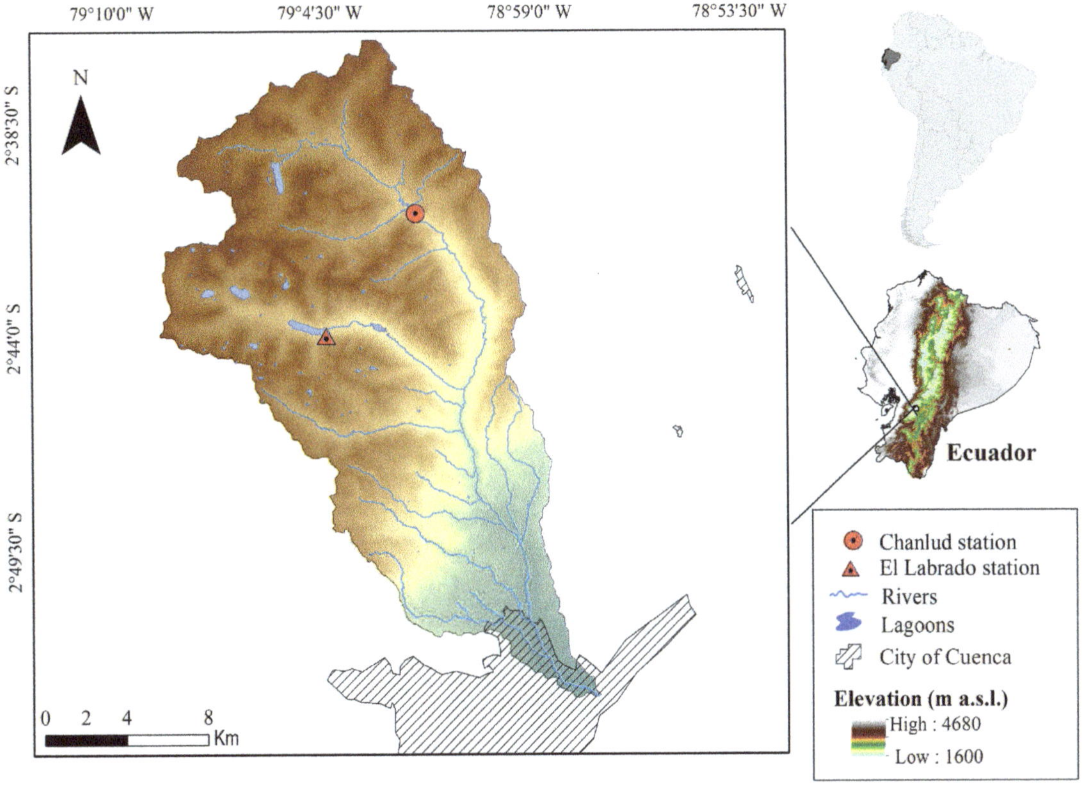

Figure 1. Location of the El Labrado and Chanlúd stations in the Machángara basin in the context of Ecuador and South America.

The integrated management of natural resources in the Machángara basin guarantees the provision of essential services, for example, water for human consumption for more than 390,000 inhabitants of Cuenca (≅60% of the population), irrigation for more than 3900 users, the generation of 39.5 MW of hydroelectricity (the first source of electric energy in Ecuador [34]), and the provision of water for various industries in the area. In the highest part of the basin, two representative stations were selected for the study, namely El Labrado and Chanlúd, which are at approximately 3335 and 3485 m a.s.l., respectively. The two stations are located in the dams that bear the same names [35] and are greatly important in the national hydroelectric generation system.

2.2. Data

The daily rainfall data of El Labrado and Chanlúd go back to 1964 and 1981, respectively. This study used monthly rainfall data from 1981 to 2021 (41 y). Therefore, the daily rainfall corresponding to each month of the period 1981–2021 was added to generate data on a monthly scale. Figure 2a shows the monthly rainfall data of El Labrado and Chanlúd in the study period. Rainfall shows a marked bimodal seasonality, which is shown in Figure 2b. A peak of heavy rain is present in April, while the driest month is August.

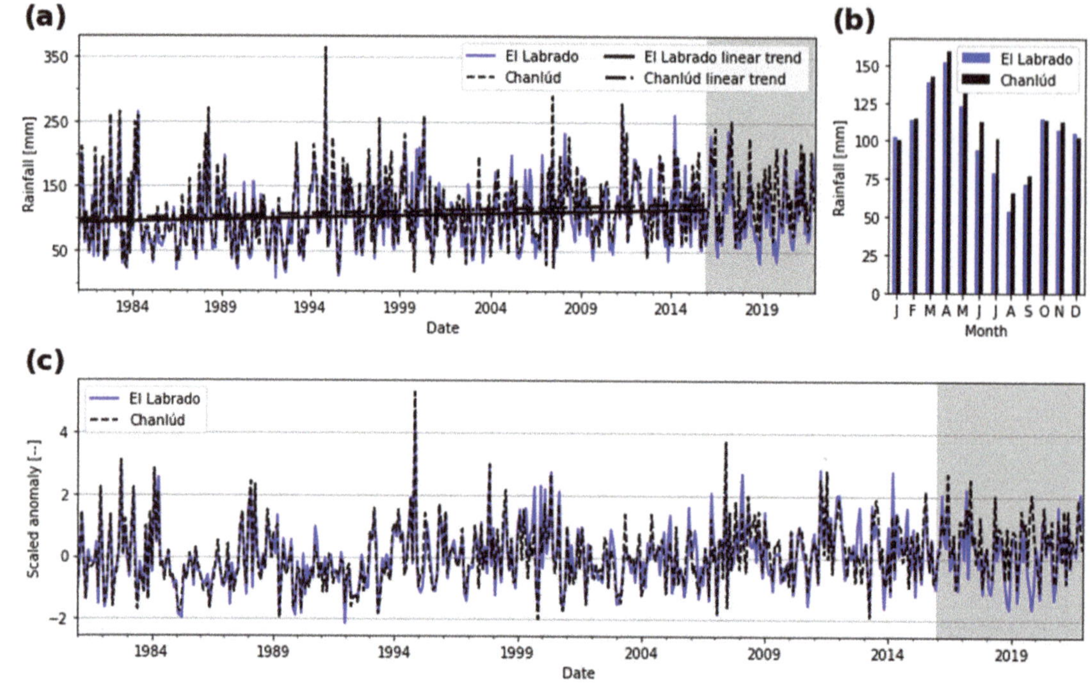

Figure 2. Rainfall in the study stations: (**a**) observations from 1981 to 2021, linear trends in the period 1981–2015 and the testing period 2016–2021 with the gray background; (**b**) seasonality in the period 1981–2015; (**c**) scaled anomalies based on the seasonality shown in (**b**).

The target variable of the forecasting models is the rainfall anomalies. Most of the predictors of such models are based on climate indices. Table 1 shows the 38 climate indices used in the study, which have a monthly resolution. Thirty-four were downloaded from the National Oceanic and Atmospheric Administration (NOAA) (https://psl.noaa.gov/data/climateindices/list/, accessed on 21 January 2022), and the sources of the rest are indicated in the table footer.

Table 1. Climate indices used in the study.

Zone	Indices
Global	Global Mean Land-Ocean Temperature (GMSST) [36].
North Hemisphere	Pacific/North American Index (PNA), East Pacific/North Pacific Oscillation (EP/NP) [37], West Pacific Index (WP), North Atlantic Oscillation (NAO) [38], Jones NAO (J.NAO *) [39], East Atlantic (EA) and Arctic Oscillation (AO) [40].
South Hemisphere	Antarctic Oscillation (AAO) [41].
Northern Pacific	North Pacific pattern (NP) [42] and Pacific Decadal Oscillation (PDO) [43].
Tropical Pacific	Pacific Warmpool Area Average (PacWarm), Extreme Eastern Tropical Pacific sea surface temperature (SST) (Niño 1+2) and its anomaly values (Niño 1+2.A), Eastern Tropical Pacific SST (Niño 3) and its anomaly values (Niño 3.A), East Central Tropical Pacific SST (Niño 3.4) and its anomaly values (Niño 3.4.A), Central Tropical Pacific SST (Niño 4) and its anomaly values (Niño 4.A), Trans-Niño Index (TNI), Southern Oscillation Index (SOI), Bivariate ENSO Timeseries (BEST), Bi-monthly Multivariate El Niño/Southern Oscillation (ENSO) index version 2 (MEIv2) and El Niño Modoki Index (EMI) [44].

Table 1. *Cont.*

Zone	Indices
Pacific	Tripole Index for the Interdecadal Pacific Oscillation (TPI.IPO) and Northern Oscillation Index (NOI) [45].
Atlantic and Eastern North Pacific	Western Hemisphere Warm Pool (WHWP) [46].
North Atlantic	Atlantic Multidecadal Oscillation UnSmoothed (AMO.US †) [47].
Tropical Atlantic	Caribbean SST Index (CAR) [48], Tropical Northern Atlantic Index (TNA) [49], Tropical Southern Atlantic Index (TSA) [49] and Atlantic Meridional Mode SST index (AMM) [50].
Tropic	Quasi-Biennial Oscillation (QBO), ENSO precipitation index (ESPI), Western Indian Ocean Dipole (IOD.W ‡), Eastern Indian Ocean Dipole (IOD.E ‡) and Dipole Mode Index (DMI ‡) [51].

* Downloaded from https://crudata.uea.ac.uk/cru/data/nao/; † downloaded from https://psl.noaa.gov/data/timeseries/AMO/; ‡ downloaded from https://psl.noaa.gov/gcos_wgsp/Timeseries/DMI/. Accessed on 24 January 2022.

Both rainfall and climate indices constitute the raw monthly dataset. The dataset was split into training and testing subsets. The former spanned from 1981 to 2015 (35 y) and the latter from 2016 to 2021 (6 y). Figure 2a shows the training subset period in a white background, while the testing period is shaded. The training subset was used to compute rainfall anomalies, standardize the dataset, generate and select predictors, and train the forecasting models. The testing subset was only used to evaluate the performance of the models. It is worth noting that the standardization of the testing subset was based on the parameters found in the training subset. The latter ensures an adequate evaluation since it simulates a scenario in which nothing is known beyond the data available for model training.

2.3. Settings and Workflow

This subsection explains the workflow followed in the study in a general manner. The following subsections describe details about data or methods in each step. So, Scheme 1 shows the workflow to assess the quarterly, semiannual and annual models to forecast monthly rainfall anomalies.

The first step shown in Scheme 1 is the deseasonalization of rainfall. The monthly rainfall climatology (Figure 2b) was computed based on the training subset period (1981–2015). Then, the climatology was subtracted from the raw rainfall signal. The second step is the standardization of the rainfall anomalies and climate indices data by removing the mean and scaling to unit variance. The Standard Scaler from the Scikit-learn library [52] was used in this step. As an example, Figure 2c shows the scaled and standardized rainfall anomalies. The shaded period in Figure 2c was not used to compute the parameters in the scaling and standardization. The third step is the generation of the candidate predictor set. For this, lagged versions of the time series of the rainfall anomalies and climate indices were generated up to a maximum τ lag (τ_{max}) which was chosen through an autocorrelation analysis. These signal delays are operationally essential to generate predictors with past information that serve the current forecast, anticipating the decision making.

The fourth step is the selection of relevant predictors for each forecasting model. For each forecasting model, different subsets of relevant predictors were selected (dashed squares in step 4) through the sequential forward selection (SFS) algorithm. There were seven different subsets for four quarterly models, two semiannual models, and one annual model. The quarterly models forecast anomaly rainfall of months belonging to each of the four seasons (e.g., the DJF model predicts rainfall in Dec–Feb), the semiannual models forecast rainfall of months belonging to each of the two semesters (e.g., the NDJFMA model forecasts rainfall in November–April), and the annual model forecasts rainfall of any month of the year. In any case, the forecasting horizon was one year. Those seven models were trained in the fifth step shown in Scheme 1. The forecasting models were based on the support vector regression (SVR) learning algorithm [53].

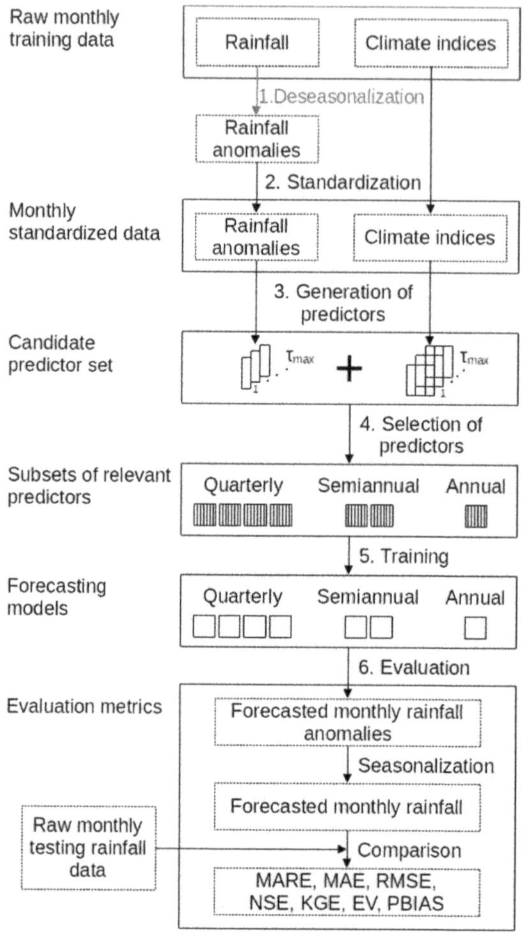

Scheme 1. Workflow followed in the study.

The last step is the evaluation. Each year of rainfall was forecasted independently for the testing subset period (2016–2021). With the entire testing period, a qualitative comparison was first made. Then, the evaluation was performed with seven evaluation metrics (the models had a horizon of one year). The comparison was performed with the raw testing rainfall data, so the results of the models were firstly converted (seasonalization) to the original scale (based on the parameters of the training subset).

2.4. *Maximum τ lag (τ_{max})*

In order to choose the $\tau_{\max}$ for the generation of candidate predictors, the autocorrelation of the raw rainfall signals was used. Each station's autocorrelation function (ACF) was plotted with 95% confidence intervals employing the statsmodels library [54]. These confidence intervals suggest that the correlation values within them are likely a statistical fluke. The standard deviation computation for the confidence intervals was performed according to Bartlett's formula [55,56].

The 38 climate indices from which the candidate predictors were derived had a particular $\tau_{\max}$, after which, the correlation with rainfall (target variable) was no longer significant. Beyond an analysis of autocorrelations, an exhaustive analysis of lagged cross-correlations (such as in [57]) would allow one to obtain a $\tau_{\max}$ for each index concerning the target variable and to possibly achieve higher-performance forecasting models. In addition, the above should have been taken into account for both El Labrado and Chanlúd rainfall. This would have led to 39 different τ_{max} for each station. However, in order not to divert the study from the objective pursued, only the autocorrelation of rainfall was taken into account. Once the autocorrelation graphs of each station were analyzed, a single reasonable τ_{max} value was taken through the study.

2.5. Generation of Candidate Predictor Sets

Four quarterly models were trained, one by each season, i.e., DJF, MAM, JJA and SON, two semiannual models related to semesters November–April (NDJFMA) and May–October (MJJASO) and one annual model (J-D). These models were schematized as blank squares in step five in Scheme 1. The forecasting horizon was one year, and as an example, if one wanted to forecast rainfall for 2016, January rainfall would be forecasted with the DJF model. Such a value would be immediately used as a possible predictor to forecast the February rainfall with the DJF model. The MAM model would be used to forecast March rainfall, and the same for April and May rainfall. The exact process would be used to forecast the rest of the months. Thus, the following lags were used for each forecasting model to generate the candidate predictor sets.

Candidate predictor set for quarterly models:

- DJF: 13 to τ_{max}.
- MAM: 6 to τ_{max}.
- JJA: 9 to τ_{max}.
- SON: 12 to τ_{max}.

Candidate predictor set for semiannual models:

- NDJFMA: 13 to τ_{max}.
- MJJASO: 11 to τ_{max}.

Candidate predictor set for annual models:

- J-D: 13 to τ_{max}.

The minimum limit of each interval shown above allows one to leverage as much information as possible. For instance, for forecasting March–May rainfall using the MAM model, information with a minimum lag of six means that information from November of the previous year could be used.

2.6. Sequential Forward Selection (SFS) of Predictors

The sequential forward selection (SFS) is a greedy approach to selecting the best new predictor iteratively from the candidate predictor set to aggregate to a subset of selected predictors [58]. The algorithm initially finds one predictor (the first) that maximizes a cross-validated score when a learning algorithm is trained on this single predictor. After the first (best) predictor is selected, the algorithm finds the second predictor that maximizes the score of the learning algorithm when it is trained on these two single predictors. The process is repeated by adding new predictors to the subset of selected predictors in each iteration. The optimum subset of relevant predictors is the one that gives the best cross-validated performance. There are different implementations of the SFS algorithm, and here, the one from the MLxtend library [59] was used. Moreover, this study used the support vector regression (SVR) learning algorithm to compute the score based on the selected predictor subsets. The SVR implementation of the Scikit-learn library [52] was used with the default hyperparameters. The cross-validated score was obtained through 5-fold cross-validation.

2.7. Support Vector Regression (SVR)

The support vector regression (SVR) model was proposed by Vapnik [60] and is a suitable model for linear and nonlinear regression. SVR is based on elements of the support vector machine (SVM), where support vectors are the closest points toward the generated hyperplane in a high-dimensional feature space [53]. As in most machine learning models, the training data are divided into two subsets: the training and validation sets [61]. The SVR model maps the training data to a high-dimensional feature space using a kernel. The radial basis function (RBF) kernel was used in this study. The hyperparameters are then optimized (i.e., model training) by fitting the model to the training data in that feature space. The formal definition of the SVR model is as follows.

Given $\{x_i, y_i\}$ denoted as a characteristic vector of sample data with $i = 1, 2, \ldots, m$ samples, where $x_i \epsilon \mathbb{R}^n$, n is the number of predictors, and $y_i \epsilon \mathbb{R}$ is the target variable (rainfall anomalies). The SVM regression estimation function is defined as

$$f(x) = W^T \phi(x) + b \tag{1}$$

where W^T is the weights matrix of the independent function, $\phi(x)$ is the nonlinear (kernel) mapping function, and b is the intercept. Then, W^T and b can be obtained by minimizing the equation

$$Min : \frac{1}{2}||W||^2 + \frac{C}{m}\sum_{i=1}^{m} R_\varepsilon[y_i, f(x_i)] \tag{2}$$

where $||W||^2$ is known as regularized term; C is the penalty parameter; and R_ε is the insensitive loss function (error control function) of the margin of tolerance ε.

The SVR model generally requires a small sample size for training, has a simple statistical structure and performs better than complex models, e.g., artificial neural networks.

In the model training, the training data were not only divided into a new training subset and a validation subset. Instead, the 5-fold cross-validation technique was used to gain generalization. A particular subset of predictors for each forecasting model was used (dashed squares of step 4 in Scheme 1). Moreover, the following set of values for each of the hyperparameters was used in the training process:

- $C = \{2^i \mid i \in \mathbb{Z}, -20 \le i \le 10\}$;
- $\varepsilon = \{2^i \mid i \in \mathbb{Z}, -20 \le i \le 10\}$;
- $\gamma = \left\{2^{-20+\frac{19}{30}i} \mid \mathrm{i} \in \mathbb{Z}, 0 \le \mathrm{i} \le 30\right\}$.

The γ hyperparameter corresponds to the RBF kernel used and relates to the inverse of the radius of influence of registers selected by the model as support vectors. The number of models trained until the best-performing one is found for each forecasting model is 31,713 × 5 = 158,565 (31,713 hyperparameter combinations, and 5 corresponds to the 5-fold cross-validation).

2.8. Evaluation Metrics

Given the raw rainfall time series $y = \{y_1, y_2, \ldots, y_m\}$ and the seasonalized forecasted rainfall $\hat{y} = \{\hat{y}_1, \hat{y}_2, \ldots, \hat{y}_m\}$, the forecasting models were evaluated with the following seven metrics. These metrics are commonly used for evaluating forecasting and prediction models [62–65].

Mean Absolute Relative Error (MARE). The MARE measures how much error exists in the forecasted rainfall relative to the observed values in absolute terms. It is computed by

$$\text{MARE} = \frac{\sum_{i=1}^{m}|y_i - \hat{y}_i|}{\sum_{i=1}^{m} y_i} \tag{3}$$

The MARE is independent of the time series scale, and its value ranges from 0 to ∞, with 0 being the measure of a perfect forecast.

Mean Absolute Error (MAE). The MAE represents the average of the absolute difference between the forecasted values and the observations. It measures the average of the residuals regardless of their sign. The MAE is defined as

$$\text{MAE} = \frac{1}{m}\sum_{i=1}^{m}|y_i - \hat{y}_i| \tag{4}$$

This metric scale depends on the scale of rainfall, and its value ranges from 0 to ∞, with 0 being the best value.

Root Mean Square Error (RMSE). The RMSE is the square root of the average of the squared difference between the forecasted values and the observations. The RMSE is defined as

$$\mathrm{RMSE} = \sqrt{\frac{1}{m}\sum_{i=1}^{m}(y_i - \hat{y}_i)^2} \quad (5)$$

The RMSE values are dependent on the time-series scale, and its value ranges from 0 to ∞, with 0 being the measure for a perfect forecast.

Nash–Sutcliffe Efficiency (NSE). The NSE [66] is widely used to evaluate the performance of hydrological models. Although the NSE is susceptible to outliers because it takes a sum over the squared values of the differences between the forecasted values and the observations, it is even better than other metrics, such as the coefficient of determination. The NSE is defined as

$$\mathrm{NSE} = 1 - \frac{\sum_{i=1}^{m}(y_i - \hat{y}_i)^2}{\sum_{i=1}^{m}\left(y_i - \bar{y}\right)^2} \quad (6)$$

where $\bar{y}$ is the average of the rainfall time series (observations). The scale of this metric is independent of the scale of the rainfall values. The values of this metric go from $-\infty$ to 1, with 1 meaning perfect forecasting, 0 meaning that the results are as good as always using $\bar{y}$ for the forecasting and negative values meaning arbitrarily bad results.

Kling–Gupta Efficiency (KGE). The KGE [67] is a robust performance measure based on three equally weighted components: variability, linear correlation, and bias ratio between forecasted and observed rainfall. The KGE is defined as

$$KGE = 1 - \sqrt{(\alpha - 1)^2 + (cc - 1)^2 + (\beta - 1)^2} \quad (7)$$

where α is the variability (the ratio between the standard deviation of forecasted rainfall over the observed rainfall), cc is the linear correlation coefficient between forecasted and observed values, and β is the division between the average of the forecasted rainfall over the average of the observed rainfall.

The KGE is independent of the rainfall scale, and its value goes from $-\infty$ to 1. The higher the value, the better the forecast.

Explained Variance (EV). The EV measures the proportion of the variance of the residuals (differences between y_i and $\hat{y}_i$) and the rainfall variance. It is computed by

$$\mathrm{EV} = 1 - \frac{\sum_{i=1}^{m}\left[(y_i - \hat{y}_i) - \frac{1}{m}\sum_{j=1}^{m}(y_j - \hat{y}_j)\right]^2}{\sum_{i=1}^{m}\left(y_i - \bar{y}\right)^2} \quad (8)$$

The EV is independent of the time series scale, and its value ranges from $-\infty$ to 1, with 1 being the optimum value and negative values indicating arbitrarily bad forecasting results. EV = 0 indicates that the model is as good as using any fixed value for the forecast.

Percent Bias (PBIAS). The PBIAS determines whether there is a tendency in the values forecasted by the model (i.e., if these are higher or lower than the observed values). A positive PBIAS indicates that the model overestimates the forecasted variable, while a negative value indicates that the variable is underestimated. The optimal value is a PBIAS equal to zero. This metric is defined as

$$\mathrm{PBIAS} = 100 \times \frac{\sum_{i=1}^{m}(\hat{y}_i - y_i)}{\sum_{i=1}^{m} y_i} \quad (9)$$

This metric is independent of the rainfall scale, and the closer the value of |PBIAS| to 0, the better the results, with 0 being the optimum value. |PBIAS| values greater than 100 indicate arbitrarily bad results.

3. Results

3.1. τ_{max} for Generating the Candidate Predictors

Figure 3 shows the ACFs of El Labrado and Chanlúd rainfall. The autocorrelation demonstrated a high seasonal rainfall signal with statistical significance until 37 months in El Labrado and 49 months in Chanlúd. A lag of 37 months was then taken as the τ_{max} to create the candidate predictor set. Since this study concentrates on comparing models differing in the number of months used in the implementation, the same number of 37 lags was taken as the maximum to create the candidate predictors based on the 38 climate indices (Table 1).

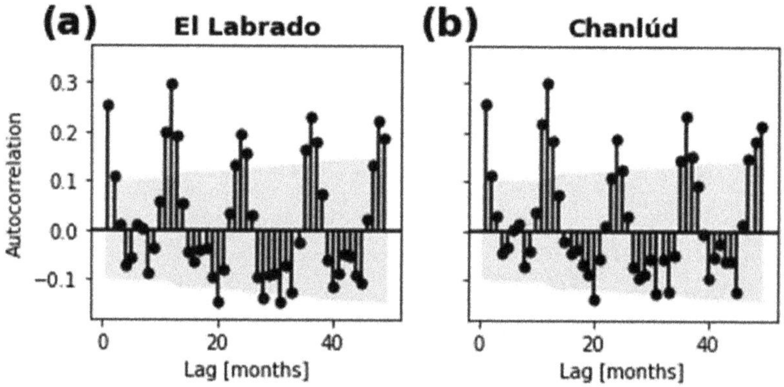

Figure 3. Autocorrelation of rainfall signals and 95% confidence intervals in gray: (**a**) El Labrado; (**b**) Chanlúd.

3.2. Optimum Number of Predictors

Figure 4 shows the results of selecting the optimum number of predictors through the SFS approach. Each column corresponds to the different model sets, i.e., quarterly, semiannual and annual models. Each row corresponds to the models for El Labrado and Chanlúd. The performance behavior of the models showed a similar tendency. The optimum number of predictors was around 81 and 68 for El Labrado and Chanlúd, respectively. The optimum number of predictors for quarterly, semiannual and annual models for El Labrado were around 93, 73 and 55, respectively. Meanwhile, for Chanlúd, the optimum numbers were around 69, 49 and 105. However, the performance of the annual model for Chanlúd (Figure 4f) had a lower variance in the approximated interval of 35–105 predictors. Thus, the greater the number of months that are considered in the models with different periods, the fewer the predictors that are needed to get the best performance. A possible explanation for this behavior is the greater number of records (instances) that the models had in the training stage of the SFS as they were built for more months (e.g., annual models). The annual models used 100% of the available records in the selection of predictors (383 of the 420 because of the candidate predictor generation with 37 lags), the semiannual models used 50% of such records (November–April: 191, May–October: 192), and the quarterly models used 25% (DJF: 95, MAM: 96, JJA: 96, SON: 96). The SFS used the SVR learning algorithm with the default hyperparameters, so varying the number of records changed the amount of relevant information for the forecasting. From a purely predictive point of view, models for more months achieve better performance with fewer predictors that have more relevant information to achieve better generalization.

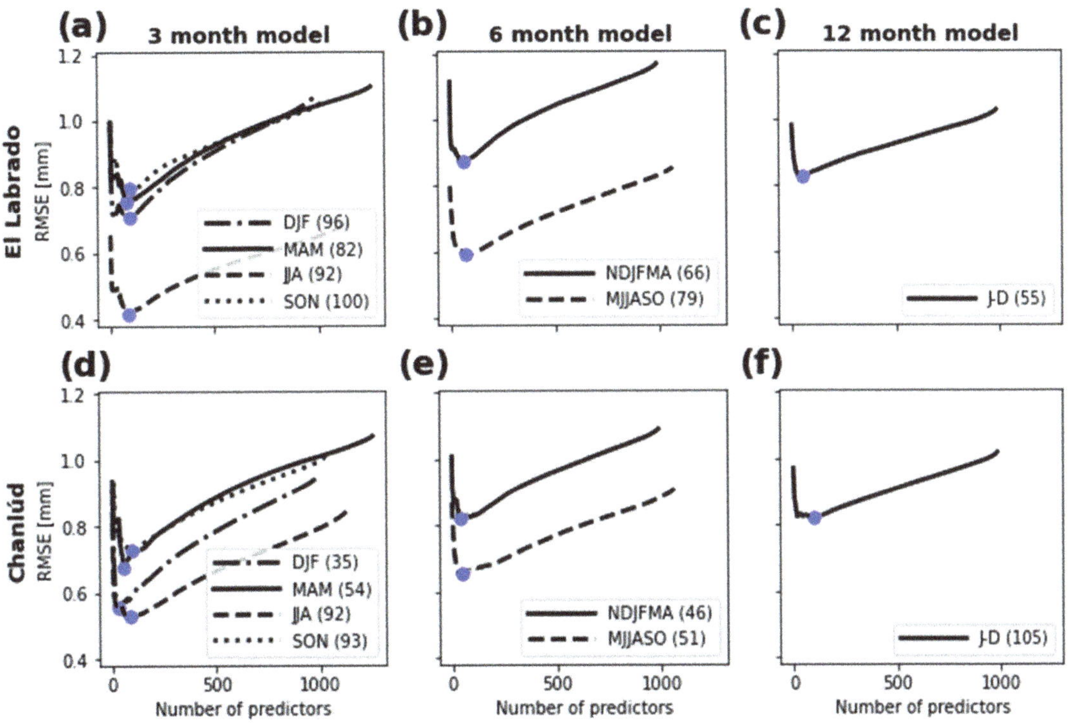

Figure 4. Sequential forward selection results: (**a**–**c**) models for El Labrado; (**d**–**f**) models for Chanlúd; (**a**,**d**) quarterly models; (**b**,**e**) semiannual models; (**c**,**f**) annual models. The dots show the best cross-validation performance, and the numbers in parentheses indicate the cardinality of the predictor subset with such best performance.

3.3. Relevant Predictors

The optimum number of predictors allows light to be shed on the more prominent indices to predict rainfall anomalies in the stations of the study zone. First, quarterly models allowed indices influencing each season to be analyzed. The analysis was conducted using the times an index with different lags was chosen. This number is labeled the frequency in Figures 5–8. The mean number of lags of the different indices chosen are shown in intervals of up to 12 months, 24 months and more than 24 months.

Figure 5 shows the climate indices providing more information for the predictor selection stage in each season for El Labrado. The most prominent feature was the EP/NP climate index in all seasons. In all the seasons, the EP/NP mean lag was within the 12 months before the rainfall value that had to be forecasted. The NP index was the second most prominent feature present in the models for the four seasons. NP was the second most prominent in DJF and SON, third in MAM, and sixth in JJA. Like EP/NP, the NP mean lag was within 12 months before the forecasted rainfall value. The Niño 3 index was present within the seven most prominent indices in DJF and MAM. DJF is a season when the climate conditions of the ENSO regions in the Pacific are more important in learning about rainfall in the Ecuadorian Andes [68]. The results showed that information 12 months before the rainfall observation is also important in learning about rainfall anomalies. It is worth noting that the Niño 3 index is a mean value index and does not refer to anomalies.

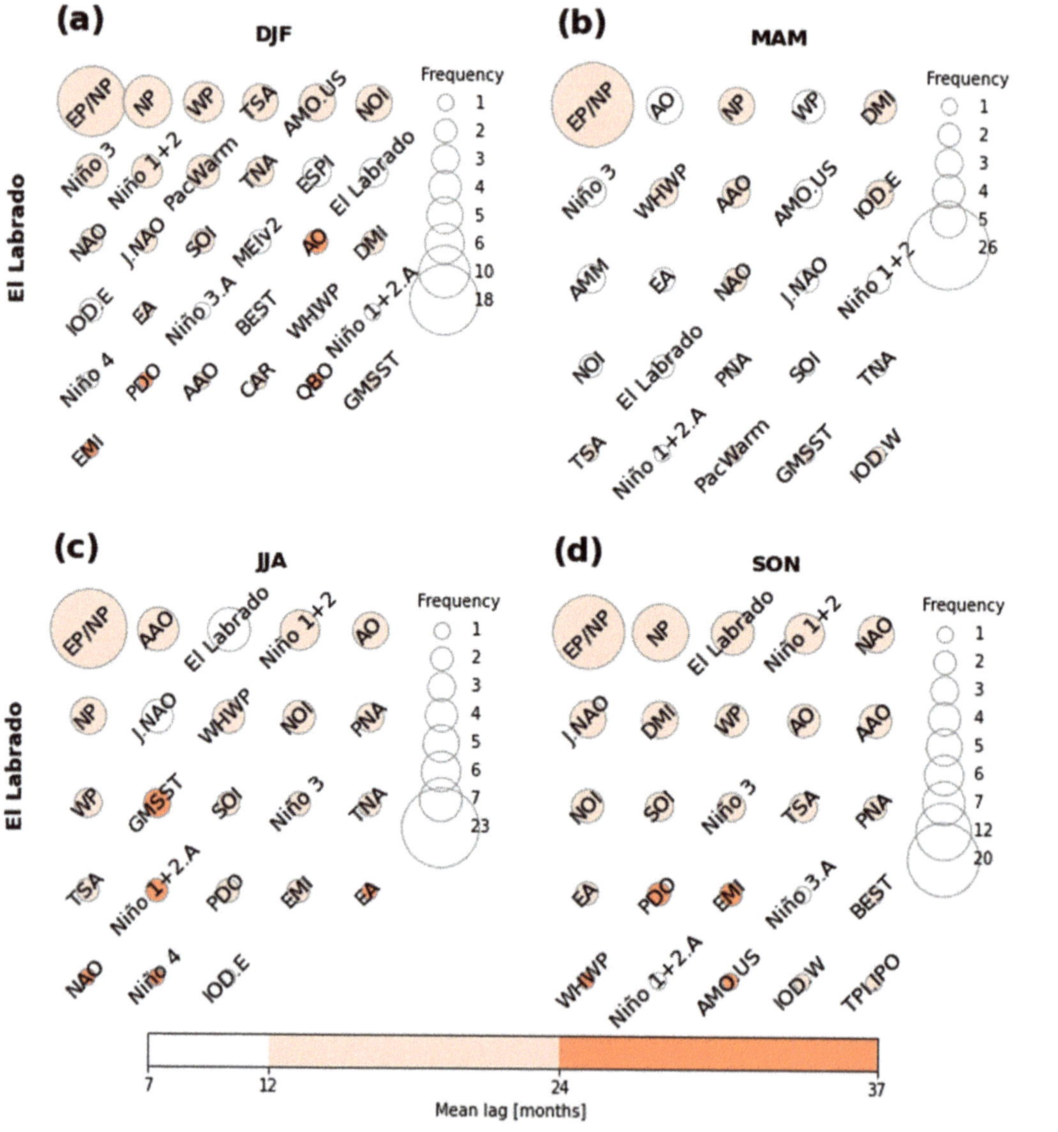

Figure 5. Predictors selected through SFS for each quarterly model for El Labrado (dots in Figure 4a): (**a**) quarter December–February; (**b**) quarter March–May; (**c**) quarter June–August; (**d**) quarter September–November. The size of each circle corresponds to the frequency with which a climate index (with different lags) appears as a predictor of the model. The color of each circle corresponds to the mean of the lags with which a climate index appears as a predictor of the model.

On the other hand, the same signal (El Labrado) and the Niño 1+2 index were prominent indices in JJA and SON models. El Labrado signal was the third most frequent index appearing in the models for JJA and SON. The Niño 1+2 index mean lag was 12 months before the rainfall value. Again, Niño 1+2 is not an anomaly index but provides information in selecting predictors.

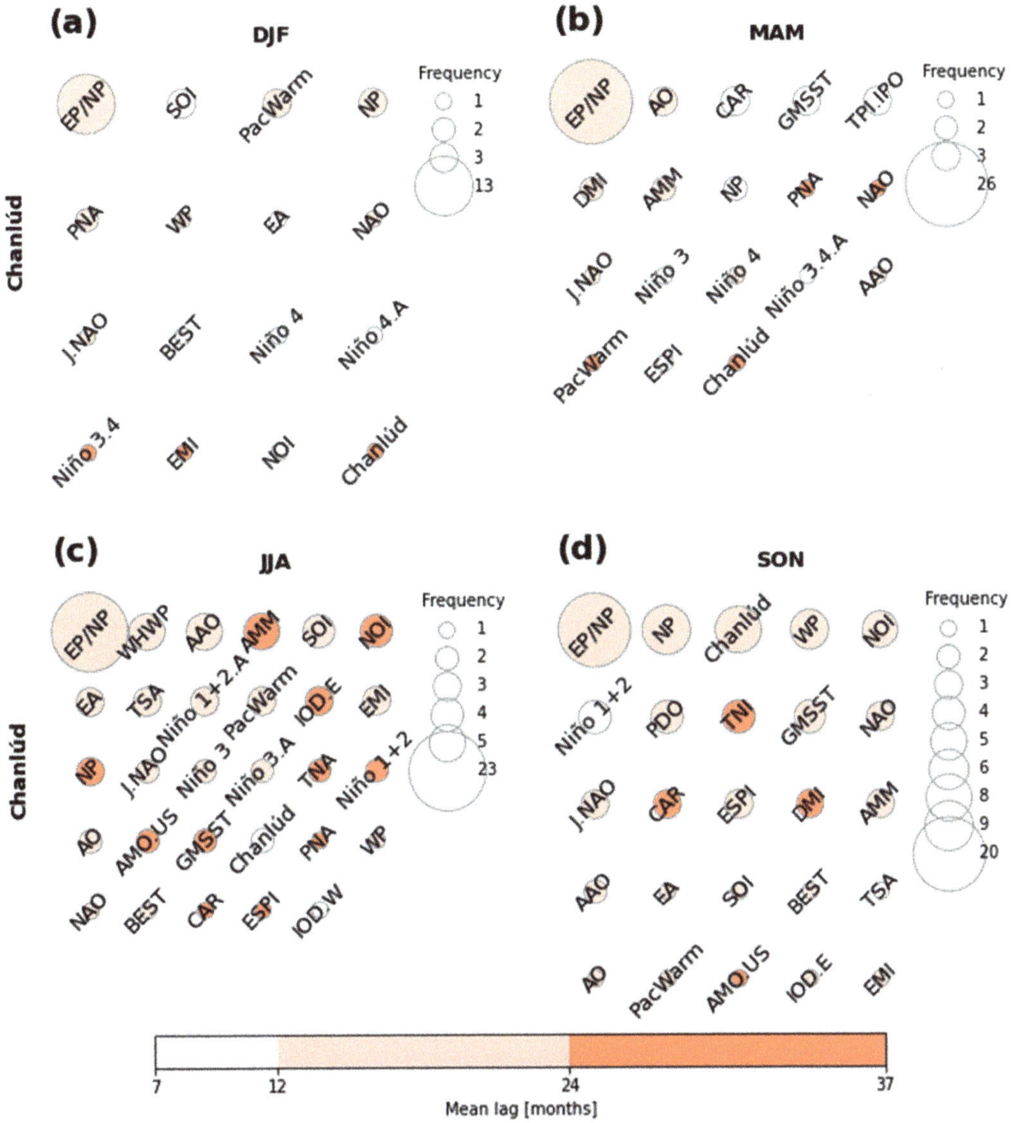

Figure 6. Predictors selected through SFS for each quarterly model for Chanlúd (dots in Figure 4d): (**a**) quarter December–February; (**b**) quarter March–May; (**c**) quarter June–August; (**d**) quarter September–November. The size of each circle corresponds to the frequency with which a climate index (with different lags) appears as a predictor of the model. The color of each circle corresponds to the mean of the lags with which a climate index appears as a predictor of the model.

GMSST was not prominent and even not present in SON. PDO was not prominent because it is a shallow frequency signal. PacWarm was only present in DJF and MAM. Niño 4 was present in DJF and JJA but with a low frequency; it was not present in MAM and SON. BEST and TPI.IPO were only present in SON but were not very prominent. CAR, QBO and ESPI were only present in DJF. TNI, MEIv2, Niño 4.A, and AMM were not present in any season.

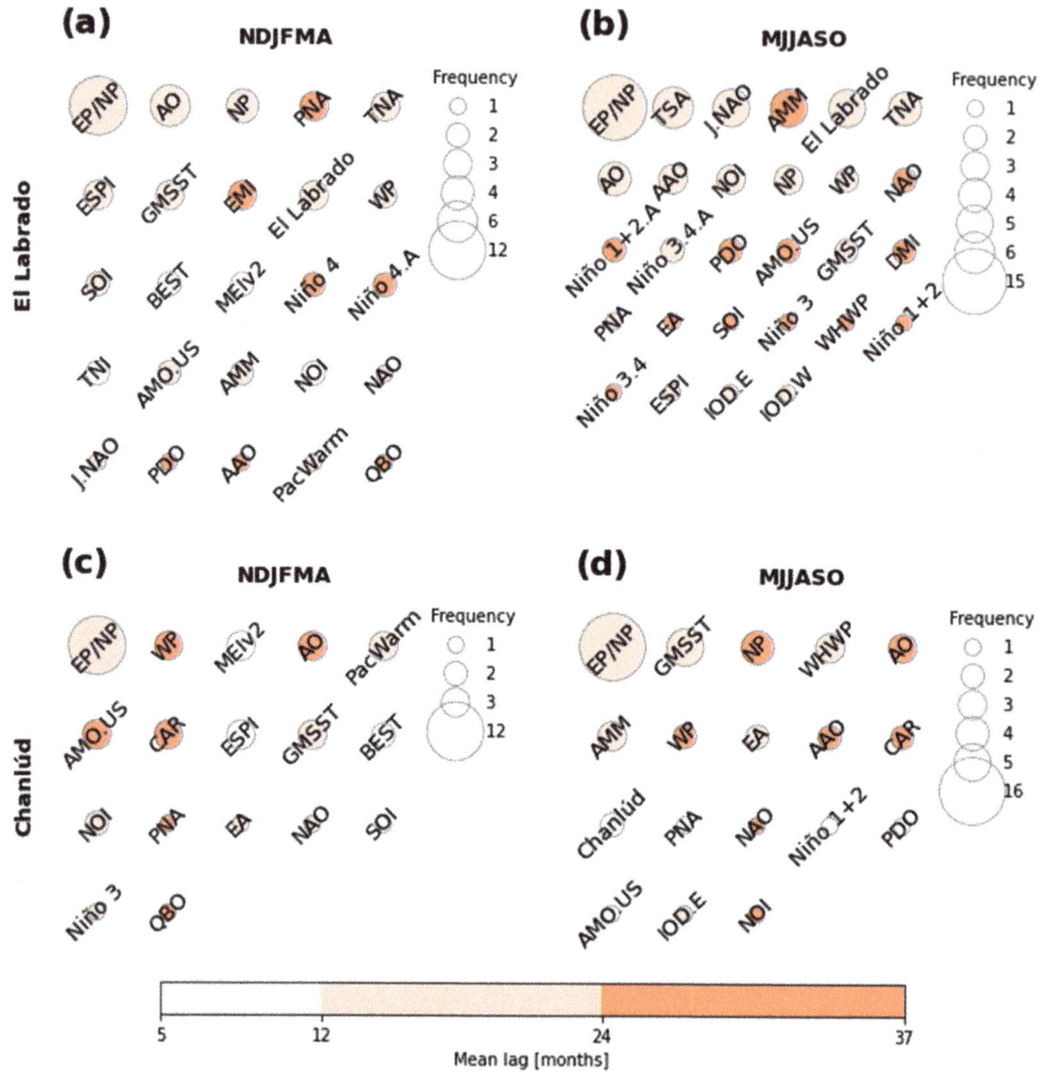

Figure 7. Predictors selected by means of SFS for each semiannual model. (**a**,**b**) Models for El Labrado (dots in Figure 4b); (**c**,**d**) models for Chanlúd (dots in Figure 4e); (**a**,**c**) semester November–April; (**b**,**d**) semester May–October. The size of each circle corresponds to the frequency with which a climate index (with different lags) appears as a predictor of the model. The color of each circle corresponds to the mean of the lags with which a climate index appears as a predictor of the model.

Figure 6 shows the climate indices providing more information in the predictor selection stage in each season for Chanlúd. Like for El Labrado, the most prominent climate index was EP/NP. Again, the mean lag of the predictors derived from EPO was within 12 months. Unlike El Labrado, NP was not found as a frequent index even though it was present in all seasons. Chanlúd appeared as the third most frequent index in SON. TNI and PDO were only present in SON. Niño 4 was only present in DJF. TPI.IPO was only present in MAM. MEIV2, QBO and AMM were not present in any season.

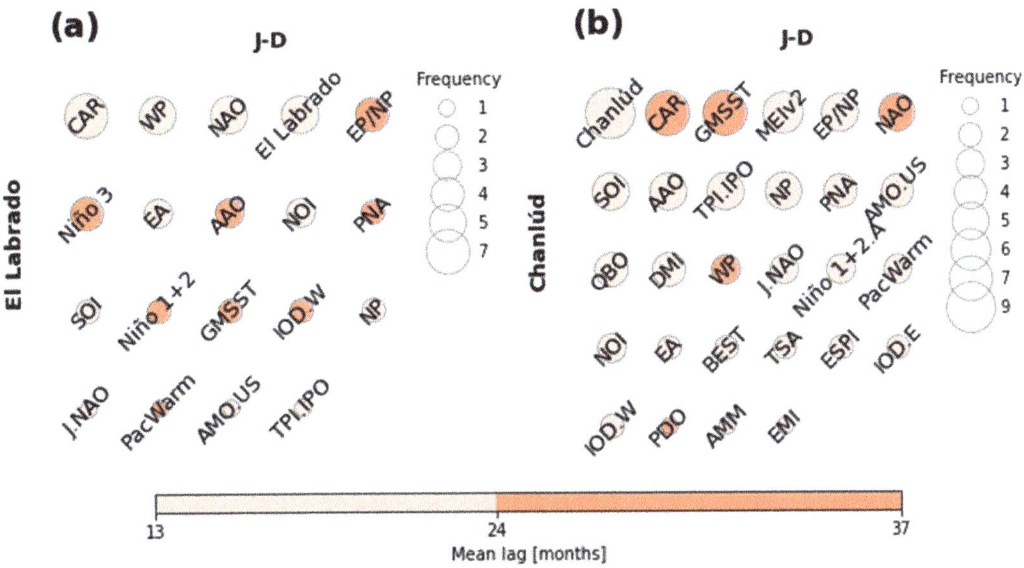

Figure 8. Predictors selected through SFS for each annual model: (**a**) model for El Labrado (dots in Figure 4c); (**b**) model for Chanlúd (dots in Figure 4f). The size of each circle corresponds to the frequency with which a climate index (with different lags) appears as a predictor of the model. The color of each circle corresponds to the mean of the lags with which a climate index appears as a predictor of the model.

Figure 7 shows the most prominent climate indices for El Labrado (Figure 7a,b) and Chanlúd (Figure 7c,d) in the semiannual models. As the number of months increased from quarterly to semiannual models, NP appeared with less relevance in El Labrado and even did not appear in Chanlúd for NDJFMA. EP/NP was the most frequent index in the subset of predictors that allowed the best performance in the models, both in El Labrado and Chanlúd. The mean lag for EP/NP was around 12 months before the forecasted month. Unlike quarterly models, in semiannual models, the higher frequency of EP/NP was 16 in MJJASO. For El Labrado, the same signal with a mean lag of 12 months appeared with more repetitions, especially in MJJASO. AO and TNA area climate indices appeared in both semiannual models for El Labrado.

Concerning El Labrado, PacWarm only appeared in NDJFMA, but its relevance was low. BEST, TNI and MEIV2 only appeared in NDJFMA with low frequency but a mean lag of 12 months. EA, Niño 1+2, Niño 1+2.A, Niño 3.4 and Niño 3.4.A were only present in MJJASO, but their frequency was low. Niño 3 only appeared in MJJASO, but its frequency was low, with a mean lag beyond 24 months. This is interesting since only Niño 4 and Niño 4.A corresponding to El Niño indices appeared in NDJFMA. EMI, Niño 4 and Niño 4.A only appeared in NDJFMA. QBO was only present in NDJFMA with the lowest frequency and a mean lag greater than 24 months. TSA, WHWP, IOD.W, IOD.E and DMI only appeared in MJJASO. Niño 3.A, TPI.IPO and CAR were not present in any semiannual models.

Concerning Chanlúd, NP, SOI, BEST, MEIv2, QBO, ESPI and PacWarm were only present in NDJFMA. Although NAO appeared in both semiannual models, J.NAO did not appear in any semiannual models. AAO, PDO, AMM, IOD.E and WHWP only appeared in MJJASO. Concerning the El Niño indices, Niño 1+2 appeared in MJJASO and Niño 3 in NDJFMA; the rest were not present in any model. TNI, EMI, TPI.IPO, TNA, TSA, IDO.W and DMI did not appear in any model.

CAR was present in both models for Chanlúd but not for El Labrado. QBO was present in NDJFMA for both El Labrado and Chanlúd but not for MJJASO.

Figure 8 shows the predictors selected for the annual models for El Labrado (Figure 8a) and Chanlúd (Figure 8b). Interestingly, the same signal of El Labrado and Chanlúd rainfall, with a mean lag within 12 months, was within the higher-frequency indices. CAR was the most frequently chosen for El Labrado. Meanwhile, lags of the Chanlúd rainfall anomalies were the most frequent in Chanlúd. NAO and EP/NP were indices within the six prominent índices in both El Labrado and Chanlúd.

In models for El Labrado AO, PDO, Niño 1+2.A, Niño 3.A, Niño 3.4, Niño 3.4.A, Niño 4, Niño 4.A, TNI, BEST, MEIV2, EMI, WHWP, TNA, TSA, AMM, QBO, ESPI, IOD.W, IOD.E and DMI were not present.

In models for Chanlúd AO, Niño 1+2, Niño 3, Niño 3.A, Niño 3.4, Niño 3.4.A, Niño 4, Niño 4.A, TNI, WHWP and TNA were not present.

3.4. Qualitative Evaluation

Figure 9 allows one to compare the rainfall forecasts from quarterly, semiannual and annual models. An outstanding feature in El Labrado (Figure 9a) is the overestimation of the semiannual model in November and December. This characteristic was prominent from 2016 to 2019. Quarterly and annual models showed similar results even though quarterly models showed better results for some specific months, for instance, from October to December 2016, and October and December 2019. Generally, quarterly models best reproduced the pattern of September–December. Likewise, semiannual models showed the best results in months such as December 2017, January 2018 and August–October 2021.

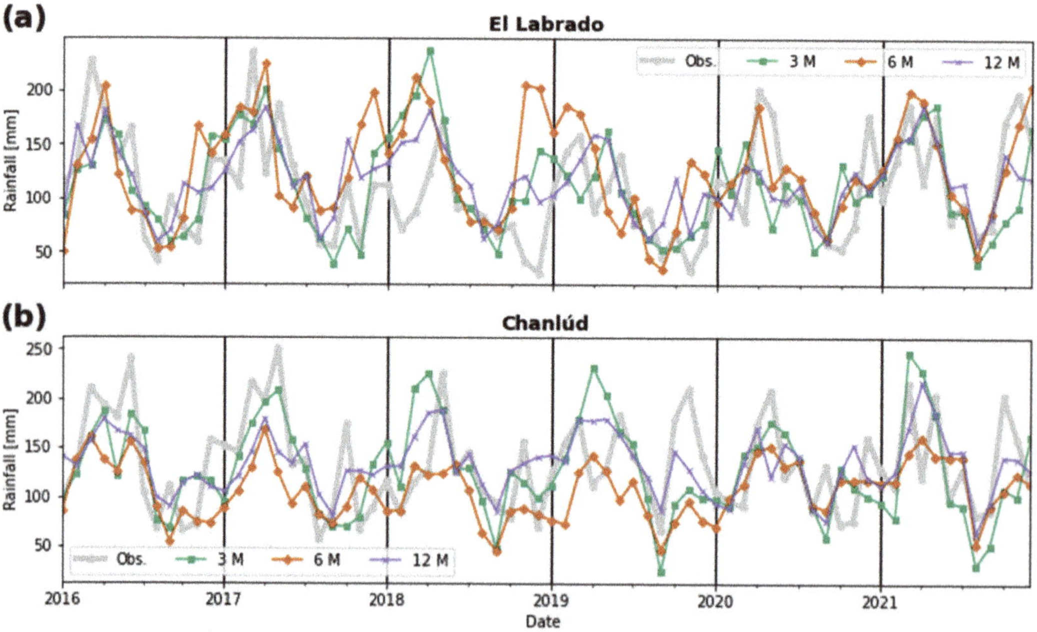

Figure 9. Qualitative evaluation of the models' performance: (**a**) for El Labrado; (**b**) for Chanlúd. The vertical lines indicate that the forecasts were made for each year individually.

Figure 9b shows that in Chanlúd, the semiannual models tended to result in mean values demonstrating the lowest performance. Annual models were better than semiannual models reproducing high values of rainfall. However, the general characteristic of quarterly models in reproducing the highest values of rainfall made them the best option. Nevertheless, it should be mentioned that quarterly models showed poor performance in some specific cases, such as in October 2017 and April 2019.

3.5. Quantitative Evaluation

Figure 10 shows the performance results for El Labrado (left-hand side of each figure panel) and Chanlúd (right-hand side of each figure panel) models that employed the seven evaluation metrics. The semiannual models showed the worst results in all the metrics, becoming the worst approach to forecast rainfall anomalies in both stations. According to the MARE, MAE, RMSE, NSE and EV metrics (Figure 10a–d,f), the best model was undoubtedly the annual model for El Labrado and Chanlúd. Moreover, the PBIAS (Figure 10g) metric confirmed the annual models as the best for Chanlúd. For El Labrado, the PBIAS indicated that the quarterly models were the best. However, the KGE metric showed that the quarterly models had the best performance.

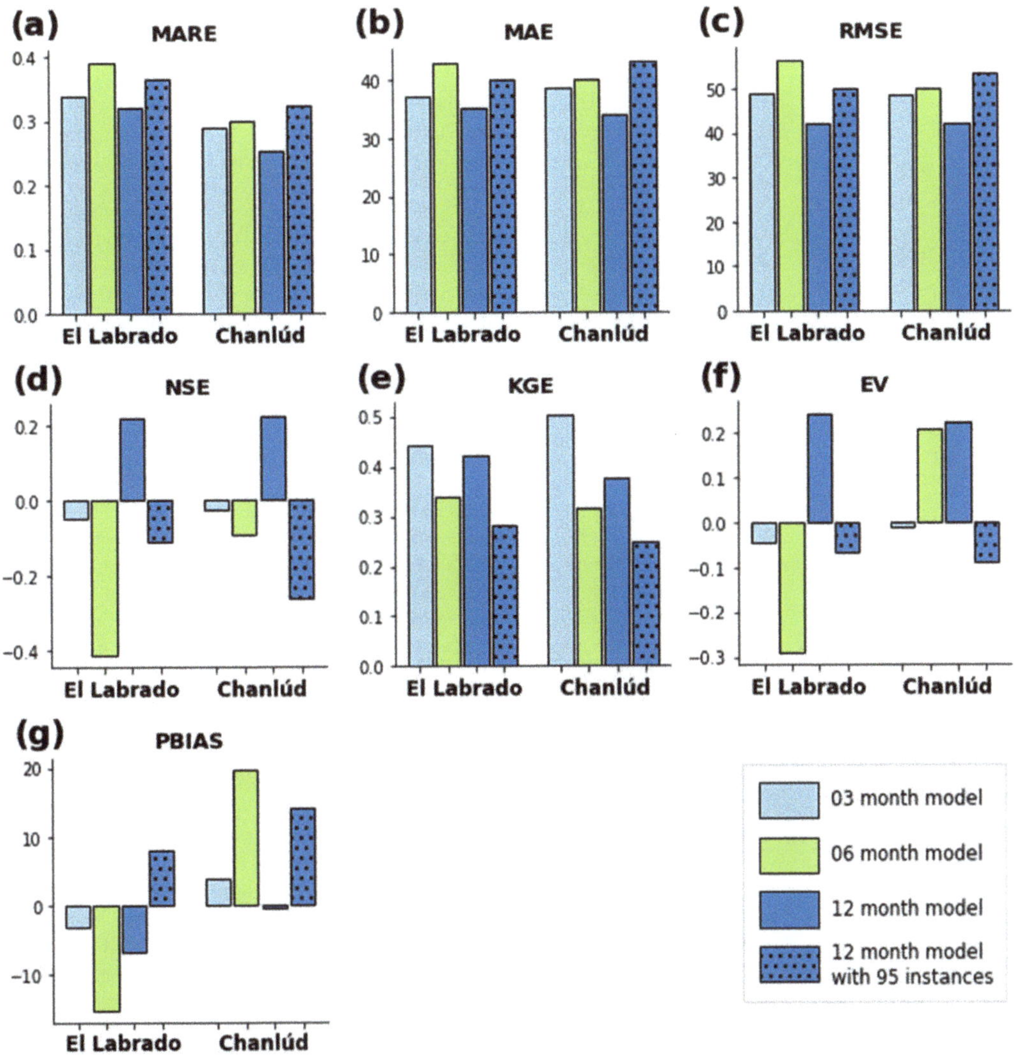

Figure 10. Quantitative evaluation of the models' performances using the metrics: (**a**) MARE; (**b**) MAE; (**c**) RMSE; (**d**) NSE; (**e**) KGE; (**f**) EV; (**g**) PBIAS. The metrics were calculated using the entire testing period.

As indicated in Section 3.2, annual models leverage a major amount of records in the model training stage, so they probably obtain better results. In order to give evidence for such conjecture, five new annual models were trained with the same followed method but by only using 95 randomly chosen records in training. This was the same number of records used when training the DJF quarterly models (in the rest of the seasons, 96 records were used). The mean values of the evaluation metrics are indicated as bars with dots in Figure 10.

Comparing the annual models trained with 95 records with the quarterly and semiannual models (Figure 10), all the metrics indicated that the quarterly models were the best models. According to KGE, the quarterly models were always the best performers.

4. Discussion

As more months were used to generate the models, the predictors chosen as the most relevant changed, especially for the annual models. EP/NP was the most prominent index in the quarterly, semiannual and annual models. This index was related to the most frequently selected predictors for quarterly and semiannual models for El Labrado and Chanlúd. In the case of annual models, EP/NP was related to the fifth most frequently selected predictors for El Labrado and Chanlúd. Except for the annual model for El Labrado, in all cases, the average lag of the predictors associated with this index was between 12 and 24 months. The EP/NP is a northern hemisphere index related to 500 hPa height anomalies over three main anomaly centers: Alaska/western Canada, central North Pacific and eastern North America. Since it is a relevant index for the climate of North America, most of the works are related to that geographical area. Córdoba Machado et al. [69] found weak but significant correlations between EP/NP and rainfall in Colombia. However, lagged correlations were not used. Mora and Willens carried out a study analyzing the relationship between the index and rainfall in a basin where the Machángara is located [70]. They found correlations around $|R^2| = 0.6$. This study showed that EP/NP also turned out to be an index with information that allows one to forecast rainfall anomalies with a horizon of approximately one year.

For the quarterly models, another prominent index was the north Pacific pattern (NP). NP is another northern hemisphere index. Specifically, it is the area-weighted sea level pressure over the region 30° N–65° N, 160° E–140° W. This confirms the relevance of information from the North Pacific in predicting rainfall anomalies in high tropical mountain areas. However, more research must be conducted to shed light on the acting mechanisms.

For the semiannual models, other prominent indices were TNA for El Labrado and AO for El Labrado and Chanlúd. TNA is a tropical Atlantic index and is defined as the anomaly of the average of the monthly SST over the region 5.5° N–23.5° N, 15° W–57.5° W. The tropical Atlantic SST is a driver of rainfall in the study zone [29,71,72]. Therefore, this study showed the applicability of the index in providing information to forecast rainfall anomalies in the study zone around 12 months in the future. AO is another northern hemisphere index that, like EP/NP, needs further study to understand the underlying mechanisms that make it a prominent index for forecasting rainfall in the study area.

For the annual models, the most prominent indices were CAR, NAO and the lagged versions of the same anomaly rainfall signal. CAR is related to the SST anomalies over the Caribbean and is not as prominent in quarterly and semiannual models. The Caribbean is known as a source of humidity that influences rainfall in the study zone [29,32]. NAO is another north hemisphere index and is a prominent teleconnection pattern in all seasons. Like EP/NP, further study is needed to understand the underlying mechanisms that make it a significant index for forecasting rainfall in the study area. Despite the known correlation between the conditions of El Niño zones and rainfall in Ecuador, these indices (Niño 3.A, Niño 3.4, Niño 3.4.A, Niño 4 and Niño 4.A) did not provide any relevant information to forecast rainfall anomalies in annual models. It should be borne in mind that the indices above are used in these models, but with delays of 13 to 37 months in order to generate forecasts with a one-year horizon. Despite the known relation between these indices and rainfall, this relation can fade when using signals with information distant in time. In addition, there may be other indices correlated with those of Niño with linear and nonlinear correlation with rainfall anomalies (standardized) that, more importantly, together with the rest of the selected predictors, are better leveraged by SVR, producing higher accuracy. Finally, another possible reason is that there are not enough events for SVR to learn the most significant patterns between the rainfall anomalies and the indices.

Another relevant result is the selection of SOI by many of the models, whether quarterlies, semiannuals or annuals. SOI had a linear correlation with the Niño 3.4 and Niño 3.4.A indices of −0.65 and −0.73, respectively. However, these last two were not selected for most models. The possible explanation is related to what was explained in the previous paragraph. Despite a high linear correlation between the indices, SOI contributed more to the learning algorithm in the context of the set of predictors chosen to produce a higher-accuracy model. In fact, the correlations between the Niño 3.4 and Niño 3.4.A indices with the standardized rainfall anomalies at Chanlúd (−0.04) were only slightly lower in magnitude than the correlation between SOI and rainfall anomalies (0.06) (in El Labrado, they were −0.07, −0.08, and 0.08, respectively). This means that SOI is more relevant to forecast rainfall when used with the other predictors in the model.

The distance between El Labrado and Chanlúd is approximately 8 km, and the difference in altitude is approximately 150 m. Despite the above, the SFS algorithm selected groups of predictors that differ for the models of these two stations, except those derived from EP/NP. EP/NP was the most relevant climatic index for the quarterly and semiannual models and was among the five most relevant in the annual one. Some reasons can explain this difference. First, note that the correlation between the standardized rainfall anomalies (Figure 2c) of the two stations (0.87) decreased compared to that of the raw data (0.92) (Figure 2a). These anomalies were used until before the evaluation (Scheme 1). As Figure 2b shows, there were systematic differences between the rainfall that were evident, for example, between May and August. Due to the above, it seems reasonable to expect certain differences between the groups of selected predictors because SVR made the best possible use of the highly nonlinear relationships in each group. Second, the relevance of the indices (size of the circles in Figures 5–8) could have been affected by the number of derived predictors that were selected (different lags). Since we used the same number of lags in climate indices and rainfall to generate the predictors, it is possible that for one of the two stations, a different τ_{max} should have been used (see Section 2.4). With that, it is possible that a larger number of predictors related to certain indices could be selected. Third, the SFS of predictors used SVR with the default hyperparameters to compute the score on the selected predictor subsets. Variations in the selection approach (e.g., sequential backward selection [73]) or in the model to calculate the score (e.g., random forests [74]) or its tuning would lead to a profoundly exhaustive sensitivity analysis of predictors. However, analyzing the influence of all the above is beyond the scope of the study and is proposed for future research.

According to almost all evaluation metrics, the annual models were the best among quarterly, semiannual and annual models. However, the KGE metric showed that quarterly models were the best. The implementation of annual models with fewer registers showed that such high performance is possibly due to the amount of information that the learning algorithm can leverage in the training stage. Not enough (or not the optimal) predictors were selected in such a case to be exploited by the SVR algorithm. When comparing such results with those from the quarterly models, all metrics demonstrated that quarterly models were the best, which KGE indicates for all models.

The semiannual models were the ones that reported the worst results. This could be related to the bimodal seasonality of rainfall (Figure 2b) that does not allow data to be separated into periods with similar rainfall characteristics. Each semiannual model contained information on both rainy months and drier months. This means that the selection of predictors was not so robust since the selection algorithm used information on transition periods.

The hypothesis that quarterly models could perform better by selecting more robust predictors was not necessarily true in practical terms. This is because, for operational reasons, the amount of information that can be used in annual models is greater. Specifically, this can be evidenced by using SVR as the learning algorithm for generating the forecast models. Depending on the learning algorithm, the negative effect of the amount of information to be used could be greater or lesser. Therefore, other studies using other learning algorithms are necessary to reach general and conclusive results.

A comprehensive comparison of the results with the predictions of South American models (e.g., the SEAS5 model) is pending, but a very brief discussion follows. Gubler et al. [68] demonstrated high precision in the highlands of Ecuador during the austral summer, which is consistent with our findings. However, Coelho et al. [75] state the low seasonal forecast skill of either empirical or coupled multimodel predictions in South America but highlight forecast assimilation's importance in obtaining better forecasts. Barnston and Tippett [10] showed that the North American Multimodel Ensemble project (NMME) [76,77] with a correction of bias with statistical methods does not improve the skill of the forecasts in South America.

5. Conclusions and Remarks

This study had the following main objectives: first, to know if the performance of the monthly rainfall forecast models improved when they were implemented for each season, by semesters or a single model for all the months of the year. These models always forecast rainfall on a monthly scale and differ in the months that are used in training and the months for which they make the forecast. Second, to analyze the main predictors that influence the improvement of the performance of the models. These predictors were generated using 38 climate indices and the same rainfall signal using lags of up to 37 months. The El Labrado and Chanlúd stations, located in the Andean Machángara basin, were used for the study.

The annual models were the best according to six of seven evaluation metrics. However, the Kling–Gupta efficiency shows that the quarterly models were the best. The study gives evidence that the performance of annual models was due to the more significant number of records (instances) that could be exploited when training them. When the annual models were trained with the same number of records as the quarterly models, the quarterly models were the best. Therefore, from a pragmatic point of view, annual models should be used to generate operational rainfall forecasting models in the study area. Studies in more areas are necessary to generalize the results obtained here.

The largest number of predictors that were chosen for the forecasting models were those derived from the EP/NP climate index. The influence of this northern hemisphere index on the Machángara rainfall has not been extensively studied, so it must be taken into account to investigate the mechanisms involved. For the quarterly models, another prominent index was NP, another northern hemisphere index. For the semiannual models, other prominent indices were the tropical Atlantic index TNA for El Labrado and the northern hemisphere index AO for El Labrado and Chanlúd. Finally, for the annual models, the most prominent indices were CAR (related to the conditions in the Caribbean), NAO (north hemisphere index) and the lagged versions of the same anomaly rainfall signal.

The results show that annual models can be operationally helpful since rainfall forecasts could be made in the current month with climatic information from twelve or more previous months. This is essential to anticipate water resources management in different sectors, e.g., agriculture and hydroelectricity.

There were some limitations in the study, which are described next. First, we selected a single $\tau_{\max}$ for the climate indices and rainfall in El Labrado and Chanlúd. Tuning a $\tau_{\max}$ for each station and index could reduce the search space of the learning algorithms and improve their performance. Second, SVR was used with the default hyperparameters in selecting predictors through SFS. Other selection methods [78–83] could be tested in future studies, as well as other learning algorithms that serve as scoring functions (e.g., random forests) performing a hyperparameter tuning. The former would help analyze the most relevant indices that influence the study area more exhaustively and with greater significance. Finally, the training of the forecast models could be carried out with other learning algorithms [17,84], also comparing their performance when combined with different selection methods.

This study shows a real approach to implementing operational forecasting models and allowing more accurate insight into the generalization of the models in a production environment.

Author Contributions: Conceptualization, A.V.-P.; methodology, A.V.-P.; software, A.V.-P.; validation, A.V.-P., M.P. and A.A.; formal analysis, A.V.-P.; investigation, A.V.-P.; resources, A.A.; data curation, A.V.-P.; writing—original draft preparation, A.V.-P.; writing—review and editing, A.V.-P., M.P. and A.A.; visualization, A.V.-P.; supervision, A.A.; project administration, A.A.; funding acquisition, A.A. All authors have read and agreed to the published version of the manuscript.

Funding: The authors would like to thank to Corporación Ecuatoriana para el Desarrollo de la Investigación y Academia—CEDIA, Empresa Electro Generadora del Austro ELECAUSTRO S.A., and Universidad de Cuenca for the financial support given to the present research work through the FONDO 11 program, especially for the project "Modelo matemático para optimización hidroenergética del complejo hidroeléctrico Machángara incluyendo criterios ambientales y de adaptación a los impactos del cambio climático". The first author was also supported by the VIUC through "Conjunto de horas 1".

Institutional Review Board Statement: Not applicable.

Informed Consent Statement: Not applicable.

Data Availability Statement: Data about climate indices are publicly available online (see Data subsection). Rainfall data analyzed during the current study are not publicly available and must be requested from the electricity generation company ELECAUSTRO S.A.

Acknowledgments: The authors thank ELECAUSTRO S.A. for sharing data.

Conflicts of Interest: The authors declare no conflict of interest.

References

1. Esquivel, A.; Llanos-Herrera, L.; Agudelo, D.; Prager, S.D.; Fernandes, K.; Rojas, A.; Valencia, J.J.; Ramirez-Villegas, J. Predictability of Seasonal Precipitation across Major Crop Growing Areas in Colombia. *Clim. Serv.* **2018**, *12*, 36–47. [CrossRef]
2. BBC News Megadrought in Southwest US Worst in a Millennium. Available online: https://www.bbc.com/news/world-us-canada-60396229 (accessed on 4 April 2022).
3. Freitas, A.A.; Drumond, A.; Carvalho, V.S.B.; Reboita, M.S.; Silva, B.C.; Uvo, C.B. Drought Assessment in São Francisco River Basin, Brazil: Characterization through SPI and Associated Anomalous Climate Patterns. *Atmosphere* **2021**, *13*, 41. [CrossRef]
4. da Rocha Júnior, R.L.; dos Santos Silva, F.D.; Costa, R.L.; Gomes, H.B.; Pinto, D.D.C.; Herdies, D.L. Bivariate Assessment of Drought Return Periods and Frequency in Brazilian Northeast Using Joint Distribution by Copula Method. *Geosciences* **2020**, *10*, 135. [CrossRef]
5. Silva, E.H.D.L.; Silva, F.D.D.S.; Junior, R.S.D.S.; Pinto, D.D.C.; Costa, R.L.; Gomes, H.B.; Júnior, J.B.C.; de Freitas, I.G.F.; Herdies, D.L. Performance Assessment of Different Precipitation Databases (Gridded Analyses and Reanalyses) for the New Brazilian Agricultural Frontier: SEALBA. *Water* **2022**, *14*, 1473. [CrossRef]
6. Cunha, A.P.M.A.; Zeri, M.; Deusdará Leal, K.; Costa, L.; Cuartas, L.A.; Marengo, J.A.; Tomasella, J.; Vieira, R.M.; Barbosa, A.A.; Cunningham, C.; et al. Extreme Drought Events over Brazil from 2011 to 2019. *Atmosphere* **2019**, *10*, 642. [CrossRef]
7. Bojinski, S.; Verstraete, M.; Peterson, T.C.; Richter, C.; Simmons, A.; Zemp, M. The Concept of Essential Climate Variables in Support of Climate Research, Applications, and Policy. *Bull. Am. Meteorol. Soc.* **2014**, *95*, 1431–1443. [CrossRef]
8. Ghil, M. Natural Climate Variability. *Encycl. Glob. Environ. Chang.* **2002**, *1*, 544–549.
9. Maslin, M. What Is Climate Change? In *Climate Change: A Very Short Introduction*; Oxford University Press: Oxford, UK, 2021; pp. 1–9, ISBN 978-0-19-886786-9.
10. Barnston, A.G.; Tippett, M.K. Do Statistical Pattern Corrections Improve Seasonal Climate Predictions in the North American Multimodel Ensemble Models? *J. Clim.* **2017**, *30*, 8335–8355. [CrossRef]
11. Ashby, S.A.; Taylor, M.A.; Chen, A.A. Statistical Models for Predicting Rainfall in the Caribbean. *Theor. Appl. Climatol.* **2005**, *82*, 65–80. [CrossRef]
12. Duan, Q.; Pappenberger, F.; Wood, A.; Cloke, H.L.; Schaake, J.C. (Eds.) *Handbook of Hydrometeorological Ensemble Forecasting*; Springer: Berlin/Heidelberg, Germany, 2019; ISBN 978-3-642-39924-4.

13. Doblas-Reyes, F.J.; García-Serrano, J.; Lienert, F.; Biescas, A.P.; Rodrigues, L.R.L. Seasonal Climate Predictability and Forecasting: Status and Prospects. *Wiley Interdiscip. Rev. Clim. Chang.* **2013**, *4*, 245–268. [CrossRef]
14. da Rocha Júnior, R.L.; Cavalcante Pinto, D.D.; dos Santos Silva, F.D.; Gomes, H.B.; Barros Gomes, H.; Costa, R.L.; Santos Pereira, M.P.; Peña, M.; dos Santos Coelho, C.A.; Herdies, D.L. An Empirical Seasonal Rainfall Forecasting Model for the Northeast Region of Brazil. *Water* **2021**, *13*, 1613. [CrossRef]
15. Amelia, R.; Dalimunthe, D.Y.; Kustiawan, E.; Sulistiana, I. ARIMAX Model for Rainfall Forecasting in Pangkalpinang, Indonesia. *IOP Conf. Ser. Earth Environ. Sci.* **2021**, *926*, 012034. [CrossRef]
16. Muñoz, P.; Orellana-Alvear, J.; Willems, P.; Célleri, R. Flash-Flood Forecasting in an Andean Mountain Catchment—Development of a Step-Wise Methodology Based on the Random Forest Algorithm. *Water* **2018**, *10*, 1519. [CrossRef]
17. Peña, M.; Vázquez-Patiño, A.; Zhiña, D.; Montenegro, M.; Avilés, A. Improved Rainfall Prediction through Nonlinear Autoregressive Network with Exogenous Variables: A Case Study in Andes High Mountain Region. *Adv. Meteorol.* **2020**, *2020*, 1828319. [CrossRef]
18. Dutta, R.; Maity, R. Time-Varying Network-Based Approach for Capturing Hydrological Extremes under Climate Change with Application on Drought. *J. Hydrol.* **2021**, *603*, 126958. [CrossRef]
19. Avilés, A.; Célleri, R.; Solera, A.; Paredes, J. Probabilistic Forecasting of Drought Events Using Markov Chain- and Bayesian Network-Based Models: A Case Study of an Andean Regulated River Basin. *Water* **2016**, *8*, 37. [CrossRef]
20. Recalde-Coronel, G.C.; Barnston, A.G.; Muñoz, Á.G. Predictability of December–April Rainfall in Coastal and Andean Ecuador. *J. Appl. Meteorol. Climatol.* **2014**, *53*, 1471–1493. [CrossRef]
21. Ghamariadyan, M.; Imteaz, M.A. Monthly Rainfall Forecasting Using Temperature and Climate Indices through a Hybrid Method in Queensland, Australia. *J. Hydrometeorol.* **2021**, *22*, 1259–1273. [CrossRef]
22. Mendoza, D.E.; Samaniego, E.P.; Mora, D.E.; Espinoza, M.J.; Campozano, L. Finding Teleconnections from Decomposed Rainfall Signals Using Dynamic Harmonic Regressions: A Tropical Andean Case Study. *Clim. Dyn.* **2019**, *52*, 4643–4670. [CrossRef]
23. Córdova, M.; Orellana-Alvear, J.; Rollenbeck, R.; Célleri, R. Determination of Climatic Conditions Related to Precipitation Anomalies in the Tropical Andes by Means of the Random Forest Algorithm and Novel Climate Indices. *Int. J. Climatol.* **2022**, joc.7519. [CrossRef]
24. Dutta, R.; Maity, R. Temporal Networks-Based Approach for Nonstationary Hydroclimatic Modeling and Its Demonstration With Streamflow Prediction. *Water Resour. Res.* **2020**, *56*, e2020WR027086. [CrossRef]
25. Pajankar, A.; Joshi, A. Preparing Data for Machine Learning. In *Hands-on Machine Learning with Python*; Apress: Berkeley, CA, USA, 2022; pp. 79–97. ISBN 978-1-4842-7920-5.
26. Vázquez-Patiño, A.; Peña, M.; Avilés, A. The Assessment of Rainfall Prediction Using Climate Models Results and Projections under Future Scenarios: The Machángara Tropical Andean Basin Case. *Int. J. Adv. Sci. Eng. Inf. Technol.* **2021**, *11*, 1903–1911. [CrossRef]
27. Campozano, L.; Trachte, K.; Célleri, R.; Samaniego, E.; Bendix, J.; Albuja, C.; Mejia, J.F. Climatology and Teleconnections of Mesoscale Convective Systems in an Andean Basin in Southern Ecuador: The Case of the Paute Basin. *Adv. Meteorol.* **2018**, *2018*, 4259191. [CrossRef]
28. Ballari, D.; Giraldo, R.; Campozano, L.; Samaniego, E. Spatial Functional Data Analysis for Regionalizing Precipitation Seasonality and Intensity in a Sparsely Monitored Region: Unveiling the Spatio-Temporal Dependencies of Precipitation in Ecuador. *Int. J. Climatol.* **2018**, *38*, 3337–3354. [CrossRef]
29. Vázquez-Patiño, A.; Campozano, L.; Ballari, D.; Córdova, M.; Samaniego, E. Virtual Control Volume Approach to the Study of Climate Causal Flows: Identification of Humidity and Wind Pathways of Influence on Rainfall in Ecuador. *Atmosphere* **2020**, *11*, 848. [CrossRef]
30. Avilés, A.; Palacios, K.; Pacheco, J.; Jiménez, S.; Zhiña, D.; Delgado, O. Sensitivity Exploration of Water Balance in Scenarios of Future Changes: A Case Study in an Andean Regulated River Basin. *Theor. Appl. Climatol.* **2020**, *141*, 921–934. [CrossRef]
31. Farfán, J.F.; Palacios, K.; Ulloa, J.; Avilés, A. A Hybrid Neural Network-Based Technique to Improve the Flow Forecasting of Physical and Data-Driven Models: Methodology and Case Studies in Andean Watersheds. *J. Hydrol. Reg. Stud.* **2020**, *27*, 100652. [CrossRef]
32. Esquivel-Hernández, G.; Mosquera, G.M.; Sánchez-Murillo, R.; Quesada-Román, A.; Birkel, C.; Crespo, P.; Célleri, R.; Windhorst, D.; Breuer, L.; Boll, J. Moisture Transport and Seasonal Variations in the Stable Isotopic Composition of Rainfall in Central American and Andean Páramo during El Niño Conditions (2015–2016). *Hydrol. Process.* **2019**, *33*, 1802–1817. [CrossRef]
33. Emck, P. A Climatology of South Ecuador—With Special Focus on the Major Andean Ridge as Atlantic-Pacific Climate Divide. Ph.D. Thesis, Friedrich-Alexander-Universität Erlangen-Nürnberg, Erlangen, Germany, 2007.
34. CENACE. *Annual Report 2020*; Management Reports; National Center for Energy Control: Quito, Ecuador, 2021; p. 209. (In Spanish)
35. Alomía Herrera, I.; Carrera Burneo, P. Environmental Flow Assessment in Andean Rivers of Ecuador, Case Study: Chanlud and El Labrado Dams in the Machángara River. *Ecohydrol. Hydrobiol.* **2017**, *17*, 103–112. [CrossRef]
36. Lenssen, N.J.L.; Schmidt, G.A.; Hansen, J.E.; Menne, M.J.; Persin, A.; Ruedy, R.; Zyss, D. Improvements in the GISTEMP Uncertainty Model. *J. Geophys. Res. Atmos.* **2019**, *124*, 6307–6326. [CrossRef]

37. Bell, G.D.; Janowiak, J.E. Atmospheric Circulation Associated with the Midwest Floods of 1993. *Bull. Am. Meteorol. Soc.* **1995**, *76*, 681–695. [CrossRef]
38. Barnston, A.G.; Livezey, R.E. Classification, Seasonality and Persistence of Low-Frequency Atmospheric Circulation Patterns. *Mon. Weather Rev.* **1987**, *115*, 1083–1126. [CrossRef]
39. Jones, P.D.; Jonsson, T.; Wheeler, D. Extension to the North Atlantic Oscillation Using Early Instrumental Pressure Observations from Gibraltar and South-West Iceland. *Int. J. Climatol.* **1997**, *17*, 1433–1450. [CrossRef]
40. Zhou, S.; Miller, A.J.; Wang, J.; Angell, J.K. Trends of NAO and AO and Their Associations with Stratospheric Processes. *Geophys. Res. Lett.* **2001**, *28*, 4107–4110. [CrossRef]
41. Lee, D.Y.; Petersen, M.R.; Lin, W. The Southern Annular Mode and Southern Ocean Surface Westerly Winds in E3SM. *Earth Space Sci.* **2019**, *6*, 2624–2643. [CrossRef]
42. Trenberth, K.E.; Hurrell, J.W. Decadal Atmosphere-Ocean Variations in the Pacific. *Clim. Dyn.* **1994**, *9*, 303–319. [CrossRef]
43. Mantua, N.J.; Hare, S.R. The Pacific Decadal Oscillation. *J. Oceanogr.* **2002**, *58*, 35–44. [CrossRef]
44. Ashok, K.; Behera, S.K.; Rao, S.A.; Weng, H.; Yamagata, T. El Niño Modoki and Its Possible Teleconnection. *J. Geophys. Res. Oceans* **2007**, *112*, C11007. [CrossRef]
45. Schwing, F.B.; Murphree, T.; Green, P.M. The Northern Oscillation Index (NOI): A New Climate Index for the Northeast Pacific. *Prog. Oceanogr.* **2002**, *53*, 115–139. [CrossRef]
46. Wang, C.; Enfield, D.B. The Tropical Western Hemisphere Warm Pool. *Geophys. Res. Lett.* **2001**, *28*, 1635–1638. [CrossRef]
47. Enfield, D.B.; Mestas-Nuñez, A.M.; Trimble, P.J. The Atlantic Multidecadal Oscillation and Its Relation to Rainfall and River Flows in the Continental U.S. *Geophys. Res. Lett.* **2001**, *28*, 2077–2080. [CrossRef]
48. Penland, C.; Matrosova, L. Prediction of Tropical Atlantic Sea Surface Temperatures Using Linear Inverse Modeling. *J. Clim.* **1998**, *11*, 483–496. [CrossRef]
49. Enfield, D.B.; Mestas-Nuñez, A.M.; Mayer, D.A.; Cid-Serrano, L. How Ubiquitous Is the Dipole Relationship in Tropical Atlantic Sea Surface Temperatures? *J. Geophys. Res. Oceans* **1999**, *104*, 7841–7848. [CrossRef]
50. Chiang, J.C.H.; Vimont, D.J. Analogous Pacific and Atlantic Meridional Modes of Tropical Atmosphere–Ocean Variability. *J. Clim.* **2004**, *17*, 4143–4158. [CrossRef]
51. Saji, N.H.; Yamagata, T. Possible Impacts of Indian Ocean Dipole Mode Events on Global Climate. *Clim. Res.* **2003**, *25*, 151–169. [CrossRef]
52. Pedregosa, F.; Varoquaux, G.; Gramfort, A.; Michel, V.; Thirion, B.; Grisel, O.; Blondel, M.; Prettenhofer, P.; Weiss, R.; Dubourg, V.; et al. Scikit-Learn: Machine Learning in Python. *J. Mach. Learn. Res.* **2011**, *12*, 2825–2830.
53. Pajankar, A.; Joshi, A. Supervised Learning Methods: Part 2. In *Hands-on Machine Learning with Python*; Apress: Berkeley, CA, USA, 2022; pp. 149–165. ISBN 978-1-4842-7920-5.
54. Seabold, S.; Perktold, J. Statsmodels: Econometric and Statistical Modeling with Python. In Proceedings of the 9th Python in Science Conference, Austin, TX, USA, 28 June–3 July 2010; pp. 92–96. Available online: http://conference.scipy.org/proceedings/scipy2010/seabold.html (accessed on 25 April 2022).
55. Brockwell, P.J.; Davis, R.A. *Time Series: Theory and Methods*; Springer Series in Statistics; Springer: New York, NY, USA, 1991; ISBN 978-1-4419-0319-8.
56. Brockwell, P.J.; Davis, R.A. *Introduction to Time Series and Forecasting*; Springer Texts in Statistics; Springer International Publishing: Cham, Switzerland, 2016; ISBN 978-3-319-29852-8.
57. Contreras, P.; Orellana-Alvear, J.; Muñoz, P.; Bendix, J.; Célleri, R. Influence of Random Forest Hyperparameterization on Short-Term Runoff Forecasting in an Andean Mountain Catchment. *Atmosphere* **2021**, *12*, 238. [CrossRef]
58. Ferri, F.J.; Pudil, P.; Hatef, M.; Kittler, J. Comparative Study of Techniques for Large-Scale Feature Selection. In *Machine Intelligence and Pattern Recognition*; Elsevier: Amsterdam, The Netherlands, 1994; Volume 16, pp. 403–413. ISBN 978-0-444-81892-8.
59. Raschka, S. MLxtend: Providing Machine Learning and Data Science Utilities and Extensions to Python's Scientific Computing Stack. *J. Open Source Softw.* **2018**, *3*, 638. [CrossRef]
60. Vapnik, V.N. Direct Methods in Statistical Learning Theory. In *The Nature of Statistical Learning Theory*; Springer: New York, NY, USA, 2000; pp. 225–265. ISBN 978-1-4419-3160-3.
61. Suykens, J.A.K.; Van Gestel, T.; De Brabanter, J.; De Moor, B.; Vandewalle, J. (Eds.) *Least Squares Support Vector Machines*; World Scientific: River Edge, NJ, USA, 2002; ISBN 981-238-151-1.
62. Costa, R.L.; Barros Gomes, H.; Cavalcante Pinto, D.D.; da Rocha Júnior, R.L.; dos Santos Silva, F.D.; Barros Gomes, H.; Lemos da Silva, M.C.; Luís Herdies, D. Gap Filling and Quality Control Applied to Meteorological Variables Measured in the Northeast Region of Brazil. *Atmosphere* **2021**, *12*, 1278. [CrossRef]
63. Fu, G.; Shen, Z.; Zhang, X.; Shi, P.; Zhang, Y.; Wu, J. Estimating Air Temperature of an Alpine Meadow on the Northern Tibetan Plateau Using MODIS Land Surface Temperature. *Acta Ecol. Sin.* **2011**, *31*, 8–13. [CrossRef]
64. Gang, F.; Wei, S.; Shaowei, L.; Jing, Z.; Chengqun, Y.; Zhenxi, S. Modeling Aboveground Biomass Using MODIS Images and Climatic Data in Grasslands on the Tibetan Plateau. *J. Resour. Ecol.* **2017**, *8*, 42–49. [CrossRef]
65. Wu, J.S.; Fu, G. Modelling Aboveground Biomass Using MODIS FPAR/LAI Data in Alpine Grasslands of the Northern Tibetan Plateau. *Remote Sens. Lett.* **2018**, *9*, 150–159. [CrossRef]
66. Nash, J.E.; Sutcliffe, J.V. River Flow Forecasting through Conceptual Models part I—A discussion of principles. *J. Hydrol.* **1970**, *10*, 282–290. [CrossRef]

67. Gupta, H.V.; Kling, H.; Yilmaz, K.K.; Martinez, G.F. Decomposition of the Mean Squared Error and NSE Performance Criteria: Implications for Improving Hydrological Modelling. *J. Hydrol.* **2009**, *377*, 80–91. [CrossRef]
68. Gubler, S.; Sedlmeier, K.; Bhend, J.; Avalos, G.; Coelho, C.A.S.; Escajadillo, Y.; Jacques-Coper, M.; Martinez, R.; Schwierz, C.; de Skansi, M.; et al. Assessment of ECMWF SEAS5 Seasonal Forecast Performance over South America. *Weather Forecast.* **2019**, *35*, 561–584. [CrossRef]
69. Córdoba Machado, S.; Palomino Lemus, R.; Castro Díez, Y.; Gámiz-Fortis, S.; Esteban Parra, M.J. Mechanisms of precipitation variability at Colombia. In *Proceedings of the VIII Congreso Internacional AEC: Cambio climático. Extremos e Impactos*; Asociación Española de Climatología: Salamanca, Spain, 2012; pp. 301–310. (In Spanish)
70. Mora, D.E.; Willems, P. Decadal Oscillations in Rainfall and Air Temperature in the Paute River Basin—Southern Andes of Ecuador. *Theor. Appl. Climatol.* **2012**, *108*, 267–282. [CrossRef]
71. Vuille, M.; Bradley, R.S.; Keimig, F. Climate Variability in the Andes of Ecuador and Its Relation to Tropical Pacific and Atlantic Sea Surface Temperature Anomalies. *J. Clim.* **2000**, *13*, 2520–2535. [CrossRef]
72. Vuille, M.; Bradley, R.S.; Keimig, F. Interannual Climate Variability in the Central Andes and Its Relation to Tropical Pacific and Atlantic Forcing. *J. Geophys. Res. Atmos.* **2000**, *105*, 12447–12460. [CrossRef]
73. Raschka, S.; Mirjalili, V. *Python Machine Learning: Machine Learning and Deep Learning with Python, Scikit-Learn, and TensorFlow 2*, 2nd ed.; Packt Publishing Ltd.: Birmingham, UK, 2019; ISBN 978-1-78995-829-4.
74. Breiman, L. Random Forests. *Mach. Learn.* **2001**, *45*, 5–32. [CrossRef]
75. Coelho, C.A.S.; Stephenson, D.B.; Balmaseda, M.; Doblas-Reyes, F.J.; van Oldenborgh, G.J. Toward an Integrated Seasonal Forecasting System for South America. *J. Clim.* **2006**, *19*, 3704–3721. [CrossRef]
76. Kirtman, B.P.; Min, D.; Infanti, J.M.; Kinter, J.L.; Paolino, D.A.; Zhang, Q.; van den Dool, H.; Saha, S.; Mendez, M.P.; Becker, E.; et al. The North American Multimodel Ensemble: Phase-1 Seasonal-to-Interannual Prediction; Phase-2 toward Developing Intraseasonal Prediction. *Bull. Am. Meteorol. Soc.* **2014**, *95*, 585–601. [CrossRef]
77. Becker, E.J.; Kirtman, B.P.; L'Heureux, M.; Muñoz, Á.G.; Pegion, K. A Decade of the North American Multimodel Ensemble (NMME): Research, Application, and Future Directions. *Bull. Am. Meteorol. Soc.* **2022**, *103*, E973–E995. [CrossRef]
78. Liu, H.; Motoda, H. (Eds.) *Computational Methods of Feature Selection*, 1st ed.; Chapman and Hall/CRC Data Mining and Knowledge Discovery Series; Chapman and Hall: London, UK; CRC: Boca Raton, FL, USA, 2008; ISBN 978-1-58488-878-9.
79. Zhang, X.; Hu, Y.; Xie, K.; Wang, S.; Ngai, E.W.T.; Liu, M. A Causal Feature Selection Algorithm for Stock Prediction Modeling. *Neurocomputing* **2014**, *142*, 48–59. [CrossRef]
80. Sun, Y.; Li, J.; Liu, J.; Chow, C.; Sun, B.; Wang, R. Using Causal Discovery for Feature Selection in Multivariate Numerical Time Series. *Mach. Learn.* **2015**, *101*, 377–395. [CrossRef]
81. Hmamouche, Y.; Casali, A.; Lakhal, L. A Causality Based Feature Selection Approach for Multivariate Time Series Forecasting. In Proceedings of the The Ninth International Conference on Advances in Databases, Knowledge, and Data Applications, IARIA, Barcelona, Spain, 21 May 2017; pp. 97–102.
82. Yu, K.; Guo, X.; Liu, L.; Li, J.; Wang, H.; Ling, Z.; Wu, X. Causality-Based Feature Selection: Methods and Evaluations. *ACM Comput. Surv.* **2020**, *53*, 1–36. [CrossRef]
83. Yu, K.; Liu, L.; Li, J. A Unified View of Causal and Non-Causal Feature Selection. *ACM Trans. Knowl. Discov. Data* **2021**, *15*, 1–46. [CrossRef]
84. Huang, J.-C.; Ko, K.-M.; Shu, M.-H.; Hsu, B.-M. Application and Comparison of Several Machine Learning Algorithms and Their Integration Models in Regression Problems. *Neural Comput. Appl.* **2020**, *32*, 5461–5469. [CrossRef]

atmosphere

Article

Plausible Precipitation Trends over the Large River Basins of Pakistan in Twenty First Century

Ammara Nusrat [1,*], Hamza Farooq Gabriel [1], Umm e Habiba [1], Habib Ur Rehman [2], Sajjad Haider [1], Shakil Ahmad [1], Muhammad Shahid [1], Saad Ahmed Jamal [3] and Jahangir Ali [4]

1 NUST Institute of Civil Engineering (NICE), School of Civil and Environmental Engineering (SCEE), National University of Sciences and Technology (NUST), Islamabad 44000, Pakistan; hamza.gabriel@nice.nust.edu.pk (H.F.G.); uhabiba@nice.nust.edu.pk (U.e.H.); sajjadhaider@nice.nust.edu.pk (S.H.); shakilahmad@nice.nust.edu.pk (S.A.); m.shahid@nice.nust.edu.pk (M.S.)

2 Civil Engineering Department, University of Engineering and Technology, Lahore 420000, Pakistan; mughalhabib@uet.edu.pk

3 Department of Geoinformatics—Z_GIS, Faculty of Digital and Analytical Sciences, Paris Lodron University of Salzburg, Schillerstraße 30, 5020 Salzburg, Austria; saad.jamal@stud.sbg.ac.at

4 Department of Geosciences, University of Arkansas, Fayetteville, AR 72701, USA; ja063@uark.edu

* Correspondence: ammara.phd14nice@student.nust.edu.pk; Tel.: +92-34-6500-1107

Abstract: Inter alia, inter-annual and spatial variability of climate, particularly rainfall, shall trigger frequent floods and droughts in Pakistan. Subsequently, a higher proportion of the country's population will be exposed to water-related challenges. This study analyzes and projects the long-term spatio-temporal changes in precipitation using the data from 2005 to 2099 across two large river basins of Pakistan. The plausible precipitation data to detect the projected trends seems inevitable to study the future water resources in the region. For, policy decisions taken in the wake of such studies can be instrumental in mitigating climate change impacts and shape water management strategies. Outputs of the Coupled Model Intercomparison Project 5 (CMIP5) climate models for the two forcing scenarios of RCP 4.5 and RCP 8.5 have been used for the synthesis of projected precipitation data. The projected precipitation data have been synthesized in three steps (1) dividing the area in different climate zones based on the similar precipitation statistics (2) selection of climate models in each climate zone in a way to shrink the ensemble to a few representative members, conserving the model spread and accounting for model similarity in a baseline period of 1971–2004 and the projected period of 2005–2099 and (3) combining the selected model's data in mean and median combinations. The future precipitation trends were detected and quantified, for the set of four scenarios. The spatial distribution of the precipitation trends was mapped for better understanding. All the scenarios produced consistent increasing or decreasing trends. Significant declining trends were projected in the warm wet season at 0.05% significance level and the increasing trends were projected in cold dry, cold wet and warm dry seasons. Framework developed to project climate change trends during the study can be replicated for any other area. The study therefore can be of interest for researchers working on climate impact modeling.

Keywords: climate change; climate model selection; spatiotemporal prediction; precipitation trends

Citation: Nusrat, A.; Gabriel, H.F.; e Habiba, U.; Rehman, H.U.; Haider, S.; Ahmad, S.; Shahid, M.; Ahmed Jamal, S.; Ali, J. Plausible Precipitation Trends over the Large River Basins of Pakistan in Twenty First Century. *Atmosphere* **2022**, *13*, 190. https://doi.org/10.3390/atmos13020190

Academic Editors: Zuohao Cao, Huaqing Cai and Xiaofan Li

Received: 16 November 2021
Accepted: 18 January 2022
Published: 24 January 2022

1. Introduction

Pakistan underwent recurring flooding during 1988, 1992, 2010, 2013, and 2014 in the Upper Catchments of Indus, Jhelum, and Chenab Rivers. Intense and devastating floods that increase fatalities and massive infrastructural damage have become somewhat annual routine in the country. Especially, heavy monsoon rains that hit the country from July to September due to the varying meteorological situations are major contributors to extreme monsoonal flooding [1,2]. Therefore, flood disaster mitigation and hazard management have become the point of concern for all stakeholders.

The intensity, variability, and frequency of temperature, floods, droughts, cyclones, and precipitation may exhibit substantial variations, thus presenting evidence of the impacts of climate change in Pakistan [3]. Northern Pakistan is the junction of three world-renowned mountain ranges known as the Karakoram, the Himalayas, and the Hindukush, producing the third largest mass of ice after the Polar Regions, located in the northern hemisphere. Westerly waves and Monsoon lows from the Mediterranean Sea, seasonal lows from the Arabian Sea, and depression caused by low pressures from the Bay of Bengal impact Pakistan [4]. The widespread perception is that this trend is part of a larger climate change phenomenon that has accelerated the hydrological cycle.

The climatic variability as a result of natural mechanisms of oceans, atmosphere, land surface, and anthropogenic forces is simulated by Global Circulation Models, commonly known as GCMs. These are the multi-dimensional numerical models, which follow the law of conservation of mass, momentum, and energy, representing the climate system. GCMs include numerous parameters related to atmospheric circulations, feedback mechanisms, moisture and wind fluxes, earth's rotational effects, and thermodynamics. Each model tends to simulate some aspects of the climate system well and some others not so well, leading to overestimation or underestimation of climate variables [5]. Therefore, these factors lead to different outputs for different GCMs with the same forcing scenarios for future projections [6–10]. The research question of the present study is how to use a variety of GCMs' outcomes to obtain plausible meteorological inputs for the climate change impact modeling for the study area, having spatiotemporal heterogeneous climate.

Many studies have consistently demonstrated that the selection of GCMs, for assessing climate change impacts, is the main contributor to uncertainty in the assessments of hydrological response to climate change. This has been proved by quantifying and comparing the uncertainties originating from different sources such as inherent errors in GCMs, forcing scenarios, downscaling and bias correction techniques, and hydrological models' parameters [11]. The Intergovernmental Panel on Climate Change [12] has proposed several GCM selection criteria such as using the latest version of GCMs' simulations, GCMs' with high temporal and spatial resolution, GCMs' producing high end and low-end climate signals [13,14], commonly known as an envelope-based selection method, or GCMs presenting realism of historical/baseline simulations [15,16].

The lack of realism of baseline simulations of some models cannot be linked with the plausibility of model projections. The correlation between past performance and future prediction is very weak; it means that model performance based on the historical period data may not be valid in an uncertain future climate [17–19]. It is necessary to consider the non-negligible probability of all the projections to use for future climate change detection, decision making, and planning [19]. Thus, the use of a multi-model ensemble (MME) is advocated and recommended over an individual model to synthesize the meteorological inputs for the climate change impact studies [20,21]. They should be synthesized, in a way to represent the full range of climate variability signals, from the available GCMs, for plausible future climate projections [22]. The subjective approach of past performance-based selection should only be used when severely unrealistic models that are not reliable for the future prediction have to be removed [14]. To attain unbiased distribution of the projected climate data, the selection of the GCMs in an ensemble should be in such a manner that they are not interdependent/correlated to each other. The correlated models gain too much weight in the larger MME [18]. The high correlation between the GCMs in the MME is responsible for the biases in the assessment of climate change impacts. It is imperative to provide the solution of biases of large MME by reducing its size using a smaller number of climate models with minimum loss of information. Effective small sub-ensembles are developed from an ensemble with a large number of models having more extensive dependencies on each other [23,24].

It is usual in the climate research community to compare or assess the spatial areal average of the climate data for selecting GCMs using any of the three methods of past performance, envelope-based approaches, and hybrid method [13,16]. This spatial average may not represent the local variances in spatial and temporal climate characteristics. Thus, strong consideration of this variation in spatial climatic conditions is required for the plausible ecological projections. To continue with the selection of a GCM for climate change impact studies, the area should be considered climatically homogeneous. The formation of homogeneous climatic zones [25] allows for a better understanding of the complex spatiotemporal variability of precipitation across an area. These climate zones are delineated using the spatial similarity/homogeneity of the precipitation statistics in an area using the dense data network.

Precipitation being the most provocative variable in terms of climate change impacts, studies have been conducted to estimate precipitation trends in various parts of Pakistan using best performing GCMs [26], single or two GCMs [27–29], hierarchical Bayesian Spatio-temporal methods [30], projections of trend line [31], artificial neural network (ANN) and Support vector regression (SVM) models [32]. Our confidence deficit in discounting any projections with a lack of realism of baseline simulations leads to the application of a novel framework of climate model selection. This framework focuses on (1) the delineation of the sub-regions having the stations with similar precipitation characteristics named as climate zone (2) selection of the climate models, based on climate signals spread in each climate zone for the possible future bandwidth of the climate change trends (3) combining the range of precipitation projections outcomes from the selected models for two climate forcing scenarios of RCP4.5 and RCP 8.5 (4) precipitation change trend detection. The selection of GCMs, from the large MME, was carried out by the formation of clusters of correlated GCMs and selecting the high and low-end producers of climate signals in each cluster in a climate zone. Therefore, a smaller MME of GCMs was formed with the larger ensemble characteristics. The daily data of the member GCMs of each MME were then combined in two ways: mean and median to investigate the future climate trend [33]. We intended to combine information from the GCMs to provide a set of scenarios for the study area that represent the uncertainty range in a credible manner. The precipitation trends were then detected for all those scenarios. The study also highlights the significance of machine learning clustering algorithms of supervised learning, for the pattern detection and grouping of long-term climate data of high temporal scale in a large study area. Subsequently, the research community is prompted to develop a framework that could analyze and predict the impacts of climate change and augment the decision support systems for a sustainable future. Moreover, the recurring demand for such research has also increased among policy-making circles and public pressure groups. Challenges such as food shortage and shelter insecurities are closely knitted with environmental degradation factors, such as floods, rising sea levels, and global warming. Availability of reliable and updated climate data trends is thus a prerequisite to forge sound policies that can reduce implications of environmental degradation upon human lives.

Section 2 of the paper describes the study area and data used, which is followed by Section 3 that explains the methodology used. Section 4 presents the results, and lastly, conclusions are drawn in Section 5.

2. Study Area and Data

2.1. Study Area

The area under study is approximately 100,845 km^2, which comprises two basins of Pakistan's rivers, namely the Jhelum and Chenab. The altitude from the mean sea level varies between 146 and 6915 m. Figure 1 presents the digital elevation model and geographic location of the study area. The precipitation gauging stations and the points on which gridded data have been sampled is also shown in Figure 1. The westerly aggravations and the southwest monsoon are the major cause of 60% and 40% of the annual precipitation, respectively [16]. The variable climatic conditions depend upon the atmospheric circulation patterns, advected moisture, and a considerable range of variations in topography. The depleting water resources and recurrence of extreme events due to high hydro-climatic variability in the region have made it an important research arena.

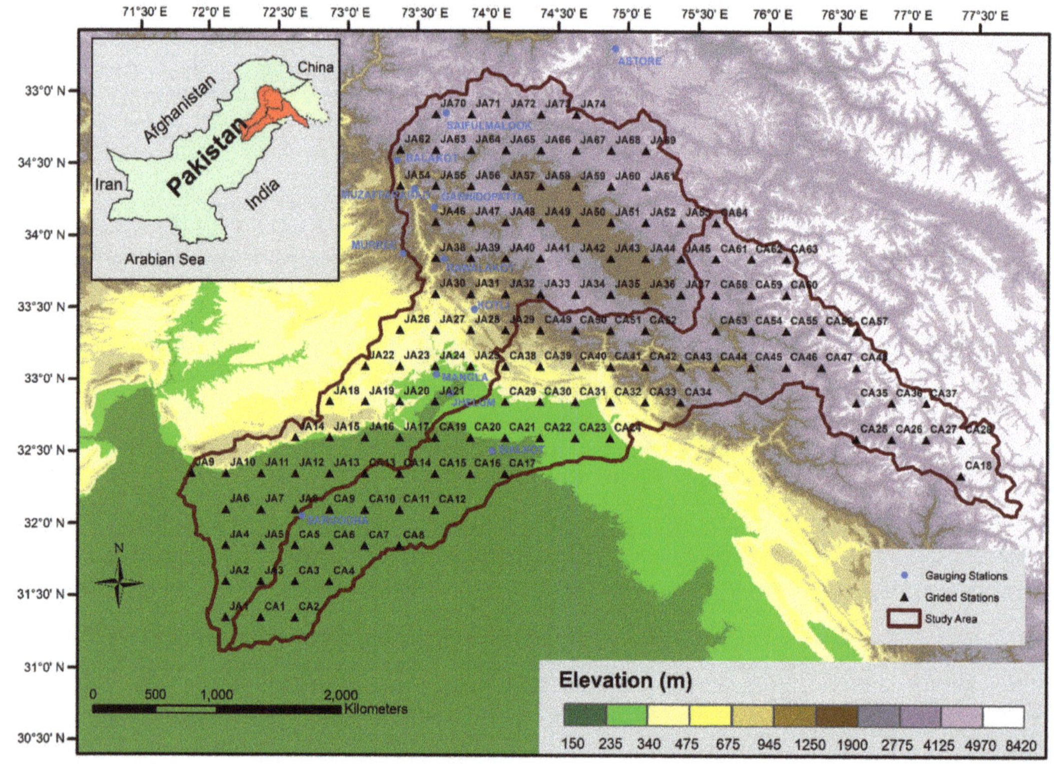

Figure 1. Study Area Digital Elevation Model (DEM) also presents the observed gauging station and Grid stations in the study area.

2.2. Climatology

The westerly aggravations and the southwest monsoon are the major cause of 60% and 40% of the annual precipitation, respectively [16]. The variable climatic conditions depend upon the atmospheric circulation patterns, advected moisture, and a considerable range of variations in topography. The depleting water resources and recurrence of extreme events due to high hydro-climatic variability in the region have made it an important research arena. Seasonal precipitation in Pakistan is affected by weather systems of three types: the monsoon depressions originating from the Bay of Bengal cause summer precipitation [34], western disturbances emanate from the Mediterranean Sea are reasons for the winter precipitation [35], and tropical cyclones from the Arabian Sea in spring and fall [36]. In Pakistan, the monsoon season lasts from June to September, with the post-monsoon season lasting from October to November [37]. On examining the mean monthly historical precipitation in Jhelum and Chenab river basins [38–40], the seasons have been defined in the present study. The seasons are warm wet (July, August, and September) Cold Dry (October, November, and December), cold wet (January, February, and March), and warm dry (April, May, and June).

For warm wet season, the average total seasonal precipitation based on APHRODITE (1970–2004) varies from 140 mm to 230 mm in north and east of the study area, 440 mm to 1050 mm in the central region, and 160 mm to 230 mm in the southwest region.

For the cold dry season, the precipitation varies between 8.93 mm and 120 mm, minimum in southwest and maximum in north and east of the study area.

For the cold wet season, in the southwest region, it varies from 120 mm to 160 mm, in the central region 230 mm to 340 mm, and in the north and northeast region 160 mm to 340 mm.

For the warm dry season, in the southwest region 100 mm to 140 mm, and the central region 140 mm to 230 mm, and 140 mm to 250 mm in the north and southwest region.

2.3. APHRODITE Data

The trend of using the gridded climate data in the climate and hydrologic assessment studies has been increasing due to its easy accessibility and reliability. A reliable gridded data network of stations is required to divide the study area into the number of homogeneous climatic zones. The reliability of the Asian Precipitation—Highly Resolved Observational Data Integration Towards Evaluation of Water Resources (APHRODITE) dataset has been advocated by many previous studies [41–43] and it is considered as the best-gridded dataset available as of yet, for the high elevated mountainous areas of Asia [44]. There are many other gridded datasets available as an open-source e.g., European Reanalysis gridded dataset (ERA5) [45] and Global Meteorological forcing dataset for land surface Modelling (GMFD) [46]. ERA5 is also famous for its good resolution and accuracy. Our decision of using the APHRODITE product (V1101) [43] for the regionalization of the study area, was based on the comparative analysis of these three datasets in a study by Nusrat et al. [25]. In this study, monthly precipitation data of APHRODITE, ERA5 and GMFD were sampled at the 11 gauging stations, mentioned in Supplementary Materials Table S1. All these gauging stations are situated at different elevations. The three datasets were compared with the monthly observed precipitation data using the Kolmogorov Smirnov Test and Pearson Correlation Coefficient test. The criteria of better performance were based on a higher correlation coefficient and p-value of KS test more than 0.05 to reject the alternative hypothesis of a dissimilar probability distribution. The results suggested that the APHRODITE dataset is more reliable than the other two datasets at nearly all the gauging stations at different altitudes.

2.4. NEX-GDDP-GCMs-CMIP5 Data

NASA Earth Exchange Global Daily Downscaled Projections (NEX-GDDP) dataset [47,48] has been used in this study for the historical and projected climate data. The outputs of the Coupled Model Intercomparison Project 5 (CMIP5) were used by National Aeronautics and Space Administration (NASA) to form NEX-GDDP. The list of 21 GCMs in the NEX-GDDP is provided in Supplementary Materials Table S1.

However, the CMIP5 experiments were meant to address the questions raised in the Assessment Report (AR4) of the Intergovernmental Panel on climate change IPCC [49]. Therefore, when fine-scale modeling is required, the climate impact modeling and Local-scale decision support system cannot rely on the coarser spatial resolution of CMIP5 GCMs. The NEX-GDDP dataset of 21 GCMs is downscaled to a finer resolution of 0.25° and bias-corrected using the bias-corrected spatial disaggregation method (BCSD) [47].

NEX-GDDP climate (maximum and minimum temperature and precipitation) datasets for the baseline period (1950–2005) and projected period (2005–2099) are publicly available. The projected data of CMIP5 GCMs have been bias-corrected and downscaled for the two forcing scenarios of representative concentration pathways (RCPs), i.e., RCP 4.5 and RCP 8.5, which were employed by IPCC for the fifth assessment report (AR5) [49].

3. Methodology

In this study, the Python module of Scikit–learn [50] has been used, assimilating many machine learning algorithms for supervised/unsupervised learning. The study area has been divided into different climate zones, of homogeneous climate, for each season, i.e., warm wet (July, August, and September), cold dry (October, November, and December), cold wet (January, February, and March) and warm dry (April, May, and June), following the methodology employed by [25].

With the help of Machine learning algorithms, large datasets of multivariate atmospheric parameters can easily be assessed and analyzed for the pattern/distribution variability study and impact modeling. Each step of the framework developed in this study has been translated into Python code. This framework of the study consists of three major steps (1) GCMs selection; (2) Combining the GCMs; and (3) climate change projections and trend detection. Figure 2 presents the flow chart of the methodology adopted in this study. Further details of all steps are discussed in the subsequent sections.

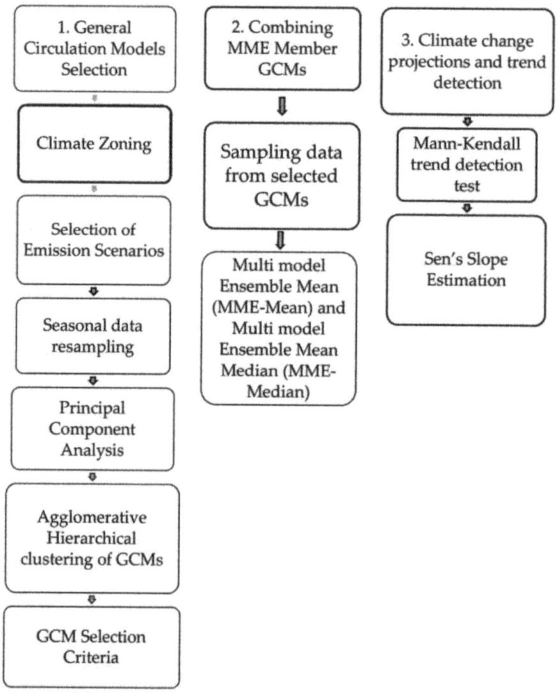

Figure 2. Methodology Flow Chart.

3.1. Climate Zoning

The first pre-requisite step in this framework is regionalization based on the climate of the region. Regionalization is the process of grouping the stations with homogeneous climate statistics and demarcating them into specific climate zones. For this process, long climate data records are required at various stations in the study area [51]. In the present study, the daily precipitation dataset of APHRODITE of 35 years was used for regionalization after its comparative performance evaluation with other gridded datasets.

There are various applications of regionalization of climate statistics e.g., in agricultural practices, hydrological extremes forecasting, and basin management [52]. The methods used to delineate climate regions are geographical convenience, subjective and objective partitioning [53,54]. In all these methods multivariate analysis techniques such as principal component analysis (PCA), correlation analysis, and clustering are widely used to demarcate the regions of homogeneous climate [55,56]. The geographical convenience method is an arbitrary and somewhat misleading approach based on the demarcation of the administrative boundaries. The subjective and objective partitioning methods are based on the demarcation of the region by grouping the meteorological sites having homogeneous climate statistics. The method employed for regionalization in the present study is agglomerative hierarchical clustering (AHC) of the principal components (PCs) of the precipitation data, at various stations of the region. Different Cluster validity indices were used to validate the number of clusters. The whole framework of regionalization step by step has been presented in the following sections.

3.1.1. Seasonal Data Resampling

The daily precipitation dataset of APHRODITE, at 138 Grid stations for the period 1975–2005, was sampled for hydrological seasons cycle: warm wet (July, August, and September), cold dry (October, November, and December), cold wet (January, February, and March) and warm dry (April, May, and June).

3.1.2. Principal Component Analysis (PCA)

The objective was to decrease the dimensions of the large matrix of the dataset of daily time series of 35 years for 138 grid stations. Principal Component Analysis (PCA) enabled us to reduce such a large matrix to a smaller sized matrix in addition to retaining as much descriptive of the data as possible. According to PCA, the data are projected onto different orthogonal axes which are called principal components. The symmetric covariance matrix is developed through the dataset, and then through the linear transformation technique principal components were identified. The eigenvectors depict the direction and the eigenvalues represent the magnitude of the extent of the axis or principal components representative of the data spread. The highly ranked principal components (PCs) which explained maximum cumulative variance in the dataset were identified through the scree plot and were used for the subsequent step. The component scores are derived by eigenvectors and the eigenvalues for all the stations in each PC. These component scores represent climate change patterns/signals in that specific site and may be considered an alternative to the meteorological parameters, which are statistically independent [14,55]. The function of scikit-learn [50] has been used to develop the code for the PCA.

3.1.3. Agglomerative Hierarchical Clustering (AHC)

Through this step, we were able to identify the clusters or groups of sites having similar climate signals. The climate change signals were estimated in the previous step of PCA, in the form of component scores. The component scores of the leading PCs were used in the clustering algorithm. The algorithm of Agglomerative Hierarchical Clustering [57,58] is an iterative process. It works on a bottom-up approach which starts from one point/self-cluster. Then, the size of the cluster keeps on increasing through nearest points one by one. In this way number of sequential combinations of clusters of the data points may be obtained. The optimum number of clusters is based upon the Euclidian distance [59] between the clusters. The dendrogram tree presents the meaningful information of different clusters and the Euclidian distances between the clusters, which forms the basis of the optimum number of clusters. The optimum number of clusters was determined with the help of cluster validity indices. Maximum Euclidian distance corresponding to the optimum number of clusters for each season was determined to truncate the dendrogram obtained through agglomerative clustering. Then number and identity of stations in each cluster were identified. The algorithm of scikit-learn [50] was used for agglomerative hierarchical clustering. The literature regarding different clustering techniques can be found in [14,60,61].

3.1.4. Formation of Climate Zones

There are different cluster validity tests, through which the number of clusters (NC) is decided. In the present study, the silhouette score (S) [62] (described in Section 3.1.5) was used to determine the optimum number of clusters/groups of sites with statistically similar climates. The number and identity of the stations in each cluster were determined by truncating the dendrogram corresponding to Euclidian distance so that the estimated optimum number of clusters could be produced.

Different clusters of stations are plotted in the map of the study area in ArcGIS Tool (Environmental Systems Research Institute, Redlands, CA, USA). To enable a clearer presentation, regions are demarcated with the visible boundaries representing the climate zone. The reference station in each climate zone was selected based on the average climate signals of all the grid stations in the respective zone. The climate of the reference station is considered representative of the climate of the respective zone.

3.1.5. Silhouette Score

The Silhouette score [62] is calculated as the average of the Euclidian distances between the clusters. The number of clusters with maximum Silhouette Score is considered optimum. The Silhouette score (S) can be calculated as

$$S = \frac{1}{NC}\sum_i \frac{1}{n_i}\sum_{r\in C_i} \frac{b(r)-a(r)}{\max[b(r),a(r)]} \quad (1)$$

and

$$a(r) = \frac{1}{n_i-1}\sum\nolimits_{s\in C_i}, s \neq r d(r,s), b(r) = min_{j,j\neq i}\left[\frac{1}{n_j}\sum\nolimits_{y\in C_j} d(r,s)\right] \quad (2)$$

where NC is the symbol of the number of clusters; the ith cluster is represented by C_i symbol for the number of objects in C_i is n_i; the center of C_i is denoted by c_i; and distance between r and s is denoted by $d(r,s)$ [62].

3.2. *GCM Selection in Climate Zones*

The climate forcing scenarios, as four Representative Concentration Pathways (RCPs), have been used for AR5 by IPCC. These RCPs are RCP 2.6, a mitigation scenario; RCP4.5 and RCP6.0, scenarios of medium stabilization, and RCP8.5, high baseline emissions scenario [63]. We used RCP 4.5 and 8.5 future scenarios to cover the wide range of greenhouse gas emissions assumed in these scenarios. The projected daily precipitation data (2005–2099) of the selected GCMs in each climate were sampled for these forcing scenarios and combined with the mean and median for the projected climate trends. The selection of GCMs out of 21 CMIP5 GCMs was done using the daily precipitation data of historical (1971–2005) and projected period (2005–2099) at every reference station. We have illustrated the method using the example of GCMs selection in the ninth climate zone of the cold dry season for RCP 8.5. The climate zones for the cold dry season have been shown in Figure 3a, highlighting the ninth climate zone in the study area. The steps are as follows:

1. For the projected period, the precipitation data were sampled for two forcing scenarios, RCP 4.5 and RCP 8.5. The set of GCMs was selected using each of the forcing scenarios separately. It means that each climate zone would have two sets of selected GCMs, each corresponding to one forcing scenario. The idea was to incorporate every possible spread and diversity of climate signals to select GCMs, which is the basis of the envelope-based selection approach.
2. After combining the data of 21 GCMs for the historical and projected periods for RCP 4.5 and RCP 8.5 for all the reference stations, Principal Component Analysis (PCA) was performed on 21 GCMs data at each of the reference stations. The objective was to reduce the dimensions of the large matrix of the daily time series of 129 years (base period: 1971–2004 and projected period: 2005–2099) for the reference stations in each of the climate zone of every season. In addition, to retain as much descriptive of the data as possible, PCA enabled us to reduce such a large matrix into a smaller-sized matrix. The Principal Components (PCs) were obtained using the data of 21 GCMs at each of the reference stations. The highly ranked PCs which explained maximum cumulative variance in the dataset were identified through the scree plot. The component scores are derived by eigenvectors and the eigenvalues for all the GCMs in each PC. These component scores represent climate change patterns/signals in that specific site and may be considered an alternative to the meteorological parameters, which are

statistically independent [14,55]. Figure 3b,c are the scree plot showing the percentage of explained variance by individual GCM and the line plot representing the cumulative percentage of explained variance, respectively. The variance explained by the individual PCs as shown in Figure 3b ranges between 4 and 6%, which means that all the PCs are equally important in deciding the hierarchy of the clusters of the GCMs for this zone 9 of the cold dry season. The gradient of the line plot in Figure 3c depicts that 18 PCs have explained the cumulative variance of 90%. All principal components have been included for the agglomerative hierarchical clustering to accommodate the maximum variance of the data. Figure 3d presents the scatter of the component scores of all the GCMs for the first two PCs, which cumulatively explained 19% variance in the data at the reference station of climate zone 9. The dendrogram tree is presented in Figure 3e, which presents the agglomerative hierarchical clustering of the GCMs for this climate zone.

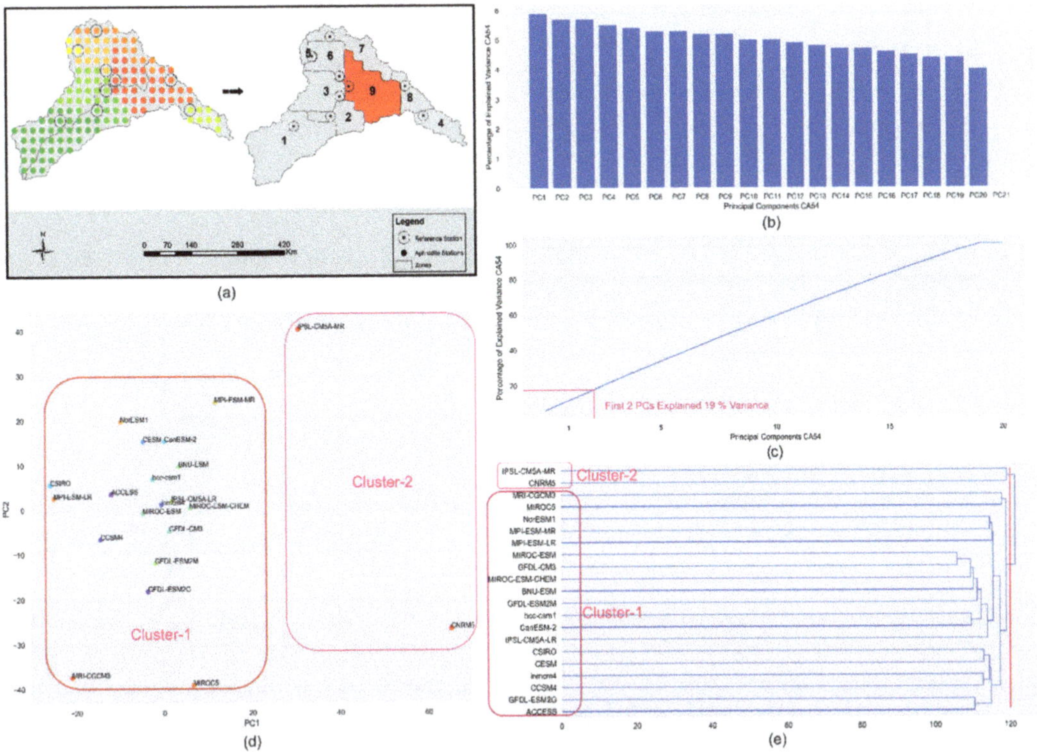

Figure 3. Climate zones for the season cold dry and indication of the zone 9 which is considered for illustration (**a**) orientation of Zone 9 in study area (stations with similar climate have been given same colored marker in first map) (**b**) cumulative percentage of explained variance by each PCs (**c**) scatter plot of PC1 and PC2 (**d**) scree plot showing the percentage of explained variance by individual PCs (**e**) dendrogram tree with cut off bar in red.

1. The climate signals in the form of component scores were obtained for the PCs, which cumulatively explained 90% of the data variance. These component scores were then used in the Agglomerative Hierarchical clustering (AHC) of the GCM. AHC would result in the clusters of GCMs having similar descriptive statistics. Through this step, we were able to identify the clusters or groups of GCMs having similar climate signals. The method has been described in Section 3.1.2; the clustering of GCMs has been illustrated with the help of a dendrogram tree of GCMs shown in Figure 3e for the

climate zone 9 of the cold dry season. The optimum number of clusters is based upon the Euclidian distance [59] between the clusters. The dendrogram tree presents the meaningful information of different clusters and the Euclidian distances between the clusters.

2. The optimum number of clusters was determined with the help of a cluster validity index called the silhouette score (described in Section 3.1.5). The silhouette scores for different numbers of clusters have been shown in Table 1 for the climate zone 9 of the cold dry season. Two clusters are optimum for this case, as they produce the highest silhouette score. Euclidian distance of 120 has been evaluated, corresponding to the optimum number of clusters. The dendrogram tree was then truncated at a value of 120 Euclidian distance, as shown in Figure 3e, to obtain the number and identity of GCMs in each cluster.

GCMs presenting the extreme climate signals, in the form of a component score, were selected in each cluster in each climate zone [18].

Figure 4 illustrates the two clusters of GCMs for climate zone 9 of the cold dry season. Cluster 1 has 19 GCMs as shown in Figure 4a, and Cluster 2 has 2 GCMs as shown in Figure 4b. MIROC5 and MIROC-ESM are selected in cluster 1 and CNRM5 is selected in cluster 2, as they are presenting the extreme climate signals in the form of component scores.

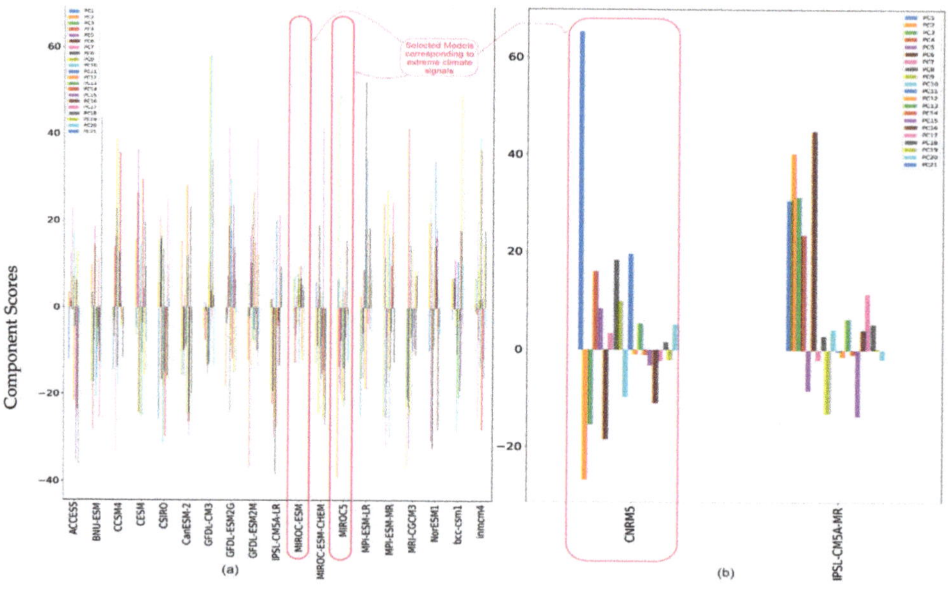

Figure 4. Agglomerative Hierarchical Clustering of GCMs at climate Zone 9 (**a**) cluster 1 with 19 GCMs and component scores (**b**) cluster 2 with 2 GCMs and respective component score.

Table 1. Silhouette score corresponding to the different number of clusters of GCMs at the reference station of Climate Zone 9 of Cold Dry Season. The highlighted cells are the optimum number of clusters and maximum Silhouette Score.

Number of Clusters	Silhouette Score	Number of Clusters	Silhouette Score
2	0.0279	6	0.0056
3	0.0249	7	0.0028
4	0.0184	8	0.0033
5	0.0171	9	0.0032

3.3. Combining GCMs and Data Sampling

Multimodel combination is a practical methodology, which is employed by the climate research community to incorporate all the model outputs (historical or projections) for the climate impact modeling to reduce the uncertainty that may originate by the use of a single model [17,64]. The approaches which are generally used to combine the models are equally weighted mean and optimum weighted mean [65], and median. In this study, two methods have been used for combining the data of the selected GCMs at each zone. The one is taking the mean and the second is the median of the data of the selected GCMs in a climate zone.

3.4. Climate Change Trends Projections

Mann–Kendall (MK) test was used to detect the trend of the precipitation change in the study for the four seasons for the century. The study period has been divided into three parts to visualize the results every three decades. The magnitudes of the trends were determined through Sen's slope. MK Test and Sen's slope test have been described as follows.

3.4.1. Mann-Kendall Test (MK Test)

The MK test [66] is used to statistically detect monotonic increasing or decreasing trends. In this study, the MK test has been used to detect the seasonal precipitation trend for the projected period to detect statistically significant trends in the chronological precipitation data. In this non-parametric distribution test, "No trend" is assumed in the Null Hypothesis (H_o) and vice versa. Equations (3)–(5) are used to calculate the test statistics Z. Equation (6) presents the test statistics.

$$T = \sum_{i=1}^{n-1} \sum_{j=i+1}^{n} sig(D_j - D_i) \tag{3}$$

$$sgn(D_j - D_i) = \begin{cases} +1\, if\ (D_j - D_i) > 0 \\ 0\, if\ (D_j - D_i) = 0 \\ -1\, if\ (D_j - D_i) < 0 \end{cases} \tag{4}$$

$$\sigma(T) = \frac{1}{18}\left[n(n-1)(2n+5) - \sum_{p=1}^{q} t_p(t_p - 1)(2t_p + 5)\right] \tag{5}$$

$$Z = \begin{cases} \frac{T-1}{\sqrt{\sigma\ (T)}}\, if\ T > 0 \\ 0\, if\ T = 0 \\ \frac{T-1}{\sqrt{\sigma\ (T)}}\, if\ T < 0 \end{cases} \tag{6}$$

where D_i and D_j are the ith and jth observations in the time series in chronological order; the length of data is n; t_p is the total number of data points in pth tied group, and the total number of tied groups is q; σ represents the variance. The negative Z value denotes the downward trend and vice versa. The Null Hypothesis of "No trend" is rejected, $|Z| > Z_{1-\alpha/2}$ indicates a statistically significant trend. The critical value $Z_{1-\alpha/2}$ corresponds to p-value 0.05. This trend has been detected at 138 stations in a region for each scenario.

3.4.2. Sen's Slope Evaluation

The magnitude/slopes of the trends in the data were obtained using Sen's method [67]. Sen's slope is the median value in the set of linear slopes in the data. Sen's slope is estimated through the following Equation (7).

$$T_i = \frac{D_j - D_k}{j - k} \text{ for } (1 \leq i < j \leq \text{n}), \tag{7}$$

where the slope is denoted by Ti, D_j and D_k are the values at time steps j and k, respectively, and n is the total number of the data points number.

4. Results and Discussion

This section presents the results of each step of the study methodology. Section 4.1 shows the results of GCM selection. In Section 4.2, the significance of precipitation trends and magnitude are presented.

4.1. Formation of Climate Zones

The two major steps involved in the regionalization of the study are (1) Principal Component analysis and (2) Agglomerative Hierarchical Clustering. Climate change patterns have been visualized through the application of PCA on the historical daily precipitation data (1971–2005) of 138 stations for every season. Then the clusters of sites/stations presenting similar climate signals were identified through agglomerative hierarchical clustering, the results of each step of the clustering procedure are presented as follows.

4.1.1. Principal Component Analysis (PCA)

After execution, 20 significant principal components were identified which explained the cumulative variance of 95% of the data for each season in the study area. The cumulative variance by five, ten, fifteen, and twenty principal components have been shown in Table 2. According to these plots, approximately the first 20 PCs explained 94–95% of the variance in every season. These 20 PCs were engaged for the agglomerative cluster analysis in the next step of regionalization. The component scores were obtained for each PC at every station. These component scores represent the climate signals generated at the respective station. The climate signals of the first two leading principal components in the study area have been visualized for each season in Figure 5.

In the warm wet season, the first two principal components explained a cumulative variance of 44.8%. In cold dry, cold wet, and warm dry seasons, the cumulative variance of 41.2%, 45.7%, and 41.5%, respectively, were explained. The higher negative or positive signals/component scores correspond to high variability in the precipitation data and the lower signals depict the low variability. The percentage variance explained by each PC in different seasons has been mentioned in each panel of Figure 5. In the southwest of the region, the high negative climate signals were obtained in the cold wet and warm dry season but low negative signals were obtained in the warm wet and cold dry season, for the first principal component. For the second principal component, low negative component scores were obtained in all the seasons in the southwest of the region. In the north, according to the first Principal component, the positive medium to high component scores were obtained in cold dry, cold wet, and warm dry season, and negative high component scores were obtained in the warm wet season. Low negative climate signals in the northern region were obtained in all the seasons for the second-highest leading Principal Component. In the southeast of the region, highest spatial heterogeneity in climate signals has been observed, as positive highest component scores were obtained for the warm wet, cold wet, and warm dry season in the second principal component. The heterogeneity of the component scores, depicting the variability of the climate in the region, forms the basis of the clustering of the sites.

Table 2. Cumulative percentage of explained variance of Principal Components.

Season	5 PCs	10 PCs	15PCs	20 PCs
Warm Wet	78	88	93	95
Cold Dry	77	87	91	94
Cold Wet	77	88	92	95
Warm Dry	77	86	92	95

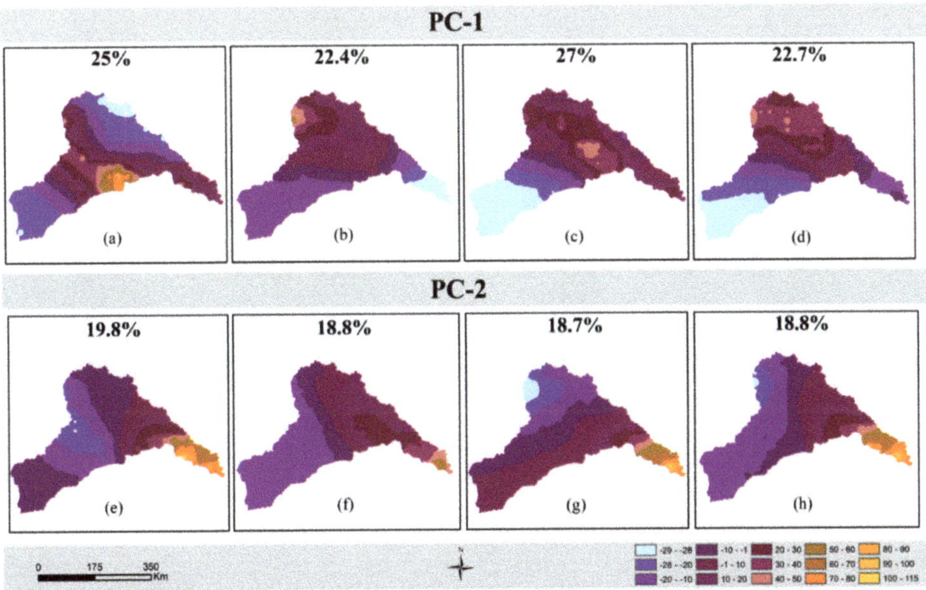

Figure 5. The spatial distribution of the component scores for the first two principal components (PC), percentage variance explained by the PCs are written in each panel (**a**) warm-wet (PC1), (**b**) cold dry (PC1), (**c**) cold wet (PC1), (**d**) warm dry (PC1), (**e**) warm wet (PC2), (**f**) cold dry (PC2), (**g**) cold wet (PC2) and (**h**) warm dry (PC2) seasons.

4.1.2. Agglomerative Hierarchical Clustering (AHC)

The clustering of the component scores of 138 stations for the first 20 leading PCs was done using Agglomerative Hierarchical Clustering. The number of optimum clusters was determined through a cluster validity test of silhouette score. The optimum clustering based on the silhouette score is decided corresponding to the highest score. The test suggested that the climate signals corresponding to the stations be optimally clustered into 17, 11, 10, and 14 for the warm wet, cold dry, cold wet seasons, and warm dry, respectively.

The maximum Euclidian distances were determined as 98, 95, 120, and 110, corresponding to the optimal number of clusters of 17, 11, 10, 14 for warm wet, cold dry, cold wet, and warm dry seasons, respectively. The dendrogram trees, as shown in Figure 6, were obtained through AHC. These trees were truncated at maximum Euclidian distances of 98, 95, 120, and 110 to obtain the optimum number of clusters and the station points in every cluster. The truncation bar in each season is also shown in Figure 6.

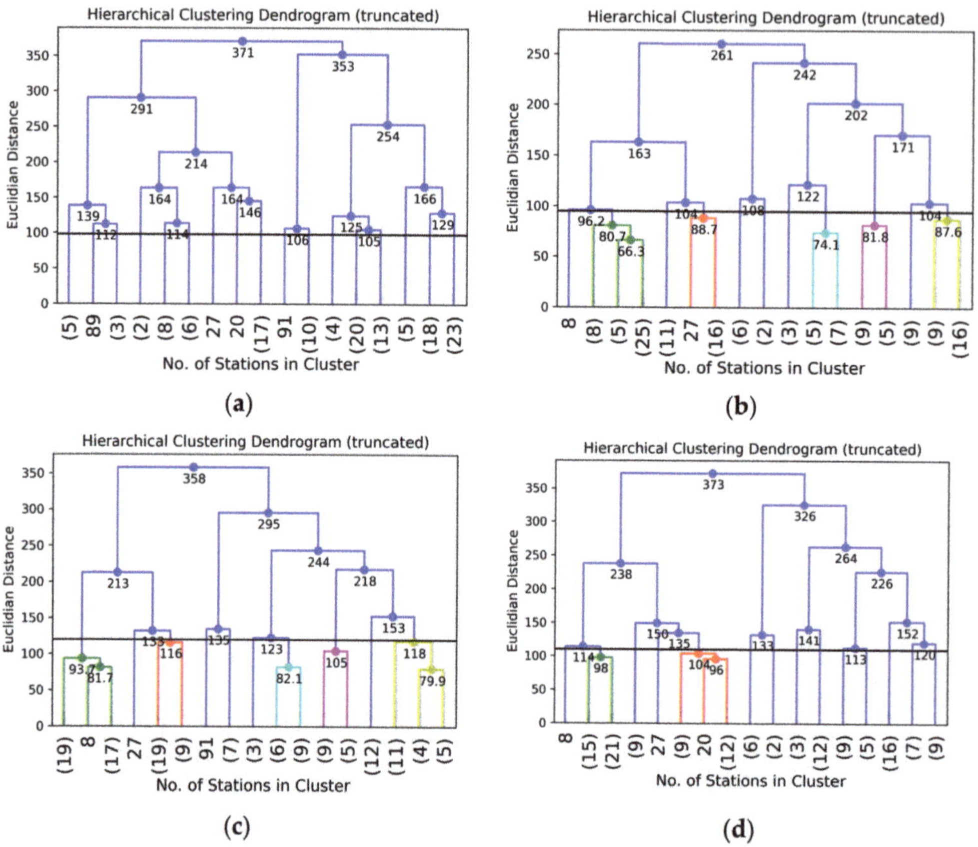

Figure 6. Dendrogram trees presenting the number of stations in each cluster (in Bracket) and cut-off bars corresponding to the optimum Euclidean distance, based on the Silhouette Score for climate zoning for: (**a**) warm wet, (**b**) cold dry, (**c**) cold wet, and (**d**) warm dry seasons.

4.1.3. Climate Zones and Reference Site

All the clusters of stations were plotted on the map of the study area. Each cluster of stations having a homogeneous climate has been given a different color to differentiate them. For clear presentation, the cluster boundaries are made to demarcate the region into several climate zones. The transformation of the clusters of stations to the climate zones for each season has been presented in Figure 7.

After merging the outliers with the nearest clusters, the river basins are apportioned into 12, 9, 9, and 10 clusters for the warm wet, cold dry, cold wet, and warm dry seasons, respectively. The sites/station, which represented approximately the average of the climate signals of the stations in the cluster, has been termed as a reference station. These reference stations were identified in every cluster/climate zone of all the seasons.

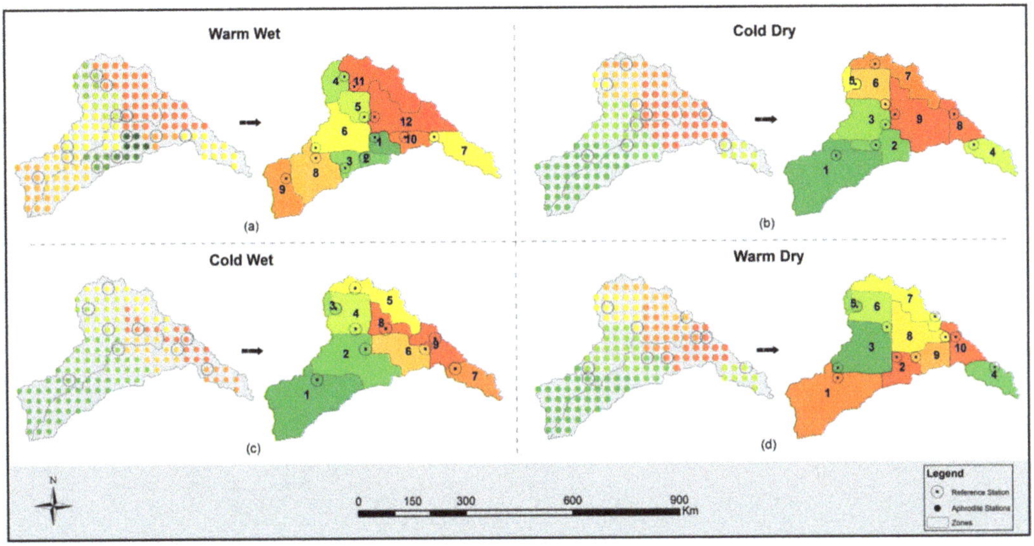

Figure 7. The transition of clusters into Climate zones in the study area in the (**a**) warm wet, (**b**) cold dry, (**c**) cold wet, and (**d**) warm dry seasons.

4.2. GCM Selection

The GCMs were selected in every climate zone in every season based on the envelope-based approach. For the study area, selected GCMs using base period data (1971–2004) and projected data of RCP 4.5 and RCP 8.5 of 21 GCMs in different climate zones and seasons have been presented in Tables S2 and S3, respectively (in Supplementary Materials).

4.3. Seasonal Precipitation Trend Projection

Although some studies have evaluated climatic variability and its implications on the hydrological regime of the Jhelum and Chenab River basins, there is no clear agreement among them regarding climate change trends projections and their effects on the hydrological regime, specifically for the next century [16,68–70].

The seasonal trends of precipitation were evaluated for the projected data for two forcing scenarios of RCP 4.5 and RCP 8.5 and two combinations (mean and median) of the Ensemble members' data. The trends and slopes have been assessed for consecutive three decadal projected periods, i.e., 2005–2040, 2041–2070, and 2071–2099 as well as the whole projected period of 2005–2099. The results show that the trends were statistically non-significant in most of the parts of the study area when the analyses were performed for the 3-decadal period. Whereas the trends were significant when the whole projected period was used in the analyses. Figures 8 and 9 present the trend analyses for the period 2005–2099 for the two ensemble combination of MME-mean and MME-median, respectively.

The spatial distributions of the significance of the trends and the slopes have been presented in Figures S1–S4 (in Supplementary Materials). MK Test p-value has been mapped spatially for a region, and the green shades depict that the trend is significant for $\alpha \leq 0.05$. When the p-value is greater than 0.05, the trend is non-significant. p-value equals unity, and depicts no trend.

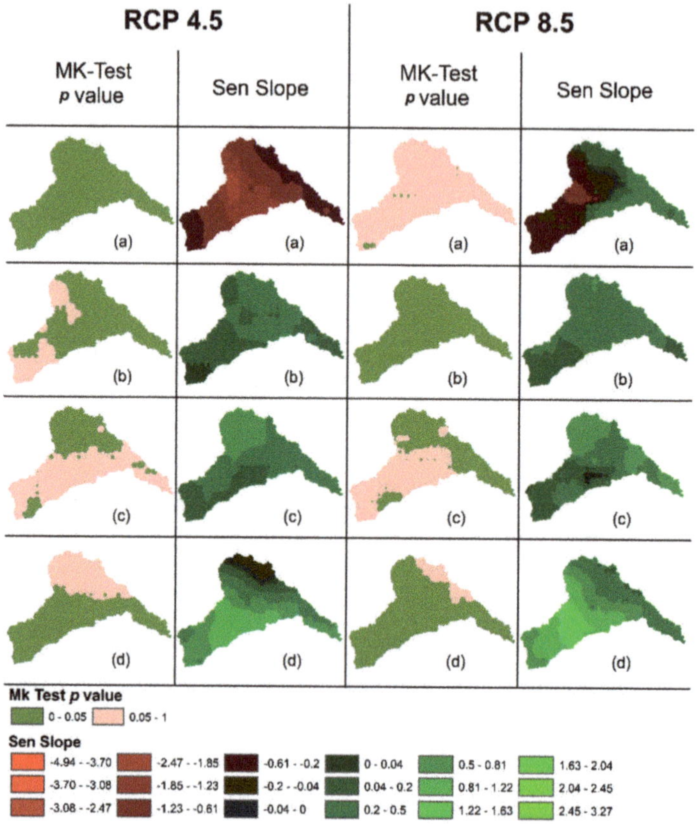

Figure 8. Mann Kendall trend detection and Sen's slope results for MME-Mean using RCP 4.5 and RCP 8.5 for analysis period (2005–2099) (**a**) Warm Wet (**b**) Cold Dry (**c**) Cold Wet (**d**) Warm Dry.

4.3.1. For 2005–2040

The data for the scenarios MME-mean and median of RCP 4.5, projected the significant increasing trends of 5–6 mm/year and 4–5 mm/year, respectively, in the warm dry season in the central part of the study area. For MME mean and median RCP 8.5, some traces of significant trends have been observed in the central part for cold dry season with magnitude ranges from 4 to 5 mm/year and 2 to 3 mm/year, respectively. In the north of the area, the data of MME-mean RCP 8.5 have projected significant positive trends of range 4–5 mm/year in east of the area.

4.3.2. For 2041–2070

The data of MME-mean RCP 8.5 have projected traces of significant decreasing trends in the warm wet season of magnitude ranging from 3 to 5 mm/year in the central part and positive significant trends ranging from 5 to 6 mm/year in the cold dry season in east. Some traces of increasing trends of magnitude 5–6 mm/year have been projected in the east of the study area for the cold dry season.

The data of MME-median RCP 8.5 have projected decreasing trends of 1–2 mm/year in east of the area for the cold dry season and increasing trends of 5–6 mm/year in the southwest for the warm dry season.

4.3.3. For 2071–2099

MME-mean and median RCP 4.5 have projected increasing trends for the warm dry season in the southeast and cold dry season in the west, respectively, with magnitude varies from 2–3 mm/year.

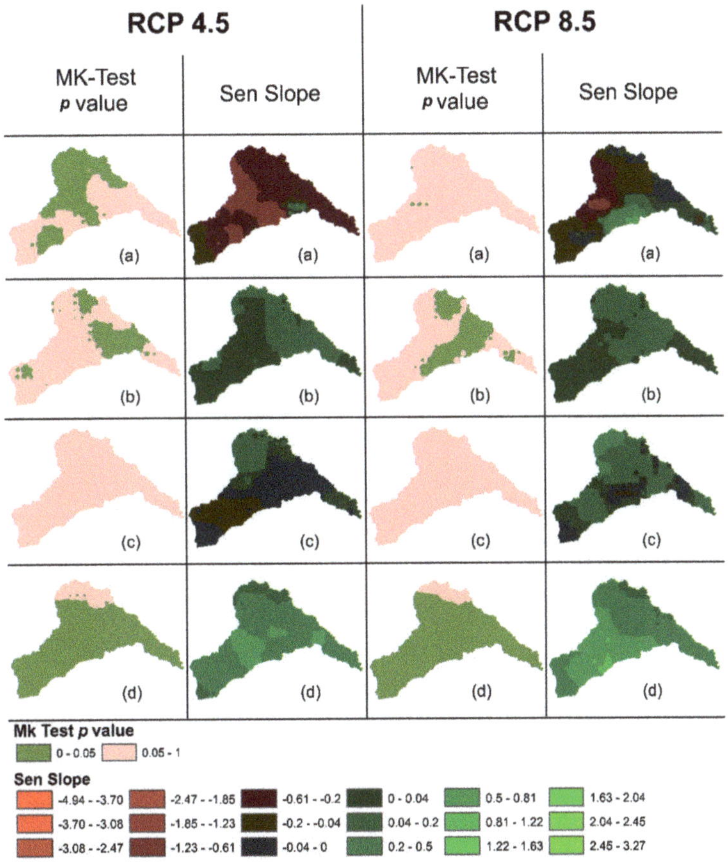

Figure 9. Mann Kendall trend detection and Sen's slope results for MME-Median using RCP 4.5 and RCP 8.5 for analysis period (2005–2099) **(a)** Warm Wet **(b)** Cold Dry **(c)** Cold Wet **(d)** Warm Dry.

4.3.4. For 2005–2099

For MME-mean RCP 4.5 (refer to Figure 8), the significant decreasing trends have been projected in the whole study area for the warm wet season with magnitude varying between 1.85 and 4.9 mm/year in the central and north, 0.6–1.2 mm/year in southeast and southwest. In the cold dry season, the significant increasing trends range between 0.04 and 1.2 mm/year have been projected in central, north, east, and southeast. In the cold wet season, significant increasing trends have been projected in the north of the region with magnitude varying between 0.8 and 1.22 mm/year. Significant increasing trends have been noticed in warm dry in nearly all parts of the area except north, with magnitude varies between 0.81 and 3.27 mm/year.

For MME-mean RCP 8.5, significant increasing trends have been projected in the study area for cold dry and warm dry seasons with magnitude varying between 0.2 and 1.2 mm/year and 0.5–3.27 mm/year, respectively. Significant increasing trends have been projected in the north and southeast for the cold wet season with magnitudes varying between 0.5 and 1.63 mm/year.

For MME-median RCP 4.5 (refer to Figure 9), significant decreasing trends have been projected in the warm wet season in the north and central parts with magnitudes varying between 1.23 and 2.47 mm/year. For the cold dry season, significant positive trends have been projected in the east with magnitude ranges between 0.81 and 1.22 mm/year. Significant positive trends have been projected in nearly all of the study areas for the warm dry season with a magnitude that varies between 0.5 and 1.63 mm/year.

For MME-median RCP 8.5, significant increasing trends have been projected in cold dry in north and east and warm dry season in nearly all the area with magnitude varying between 0.5 and 1.22 mm/year.

5. Conclusions

There is currently no consensus on widely agreed criteria and approaches for GCM selection in previous research [71,72]. Studies continue to explore ways, be it statistical or dynamic, to minimize the uncertainty in the climate change projections [25]. To minimize uncertainty associated with the GCMs suitability across various regions with heterogeneous climates, this research proposed a novel method for the selection of GCMs in homogeneous climate zones based on daily seasonal precipitation statistics. These statistics are reflected by the reanalysis gridded data series spanning throughout 1970–2005 for the Jhelum and Chenab River basins. The GCMs were selected by agglomerative hierarchical clustering of PCs obtained through the data spanning over the baseline period of 1970–2005 and the projected data for two forcing scenarios spanning between 2005 and 2099. The PCs represented the climate variability signals produced by various GCMs' data in the specific climate zone. Agglomerative hierarchical clustering of these climate signals produced the clusters of GCMs having the homogeneous variability of climate signals. The GCMs, producing the extreme climate signals in every cluster, were selected for the specific climate zone and season, thus fulfilling the criteria of envelope-based selection.

We sampled the daily precipitation data for the projected period using the selected GCMs for the two radiative forcing scenarios of RCP 4.5 and 8.5 and combining the data as mean and the median at every grid point, to detect the trends in the precipitation variability, in the Jhelum and Chenab River basins for the period spanning 2005 to 2099.

The machine learning algorithms modules of the Scikit-learn library of Python were used to develop a program for GCM selection, Data sampling, and trend detection. The program can be used to augment the decision support system for water resource management, even with the data of the new versions of the GCMs. However, the assessment of the future projections derived from the GCMs outputs are based on the forcing scenarios, which are unknown in the future, thus are fundamentally uncertain. So, it is important to understand the uncertainty associated with the GCMs' outputs to use such simulations for the climate change impact assessment [73–76]. The following conclusions can be drawn from the results presented in the study:

(1) The high variability in the climate in Pakistan poses a major challenge to the scientific community to project the plausible trends in climate, specifically precipitation change, which is considered to be the basic representative of climate and covariates. It was intended to select the suitable GCMs across the multiple homogeneous climate zones, which are representative of spatiotemporal variability of the climate in a region. The conventional method of using the spatiotemporal area average [13,31,77] of the climate data or various spatial metrics after analyzing the individual grid point data [66,78,79] is very common among the climate research community. However, the selected GCMs, through these methods, may not represent variability and range of climate signals in the region having spatially heterogeneous climate statistics, which poses uncertainty in projecting the climate data using these GCMs. Therefore, the entire study area was divided into 12, 9, 9, and 10 homogeneous precipitation regions for the warm wet,

cold dry, cold wet, and warm dry season, respectively. The selection of GCMs was made in each homogeneous climate zone.

(2) The precipitation trends were projected using the selected GCMs data on two forcing scenarios, RCP 4.5 and RCP 8.5, and two ensemble combinations; mean and median, thus making the total of four scenarios (RCP 4.5 Mean, RCP 4.5 Median, RCP 8.5 Mean, and RCP 8.5 Median). The trends projected using these scenarios provide the details of the range of trend variability of climate change in the region, with the knowledge of maximum increasing and decreasing trend quantification in the region seasonally, which is the purpose of envelope-based selection of GCMs.

Statistically significant trends were projected when the analyses were performed using the study period of 2005–2099. Significant negative trends were projected in the warm wet season and significant positive trends were projected in warm dry seasons for RCP 4.5. For RCP 8.5, statistically significant positive trends were projected in cold dry and warm dry seasons. The high evaporation and convection rate over the agro-economic zones is anticipated to be the cause of increasing trends in high emission scenarios.

Further research avenues that can be explored include a redefinition of the homogeneous climate zones based on GCMs' output and selection of GCMs based on spatial coherence of these climate regions with the regions derived through observed or high-resolution reanalysis data.

Supplementary Materials: The following supporting information can be downloaded at: https://www.mdpi.com/article/10.3390/atmos13020190/s1. Table S1. Comparative analysis of APHRODITE and ERA5 monthly dataset. Pearson correlation coefficients and Kolmogorov Smirnov Test results (KS Test). (The shaded *p*-values are >0.05, depicting the null hypothesis of similar distribution, is not rejected). Table S2. Descriptions of the general circulation models (GCMs) used in the study. Table S2. Selected GCMs using RCP 4.5 and climate Zones for seasons warm wet cold dry cold wet and warm dry. Table S3: Selected GCMs using RCP 8.5 scenarios for seasons Warm Wet, Cold Dry, Cold Wet and Warm Dry. Figure S1. Mann Kendall trend detection and Sen's slope results for MME-Mean using RCP 4.5 (a) Warm Wet (b) Cold Dry (c) Cold Wet (d) Warm Dry. Figure S2. Mann Kendall trend detection and Sen's Slope results for MME-Median using RCP 4.5 (a) Warm Wet (b) Cold Dry (c) Cold Wet (d) Warm Dry. Figure S3. Mann Kendall trend detection and Sen's slope results for MME-Mean using RCP 8.5 (a) Warm Wet (b) Cold Dry (c) Cold Wet (d) Warm Dry. Figure S4. Mann Kendall trend detection and Sen's slope results for MME-Median using RCP 8.5 (a) Warm Wet (b) Cold Dry (c) Cold Wet (d) Warm Dry.

Author Contributions: Conceptualization: A.N. and H.U.R.; Data curation: U.e.H., S.A.J., J.A. and M.S., Formal analysis: A.N., S.H. and S.A.; Investigation: U.e.H., H.F.G. and A.N.; Methodology: S.A. and J.A.; Project administration: S.A. and H.U.R.; Software: S.A.J. and A.N.; Supervision: H.F.G.; Validation: A.N. and H.U.R.; Visualization: S.A.J., A.N.; Writing—S.H., S.A. and H.F.G. All authors have read and agreed to the published version of the manuscript.

Funding: This research received no external funding.

Data Availability Statement: The data presented in this study are openly available in [GitHub—SaadAhmedJamal/PlausiblePrecipitationTrends: for jehlum and chenab basin https://github.com/SaadAhmedJamal/PlausiblePrecipitationTrends, accessed on 2 November 2021].

Acknowledgments: The authors highly acknowledge the resources and facilities provided by NUST Institute of Civil Engineering for the successful compliance of the present study.

Conflicts of Interest: The authors declare no conflict of interest.

References

1. Shah, S.M.H.; Mustaffa, Z.; Teo, F.Y.; Imam, M.A.H.; Yusof, K.W.; Al-Qadami, E.H.H. A review of the flood hazard and risk management in the South Asian Region, particularly Pakistan. *Sci. Afr.* **2020**, *10*, e00651. [CrossRef]
2. Tariq, M.A.U.R.; Van De Giesen, N. Floods and flood management in Pakistan. *Phys. Chem. Earth Parts A/B/C* **2012**, *47*, 11–20. [CrossRef]

3. Ahmed, K.; Shahid, S.; Chung, E.-S.; Wang, X.; Harun, S.B. Climate Change Uncertainties in Seasonal Drought Severity-Area-Frequency Curves: Case of Arid Region of Pakistan. *J. Hydrol.* **2019**, *570*, 473–485. [CrossRef]
4. Anjum, M.N.; Ding, Y.J.; Shangguan, D.H. Simulation of the projected climate change impacts on the river flow regimes under CMIP5 RCP scenarios in the westerlies dominated belt, northern Pakistan. *Atmos. Res.* **2019**, *227*, 233–248. [CrossRef]
5. Bosshard, T.; Carambia, M.; Goergen, K.; Kotlarski, S.; Krahe, P.; Zappa, M.; Schär, C. Quantifying uncertainty sources in an ensemble of hydrological climate-impact projections. *Water Resour. Res.* **2013**, *49*, 1523–1536. [CrossRef]
6. Sood, A.; Smakhtin, V. Global hydrological models: A review. *Hydrol. Sci. J.* **2015**, *60*, 549–565. [CrossRef]
7. Pradhan, P.; Shrestha, S.; Sundaram, S.M.; Virdis, S.G.P. Evaluation of the CMIP5 general circulation models for simulating the precipitation and temperature of the Koshi River Basin in Nepal. *J. Water Clim. Chang.* **2021**, *12*, 3282–3296. [CrossRef]
8. Almazroui, M.; Islam, M.N.; Saeed, F.; Saeed, S.; Ismail, M.; Ehsan, M.A.; Diallo, I.; O'Brien, E.; Ashfaq, M.; Martínez-Castro, D.; et al. Projected Changes in Temperature and Precipitation Over the United States, Central America, and the Caribbean in CMIP6 GCMs. *Earth Syst. Environ.* **2021**, *5*, 1–24. [CrossRef]
9. Jang, D. An Application of ANN Ensemble for Estimating of Precipitation Using Regional Climate Models. *Adv. Civ. Eng.* **2021**, *2021*, 7363471. [CrossRef]
10. Shiru, M.S.; Chung, E.-S. Performance evaluation of CMIP6 global climate models for selecting models for climate projection over Nigeria. *Theor. Appl. Climatol.* **2021**, *146*, 599–615. [CrossRef]
11. Chen, J.; Brissette, F.P.; Poulin, A.; Leconte, R. Overall uncertainty study of the hydrological impacts of climate change for a Canadian watershed. *Water Resour. Res.* **2011**, *47*. [CrossRef]
12. IPCC. *IPCC Expert Meeting on Emission Scenarios*; IPCC: Washington, DC, USA, 2005.
13. Lutz, A.F.; ter Maat, H.W.; Biemans, H.; Shrestha, A.B.; Wester, P.; Immerzeel, W.W. Selecting Representative Climate Models for Climate Change Impact Studies: An Advanced Envelope-Based Selection Approach. *Int. J. Climatol.* **2016**, *36*, 3988–4005. [CrossRef]
14. Mendlik, T.; Gobiet, A. Selecting Climate Simulations for Impact Studies Based on Multivariate Patterns of Climate Change. *Clim. Chang.* **2016**, *135*, 381–393. [CrossRef] [PubMed]
15. Chhin, R.; Yoden, S. Ranking CMIP5 GCMs for Model Ensemble Selection on Regional Scale: Case Study of the Indochina Region. *J. Geophys. Res. Atmos.* **2018**, *123*, 8949–8974. [CrossRef]
16. Azmat, M.; Qamar, M.U.; Huggel, C.; Hussain, E. Future Climate and Cryosphere Impacts on the Hydrology of a Scarcely Gauged Catchment on the Jhelum River Basin, Northern Pakistan. *Sci. Total Environ.* **2018**, *639*, 961–976. [CrossRef]
17. Weigel, A.P.; Knutti, R.; Liniger, M.A.; Appenzeller, C. Risks of Model Weighting in Multimodel Climate Projections. *J. Clim.* **2010**, *23*, 4175–4191. [CrossRef]
18. Knutti, R.; Furrer, R.; Tebaldi, C.; Cermak, J.; Meehl, G. Challenges in Combining Projections from Multiple Climate Models. *J. Clim.* **2010**, *23*, 2739–2758. [CrossRef]
19. McSweeney, C.F.; Jones, R.G.; Lee, R.W.; Rowell, D.P. Selecting CMIP5 GCMs for downscaling over multiple regions. *Clim. Dyn.* **2015**, *44*, 3237–3260. [CrossRef]
20. Zhang, H.; Huang, G.H.; Wang, D.; Zhang, X. Uncertainty assessment of climate change impacts on the hydrology of small prairie wetlands. *J. Hydrol.* **2011**, *396*, 94–103. [CrossRef]
21. Cannon, A.J. Selecting GCM Scenarios that Span the Range of Changes in a Multimodel Ensemble: Application to CMIP5 Climate Extremes Indices. *J. Clim.* **2015**, *28*, 1260–1267. [CrossRef]
22. Carter, T.R. *General Guidelines on the Use of Scenario Data for Climate Impact and Adaptation Assessment*, 2nd ed.; Task Group on Data and Scenario Support for Impact and Climate Assessment (TGICA); Intergovernmental Panel on Climate Change (IPCC): Helsinki, Finland, 2007; p. 66.
23. Pennell, C.; Reichler, T. On the Effective Number of Climate Models. *J. Clim.* **2011**, *24*, 2358–2367. [CrossRef]
24. Masson, D.; Knutti, R. Climate model genealogy. *Geophys. Res. Lett.* **2011**, *38*, 1–4. [CrossRef]
25. Nusrat, A.; Gabriel, H.F.; Haider, S.; Ahmad, S.; Shahid, M.; Ahmed Jamal, S. Application of Machine Learning Techniques to Delineate Homogeneous Climate Zones in River Basins of Pakistan for Hydro-Climatic Change Impact Studies. *Appl. Sci.* **2020**, *10*, 6878. [CrossRef]
26. Khan, F.; Pilz, J.; Ali, S. Evaluation of CMIP5 models and ensemble climate projections using a Bayesian approach: A case study of the Upper Indus Basin, Pakistan. *Environ. Ecol. Stat.* **2021**, *28*, 383–404. [CrossRef]
27. Yaseen, M.; Ahmad, I.; Guo, J.L.; Azam, M.I.; Latif, Y. Spatiotemporal Variability in the Hydrometeorological Time-Series over Upper Indus River Basin of Pakistan. *Adv. Meteorol.* **2020**, *2020*. [CrossRef]
28. Asmat, U.; Athar, H.; Nabeel, A.; Latif, M. An AOGCM based assessment of interseasonal variability in Pakistan. *Clim. Dyn.* **2018**, *50*, 349–373. [CrossRef]
29. Azmat, M.; Wahab, A.; Huggel, C.; Qamar, M.U.; Hussain, E.; Ahmad, S.; Waheed, A. Climatic and hydrological projections to changing climate under CORDEX-South Asia experiments over the Karakoram-Hindukush-Himalayan water towers. *Sci. Total Environ.* **2020**, *703*, 135010. [CrossRef]
30. Ahmad, M.; Chand, S.; Yaseen, M. High resolution bayesian spatio-temporal precipitation modelling in pakistan for the appraisal of trends. *Pak. J. Agric. Sci.* **2020**, *57*, 1669–1680. [CrossRef]
31. Mahmood, R.; Jia, S.; Tripathi, N.K.; Shrestha, S. Precipitation Extended Linear Scaling Method for Correcting GCM Precipitation and Its Evaluation and Implication in the Transboundary Jhelum River Basin. *Atmosphere* **2018**, *9*, 160. [CrossRef]

32. Ateequr, R.; Ghumman, A.R.; Ahmad, S.; Hashmi, H.N. Performance assessment of artificial neural networks and support vector regression models for stream flow predictions. *Environ. Monit. Assess.* **2018**, *190*. [CrossRef]
33. Lenderink, G.; van Ulden, A.; van den Hurk, B.; Keller, F. A study on combining global and regional climate model results for generating climate scenarios of temperature and precipitation for the Netherlands. *Clim. Dyn.* **2007**, *29*, 157–176. [CrossRef]
34. Wang, B. *The Asian Monsoon*; Springer Science & Business Media: Cham, Switzerland, 2006.
35. Dimri, A.; Niyogi, D.; Barros, A.; Ridley, J.; Mohanty, U.; Yasunari, T.; Sikka, D. Western disturbances: A review. *Rev. Geophys.* **2015**, *53*, 225–246. [CrossRef]
36. Khatri, W.D.; Xiefei, Z.; Ling, Z. Interannual and Interdecadal Variations in Tropical Cyclone Activity over the Arabian Sea and the Impacts over Pakistan. In *High-Impact Weather Events over the SAARC Region*; Springer: Berlin/Heidelberg, Germany, 2015; pp. 129–145.
37. Rasul, G.; Chaudhry, Q. Review of advance in research on Asian summer monsoon. *Pak. J. Meteorol.* **2010**, *6*, 1–10.
38. Parvaze, S.; Ahmad, D. Meteorological Drought Quantification with Standardized Precipitation Index for Jhelum Basin in Kashmir Valley. *Int. J. Adv. Res. Comput. Sci. Manag.* **2018**, *7*, 688–697.
39. Mahmood, R.; Babel, M.S.; Jia, S. Assessment of temporal and spatial changes of future climate in the Jhelum river basin, Pakistan and India. *Weather Clim. Extrem.* **2015**, *10*, 40–55. [CrossRef]
40. Rizwan, M.; Jamal, K.; Chen, Y.; Chauhdary, J.N.; Zheng, D.; Anjum, L.; Youhua, R.; Pan, X. Precipitation Variations under a Changing Climate from 1961-2015 in the Source Region of the Indus River. *Water* **2019**, *11*, 1366. [CrossRef]
41. Immerzeel, W.W.; Wanders, N.; Lutz, A.F.; Shea, J.M.; Bierkens, M.F.P. Reconciling High-Altitude Precipitation in the Upper Indus Basin with Glacier Mass Balances and Runoff. *Hydrol. Earth Syst. Sci.* **2015**, *19*, 4673–4687. [CrossRef]
42. Lutz, A.; Immerzeel, W.; Kraaijienbrink, P.D.A. *Gridded Meteorological Datasets and Hydrological Modelling in the Upper Indus Basin*; Future Water: Wageningen, The Netherlands, 2014; p. 83.
43. Yatagai, A.; Kamiguchi, K.; Arakawa, O.; Hamada, A.; Yasutomi, N.; Kitoh, A. APHRODITE: Constructing a Long-Term Daily Gridded Precipitation Dataset for Asia Based on a Dense Network of Rain Gauges. *Bull. Am. Meteor. Soc.* **2012**, *93*, 1401–1415. [CrossRef]
44. Lutz, A.; Immerzeel, W. *Water Availability Analysis for the Upper Indus, Ganges and Brahmaputra River Basins*; Future Water: Wageningen, The Netherlands, 2013; p. 85.
45. Hersbach, H.; Bell, B.; Berrisford, P.; Biavati, G.; Horányi, A.; Muñoz Sabater, J.; Nicolas, J.; Peubey, C.; Radu, R.; Rozum, I.; et al. ERA5 Hourly Data on Single Levels from 1979 to Present. Available online: https://cds.climate.copernicus.eu/cdsapp#!/dataset/reanalysis-era5-single-levels?tab=form (accessed on 2 November 2021).
46. Department of Civil and Environmental Engineering Princeton University. *Global Meteorological Forcing Dataset for Land Surface Modeling*; Computational and Information Systems Laboratory, Research Data Archive at the National Center for Atmospheric Research: Boulder, CO, USA, 2006. [CrossRef]
47. Chen, H.-P.; Sun, J.-Q.; Li, H.-X. Future changes in precipitation extremes over China using the NEX-GDDP high-resolution daily downscaled data-set. *Atmos. Ocean. Sci. Lett.* **2017**, *10*, 403–410. [CrossRef]
48. Kumar, P.; Kumar, S.; Barat, A.; Sarthi, P.P.; Sinha, A.K. Evaluation of NASA's NEX-GDDP-simulated summer monsoon rainfall over homogeneous monsoon regions of India. *Theor. Appl. Climatol.* **2020**, *141*, 525–536. [CrossRef]
49. Taylor, E.K.; Ronald, S.; Meehl, G. An Overview of CMIP5 and the Experiment Design. *Bull. Am. Meteor. Soc.* **2011**, *93*, 485–498. [CrossRef]
50. Pedregosa, F.; Varoquaux, G.; Gramfort, A.; Michel, V.; Thirion, B.; Grisel, O.; Blondel, M.; Prettenhofer, P.; Weiss, R.; Dubourg, V.; et al. Scikit-learn: Machine Learning in Python. *J. Mach. Learn. Res.* **2011**, *12*, 2825–2830.
51. Asong, Z.E.; Khaliq, M.N.; Wheater, H.S. Regionalization of Precipitation Characteristics in the Canadian Prairie Provinces Using Large-scale Atmospheric Covariates and Geophysical Attributes. *Stoch. Environ. Res. Risk Assess.* **2015**, *29*, 875–892. [CrossRef]
52. Liu, M.; Huang, Y.; Li, Z.; Tong, B.; Liu, Z.; Sun, M.; Jiang, F.; Zhang, H. The Applicability of LSTM-KNN Model for Real-Time Flood Forecasting in Different Climate Zones in China. *Water* **2020**, *12*, 440. [CrossRef]
53. Zeraatpisheh, M.; Bakhshandeh, E.; Emadi, M.; Li, T.; Xu, M. Integration of PCA and Fuzzy Clustering for Delineation of Soil Management Zones and Cost-Efficiency Analysis in a Citrus Plantation. *Sustainability* **2020**, *12*, 5809. [CrossRef]
54. Chen, Y.; Zheng, B.; Hu, Y. Mapping Local Climate Zones Using ArcGIS-Based Method and Exploring Land Surface Temperature Characteristics in Chenzhou, China. *Sustainability* **2020**, *12*, 2974. [CrossRef]
55. Benestad, R.E.; Chen, D.; Mezghani, A.; Fan, L.; Parding, K. On Using Principal Components to Represent Stations in Empirical–Statistical Downscaling. *Tellus A* **2015**, *67*, 28326. [CrossRef]
56. Liu, Q.; Huang, C.; Li, H. Quality Assessment by Region and Land Cover of Sharpening Approaches Applied to GF-2 Imagery. *Appl. Sci.* **2020**, *10*, 3673. [CrossRef]
57. Abbasi, F.; Bazgeer, S.; Kalehbasti, P.R.; Oskoue, E.A.; Haghighat, M.; Kalebasti, P.R. New climatic zones in Iran: A comparative study of different empirical methods and clustering technique. *Theor. Appl. Climatol.* **2021**, *147*, 47–61. [CrossRef]
58. Sammour, M.; Othman, Z.A.; Muda, Z.; Ibrahim, R. An agglomerative hierarchical clustering with association rules for discovering climate change patterns. *Int. J. Adv. Comput. Sci. Appl.* **2019**, *10*, 233–240. [CrossRef]
59. Mimmack, G.M.; Mason, S.J.; Galpin, J.S. Choice of Distance Matrices in Cluster Analysis: Defining Regions. *J. Clim.* **2001**, *14*, 2790–2797. [CrossRef]

60. Nam, W.; Shin, H.; Jung, Y.; Joo, K.; Heo, J.-H. Delineation of the Climatic Rainfall Regions of South Korea Based on a Multivariate Analysis and Regional Rainfall Frequency Analyses. *Int. J. Climatol.* **2015**, *35*, 777–793. [CrossRef]
61. Carvalho, M.J.; Melo-Gonçalves, P.; Teixeira, J.C.; Rocha, A. Regionalization of Europe Based on a K-Means Cluster Analysis of the Climate Change of Temperatures and Precipitation. *Phys. Chem. Earth Parts A/B/C* **2016**, *94*, 22–28. [CrossRef]
62. Rousseeuw, P.J. Silhouettes: A graphical aid to the interpretation and validation of cluster analysis. *J. Comput. Appl. Math.* **1987**, *20*, 53–65. [CrossRef]
63. Van Vuuren, D.P.; Edmonds, J.; Kainuma, M.; Riahi, K.; Thomson, A.; Hibbard, K.; Hurtt, G.C.; Kram, T.; Krey, V.; Lamarque, J.-F.; et al. The Representative Concentration Pathways: An Overview. *Clim. Change* **2011**, *109*, 5–31. [CrossRef]
64. Rozante, J.; Moreira, D.; Godoy, R.C.M.; Fernandes, A. Multi-model ensemble: Technique and validation. *Geosci. Model Dev. Discuss.* **2014**, *7*, 2933–2959. [CrossRef]
65. Stephenson, D.B.; Coelho, C.A.S.; Doblas-Reyes, F.J.; Balmaseda, M. Forecast assimilation: A unified framework for the combination of multi-model weather and climate predictions. *Tellus A Dyn. Meteorol. Oceanogr.* **2005**, *57*, 253–264. [CrossRef]
66. Salman, S.A.; Shahid, S.; Ismail, T.; Ahmed, K.; Wang, X.-J. Selection of Climate Models for Projection of Spatiotemporal Changes in Temperature of Iraq with Uncertainties. *Atmos. Res.* **2018**, *213*, 509–522. [CrossRef]
67. Sen, P.K. Estimates of the Regression Coefficient Based on Kendall's Tau. *J. Am. Stat. Assoc.* **1968**, *63*, 1379–1389. [CrossRef]
68. Mahmood, R.; Jia, S.F. Spatial and temporal hydro-climatic trends in the transboundary Jhelum River basin. *J. Water Clim. Change* **2017**, *8*, 423–440. [CrossRef]
69. Jasrotia, A.S.; Baru, D.; Kour, R.; Ahmad, S.; Kour, K. Hydrological modeling to simulate stream flow under changing climate conditions in Jhelum catchment, western Himalaya. *J. Hydrol.* **2021**, *593*. [CrossRef]
70. Dahri, Z.H.; Ludwig, F.; Moors, E.; Ahmad, S.; Ahmad, B.; Ahmad, S.; Riaz, M.; Kabat, P. Climate change and hydrological regime of the high-altitude Indus basin under extreme climate scenarios. *Sci. Total Environ.* **2021**, *768*, 144467. [CrossRef] [PubMed]
71. McMahon, T.A.; Peel, M.C.; Karoly, D.J. Assessment of Precipitation and Temperature Data from CMIP3 Global Climate Models for Hydrologic Simulation. *Hydrol. Earth Syst. Sci.* **2015**, *19*, 361–377. [CrossRef]
72. Smith, I.; Chandler, E. Refining Rainfall Projections for the Murray Darling Basin of South-East Australia—The Effect of Sampling Model Results Based on Performance. *Clim. Chang.* **2010**, *102*, 377–393. [CrossRef]
73. Xu, C.-y.; Widén, E.; Halldin, S. Modelling Hydrological Consequences of Climate Change—Progress and Challenges. *Adv. Atmos. Sci.* **2005**, *22*, 789–797. [CrossRef]
74. Kay, A.L.; Davies, H.N.; Bell, V.A.; Jones, R.G. Comparison of Uncertainty Sources for Climate Change Impacts: Flood Frequency in England. *Clim. Change* **2009**, *92*, 41–63. [CrossRef]
75. Woldemeskel, F.M.; Sharma, A.; Sivakumar, B.; Mehrotra, R. An Error Estimation Method for Precipitation and Temperature Projections for Future Climates. *J. Geophys. Res.* **2012**, *117*, D22. [CrossRef]
76. Zhang, X.; Xu, Y.-P.; Fu, G. Uncertainties in SWAT Extreme Flow Simulation under Climate Change. *J. Hydrol.* **2014**, *515*, 205–222. [CrossRef]
77. Latif, M.; Hannachi, A.; Syed, F.S. Analysis of Rainfall Trends over Indo-Pakistan Summer Monsoon and Related Dynamics Based on CMIP5 Climate Model Simulations. *Int. J. Climatol.* **2018**, *38*, e577–e595. [CrossRef]
78. Srinivasa Raju, K.; Sonali, P.; Nagesh Kumar, D. Ranking of CMIP5-based global climate models for India using compromise programming. *Theor. Appl. Climatol.* **2017**, *128*, 563–574. [CrossRef]
79. Khan, N.; Shahid, S.; Ahmed, K.; Ismail, T.; Nawaz, N.; Son, M. Performance Assessment of General Circulation Model in Simulating Daily Precipitation and Temperature Using Multiple Gridded Datasets. *Water* **2018**, *10*, 1793. [CrossRef]

atmosphere

MDPI

Article

Assessment of Satellite-Based Rainfall Products Using a X-Band Rain Radar Network in the Complex Terrain of the Ecuadorian Andes

Nazli Turini [1,*], Boris Thies [1], Rütger Rollenbeck [1], Andreas Fries [2], Franz Pucha-Cofrep [3], Johanna Orellana-Alvear [1,4], Natalia Horna [5] and Jörg Bendix [1]

1 Laboratory for Climatology and Remote Sensing, Faculty of Geography, Philipps-Universität Marburg, 35032 Marburg, Germany; thies@staff.uni-marburg.de (B.T.); rollenbe@staff.uni-marburg.de (R.R.); johanna.orellana@ucuenca.edu.ec (J.O.-A.); bendix@staff.uni-marburg.de (J.B.)
2 Unidad Ingeniería Civil y Geología (UCG), Hidrología y Climatología Working Group, Universidad Técnica Particular de Loja, Loja 1101608, Ecuador; aefries@utpl.edu.ec
3 Institute for Environmental Sciences, Brandenburg University of Technology (BTU) Cottbus-Senftenberg, Burger Chaussee 2, 03044 Cottbus, Germany; puchacof@b-tu.de
4 Departamento de Recursos Hídricos y Ciencias Ambientales, Universidad de Cuenca, Cuenca 010207, Ecuador
5 Department of Geography and Environmental Management, University of Waterloo, Waterloo, ON N2L 3G1, Canada; nhornaku@uwaterloo.ca
* Correspondence: turini@staff.uni-marburg.de; Tel.: +49-2824204

Abstract: Ground based rainfall information is hardly available in most high mountain areas of the world due to the remoteness and complex topography. Thus, proper understanding of spatio-temporal rainfall dynamics still remains a challenge in those areas. Satellite-based rainfall products may help if their rainfall assessment are of high quality. In this paper, microwave-based integrated multi-satellite retrieval for the Global Precipitation Measurement (GPM) (IMERG) (MW-based IMERG) was assessed along with the random-forest-based rainfall (RF-based rainfall) and infrared-only IMERG (IR-only IMERG) products against the quality-controlled rain radar network and meteorological stations of high temporal resolution over the Pacific coast and the Andes of Ecuador. The rain area delineation and rain estimation of each product were evaluated at a spatial resolution of 11 km^2 and at the time of MW overpass from IMERG. The regionally calibrated RF-based rainfall at 2 km^2 and 30 min was also investigated. The validation results indicate different essential aspects: (i) the best performance is provided by MW-based IMERG in the region at the time of MW overpass; (ii) RF-based rainfall shows better accuracy rather than the IR-only IMERG rainfall product. This confirms that applying multispectral IR data in retrieval can improve the estimation of rainfall compared with single-spectrum IR retrieval algorithms. (iii) All of the products are prone to low-intensity false alarms. (iv) The downscaling of higher-resolution products leads to lower product performance, despite regional calibration. The results show that more caution is needed when developing new algorithms for satellite-based, high-spatiotemporal-resolution rainfall products. The radar data validation shows better performance than meteorological stations because gauge data cannot correctly represent spatial rainfall in complex topography under convective rainfall environments.

Keywords: complex terrain; Ecuador; GPM IMERG; rainfall; radar network; satellite retrieval

Citation: Turini, N.; Thies, B.; Rollenbeck, R.; Fries, A.; Pucha-Cofrep, F.; Orellana-Alvear, J.; Horna, N.; Bendix, J. Assessment of Satellite-Based Rainfall Products Using a X-Band Rain Radar Network in the Complex Terrain of the Ecuadorian Andes. *Atmosphere* **2021**, *12*, 1678. https://doi.org/10.3390/atmos12121678

Academic Editors: Zuohao Cao, Huaqing Cai and Xiaofan Li

Received: 28 October 2021
Accepted: 8 December 2021
Published: 14 December 2021

1. Introduction

Understanding precipitation amounts and patterns is essential for sustainable water management and monitoring the hydrological cycle [1]. In complex mountainous regions characterized by high spatiotemporal variability, coarse networks of operational precipitation gauge stations are often lacking. The spatiotemporal variability, combined with lack of gauge data, makes the time series and area-averaged rainfall analysis more complicated in these regions [2]. This also applies to the complex topography of the Andes in Ecuador.

Early satellite-based rainfall retrieval efforts estimated rainfall from geostationary infrared (IR) data, using the indirect relationship between precipitation rate and the temperature of cloud on top [3]. Hence, the algorithms and the product accuracy were limited to the top of the cloud's characteristics. Unlike IR, microwave (MW) sensors measure thermal radiance from actual precipitation particles in the clouds; consequently, MW retrieval generally provides superior precipitation information [4].

A recent result of the continuous technological improvement of low-Earth-orbiting passive MW satellites and spaceborne radars in the MW band is the Global Precipitation Measurement (GPM) mission [5]. GPM was launched in 2014 as post Tropical Rainfall Measuring Mission (TRMM) [6]. Compared with TRMM, the GPM improved sensitivity to light precipitation and distribution of rain and snow. These improvements have achieved by a two-frequencies precipitation radar (Ku band (13.6 GHz) and Ka-band (35.5 GHz)) as well as the GPM multi-channel microwave imager (GMI) that accommodates higher spectral resolution at frequencies of 10.65, 18.7, 23.8, 26.5, 89, 165.5, and 183.3 GHz [5,7,8].

However, several studies showed that machine learning could improved the regionally calibrated retrievals using simply passive IR data from geostationary orbit (GEO) [3,8–13]. Compared to the passive MW and radar sensors, the GEO systems provide the high temporal (10–30 min) and spatial (2–4 km^2) resolution. It is essential to capture the short-term characteristics of rainfall systems in the retrieval [8]

A few studies have investigated the performance of satellite-based rainfall products over Ecuadorian areas. The Precipitation Estimation from Remotely Sensed Information using Artificial Neural Networks (PERSIANN) [14] shows low agreement with rain gauge in daily resolution [2] in rain area detection. Manz et al. [15] investigated the performance of the integrated multi-satellite retrievals for GPM (IMERG) [5] and TRMM multi-satellite precipitation analysis (TMPA) [6] against gauge data with different temporal resolutions (hourly, 3 h, and daily). In their study, IMERG showed better agreement than TMPA, especially on the high elevation of Andes. Erazo et al. [16] reported that at high elevations in the Andes, TRMM 3B43 Version 7 retrievals showed a higher correlation (R^2 = 0.82) on monthly compared with interpolated gauge data at a spatial resolution of 27.75 km^2. The result of the validation of the regionally developed algorithm in Ecuador, the random forest-based rainfall (RF-based rainfall) of Turini et al. [3] with an 11 km^2 resolution, obtained a median Heike skill score (HSS) around 0.35 for daily gauge stations, meanwhile the lower performance of the IR-only from the IMERG (IR-only IMERG) showed by HSS = 0.2. In their method, they used the Random forest algorithm to retrieve rainfall. In this text, the RF-based rainfall stands for the rainfall retrieval from random forest algorithm [3]. The RF-based rainfall retrieval performed in estimating the rainfall rate with correlation coefficient (r) values 0.34 [3].

To improve satellite-based products' overall performance, understanding the sources of error on the highest possible temporal resolution is crucial [6,17]. Given the high spatiotemporal variability of rainfall in Ecuador, spatiotemporally high-resolution validation sources for rainfall are lacking. Therefore, as stated before, only a couple of studies have investigated the performance of satellite-based rainfall products at higher spatiotemporal resolution [15,18].

Different studies have found that, due to the variability of weather and climate in complex terrain, the satellite retrievals are posed to challenges both in IR and MW products [3,8,12,13,19]. Dinku et al. [19] evaluated the impact of topography on IR-based Tropical Applications of Meteorology using Satellite and ground-based observation (TAMSAT) [20] in East Africa for 1998–2012, comprising five different countries: Uganda, Kenya, Tanzania, Rwanda, and Burundi. In the study, the elevation varied between 1500 and 4500 m [19]. TAMSAT showed an underestimation. Dinku et al. [19] argued that the underestimation corresponded mainly to convective and orographic rainfall during the rainy season (March, April, and May), mostly in the windward exposition.

In this work, we aimed to validate different satellite-based rainfall products to identify and understand sources of errors in the complex elevation of the Andes in Ecuador on a sub-daily time scale. Our aim was not just to compare satellite-based rainfall products with ground measurements but also to identify the sources of the differences between the satellite-based rainfall products and ground measurements. Therefore, in this study, we evaluated the performance of MW-based IMERG in comparison with RF-based rainfall and IR-only IMERG against high-spatiotemporal-resolution data from ground based radar network and high temporal resolution of meteorological stations to characterize the impact of climatic and topographic conditions on satellite-based rainfall products at the time of MW overpass. We also assessed the performance of regionally trained RF-based rainfall in Ecuador on the subdaily time scale (30 min) and high spatial resolution (2 km^2) with the aim of finding the source of possible errors for further development. Following a description of the climatology of the study area, the satellite-based rainfall products, ground based radar data and meteorological stations are described in Section 2.1. Section 2.2 introduces the evaluation methodology with a focus on rain area detection and rain estimation. The results are presented in Section 3 and discussed in Section 4. Finally, the important findings are summarized in Section 5.

2. Materials and Methods

2.1. Data

2.1.1. Radar

In the current study, the data from two rainfall radars, which are part of the Radarnet-Sur network in Southern Ecuador, were used. The westernmost radar system is located on Cerro Guachaurco (3100 m above sea level (m.a.s.l) (GUAXX radar)). Another radar system is located at 4450 m.a.s.l (to the best of our knowledge, this is the highest worldwide) on the Paragüillas peak on the north border of the Cajas National Park in Southern Ecuador (CAXX radar). The radars have a maximum range of 100 km^2 and provide images with spatial resolutions of 500 m every 5 min. For more information about the Radarnet-Sur network (Figure 1a) infrastructure, please refer to Bendix et al. [21]. The coverage of radars in this study is shown in Figure 1a.

Radarnet-Sur calibration strategies have been continuously developed since 2006. The calibration strategy is based on a statistical procedure that uses the available rain gauge data. The data processing and correction algorithms in this empirical calibration consisted of four steps: (i) clutter and noise removal; (ii) atmospheric and geometric attenuation correction; (iii) interpolation of blind sectors; (vi) application of the empirically derived daily variable Z/R relationship. In this equation Z means radar reflectivity factor and R stands for rainfall intensity. For more information about the calibration algorithm, please refer to [22]. The final product from the radars used a blending technique for overlapping areas and temporal data gaps were completed using additional data from the rain gauges. For further information about the extended calibration strategy, please refer to [23].

The observed rainfall data from the radars were quality-controlled for detecting possible inconsistencies and selecting high-quality data. All the scenes from the radars were visually inspected. For this, the gauge data from the National Institute of Meteorology and Hydrology (INAMHI) (daily), the Universidad TécnicaParticular de Loja (UTPL) gauge network (10 min resolution), and the Cuenca University gauge network were used as references. The scenes in which there were no rain in the radar but rain in each of the gauges and vice versa were removed. Additionally, obviously failed recordings were manually removed. Furthermore, we used the infrared channel IR 3.9 from GOES-16 to detect the movement of cold clouds and radar rainfall rate. Although we have enough data available in our observation period, electronic technical problems and other issues caused data failure.

We delivered the final products of radar reflectivity and rainfall rate after attenuation and clutter correction for the time period between April 2017 and the end of January 2018 (GUAXX: 16 June 2017 to 1 February 2018; CAXX: 19 April 2017 to 1 July 2017). The reflectivity ranged from −31.5 to 91.5 dBZ with a total of 256 possible values.

The spatial distribution of the sum of the radar rainfall for the observation period is shown in Figure 1b. The rainfall sum, showing totals between 250 and 4492 mm. The rainfall pattern is different over the study region, covering a climatically diverse area. The spatiotemporal rainfall distribution in the radar coverage is generally affected by the Andes mountains, the El Niño–Southern Oscillation (ENSO), the biannual migration of the intertropical convergence zone (ITCZ), and also the cold von Humboldt current in the Pacific Ocean, [15,24,25]. On the eastern sides of the Andes, the strong topographic slopes and easterly winds result in orographic effects [26,27], which is causing the cyclical spatiotemporal rainfall behavior and deep convection [28].

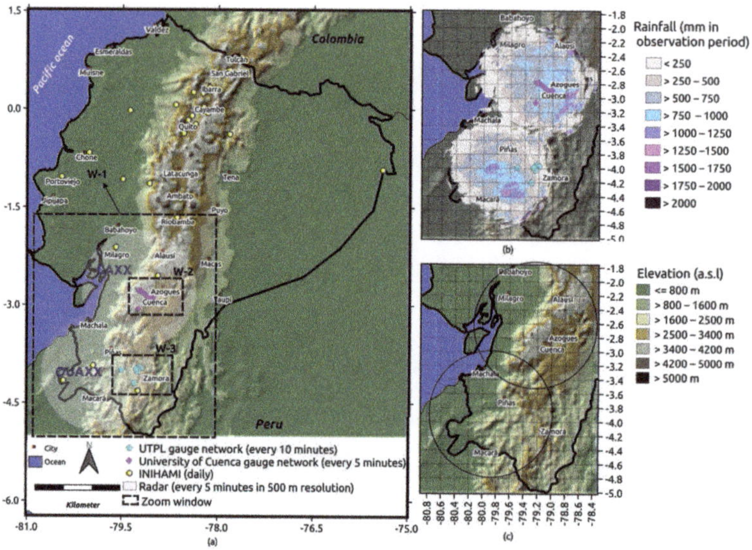

Figure 1. The distribution of (**a**) meteorological stations (19 April 2017 to 28 February 2018) and spatial coverage of radars (GUAXX: 16 June 2017 to 1 February 2018; CAXX: 19th April 2017 to 1 July 2017) used in this study, (**b**) the radars in the study period (GUAXX: 16 June 2017 to 1 February 2018; CAXX: 19th April 2017 to 1 July 2017). For validation purposes, we excluded the radar data in the very near range (<10 km distance from the radar site) to avoid contamination through noise. We also excluded the far range >50 km due to possible attenuation errors. Nevertheless, we show the rainfall amount in the entire radar range for better illustration. The extent of study area is shown in windows (W)-1. (**c**) Spatial distribution of the elevation in the radar coverage area. W-2 and W-3 rectangles outline the extent of Figure 2a,b.

2.1.2. Meteorological Stations

A meteorological station network, comprising 21 high-temporal-resolution rain stations, was used in this study. Meteorological station data were obtained from UTPL and University of Cuenca. Meteorological stations from UTPL and University of Cuenca provide rain data every 10 and 5 min, respectively. Daily rainfall information was acquired from INAMHI. Meteorological station data from 19 April 2017 to 28 February 2018 were used as validation information to examine radar quality. The high temporal resolution meteorological stations from UTPL and the University of Cuenca are used to validate the satellite-based products at the time of MW-overpasses. We obtained the data from all organizations after quality checks.

The quality check for the station data from the University of Cuenca is performed by drawing a cumulative precipitation curve that identifies abnormal records (outliers and wrong measurements). These measurements are disregarded from the time series. In addition, correlation to nearby stations is also performed as a double check if necessary. In order to maximize the quality of the measurements, regular maintenance of the stations in the field (every three weeks or fewer) is performed. For the INAMHI data, it is checked if daily values are between 0 and 250 mm, which is the maximum daily precipitation value registered at a national scale.

Figure 1a shows the distribution of the meteorological stations used in this study.

It should be noted that these data are not included in the Global Precipitation Climatology Center (GPCC) network and therefore not used for the gauge-calibrated final IMERG product.

2.1.3. Integrated Multi-Satellite Retrievals for GPM

IMERG is a level 3 product which integrates all MW sensors, MW-calibrated IR estimates, and rain gauge measurements on a global scale [29]. All MW estimates, after calibration, were subjected to the Climate Prediction Center MORPHing technique (CMORPH) [30] to calculate the motion vectors from the IR measurements and the different atmospheric variables from numerical models. In regions without direct PMW overpasses, the algorithm uses the retrieved rainfall from PERSIANN-CCS [14] and GEO IR (IR-only IMERG) to complete the gridded product. In the last step, the monthly rain data from the GPCC were used to as a bias correction of the rainfall estimate [29].

In this study, the latest available version of IMERG (IMERG-V06 [29]), which displayed an overall improvement in the precipitation estimation compared with version-05 [31], was used.

The IMERG provides rainfall estimates with the spatial resolution of 0.1° (11 km^2) in every 30 min. We focused on the final product of IMERG Version 06 (IMERG-V06), gauge-adjusted retrievals for the study period. NASA also provided the quality index (QI) as a variable in 30 min resolution [32]. The QI indicates the relative quality of rainfall estimates in half-hourly IMERG products, fluctuating temporally between passive MW (PMW) and IR-based rainfall estimates. Additionally, the time of the overpass of each MW swath is provided in metadata with the name of 'HQobservationTime'.

For our validation, the multi-satellite precipitation estimates with the gauge calibration subdata set of IMERG (precipitationCal), as well as "IRprecipitation" was used. In this study, IRprecipitation and IR-only IMERG are equivalent.

2.1.4. Random Forest-Based Rainfall

The random forest-based rainfall (RF-based rainfall) product is the regionally calibrated rainfall retrieval scheme developed Ecuador by Turini et al. [3]. The algorithm uses random forest (RF) to calculate rainfall rates in surface level by means of multi-spectral IR data from Geostationary Operational Environmental Satellite 16 (GOES-16). The algorithm is trained based on MW-only precipitation data from IMERG-V06. The RF-based rainfall product was implemented by (i) delineating the rain area, and (ii) assigning of the rainfall rate at 11 km^2 spatial resolution and for the time of a MW overpass. As predictors, GOES IR bands, band combinations, geostatistical texture features calculated from the original GOES IR bands, and ancillary data were used. Turini et al. [3] used the geostatistical texture features to capture the clouds' heterogeneity. They calculated the texture features using a 5 × 5 pixel moving window method. First, for each GOES IR band, variograms (VARs), madograms (MADs), and rodograms (RODs) and then, for each possible bands combination, cross-variograms (CVs) and pseudo cross-variograms (PCVs) were calculated. Please refer to Schulz et al. [33] for more information about definitions and equations of texture features. The most important features were obtained monthly for each of the steps (rain area delineation and rainfall rate assignment) separately. The model tuning and feature selection results showed that, in addition to the ancillary data, the information recorded in the geostatistical texture features was the most important for rain area delineation and rainfall rate assignment [3].

The PCV was the dominant texture feature selected in almost all months, both for rain area delineation and rain rate assignment [3].

After training the models, the RF-based rainfall at a high spatiotemporal resolution (2 km^2, 15 min) was estimated. In this step, the models were applied to the GOES-16 scenes where MW-IMERG was available and the following scenes until the next model was present in Turini et al. [3]. The product is available from 19 April 2017 to 19 April 2018.

2.2. Methods

Three different validations were employed in this study to assess satellite-based rainfall product performance. Due to the different availabilities of the slots of the products, the period for this study ranged from 19 April 2017 to 1 February 2018 in the time slots where radar data are available.

- The performance of the satellite-based rainfall products are investigated against the X-band rain radar network at the time of MW overpass at a spatial resolution of 11 km^2.
- The performance of the satellite-based rainfall products is investigated against the ground based meteorological station network at the time of MW overpass at a spatial resolution of 11 km^2.
- The RF-based rainfall product is validated against the X-band rain radar network in the temporal resolution of 30 min and spatial resolution of 2 km^2.

2.2.1. Validation of Satellite-Based Rainfall Products at the Time of MW Overpass from IMERG

The first validation was performed to investigate satellite-based rainfall products' performance against X-band rain radar network when MW overpass sensors from IMERG are present. This is essential since the IMERG data set has been widely used to develop satellite-based rainfall products [8,12,13,34].

We used different subdata sets in the IMERG product. We first considered the pixels from "precipitationCal" when the PMWs swat was available ("HQobservation"). Then, the pixels with the "PrecipitationQualityIndex" >0.6 (which indicates the current half-hour microwave swath data) [32] were picked out. "IRprecipitation" were also selected in the same pixels from IMERG. This data set (IR only) was retrieved from the PERSIANN-CCS in IMERG, which are calibrated regionally to the PMW-only measures [29]. Therefore, in this study, we named this product "IR-only IMERG".

To compile the most robust data set for the first validation of satellite-based rainfall products against the radars at the time of MW overpass in IMERG, we defined the following criteria: (i) For temporal matching, we used "HQobservationTime" for IMERG to determine the exact time of MW overpass in each pixel. Then, we rounded the WM overpass time to the closest 5 min to be compatible with the temporal resolution of the radar (every 5 min). In this step, we assumed that the RF-based rainfall and IR-only IMERG have the same timing as the time of MW overpass. (ii) To ensure the high-quality rainfall information from IMERG (merged MW-only precipitation estimates), we used the "PrecipitationQualityIndex". (iii) Sensitivity to light rain continuously degrades with increasing distance from the radar. To only assess the near range, we applied a circular mask with a radius of 50 km from the center of each radar. (iv) A mask for filtering the radar data for plausibility was also applied. A value of 1 indicates reliable data from radars. (v) There was some noise in the center of the radar due to the cross-talk from the antenna's side-lobes. Therefore, we omitted the inner pixels with a radius of 10 km from the center for the validation. (vi) Due to the different spatial resolutions of the RF-based rainfall (2 km^2), radar (0.5 km^2), radar quality index (0.5 km^2), DEM (1 km^2), and IMERG (11 km^2), the average resampling techniques in gdal [35] were used to guarantee spatial matching between the different data sets. In our study, we used the WGS84 projection coordinate system and all dataset were resampled to the spatial resolution of IMERG (11 km^2). (vi) The 0.5 mm/h was used as a threshold between rainy and non-rainy pixels for validation. (vii) The pixels in the radar considered rainy (>0.5 mm/h) but has a dBZ lower than −15 were considered false and filtered out from the validation data set.

By applying above criteria and withdraw the data pairs of the first validation against radar on a pixel basis, a total of 117,183 pixels of radar and MW-based IMERG, RF-based rainfall, and IR-only IMERG were made available at a half-hourly resolution for validation.

In the second validation, the overall performance of the rainfall area delineation and rainfall rate assignment was investigated for each product against data from ground based meteorological stations at the time of the MW overpass.

For comparison with the meteorological ground based station network, we only considered pixels with a minimum number of three gauges (see Figure 2). Tang et al. [36] underline that gauge networks with limited numbers of gauges in each pixel leads to underestimation of the performance of satellite-based rainfall products. This is because the point observations of gauges cannot represent pixel-based precipitation. Therefore, for this validation the stations from University of Cuenca with the temporal resolution of 5 min (Ana Davis, Zona Militar Davis and Balzay) and from UTPL (UTPL Militar, UTPL Tecnico and UTPL Villonaca) with the temporal resolution of 10 min are considered.

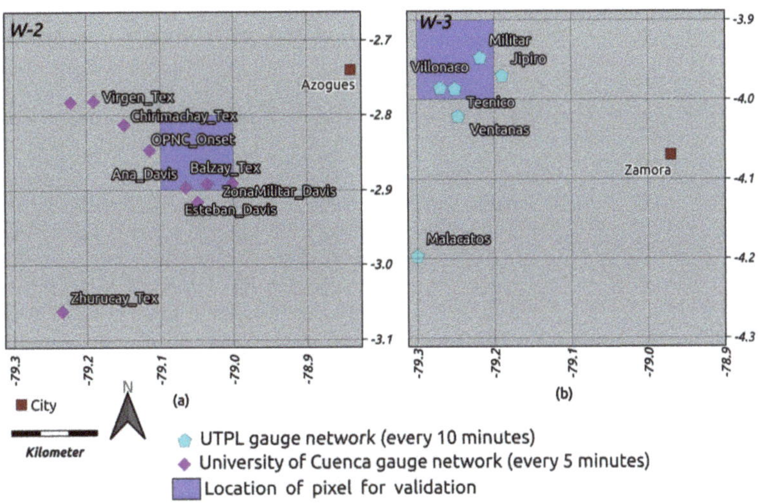

Figure 2. Location of pixels with a minimum number of three gauges for (**a**) the University of Cuenca gauge network and (**b**) the UTPL gauge network. In Figure 1a, W-2 and W-3 rectangles outline the extent of (**a**) and (**b**), respectively.

To generate the dataset for ground truth validation of the three satellite-based products against the gauge network, we proceeded as follows: (i) for temporal matching, we used "HQobservationTime" for IMERG to determine the exact time of MW overpass in each pixel. Then, we rounded the WM overpass time to the closest 5 min to be compatible with the temporal resolution of the radar (every 5 min). In this step, we assumed that the RF-based rainfall and IR-only IMERG have the same timing as the time of MW overpass. (ii) To ensure the high-quality rainfall information from IMERG (merged MW-only precipitation estimates), we used the "PrecipitationQualityIndex". In the next step (iii), the spatial matching were done using the average resampling techniques in gdal [35] to resample the products to the spatial resolution of IMERG (11 km^2). (iv) The threshold of 0.5 mm/h was used to distinguish between rainy and non-rainy events. (v) After selecting pixels, the arithmetic mean rainfall from station data was computed in these pixels, given that every pixel includes three stations at minimum.

2.2.2. Validation of RF-Based Rainfall Products in Native Resolution

In the third validation, we investigated the general behavior RF-based rainfall in rainfall area delineation and rainfall estimation in the native spatial resolution (2 km^2) and every 30 min in the entire study area for the study period. To prepare the data set for this validation strategy, we defined the following criteria: (i) In our study, area, subscale convective rainfall systems in the transition zones and valleys [37] are dominant. To understand satellite-based rainfall products' capability to capture these events, we kept the original spatial resolutions of the RF-based rainfall, 2 km^2; (ii) to minimize the uncertainties caused by the potential temporal offset between RF-based rainfall products, the radar and RF-based rainfall were aggregated in time to 30 min. For the temporal aggregation of the radar and the RF-based rainfall, we considered a unit conversion between mm/h and mm/30 min. (iii) We used a threshold of 0.2 mm/30 min to distinguish between rainy and non-rainy pixels for validation; (iv) equal to the first validation strategy, the pixels of the radar considered rainy (>0.2 mm/30 min) at a dBz lower than −15 dBz were considered false and were removed from the validation data set. (v) We omitted the inner pixels within a radius of 10 km^2 from the center; (vi) a mask for filtering the radar data for plausibility was also applied. (vii) In the next step, the RF-based rainfall was aggregated for the observation period in 1 h, 3 h, and daily for evaluation against the radar.

2.2.3. Validation Metrics for Rainfall Area Delineation and Rainfall Estimate

We considered all pixels from the validation data set in each validation strategy for the validation of rainfall area delineation. First, we calculated the cross-table's respective satellite-based rainfall products in comparison with the radar as a reference. Therefore we calculated the misses (M), hits (H), false alarms (F), and correct negatives (C). We define hit when the satellite-rainfall product and the radar are both raining in the same location. A miss occurs when the satellite-rainfall product is not raining but the radar shows rain, a false alarm holds when the satellite-rainfall product is raining but the radar is not and a correct negative is when both, the satellite-rainfall product and radar are showing cloudy but not rainy conditions (Figure 3).

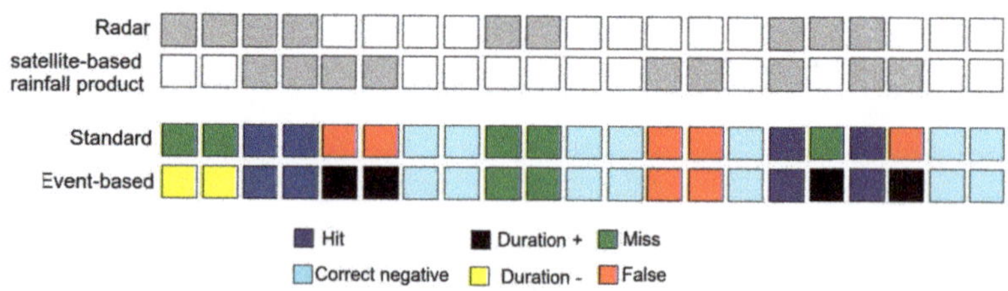

Figure 3. Schematic view of how H, M, and F were designated in the rain area validation. The dry pixels are shown in white, and the rainy pixels are shown in grey. The standard approach defines M (F) when a rainy pixel in the radar (satellite-based rainfall product) is related to a dry pixel in the satellite-based rainfall product (radar) at the same time. In the temporal event-based approach (fourth row), the M (F) in the vicinity time of hits are defined as a reduction (continuous) in the event duration. Thus, the terms Duration+ (Duration-) are described. True misses and true false alarms are the errors occurring simultaneously or in the same pixel, respectively [17].

We also defined temporal and spatial events. Schematic images of temporal and spatial events are illustrated in Figures 3 and 4, respectively.

Temporal events were defined to check the time lag effect of satellite scanning. You et al. [38] stressed this aspect for PMW observation. Later, Maranan et al. [17] investigated the time lag effect in IMERG, where false alarms were reduced through the temporal shift in IMERG relative to surface observations.

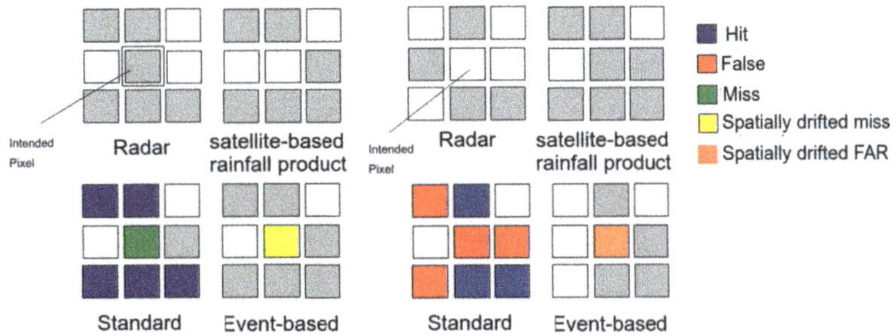

Figure 4. Schematic view of how hits, misses, and false alarms are designated in the rain area validation. The dry pixels are shown in white, and the rainy pixels are shown in grey. The standard approach defines M (F) when a rainy pixel in the radar (satellite-based rainfall product) is related to a dry pixel in the satellite-based rainfall product (radar) at the same time. In the spatial event-based approach (second row), the M (F) in the neighboring pixels are defined as a spatially drifted miss (false alarm) of the event. The errors simultaneously and in the same pixel are called true misses and false alarms, respectively.

We calculated the probability of detection (POD), false alarm ratio (FAR), and Heike skill score (HSS) as validation metrics from the H, M, F, and C.

To evaluate the accuracy of estimated rainfall from each satellite-based rainfall product, we used the mean absolute error (MAE), root mean square error (RMSE) and mean error (ME), and their normalized counterparts. These metrics were calculated when it was rainy for both radar and satellite-based rainfall products. Table 1 shows the detailed equations and the range of these metrics.

Table 1. List of validation metrics used in this study for rain area delineation and rain estimate.

Name	Metrics Equation	Range	Optimum
Probability of detection	$\text{POD} = \frac{H}{H-M}$	(0,1)	1
False alarm ratio	$\text{FAR} = \frac{F}{F-C}$	(0,1)	0
Heike skill score	$\text{HSS} = \frac{2(H\times C-F\times M)}{(H-M)(M+C)+(H+F)(F+C)}$	(0,1)	1
Mean absolute error	$\text{MAE} = \frac{1}{n}\sum_{i=1}^{n}\lvert P_i-O_i\rvert$	-	-
Normalized mean absolute error	$\text{NMAE} = \frac{\frac{1}{n}\sum_{i=1}^{n}\lvert P_i-O_i\rvert}{\frac{1}{n}\sum_{i=1}^{n}O_i}$	-	-
Root mean square error	$\text{RMSE} = \sqrt{\frac{\sum_{i=1}^{n}(P_i-O_i)^2}{n}}$	-	-
Normalized root mean square error	$\text{NRMSE} = \frac{\sqrt{\frac{\sum_{i=1}^{n}(P_i-O_i)^2}{n}}}{\frac{1}{n}\sum_{i=1}^{n}O_i}$	-	-
Mean error	$\text{ME} = \frac{1}{n}\sum_{i=1}^{n}(P_i-O_i)$	-	-
Normalized mean error	$\text{NME} = \frac{\frac{1}{n}\sum_{i=1}^{n}(P_i-O_i)}{\frac{1}{n}\sum_{i=1}^{n}O_i}$	-	-

3. Results

3.1. Validation Metrics for Satellite-Based Rainfall Products at the Time of MW Overpass against X-Band Rain Radar Network

3.1.1. Rain Area Delineation

The frequency of occurrence of the cross-table components formed on all available MW overpass timing (n = 51,384) is presented in Figure 5. Less than 5% of the MW overpass times in either radar or satellite-based rainfall products contain rainfall and a total of 0.73%, 0.58%, and 0.39% are hits for MW-based IMERG, RF-based rainfall, and IR-only IMERG, respectively. Successively, false alarms dominated the error with a fraction of 2.53% for MW-based IMERG, 2.08% for RF-based rainfall, and 2.24% for IR-only IMERG. All three product show reasonable agreement with the radar at the time of MW overpass Table 2. All products have a high FAR (0.78 for MW-based IMERG and RF-based rainfall, and 0.85 for IR-only IMERG).

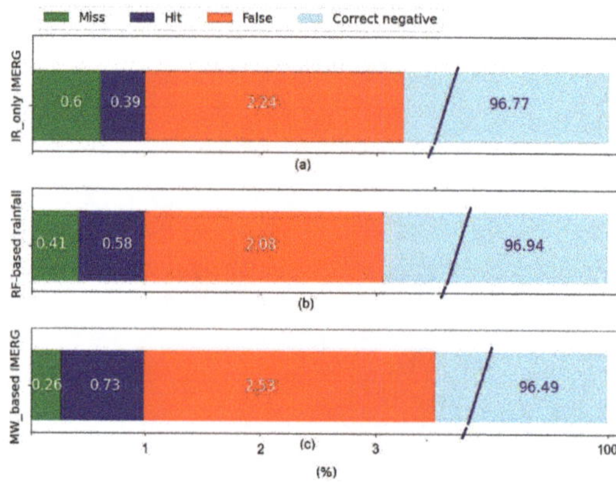

Figure 5. Standard cross-table approach for all available MW overpass times for the validation of rain area delineation for (**a**) IR-only IMERG, (**b**) RF-based rainfall, and (**c**) MW-based IMERG. Note that the correct negative fraction extends to 100%.

Table 2. The rain area delineation performance of satellite-based rainfall over the MW overpass time compared to ground radar network.

Satellite-Based Rainfall Products	POD	FAR	HSS
MW-based IMERG	0.74	0.78	0.33
RF-based rainfall	0.58	0.78	0.31
IR-only IMERG	0.39	0.85	0.2

Overall, MW-based IMERG exhibits relatively better performance (HSS = 0.33), RF-based rainfall performs somewhat the same as MW-based IMERG (HSS = 0.3), whereas IR-only IMERG performs the worst (HSS = 0.2). This shows the higher potential of using multispectral GEO data (RF-based rainfall) compared with only one IR channel rainfall retrieval, as is the case for IR-only IMERG [3,8,12,13].

Figure 6 reveals the spatial performance of the satellite-based rainfall products at the time of MW overpass during the study period. Figure 6c,f,i shows the spatial distribution of HSS for MW-based IMERG, IR-only IMERG, and RF-based rainfall, respectively. The HSS share similarities in the spatial distribution for all products, with the maximum occurring at the north and northeast of the study region (0.4–0.7 for MW-based IMERG, and 0.4–0.8 for IR-only IMERG and RF-based rainfall). However, in the northwestern part of the region, the ability to capture precipitation is almost lost due to the lower POD and higher FAR (0.7–1 for MW-based IMERG, IR-only IMERG, and RF-based rainfall) in all the products.

The GUAXX radar performs better in terms of POD in general but with a relatively higher FAR (0.6–1 for MW-based IMERG, and 0.7–1 for IR-only IMERG and RF-based rainfall), and this phenomenon illustrates that the products have difficulties in capturing the rainfall in these region (HSS of 0.1–0.6 for MW-based IMERG, 0.1–0.3 for IR-only IMERG, and 0.1–0.6 for RF-based rainfall). Please note that the time periods of available data for GUAXX and CAXX are different.

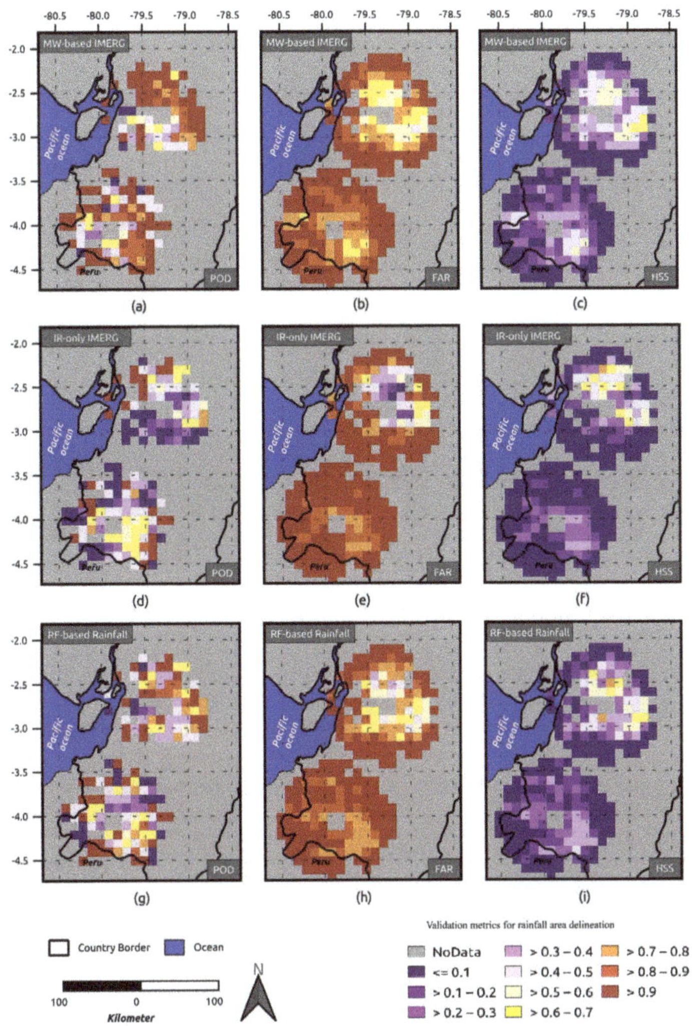

Figure 6. Spatial distribution of the validation metrics for rain area delineation at the time of MW overpass. (**a**) POD, (**b**) FAR, and (**c**) HSS showing the matrics for MW. The variables were calculated for MW-based IMERG. (**d**) POD, (**e**) FAR, and (**f**) HSS illustrating the performance of IR-only IMERG. (**g**) POD, (**h**) FAR, and (**i**) HSS showing the RF-based rainfall performance. The variables were calculated for each grid point of the validation data set over the stated period. For better illustration, we show the results up to 75 km distance from the center of each radar.

Figure 7 provides an overview of the validation metrics of the three satellite-based rainfall products for rain area delineation, along with the altitude. All products have a high FAR and a convincing POD. The performance at a terrain elevation of approximately 0–1500 m.a.s.l is relatively lower for all of the three products with HSS of 0.2–0.29 for MW-based IMERG, 0.1–0.22 for IR-only IMERG, and 0.2–0.25 for RF-based rainfall. The rain area delineation performance increased until 3000 m.a.s.l. At 0–750 m.a.s.l, the RF-based rainfall (HSS = 0.25) performs the best of all products.

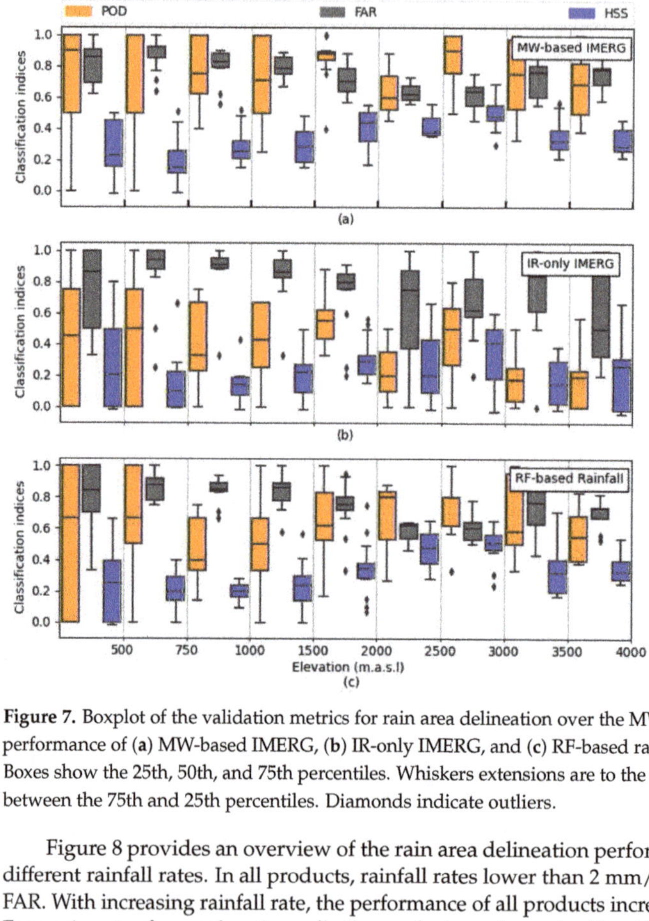

Figure 7. Boxplot of the validation metrics for rain area delineation over the MW overpass time. The performance of (**a**) MW-based IMERG, (**b**) IR-only IMERG, and (**c**) RF-based rainfall along elevation. Boxes show the 25th, 50th, and 75th percentiles. Whiskers extensions are to the maximum data value between the 75th and 25th percentiles. Diamonds indicate outliers.

Figure 8 provides an overview of the rain area delineation performance, along with different rainfall rates. In all products, rainfall rates lower than 2 mm/h have the highest FAR. With increasing rainfall rate, the performance of all products increases until 6 mm/h. For a rain rate of more than 6 mm/h, the products perform steadily. Altogether, the graph confirms (i) the poor rain area delineation performance at lower rainfall rates in Ecuador, and (ii) the WM-based IMERG shows the best performance with different rain rates in Ecuador, followed by RF-based rainfall.

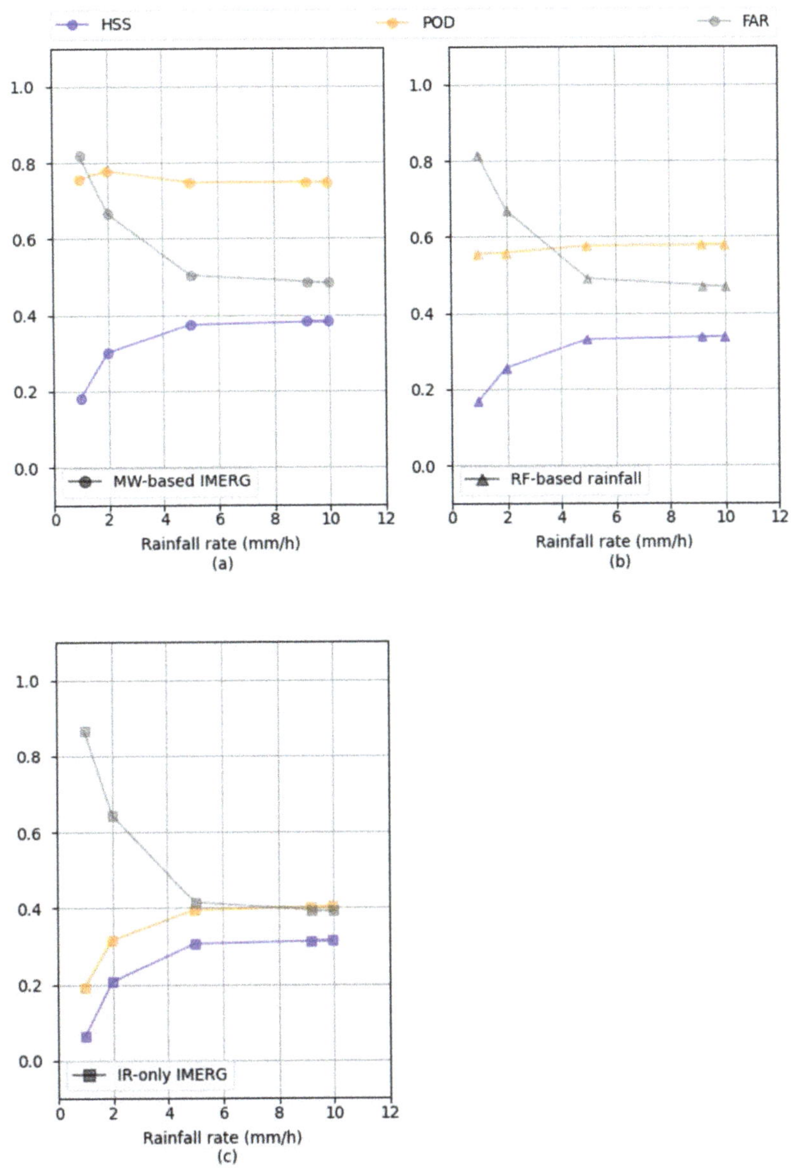

Figure 8. The rain area delineation performance over the MW overpass time and at 11 km^2 for different rainfall rates for (**a**) MW-based IMERG, (**b**) RF-based rainfall, and (**c**) IR-only IMERG.

3.1.2. Rainfall Estimation

Table 3 exhibits the ability of satellite-based rainfall products to estimate rainfall at the time of MW overpass. RF-based rainfall shows the best performance compared with the two other products. All three products underestimate rainfall, indicated by their negative ME and NME.

Table 3. The rainfall estimation performance of satellite-based rainfall over the MW overpass time compared to the ground radar network.

Satellite-Based Rainfall Products	RMSE (mm/h)	NRMSE	MAE (mm/h)	NMAE	ME (mm/h)	NME
MW-based IMERG	2.96	0.82	1.86	1.35	−1.48	−1.07
RF-based rainfall	2.47	0.8	1.66	1.16	−1.34	−0.93
IR-only IMERG	4.65	1.27	2.5	1.63	−2.13	−1.39

The scatter plots in Figure 9 illustrate how the rainfall rate at the time of MW overpass is distributed for each of the satellite-based rainfall products against the radar. Only pixels with hits are considered, therefore the number of hits (n) differs for each product (Figure 9a,d,g). The overall variability in all the products is high, which might be due to issues in timing or/and rainfall estimation (Figure 9a,d,g) [17]. Overall, IR-only rainfall shows the best correlation line close to 1:1. The regression line also indicates the underestimation by RF-based rainfall. MW-based IMERG and IR-only rainfall overestimate the rainfall rate. Figure 9b,e,h shows the rainfall rate for each product against radar in quantile–quantile (Q–Q) plots.

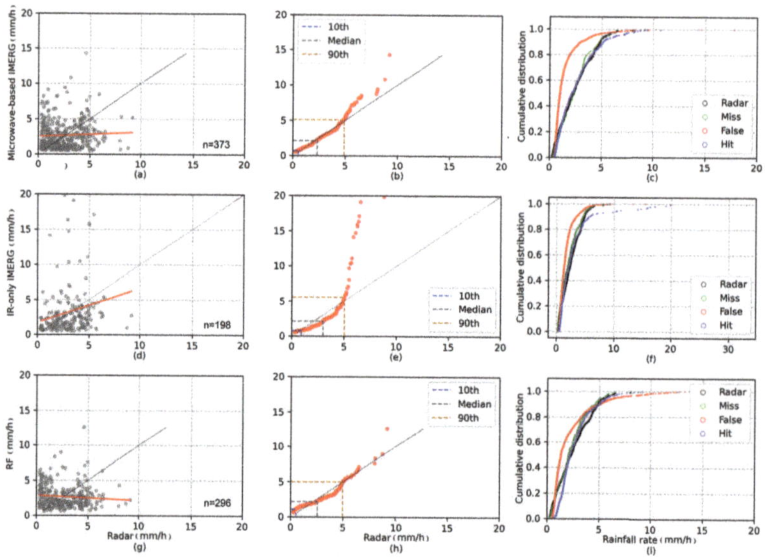

Figure 9. Comparison of rainfall rates estimated by the radar and satellite-based products. **(a,d,g)** Scatter plot with radar rainfall rates (x-axis) and microwave-based IMERG, IR-only IMERG, and RF-based rainfall rates (y-axis), respectively. Only pixels with hits are considered. The parameters *n* show the total number of hits. **(b,e,h)** Quantile–quantile (Q–Q) plot of the radar (x-axis) and microwave-based IMERG (y-axis), IR-only IMERG (y-axis), and RF-based (y-axis) rainfall rates. The 10th, 50th, and 90th percentiles are illustrated. **(c,f,i)** The distribution of cumulative rainfall rate for the contingency table of each satellite-based product. The radar rain rate is displayed in black as a reference.

The Q–Q plot ignores the corresponding time steps in order to underline the differences between the radar and each product in a more comprehensive manner [17]. In MW-based IMERG (Figure 9b) the rainfall rate is almost evenly distributed up to 5 mm/h, the positive values for MW-based IMERG at higher rainfall rates are more evident. The distribution of the rainfall rate between radar and IR-only IMERG shows more discrepancies (Figure 9e). IR-only IMERG shows negative biases until the 90th percentile and shows high positive bias for the higher rainfall rates. RF-based rainfall is distributed relatively even for all rainfall rates, with a slight negative bias between 3 and 5 mm/h. Overall, IR-only IMERG and MW-based IMERG are unable to model the most extreme rainfall rates. For extreme rainfall rates, RF-based rainfall shows better performance. The cumulative distribution of the rainfall rates for hits and the other contingency table elements is compared in Figure 9c,f, i, for MW-based IMERG, IR-only IMERG, and RF-based rainfall, respectively. In MW-based IMERG (Figure 9c) and RF-based rainfall (Figure 9i), around 60% of the FARs is equal to or less than 1 mm/h. This is also true for IR-only IMERG (Figure 9f). The FAR is also shown for higher rainfall rates in the RF-based rainfall product. This underlines that the algorithm is flawed for low-intensity rainfall in these products [17]. The misses show the same distribution as the radar's distribution for all three products.

Figure 10 provides an overview of the validation metrics of the three satellite-based rainfall products for rain estimation along with altitude. MW-based IMERG and IR-only IMERG have difficulty estimating rainfall at lower elevations (0–500 m.a.s.l), which is shown by the extension of the boxplot for NRMSE and NMAE in this elevation range. RF-based rainfall has relatively lower values of NRMSE, NMAE, and NME at an elevation of 0–500 m.a.s.l. With increasing elevation, the rain estimation performance is relatively moderate until 2500 m.a.s.l. For high terrain elevations of approximately 2500–4000 m.a.s.l, all products show a significant uncertainty, mainly in NME. All the products underestimate the rainfall rate at high elevation (2000–4000 m.a.s.l).

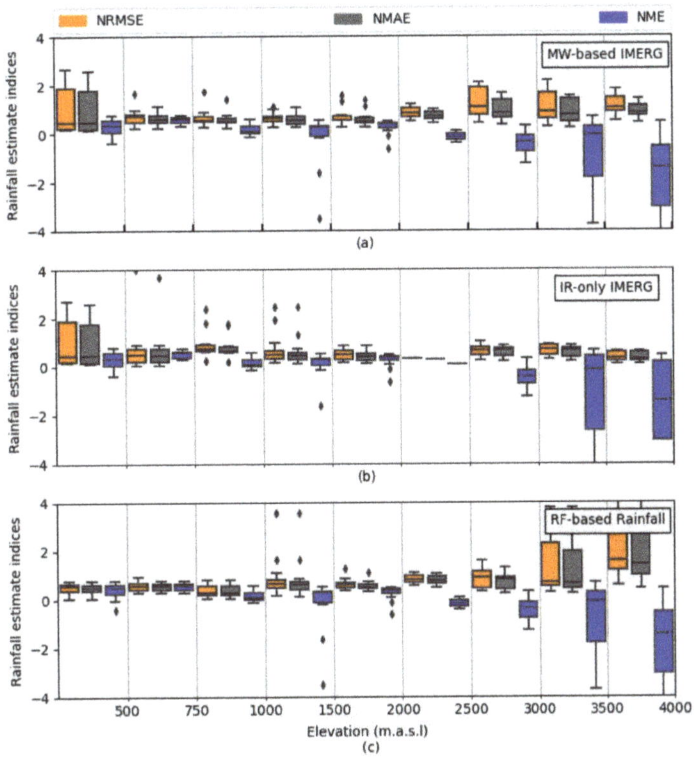

Figure 10. Boxplot of the validation metrics for rain estimation at the MW overpass time. The performance of (**a**) MW-based IMERG, (**b**) IR-only IMERG, and (**c**) RF-based rainfall are shown along with elevation. Boxes show the 25th, 50th, and 75th percentiles. Whiskers extensions are to the maximum data value between the 75th and 25th percentiles. Diamonds indicate outliers.

3.2. Validation Metrics for Satellite-Based Rainfall Products at the Time of MW Overpass from IMERG against Meteorological Stations

Table 4 summarizes the performance of satellite-based rainfall products for rain area delineation against meteorological stations at the time of MW overpasses for the pixel in W-2 (Figure 2a) and W-3 (Figure 2b).

Table 4. Rain area delineation performance of satellite-based rainfall products at the time of MW overpass compared to the meteorological station network. Pixel W-2 and W-3 are shown in Figure 2a and Figure 2b, respectively.

Pixel Number	Satellite-Based Rainfall Products	POD	FAR	HSS
W-3	MW-based IMERG	0.33	0.83	0.21
W-3	RF-based rainfall	0.25	0.83	0.19
W-3	IR-only IMERG	0.17	0.89	0.12
W-2	MW-based IMERG	0.29	0.82	0.19
W-2	RF-based rainfall	0.35	0.78	0.24
W-2	IR-only IMERG	0.23	0.82	0.19

The validation scores show the superior performance of the IMERG-MW-based and RF-based rainfall products in comparison to IMERG-IR-only in W3. W2 shows a slightly better performance for RF-based rainfall while IR-only IMERG and MW-based IMERG are more or less the same. Still, all of the products overestimate precipitation area. These behaviors are similar to the validation of the rainfall products at the MW overpass time against the X-band rain radar network (Table 2). However, the validation scores indicate a lower performance of the satellite-based rainfall products by using the radar data compared to higher scores by using the station data. This is not surprising, since a low number of the gauges in a pixel (3 gauges in 11 km^2) is not representative for the spatial distribution of rain. Therefore, the assessment of satellite-based rainfall products against a low number of gauges in each pixel underestimates their performance [36].

Table 5 shows the satellite-based rainfall products' ability to estimate the rainfall at the time of MW overpass against ground truth data. The behavior of satellites is different in two pixels. In W-2, IR-only IMERG shows the best performance compared to the other two products. Meanwhile, in W-3, the RF-based rainfall capture the rain estimate more accurately compared to other products. In general, all of the products overestimate rainfall slightly (positive ME).

Table 5. Rainfall estimation performance of satellite-based rainfall over the MW overpass time compared to ground radar network.

Pixel Number	Satellite-Based Rainfall Products	RMSE (mm/h)	NRMSE	MAE (mm/h)	NMAE	ME (mm/h)	NME
W-3	MW-based IMERG	1.10	1.03	0.96	0.91	0.83	0.78
W-3	RF-based rainfall	0.48	0.45	0.85	0.80	0.83	0.78
W-3	IR-only IMERG	3.32	3.11	3.27	3.07	0.83	0.80
W-2	MW-based IMERG	1.03	0.87	0.68	0.57	0.68	0.57
W-2	RF-based rainfall	1.11	0.94	1.05	0.89	0.60	0.71
W-2	IR-only IMERG	0.68	0.58	0.49	0.41	0.77	0.67

3.3. Validation Metrics for RF-Based Rainfall Products in Native Resolution

3.3.1. Rain Area Delineation

Using the analysis techniques described in Section 2.2.3, the ability of RF-based rainfall to estimate rainfall in comparison with the radar at 2 km^2 spatial resolution and 30 min temporal resolution is shown in Figure 11 (n = 1,048,575). Less than 3% of the time steps in either radar or RF-based rainfall contain rainfall including 0.31% of hits (Figure 11a). The errors are dominated by false alarms at 1.57%. The decomposition of misses using the temporal event-based approach shows that almost 12% of the misses occur in the coincidental timing from radar precipitation (Figure 11b, Duration −, yellow bar; Duration +, black bar), whereas the spatially drifted misses are not recognizable (Figure 11d). Almost 4% of the overestimation occurs by overestimating event duration (Figure 11c), and 8.5% by overestimating events in the neighboring pixel (Figure 11e).

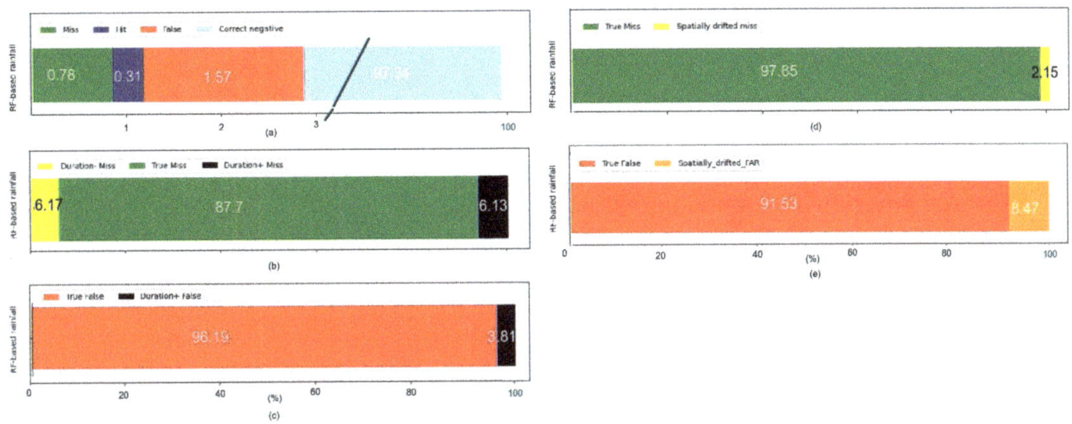

Figure 11. (**a**) Standard contingency table approach for all available RF-based rainfall products for both radars at 2 km^2 and 30 min. Note that the correct negative fraction extends to 100%. (**b**,**c**) The temporal event-based approach of the contingency table was evaluated in the M and F subsets, respectively. (**d**,**e**) The spatial event-based approach of the contingency table was evaluated in the M and F subsets, respectively. The numbers in the bars show the percentage.

The performance is summarized in Table 6. As expected, a noticeable result is the high FAR of 83%, showing that 83% of rainy events are false alarms. This is almost similar behaviour as for RF-based rainfall at the MW overpass in 11 km^2 temporal resolution (Table 2). By applying the algorithm in 2 km^2 spatial and 30 min temporal resolution, the ability of rain detection by RF-based rainfall has reduced compared to the RF-based rainfall in MW-overpasses and at 11 km^2 spatial resolution (HSS = 0.31).

Table 6. Performance evaluation of RF-based rainfall at rainfall area delineation for 2 km^2 spatial and 30 min temporal resolution.

Satellite-Based Rainfall Products	POD	FAR	HSS
RF-based rainfall	0.28	0.83	0.20

3.3.2. Rain Estimation

Table 7 summarizes the performance of RF-based rainfall in estimating rain at 2 km^2 spatial and 30 min temporal resolution. The RF-based rainfall shows better performance in estimating rainfall at higher resolution compared with lower resolution (Table 3).

Table 7. Performance evaluation of RF-based rainfall for rainfall estimation 2 km^2 spatial and 30 min temporal resolution.

Satellite-Based Rainfall Products	RMSE (mm/h)	NRMSE	MAE (mm/h)	NMAE	ME (mm/h)	NME
RF-based rainfall	2.39	1.07	1.72	0.77	0.51	0.22

Focusing on hits, Figure 12 shows the rain estimation retrieval ability of RF-based rainfall in comparison with the radar. The scatter plot in Figure 12a shows the distribution of the half-hourly rain rates. The rain rates illustrates high variability, suggesting problems in rain estimation retrieval and/or timing. This is also shown in Figure 9g at the time of

MW overpass. Figure 12b shows the Q–Q plot for RF-based rainfall. The overall estimation of the rainfall is placed along the 1:1 line to the 90th percentile. However, the curve deviates towards the left after the 90th percentile, showing an overestimation of rain intensities in the outliers. Figure 12c decomposes the results in more detail. Overall, RF-based rainfall is unable to detect the most extreme rainfall rates, as reported by Turini et al. [3]. The cumulative distribution of rainfall rates for hits, misses, radar, and false alarms are compared in Figure 12a. Around 60% of false alarms and misses are less than or equal to 1 mm/h. This is also true for 60% of event-based (temporally and spatially) false alarms Figure 12d. The event-based misses are evenly distributed over the different rainfall rates.

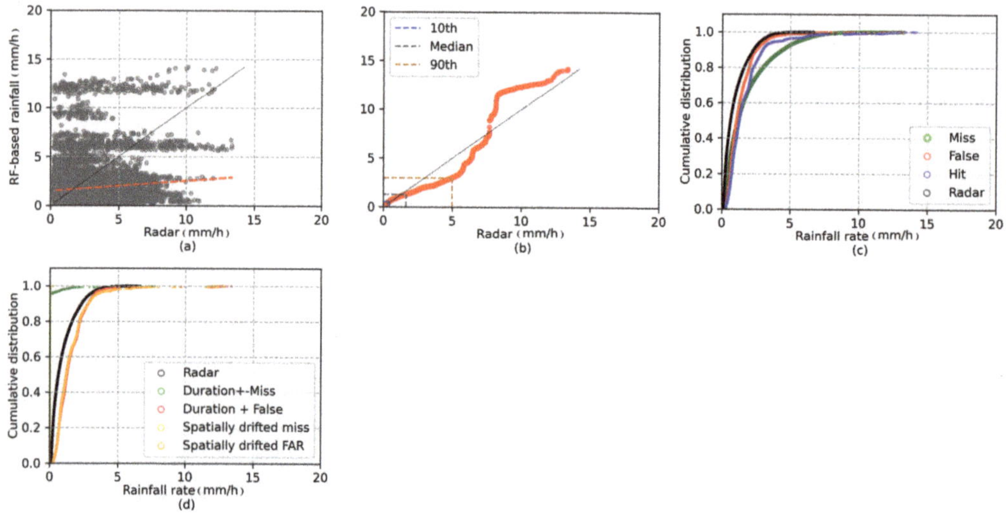

Figure 12. Comparison of rain rates estimated by the radar and RF-based rainfall at 2 km^2 and 30 min. (**a**) Scatter plot with radar rainfall (x-axis) and RF-based rainfall (y-axis). Only the pixels with hit are considered. (**b**) Q–Q plot of radar (x-axis) and RF-based rainfall rates (y-axis). The 10th, 50th, and 90th percentiles are illustrate. (**c**) The distribution of cumulative rainfall rate for the contingency table. (**d**) The distribution of cumulative rainfall rate based on the event-based (spatial and temporal) contingency table.

3.4. Validation Metrics for RF-Based Rainfall Products at Different Temporal Resolutions

To validate the results of rain area delineation and rain estimation in different temporal resolutions, Figure 13a,b presents the validation metrics with the radar for the whole study region and observation period. The results show the best agreements regarding rain area delineation in daily resolution (POD 0.68, HSS 0.4, and FAR 0.6).

The rain estimation indices for RF-based rainfall do not show a significant improvement for the different temporal resolutions. The NME suggests the overestimation of precipitation by RF-based rainfall at lower resolution (after 3 h) and an underestimation at higher temporal resolutions. Note that in this step, we considered rainfall at a rate of more than 0.5 mm/h as rainy.

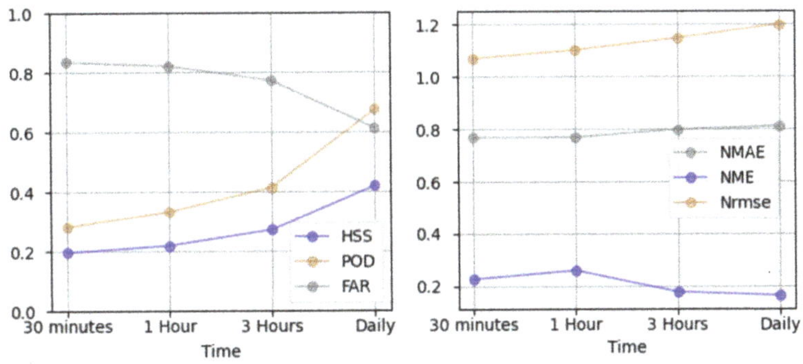

(a) (b)

Figure 13. Comparison of the validation metrics between the radar and RF-based rainfall at 2 km^2 and 30 min, 1 h, 3 h, and daily. The performance of RF-based (**a**) rain area delineation and (**b**) rain estimation is shown for different temporal resolutions.

4. Discussion

In Section 3.1, satellite-based rainfall products at the time of MW overpasses from IMERG were assessed using radar data. We evaluated the satellite-based products in grid cells at the time of MW overpasses and a spatial resolution of 11 km^2.

The verification scores for rain area delineation revealed that the MW-based IMERG has superior performance in estimating rain area (POD = 0.74, HSS = 0.33). RF-based rainfall, which is trained based on MW-based IMERG, has slightly lower performance compared to MW-based IMERG data (HSS = 0.31). IR-only IMERG performed the worst in Ecuador. This is in line with the findings of Kolbe et al. [12], Kolbe et al. [13], Turini et al. [8], and Turini et al. [3]. It shows that multispectral GEO data has more potential than using one IR channel only for rainfall retrieval.

The frequent false alarm is one of the most noticeable issues identified in the present study. This agrees well with the result of IMERG-V06 validation in the west African forest zone [17] and confirms the previous investigation of IMERG-v05 by Manz et al. [15] in the Andes region. In our study, around 60% of the false alarms were related to rain rates less than 1 mm/h for all products (Figure 9), which was found to be the dominant rainfall intensity in this region of the world [39]. We also note that the radar potentially underestimated rainfall [40–43]. This was also reported elsewhere for the radars in Ecuador [23]. In MW-based IMERG and RF-based rainfall, with increasing the rainfall rate, the FAR decreases while POD does not change (Figure 8).

The results of the topography-based evaluation indicated the high detection accuracy of MW-based IMERG and RF-based rainfall in different topographical regions. Moreover, the highest errors occurred for coastal areas and foothills (0–1500 m.a.s.l) and high mountains regions (>3000 m.a.s.l) compared to the other topographical regions. All the products experienced challenges in estimating rainfall at high elevation in the Andes (Figure 10). In Ecuador, high-elevation areas and volcanoes have two issues for rainfall retrieval algorithms: (i) They are regularly covered by ice, which generates errors in MW-based IMERG [29,44]; (ii) the drizzle on the high elevation is hard to be captured by MW and IR channels. This conclusion is in agreement with the findings of study conducted by Prakash et al. [45], who assessed the performance of IMERG products in monsoon-dominated regions in India. Their results showed that IMERG was affected by the orographic process, which leads to higher errors in mountainous areas. Another study by Kim et al. [46] revealed the disadvantage of IMERG products over mountainous and coastal regions. Similar results were obtained by Turini et al. [3] in Ecuador for RF-based rainfall. They argued that because of local topography, the subscale convective rainfall systems probably could not be captured by GOES data and IMERG [3,37]. Altogether, at the elevation of 0–750 m.a.s.l, RF-based rainfall showed the best performance of all products (Figures 7 and 10).

Concerning rainfall rate validation, the overall variability in all the products is high, suggesting rainfall rate estimation and/or timing issues. Different studies discuss a possible time lag between the satellite-based rainfall products and the ground-based rainfall measurements as a source of degrading validation results [17,38,47–49]. The time lag is defined as the time shift when satellite observation and surface precipitation rate from ground data obtain to their optimum correlation. This time lag might be due to the time it takes for the precipitation detected by the satellite to reach the ground [17,47]. You et al. [38] related the precipitation time from GMI to the environmental temperature and storm top height. They found that when the storm is taller, the lag time increases to obtain the optimum correlation between the GMI and ground truth data. This is due to the long way of raindrops from the storm top to the gauge.

Ignoring the corresponding time steps in the Q–Q plots shows that the MW-based IMERG and RF-based (Figure 9b,h) rainfall rates are distributed up to 5 mm/h evenly. The positive values in MW-based IMERG at higher rainfall rates are more evident. Conversely, the rainfall rate distribution between the radar and IR-only IMERG shows more discrepancies (Figure 9e). The validation of satellite-based rainfall products against the gauges show lower consistency (Table 5). However, in the term of rain area delineation (Table 4), the RF-based rainfall product shows better performance than IMERG-IR-only, which confirms the potential to use multispectral GEO data.

The validation of satellite-based rainfall show a slight overestimation of rainfall totals for all products (Table 5).

It should be noted that the evaluation of satellite-based products against only a few gauges has high uncertainties [8,36], especially in areas with high small-scale precipitation variability in mainly convective environments, like the Ecuadorian Andes, where point based observations at weather stations cannot properly represent the spatial rainfall distribution.

The validation of RF-based rainfall retrieval at high spatiotemporal resolution for all the available rain events is shown in Table 6. The RF-based rainfall is calibrated locally for Ecuador. The importance of local calibration, which involves determining relevant climatic parameters, including the selection of appropriate temperature thresholds for clouds and a local correlation systematic biases that may not have been adjusted in global products, have been mentioned in different studies [50–52].

RF-based rainfall for 2 km^2 and 30 min shows a lower HSS compared to the RF-based rainfall for 11 km^2 at the time of MW overpass. This was expected because the errors at higher temporal resolutions may cancel each other out following the aggregation to a lower temporal resolution [50]. However, in terms of rainfall estimation, RF-based rainfall performs better at higher spatial resolution (Table 3). This result needs to be interpreted with caution, since the rainfall events at the time of MW overpasses differ from the validation of the RF-based rainfall at 2 km^2 and 30 min.

An event-based analysis was then used to investigate the source of error in the RF-based rainfall product. Shifting the RF-based rainfall backward by one to two time steps (i.e., 30 min) resulted in the more accurate detection of rainfall around 10% (Figure 11b) by lowering the misses. RF-based rainfall rates are lower than their counterparts in radar, as shown in Figure 12d. We speculate that this lag appears due to the lag time between the time of MW overpass and the GOES-16 scan time. The RF-based rainfall algorithm relies on the precipitation information from MW-based IMERG and IR data from GOES-16.

However, RF-based rainfall also has a high FAR. The event-based spatial analysis reduced the FAR by 8.5% (Figure 11e), but the challenge remains the same. High FAR values occur for all the different types of rain with different intensities (Figure 12c,d). The reason for the high FAR in RF-based rainfall might be (i) the high amount of FAR from MW-based IMERG in Ecuador (Table 2), which is used as a reference for calibrating of RF-based rainfall; (ii) A bias in IR retrievals that classify cold cloud pixels as rainy. They experience difficulties in defining the correct rainfall cloud and profile, thus producing error in statistical-physical rainfall algorithms.

By increasing the temporal resolution of the RF-based rainfall product, the performance of the product increased. However, the FAR (60% in daily resolution) remains a main challenge.

5. Conclusions

In this study, we evaluated and compared the performance of different satellite-based rainfall products over the Pacific coast and Andes of Ecuador. A mesoscale quality-controlled rain radar network was used as the rainfall reference. Statistical comparison indices were used to analyze the performance and to describe different aspects of the satellite-based rainfall products. The first validation was performed at 11 km^2 spatial resolution and at the time of MW overpass for MW-based IMERG, RF-based rainfall, and IR-only IMERG products. Based on the validation, MW-based IMERG and RF-based rainfall provided better rainfall estimates in Ecuador than IR-only IMERG during MW overpasses. The distribution of the evaluation metrics spatially shows the impact of topography and the complex climate zonations in the study region. High precipitation values were better captured by the MW-based IMERG and the RF-based rainfall algorithms. The frequent false alarms are one of the most important issues in all products; FAR decreases with an increasing rainfall rate. Future studies on the lag time are therefore required in order to elucidate the high FAR in the satellite-based products. In the third validation, we investigated regionally calibrated RF-based rainfall products for Ecuador. RF-based rainfall is trained by MW-based IMERG. Although the product shows convincing results at a MW overpass of 11 km^2, the performance decreased by increasing the resolution to 2 km^2 spatial and 30 min temporal resolution. Furthermore, RF-based rainfall is trained to the available microwave-only data from IMERG. Consequently, due to the low temporal resolution of the data from MW satellites, some rainfall events might not have been considered [8].

Author Contributions: N.T., B.T. and J.B. developed the theoretical methodology; N.T. performed the analytic calculations; N.T., B.T., J.B. and R.R. analyzed and discussed the results; N.T. wrote the manuscript; N.T., B.T., N.H., R.R., A.F., F.P.-C., J.O.-A. and J.B. reviewed the paper. All authors have read and agreed to the published version of the manuscript.

Funding: This research was funded by the Federal Ministry for Education and Research of Germany within the projects "Seasonal Water Resource Management in Semi-arid regions" (02WGR1421C: http://www.grow-sawam.org, accessed on 6 October 2021); DFG research unit "RESPECT" (BE1780/51-1) and DFG "High-resolution Radar analysis of precipitation extremes in Ecuador and North Peru and implications of the ENSO-dynamics" (RO3815/2-1).

Data Availability Statement: The RF-based rainfall data presented in this study are available online: http://lcrs.geographie.uni-marburg.de/lcrs/data_pre.do?citid=448 (accessed on 6 October 2021). Restriction applies for the availability of ground radar data and gauge data. Because the data were obtained from third parties.

Acknowledgments: This work was carried out within the "Seasonal Water Resource Management" (SaWaM) research program by the Federal Ministry for Education and Research of Germany's (BMBF). The work was done in the subproject "Remote Sensing of Precipitation" (02WGR1421C). We gratefully acknowledge the DFG research unit FOR2730 RESPECT, subproject A1 (BE1780/51-1) and the research project "High-resolution Radar analysis of precipitation extremes in Ecuador and North Peru and implications of the ENSO-dynamics" (DFG RO3815/2-1). The authors also thank National Weather Service of Ecuador (INAMHI), Universidad TécnicaParticular de Loja (UTPL), and University of Cuenca (Proyecto "Desarrollo de modelos para pronóstico hidrológico a partir de datos de radar meteorológico en cuencas de montaña") for providing their meteorological station data and NOAA for produce the GOES-16 satellite data used in this study. Additionally, the IMERG-V06 data were provided by the NASA/Goddard Space Flight Centers and PPS, which develop and compute IMERG as a contribution to GPM, archived at the NASA GES DISC.

Conflicts of Interest: The authors declare no conflict of interest.

References

1. Seidel, J.; Trachte, K.; Orellana-Alvear, J.; Figueroa, R.; Célleri, R.; Bendix, J.; Fernandez, C.; Huggel, C. Precipitation characteristics at two locations in the tropical Andes by means of vertically pointing micro-rain radar observations. *Remote Sens.* **2019**, *11*, 2985. [CrossRef]
2. Ward, E.; Buytaert, W.; Peaver, L.; Wheater, H. Evaluation of precipitation products over complex mountainous terrain: A water resources perspective. *Adv. Water Resour.* **2011**, *34*, 1222–1231. [CrossRef]
3. Turini, N.; Thies, B.; Horna, N.; Bendix, J. Random forest-based rainfall retrieval for Ecuador using GOES-16 and IMERG-V06 data. *Eur. J. Remote Sens.* **2021**, *54*, 117–139. [CrossRef]
4. Tan, J.; Petersen, W.A.; Tokay, A. A novel approach to identify sources of errors in IMERG for GPM ground validation. *J. Hydrometeorol.* **2016**, *17*, 2477–2491. [CrossRef]
5. Huffman, G.; Bolvin, D.; Braithwaite, D.; Hsu, K.; Joyce, R.; Kidd, C.; Nelkin, E.; Xie, P. *NASA Global Precipitation Measurement Integrated Multi-satellitE Retrievals for GPM (IMERG). Algorithm Theoretical Basis Doc.*; version 4.5; National Aeronautics and Space Administration: Washington, DC, USA, 2015; 30p.
6. Huffman, G.J.; Bolvin, D.T.; Nelkin, E.J.; Wolff, D.B.; Adler, R.F.; Gu, G.; Hong, Y.; Bowman, K.P.; Stocker, E.F. The TRMM multisatellite precipitation analysis (TMPA): Quasi-global, multiyear, combined-sensor precipitation estimates at fine scales. *J. Hydrometeorol.* **2007**, *8*, 38–55. [CrossRef]
7. Olson, W.S.; Masunaga, H.; GPM, C.R.R.A.T. *GPM Combined Radar-Radiometer Precipitation Algorithm Theoretical Basis Document (Version 4)*; NASA: Washington, DC, USA, 2016.
8. Turini, N.; Thies, B.; Bendix, J. Estimating High Spatio-Temporal Resolution Rainfall from MSG1 and GPM IMERG Based on Machine Learning: Case Study of Iran. *Remote Sens.* **2019**, *11*, 2307. [CrossRef]
9. Kimani, M.W.; Hoedjes, J.C.B.; Su, Z. An Assessment of Satellite-Derived Rainfall Products Relative to Ground Observations over East Africa. *Remote Sens.* **2017**, *9*, 430. [CrossRef]
10. Kühnlein, M.; Appelhans, T.; Thies, B.; Nauss, T. Improving the accuracy of rainfall rates from optical satellite sensors with machine learning—A random forests-based approach applied to MSG SEVIRI. *Remote Sens. Environ.* **2014**, *141*, 129–143. [CrossRef]
11. Kühnlein, M.; Appelhans, T.; Thies, B.; Nauß, T. Precipitation Estimates from MSG SEVIRI Daytime, Nighttime, and Twilight Data with Random Forests. *J. Appl. Meteorol. Climatol.* **2014**, *53*, 2457–2480. [CrossRef]
12. Kolbe, C.; Thies, B.; Egli, S.; Lehnert, L.; Schulz, H.M.; Bendix, J. Precipitation Retrieval over the Tibetan Plateau from the Geostationary Orbit—Part 1: Precipitation Area Delineation with Elektro-L2 and Insat-3D. *Remote Sens.* **2019**, *11*, 2302. [CrossRef]
13. Kolbe, C.; Thies, B.; Turini, N.; Liu, Z.; Bendix, J. Correction: Kolbe, C., et al. Precipitation Retrieval over the Tibetan Plateau from the Geostationary Orbit—Part 2: Precipitation Rates with Elektro-L2 and Insat-3D. Remote Sensing 2020, 12, 2114. *Remote Sens.* **2020**, *12*, 3594. [CrossRef]
14. Hong, Y.; Hsu, K.L.; Sorooshian, S.; Gao, X. Precipitation estimation from remotely sensed imagery using an artificial neural network cloud classification system. *J. Appl. Meteorol.* **2004**, *43*, 1834–1853. [CrossRef]
15. Manz, B.; Páez-Bimos, S.; Horna, N.; Buytaert, W.; Ochoa-Tocachi, B.; Lavado-Casimiro, W.; Willems, B. Comparative ground validation of IMERG and TMPA at variable spatiotemporal scales in the tropical Andes. *J. Hydrometeorol.* **2017**, *18*, 2469–2489. [CrossRef]
16. Erazo, B.; Bourrel, L.; Frappart, F.; Chimborazo, O.; Labat, D.; Dominguez-Granda, L.; Matamoros, D.; Mejia, R. Validation of satellite estimates (Tropical Rainfall Measuring Mission, TRMM) for rainfall variability over the Pacific slope and Coast of Ecuador. *Water* **2018**, *10*, 213. [CrossRef]
17. Maranan, M.; Fink, A.H.; Knippertz, P.; Amekudzi, L.K.; Atiah, W.A.; Stengel, M. A process-based validation of GPM IMERG and its sources using a mesoscale rain gauge network in the West African forest zone. *J. Hydrometeorol.* **2020**, *21*, 729–749. [CrossRef]
18. Bulovic, N.; McIntyre, N.; Johnson, F. Evaluation of IMERG V05B 30-Minute Rainfall Estimates over the High Elevation Tropical Andes Mountains. *J. Hydrometeorol.* **2020**, *21*, 1–53. [CrossRef]
19. Dinku, T.; Ceccato, P.; Grover-Kopec, E.; Lemma, M.; Connor, S.; Ropelewski, C. Validation of satellite rainfall products over East Africa's complex topography. *Int. J. Remote Sens.* **2007**, *28*, 1503–1526. [CrossRef]
20. Maidment, R.I.; Grimes, D.; Black, E.; Tarnavsky, E.; Young, M.; Greatrex, H.; Allan, R.P.; Stein, T.; Nkonde, E.; Senkunda, S.; et al. A new, long-term daily satellite-based rainfall dataset for operational monitoring in Africa. *Sci. Data* **2017**, *4*, 170063. [CrossRef]
21. Bendix, J.; Fries, A.; Zárate, J.; Trachte, K.; Rollenbeck, R.; Pucha-Cofrep, F.; Paladines, R.; Palacios, I.; Orellana, J.; Oñate-Valdivieso, F.; et al. RadarNet-Sur first weather radar network in tropical high mountains. *Bull. Am. Meteorol. Soc.* **2017**, *98*, 1235–1254. [CrossRef]
22. Rollenbeck, R.; Bendix, J. Experimental calibration of a cost-effective X-band weather radar for climate ecological studies in southern Ecuador. *Atmos. Res.* **2006**, *79*, 296–316. [CrossRef]
23. Rollenbeck, R.; Bendix, J. Rainfall distribution in the Andes of southern Ecuador derived from blending weather radar data and meteorological field observations. *Atmos. Res.* **2011**, *99*, 277–289. [CrossRef]
24. Boers, N.; Bookhagen, B.; Marwan, N.; Kurths, J.; Marengo, J. Complex networks identify spatial patterns of extreme rainfall events of the South American Monsoon System. *Geophys. Res. Lett.* **2013**, *40*, 4386–4392. [CrossRef]
25. Vuille, M.; Bradley, R.S.; Keimig, F. Climate variability in the Andes of Ecuador and its relation to tropical Pacific and Atlantic sea surface temperature anomalies. *J. Clim.* **2000**, *13*, 2520–2535. [CrossRef]

26. Espinoza, J.C.; Chavez, S.; Ronchail, J.; Junquas, C.; Takahashi, K.; Lavado, W. Rainfall hotspots over the southern tropical Andes: Spatial distribution, rainfall intensity, and relations with large-scale atmospheric circulation. *Water Resour. Res.* **2015**, *51*, 3459–3475. [CrossRef]
27. Bookhagen, B.; Strecker, M.R. Orographic barriers, high-resolution TRMM rainfall, and relief variations along the eastern Andes. *Geophys. Res. Lett.* **2008**, *35*. [CrossRef]
28. Romatschke, U.; Houze, R.A., Jr. Extreme summer convection in South America. *J. Clim.* **2010**, *23*, 3761–3791. [CrossRef]
29. Huffman, G.; Bolvin, D.; Braithwaite, D.; Hsu, K.; Joyce, R.; Kidd, C.; Nelkin, E.; Sorooshian, S.; Tan, J.; Xie, P. *NASA Global Precipitation Measurement (GPM) Integrated Multi-satellitE Retrievals for GPM (IMERG)*; Version 06; Algorithm Theoretical Basis Document (ATBD): Greenbelt, MD, USA, 2019.
30. Joyce, R.J.; Janowiak, J.E.; Arkin, P.A.; Xie, P. CMORPH: A method that produces global precipitation estimates from passive microwave and infrared data at high spatial and temporal resolution. *J. Hydrometeorol.* **2004**, *5*, 487–503. [CrossRef]
31. Tan, J.; Huffman, G.J.; Bolvin, D.T.; Nelkin, E.J. IMERG V06: Changes to the morphing algorithm. *J. Atmos. Ocean. Technol.* **2019**, *36*, 2471–2482. [CrossRef]
32. Huffman, G. IMERG V06 Quality Index 2019. Available online: https://gpm.nasa.gov/resources/documents/imerg-v06-quality-index (accessed on 15 March 2019).
33. Schulz, H.M.; Li, C.F.; Thies, B.; Chang, S.C.; Bendix, J. Mapping the montane cloud forest of Taiwan using 12 year MODIS-derived ground fog frequency data. *PLoS ONE* **2017**, *12*, e172663. [CrossRef]
34. Min, M.; Bai, C.; Guo, J.; Sun, F.; Liu, C.; Wang, F.; Xu, H.; Tang, S.; Li, B.; Di, D.; et al. Estimating summertime precipitation from Himawari-8 and global forecast system based on machine learning. *IEEE Trans. Geosci. Remote Sens.* **2018**, *57*, 2557–2570. [CrossRef]
35. GDAL; OGR Contributors. Geospatial Data Abstraction Software Library 2020. Open Source Geospatial Foundation 2021. Available online: https://gdal.org (accessed on 6 October 2021).
36. Tang, G.; Behrangi, A.; Long, D.; Li, C.; Hong, Y. Accounting for spatiotemporal errors of gauges: A critical step to evaluate gridded precipitation products. *J. Hydrol.* **2018**, *559*, 294–306. [CrossRef]
37. Bendix, J.; Trachte, K.; Cermak, J.; Rollenbeck, R.; Nauß, T. Formation of convective clouds at the foothills of the tropical eastern Andes (South Ecuador). *J. Appl. Meteorol. Climatol.* **2009**, *48*, 1682–1695. [CrossRef]
38. You, Y.; Meng, H.; Dong, J.; Rudlosky, S. Time-Lag Correlation Between Passive Microwave Measurements and Surface Precipitation and Its Impact on Precipitation Retrieval Evaluation. *Geophys. Res. Lett.* **2019**, *46*, 8415–8423. [CrossRef]
39. Orellana-Alvear, J.; Célleri, R.; Rollenbeck, R.; Bendix, J. Analysis of rain types and their Z–R relationships at different locations in the high Andes of southern Ecuador. *J. Appl. Meteorol. Climatol.* **2017**, *56*, 3065–3080. [CrossRef]
40. Hazenberg, P.; Leijnse, H.; Uijlenhoet, R. Radar rainfall estimation of stratiform winter precipitation in the Belgian Ardennes. *Water Resour. Res.* **2011**, *47*. [CrossRef]
41. Goudenhoofdt, E.; Delobbe, L. Generation and verification of rainfall estimates from 10-yr volumetric weather radar measurements. *J. Hydrometeorol.* **2016**, *17*, 1223–1242. [CrossRef]
42. Fairman, J.G., Jr.; Schultz, D.M.; Kirshbaum, D.J.; Gray, S.L.; Barrett, A.I. A radar-based rainfall climatology of Great Britain and Ireland. *Weather* **2015**, *70*, 153–158. [CrossRef]
43. Smith, J.A.; Baeck, M.L.; Villarini, G.; Welty, C.; Miller, A.J.; Krajewski, W.F. Analyses of a long-term, high-resolution radar rainfall data set for the Baltimore metropolitan region. *Water Resour. Res.* **2012**, *48*. [CrossRef]
44. Petkovi, V.; Kummerow, C.D. Understanding the sources of satellite passive microwave rainfall retrieval systematic errors over land. *J. Appl. Meteorol. Climatol.* **2017**, *56*, 597–614. [CrossRef]
45. Prakash, S.; Mitra, A.K.; AghaKouchak, A.; Liu, Z.; Norouzi, H.; Pai, D. A preliminary assessment of GPM-based multi-satellite precipitation estimates over a monsoon dominated region. *J. Hydrol.* **2018**, *556*, 865–876. [CrossRef]
46. Kim, K.; Park, J.; Baik, J.; Choi, M. Evaluation of topographical and seasonal feature using GPM IMERG and TRMM 3B42 over Far-East Asia. *Atmos. Res.* **2017**, *187*, 95–105. [CrossRef]
47. Tan, J.; Petersen, W.A.; Kirchengast, G.; Goodrich, D.C.; Wolff, D.B. Evaluation of Global Precipitation Measurement Rainfall Estimates against Three Dense Gauge Networks. *J. Hydrometeorol.* **2018**, *19*, 517–532. [CrossRef]
48. Foelsche, U.; Kirchengast, G.; Fuchsberger, J.; Tan, J.; Petersen, W.A. Evaluation of GPM IMERG Early, Late, and Final rainfall estimates using WegenerNet gauge data in southeastern Austria. *Hydrol. Earth Syst. Sci.* **2017**, *21*, 6559–6572.
49. Villarini, G.; Krajewski, W.F. Evaluation of the research version TMPA three-hourly 0.25 rainfall estimates over Oklahoma. *Geophys. Res. Lett.* **2007**, *34*, 2007. [CrossRef]
50. Ayehu, G.T.; Tadesse, T.; Gessesse, B.; Dinku, T. Validation of new satellite rainfall products over the Upper Blue Nile Basin, Ethiopia. *Atmos. Meas. Tech.* **2018**, *11*, 1921–1936. [CrossRef]
51. Ghajarnia, N.; Liaghat, A.; Arasteh, P.D. Comparison and evaluation of high resolution precipitation estimation products in Urmia Basin-Iran. *Atmos. Res.* **2015**, *158*, 50–65. [CrossRef]
52. Katiraie-Boroujerdy, P.S.; Nasrollahi, N.; Hsu, K.l.; Sorooshian, S. Evaluation of satellite-based precipitation estimation over Iran. *J. Arid. Environ.* **2013**, *97*, 205–219. [CrossRef]

atmosphere

Article

Time Series Analysis of Atmospheric Precipitation Characteristics in Western Siberia for 1979–2018 across Different Datasets

Elena Kharyutkina [1,2,*], Sergey Loginov [1], Yuliya Martynova [1] and Ivan Sudakov [2,3]

1 Institute of Monitoring of Climatic and Ecological Systems of Siberian Branch of Russian Academy of Sciences, 634055 Tomsk, Russia; logsv13@gmail.com (S.L.); foxyj13@gmail.com (Y.M.)
2 The Center for Research and Invention, 173008 Veliky Novgorod, Russia; ivan.sudakov@gmail.com
3 Department of Physics, University of Dayton, Dayton, OH 45469, USA
* Correspondence: kh_ev@mail2000.ru; Tel.: +7-(3822)49-15-65

Abstract: A comparative statistical analysis of the spatiotemporal variability of atmospheric precipitation characteristics (mean and extreme values) in Western Siberia was performed based on data acquired from meteorological stations, global precipitation datasets such as the project of Asian Precipitation—Highly-Resolved Observational Data Integration Towards Evaluation (APHRODITE) and from Global Precipitation Climatology Centre (GPCC), and reanalysis archives, including from National Centers of Environmental Prediction (NCEP-DOE) and the European Center for Medium Range Weather Forecasts (ERA5) for the period 1979–2018. The best agreement of the values from the observational data was observed with the values from GPCC. This archive also represented the periodicities in the time series of observational data from meteorological stations, especially in the short-period part of the spectrum. Underestimated values were revealed for the APHRODITE archive, while overestimated ones were found for the NCEP reanalysis data. In comparison with GPCC, the ERA5 dataset reproduced the general variability but with a smaller amplitude (the correlation coefficient was up to 0.9). In general, the median estimates of the precipitation amount derived from the meteorological stations' data, as well from the reanalysis data, were in better agreement with each other rather than their extreme values. However, their temporal variability can be effectively described by other datasets.

Keywords: atmospheric precipitation; time series; extreme values; periodicities; correlation analysis; spatiotemporal variability; Western Siberia

Citation: Kharyutkina, E.; Loginov, S.; Martynova, Y.; Sudakov, I. Time Series Analysis of Atmospheric Precipitation Characteristics in Western Siberia for 1979–2018 across Different Datasets. *Atmosphere* **2022**, *13*, 189. https://doi.org/10.3390/atmos13020189

Academic Editors: Zuohao Cao, Huaqing Cai and Xiaofan Li

Received: 30 December 2021
Accepted: 20 January 2022
Published: 24 January 2022

1. Introduction

Climate temperature warming, which has been observed across the planet in recent decades, is also typical for the territory of Russia. The most important climate variables that are often used as climate change indicators are surface air temperature and precipitation [1]. The growth rate of the average annual temperature in Russia in 1976–2019 was 0.47 °C/decade, which is more than two and a half times higher than the rate of global temperature increase over the same time interval (0.18 °C/decade) [2]. As for atmospheric precipitation for the same time interval, we see (according to Reference [2]) that there is a tendency toward an increase in the annual precipitation in several regions of Siberia and the Russian Far East; however, precipitation decreases in the northeast of the country.

The most significant linear trend coefficients are observed in the regions of Siberia in the spring. Moreover, in Northern Eurasia, there is a moderate increase in the total amount of precipitation accompanied by a relatively strong increase in heavy rainfall and a simultaneous decrease in stratiform precipitation for the period 1966–2016 [3]. According to the CMIP5 projections, in the 21st century, annual and seasonal precipitation will increase everywhere, especially in the arctic region of Russia [4]. The frequency of heavy precipitation is influenced by changes in the characteristics of air humidity and atmospheric circulation that can lead to the development of extreme climatic events. First of all, this increases the frequency of severe floods and droughts [5]. The trends in extreme precipitation are also correlated with the geographic features of the region and the used data source quality [6,7].

In general, there is an increase in the frequency and intensity of extreme precipitation across the territory of Russia [7–11]; the number of days without precipitation increases in the winter and decreases in the summer [2]. Moreover, most of the territory is characterized by an increase in the number of days with heavy snowfall [9]. Thus, in Western Siberia, there is an increase in extreme precipitation in the winter and a tendency toward dry periods in the summer [12]. In the south of the region, the risks of heavy and long-term precipitation increase [6,13].

The most accurate source of information on atmospheric precipitation is data of meteorological observations, but they are irregular in space and time [14]. Satellite data provide regular observations in space, with almost global coverage. However, they have significant errors associated, among other things, with the inaccuracy of the algorithms for calculating precipitation based on the intensity of direct or scattered radiation [15]. The reanalysis data are given at regular grids, but they may differ from the data of meteorological stations due to various simplifications, parameterizations, and numerical schemes that introduce errors into the calculations [14].

A comparative analysis of the precipitation extremality indices based on the station observational data, reanalyses, and satellite measurements for the territory of Eurasia was carried out in Reference [16]. It showed that the reanalysis data, in comparison with the observational ones, significantly underestimated the extreme precipitation values by 30–35% on average, and the satellite data overestimated the extreme values by 30–50% in the winter and underestimated them by 40–60% in the summer. The comparison of ERA5 reanalysis data and station observational data revealed a high linear correlation for the southern part of Siberia for the period from 1979 to 2015. In this case, the maximum differences are typical for mountainous regions [17].

We would like to mention that average precipitation values are calculated with much greater accuracy than extreme precipitation, since the latter are observed less frequently and have a large uncertainty in magnitude, especially in regions with a scattered observational grid (for example, in the north of Siberia) [8,18]. Therefore, the analysis of precipitation characteristics derived from different data sources can lead to contradictory conclusions. In this regard, an urgent question arises about the accuracy in assessing precipitation data, the importance of which is obvious not only for weather and climate change monitoring but also for solving forecasting problems.

The goal of this work is to investigate the variability and to compare the characteristics of atmospheric precipitation in Western Siberia over recent decades across different datasets.

The paper is organized as follows. In Section 2, we present the study region and datasets and briefly describe the used methods. In Section 3, the results of a time series analysis of precipitation characteristics (including their extremes) are presented. In Section 4 we discuss the obtained results and compare them with studies from other works. The conclusion, limitations, and future direction of this research are briefly summarized in Section 5.

2. Materials and Methods

2.1. Study Region

The analysis was conducted for the territory of Western Siberia (50°–70° N, 60°–90° E) (Figure 1).

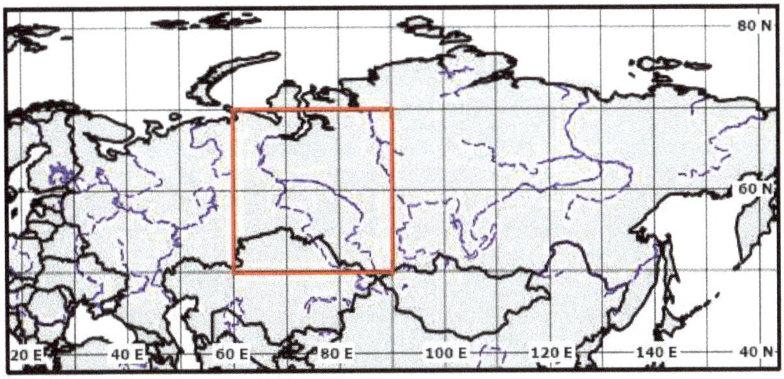

Figure 1. Geographical location of the study region (marked with a red box).

2.2. Datasets and Preprocessing

The following data sources on daily and monthly mean values of atmospheric precipitation were used: (1) observational data at 57 meteorological stations from RIHMI-WDC (All-Russian Research Institute of Hydrometeorological Information—World Data Center) from 1979 to 2018 [19], (2) gridded observational data from APHRODITE (Asian Precipitation—Highly-Resolved Observational Data Integration Towards Evaluation, hereinafter APHRO) with spatial resolution 0.25° × 0.25° for 1979–2007 [20] and from GPCC (Global Precipitation Climatology Centre Full Data Monthly Product Version 2018) with spatial resolution 0.50° × 0.50° for 1979–2018 [21], and (3) reanalysis data from NCEP-DOE Reanalysis 2 (NOAA National Center for Environmental Prediction, hereinafter NCEP) with spatial resolution 1.90° × 1.90° [22] and from ERA5 (the fifth generation of the European Centre for Medium-Range Weather Forecasts reanalysis) with spatial resolution 0.25° × 0.25° from 1979 to 2018 [23]. Reanalysis combines the model data with observations from across the world. We consider that observational data at stations, in spite of their spatial inhomogeneity, provide more objective information due to being based on measurements, while a reanalysis applies different kinds of data assimilation, which can lead to some uncertainties.

2.3. Methodology

The first step in a time series analysis of the meteorological parameters is the derivation of their statistical estimations that characterize the variability of the processes. To calculate these characteristics, we used stochastic processes methods.

We consider that a random quantity ξ is defined by probability density $p_\xi(x)$, satisfied by the following conditions [24]:

$$p_\xi(x) \geq 0,\ \forall x,\ \int_{-\infty}^{\infty} p_\xi(x)dx = 1 \tag{1}$$

where x is the dimensionless quantity in the definition domain of ξ.

Then, we define $f(\xi)$ as a function of the random quantity ξ. Hence, the mean statistical value of this function can be written as follows:

$$\langle \mathrm{f}(\xi)\rangle \equiv \int_{-\infty}^{\infty} \mathrm{f}(\xi)p_\xi(x)dx \tag{2}$$

Distribution moments of the random value (ξ), such as α_n^ξ and μ_n^ξ, are defined as:

$$\alpha_n^\xi \equiv \langle \xi^n \rangle \equiv \int_{-\infty}^{\infty} \mathrm{x}^n p_\xi(x)dx \text{ and } \mu_n^\xi \equiv \langle (\xi - \langle \xi \rangle)^n \rangle \tag{3}$$

where n is a moment order.

According to Equation (1), for each distribution, the p_ξ (x) moments are identically determined and the characteristics of the distribution. Then, we define the mean value of a random quantity (m) and its variability, called a dispersion (D):

$$m \equiv \alpha_1 \ i \ D \equiv \sigma^2 \equiv \mu_2 =< (\xi - m)^2 > \tag{4}$$

These values are enough to describe normal (Gaussian) processes, the probability density of which can be written as:

$$p(x) = \frac{1}{\sigma\sqrt{2\pi}} e^{-\frac{1}{2}\left(\frac{x-m}{\sigma}\right)^2} \tag{5}$$

where σ is a standard deviation.

In that case, if processes do not follow the normal distribution law (nonstationary), it is necessary to use additional characteristics—for instance, coefficients of skewness (A_s) and kurtosis (K_s):

$$A_s = \frac{\mu_3}{\sigma^3}, \ K_s = \frac{\mu_4}{\sigma^4} - 3 \tag{6}$$

The coefficient of A_s allows to view the symmetry of the probability density with respect to m and the presence of heavy tails of the probability distribution density, i.e., about the existence of extreme values. The K_s value can represent the $p(x)$ deformation.

The median of the sample probability distribution function (PDF) is used as the average characteristic for the territory. The extreme values of precipitation are determined using PDF threshold percentiles such as 1% and 5%—extremely low and 95% and 99%—extremely high.

If there are periodic components in the time series, then it is better to use trigonometric functions as basic ones to find the sample spectral density C_{XX} (ω) for a stochastic process $X(t)$ that is defined on a finite interval $[-T, T]$ [25,26]:

$$C_{XX}(\omega) = |A_{XX}(\omega)|^2 = \frac{1}{T}\left(\int_{-T/2}^{T/2} X(t) e^{-i\omega t} dt\right)^2 \tag{7}$$

where A_{XX} (ω)—the amplitude spectral density of the process $X(t)$, i—imaginary number, ω—the frequency of oscillation, and t—time. C_{XX} (ω) shows the distribution of D by the frequency.

The dispersion of C_{XX} (ω) can be reduced if a smoothed window $W(\omega)$ is used [25]:

$$\overline{C_{XX}}(\omega) = \int_{-M}^{M} W(g) C_{XX}(\omega - g) dg \tag{8}$$

$$\int_{-M}^{M} W(\omega) d\omega = 1; \ W(\omega) = W(-\omega) \tag{9}$$

where M is a cutoff point and g—fictitious frequency.

The type of the spectral window can be found as a compromise between reducing the spectrum shift (i.e., narrow window $W(\omega)$) and the C_{XX} (ω) variance (wide window $W(\omega)$). Reference [25] proposed to use the mean square error minimization procedure to find a compromise. In Reference [27], the analysis of different spectral windows was provided in detail, and their characteristics were given there. In the framework of our study, a rectangular window was applied to obtain a small estimate bias.

Data smoothing was conducted using a low-pass filter (LPF) with a reference point of ten years.

To provide a comparative analysis of different data sources, two interpolation methods were used. In the first method, the reanalysis data are interpolated to the meteorological station coordinates with the subsequent calculation of the statistical characteristics for the derived time series. In the framework of this method, two types of interpolation were used: bilinear and cubic (makima) in MATLAB software (MATLAB and Statistics Toolbox Release 2015b, The MathWorks, Inc., Natick, MA, USA), a simplified version of the modified Akima piecewise cubic Hermite interpolation [28]. In the second method, we conducted a spatial interpolation of the calculated average monthly and average annual values of the observational data to the reanalysis data grid according to the Kriging algorithm [29]. The spatial interpolation error could be estimated using the cross-validation procedure [30].

A comparative analysis of the amount of precipitation from different datasets was also performed using Taylor diagrams [31]. The method can provide concise statistical values of how well patterns match each other in terms of their correlation (Pearson coefficient correlation [24]), their root mean square difference (RMS) [24] between processes X and Y, and the ratio of their variances (σ_X and σ_Y). These statistics make it easy to determine how much of the overall RMS difference in patterns is attributable to a difference in variance and how much is due to poor pattern correlation [31].

The estimates are calculated for the warm (April–October) and cold (November–March) seasons of a year. To identify the dynamics of the precipitation characteristics, a comparison of the estimates over the time intervals from 1979 to 2018 was made, as well as from 1979 to 2007 when we made a comparison with APHRO.

3. Results

The interannual variability of the smoothed average seasonal median estimates of precipitation amount in different datasets for the period from 1979 to 2018 is presented in Figure 2. The figure shows that the GPCC data has the maximum approximate annual values of the observational data. The APHRO data archive underestimates the values, and the NCEP reanalysis data overestimates ones that are especially pronounced in the warm season. At the same time, in the cold season, and only in the first half of the time interval (from 1979 to 1995), the highest values are observed for the ERA5 reanalysis dataset.

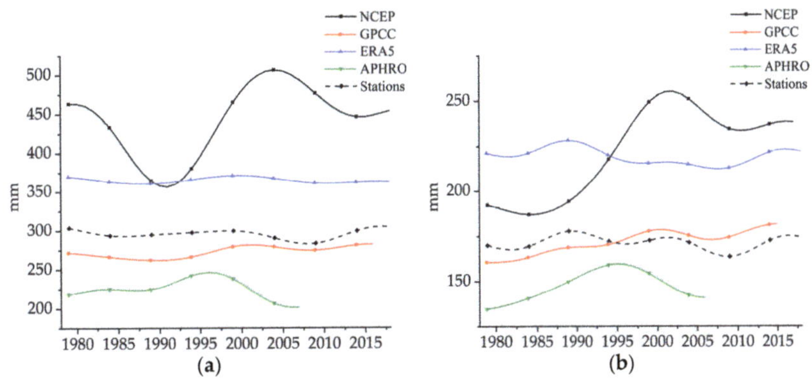

Figure 2. The interannual variability of the precipitation median values smoothed with LPF: (**a**) warm season and (**b**) cold season.

The NCEP reanalysis, in general, is characterized by the greatest variability in precipitation time series compared to other datasets. The discrepancies between the time series at the beginning of the 21st century are probably associated with using different calculation methods in data assimilation.

A spectral analysis of the time series showed that the periodic structure in the precipitation time series constructed from the observational data was well-explicit (significance level $p < 0.01$) in ERA5 and in GPCC data, especially in the short-period part of the amplitude spectrum (fluctuation scale < 10 years), particularly in the cold season (Figure 3).

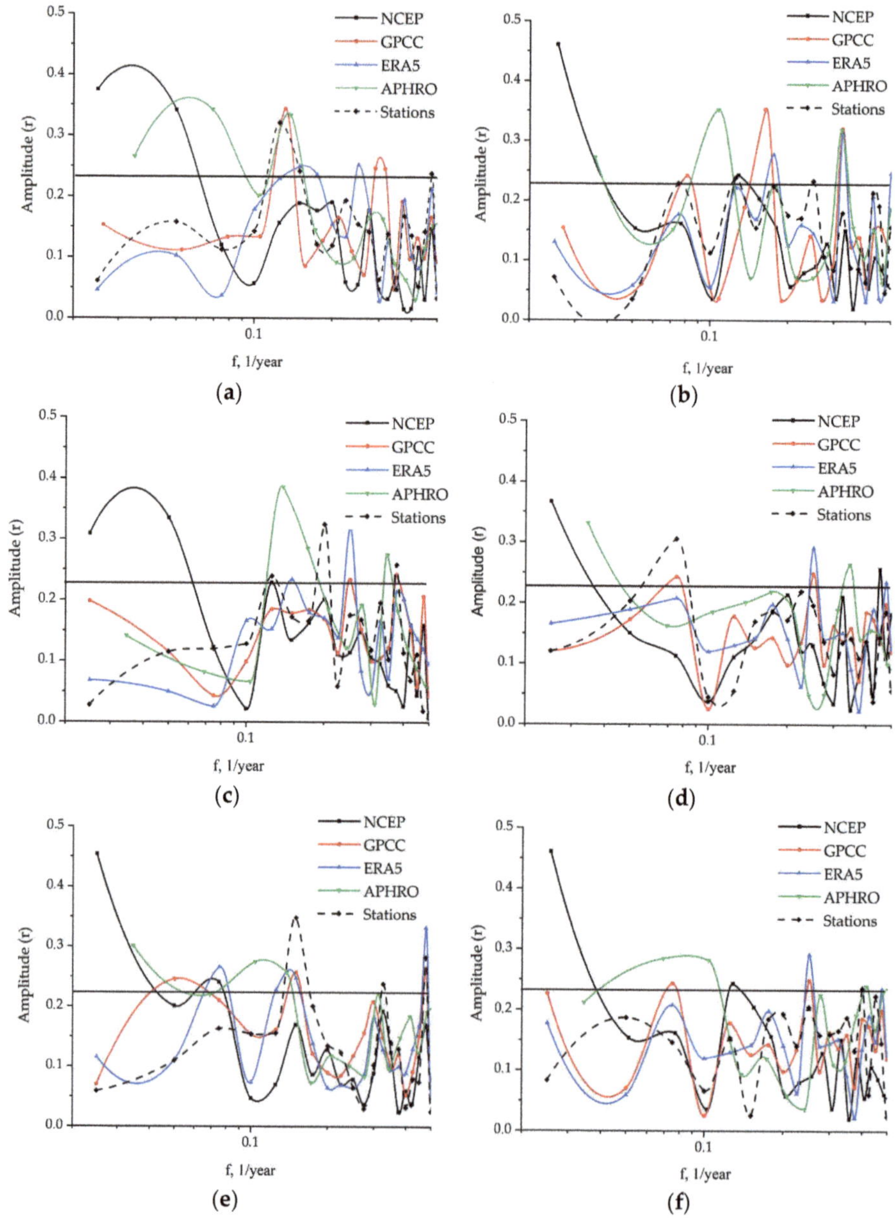

Figure 3. The Fourier amplitude spectrum of the precipitation time series: (**a**,**b**) median values, (**c**,**d**) extremely high (95%) values, and (**e**,**f**) extremely low (5%) values. Left panel: (**a**,**c**,**e**) warm season; right panel: (**b**,**d**,**f**) cold season. A straight horizontal line defines the significance level ($p < 0.01$).

Statistically significant fluctuations ($p < 0.01$) in the seasonal series obtained from observational data, as well as from GPCC and APHRO datasets, were discovered at periods of 7 to 8 years in the warm season, while shorter-period fluctuations also appeared in the extreme values series (Figure 3c,e). In the cold season, the indicated periodicities were not distinguished in the observational data (Figure 3b,d,f). However, as for the long-period part of the spectrum, a period of 12 to 13 years was determined for extremely high precipitation characteristics that also exist in the GPCC dataset (Figure 3f). No statistically significant fluctuations were found in the data series from meteorological stations; however, they were found in the NCEP and APHRO datasets. This could result from the influence of the trend components on the variability during time series processing.

Analyzing the calculated values by the threshold quantiles (see Table 1 and Figure 4), we found that the average annual median estimates based on the observational data were closest to the corresponding estimates derived from the APHRO dataset (274.0 mm and 291.2 mm, respectively). The NCEP reanalysis data had the highest median values (490.9 mm).

Table 1. The distribution of the precipitation values by the threshold quantiles.

Quantile	0.01	0.05	0.10	0.25	0.50	0.75	0.90	0.95	0.99
Season					APHRO (1979–2007)				
Warm	58.9	119.8	158.5	213.3	274.0	332.9	380.6	414.2	500.4
Cold	37.4	50.9	64.2	87.8	115.6	143.3	170.5	189.9	227.2
Year	106.7	187.7	236.7	314.3	391.9	468.1	531.4	577.6	693.2
					ERA5				
Warm	163.0	235.0	275.0	342.0	414.0	484.0	557.0	621.0	867.0
Cold	75.0	95.0	107.0	130.0	156.0	189.0	229.0	265.0	376.0
Year	283.0	358.0	407.0	486.0	570.0	660.0	761.0	853.0	1000.0
					NCEP				
Warm	134.2	194.3	242.6	344.2	490.9	633.2	744.6	809.3	912.5
Cold	57.9	81,6	96.9	127.5	165.6	208.1	250.3	277.5	321.3
Year	227.9	306.2	361.9	486.8	659.3	832.7	969.4	1047.7	1174.9
					GPCC				
Warm	105.5	164.9	198.5	264.4	337.9	403.0	466.6	509.5	609.5
Cold	41.6	62.5	75.6	99.9	133.4	168.6	206.6	233.4	302.2
Year	181.3	248.5	290.2	377.8	476.2	565.4	647.9	699.9	828.8
					Stations				
Warm	165.0	200.0	219.6	252.1	291.2	328.9	369.0	395.8	463.7
Cold	94.0	110.9	121.8	140.8	168.8	205.9	242.9	267.0	321.0
Year	269.0	315.9	342.9	395.1	464.3	531.1	599.6	641.4	760.9

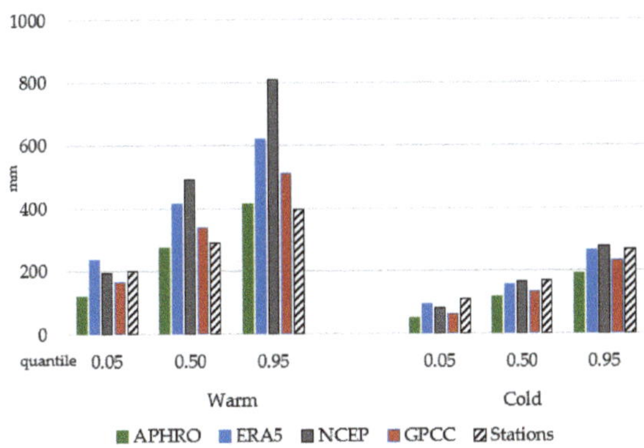

Figure 4. The distribution of the precipitation values in the warm and cold seasons by the threshold quantiles. Visualization of the data presented in Table 1.

In the warm season, the closest values of extremely low precipitation to the estimates in the observational data belonged to the ERA5 dataset and the NCEP dataset of 1% and 5%, respectively. The extremely high values (95% and 99%) of precipitation at the stations were in good agreement with the APHRO and the GPCC datasets. For the cold season, in the main, a good agreement was observed with the ERA5 reanalysis data and with the NCEP reanalysis data for extremely high precipitation (99%). The greatest discrepancy in the cold season with the observational data at the stations was observed with the APHRO data archive. Thus, the annual average median extreme high values from the GPCC data showed the best agreement with these values from observational data at the stations.

The range of variability can be estimated using the coefficients of kurtosis (Ks) and skewness (As). In the first approximation, their analysis allowed us to describe the form of the PDF, taking into account its deviation from the normal distribution. Analyzing the density function, we concluded that, for the cold season (as well as for an entire year in general), all datasets (except the APHRO dataset) were characterized by positive skewness ($As > 0$ is a positive skew). That means that the distribution was right-skewed, right-tailed, or skewed to the right, which may indicate a decrease in precipitation values for the considered period and an increase in the frequency of extreme events with increased precipitation. This confirmed the results that were obtained earlier in Reference [12]. In the warm season, only the GPCC dataset represented a positive skewness. The analysis of the derived kurtosis coefficients showed that the warm season was characterized by a flat-topped distribution ($Ks < 0$) that was the amount of precipitation over the studied time interval varying in a wide range of values. In the cold season, data from the stations and ERA5 dataset had a narrow peak distribution ($Ks > 0$). This indicated that the values varied within a narrow range of values. For an entire year, the kurtosis coefficient was positive and determined mainly by the cold season.

Figure 5 represents the spatial distribution of precipitation based on the observational data. We reveal that the amount of precipitation in the northern regions of Western Siberia is more than in the southern ones.

This may happen due to geographical reasons (latitude and relief) or may be related to the dominant form of atmospheric circulation and the influence of the ocean. Both in the warm and cold seasons, the maximum amount of precipitation is was at the station located in the mountain area, the minimum found in the Chuy River Basin.

Figure 5 also shows the spatial distribution of the precipitation characteristics for all the datasets, with the incorporation of the values from the considered weather stations. The maximum values of extreme precipitation were also indicated in the Altay Mountains region in all the considered datasets. We found an increase in the values of both extremely low and extremely high values of precipitation that were observed in the northeast, northwest (the arctic zone), and in the central part (the Siberian Ridges) of the territory. This is consistent with the observational data that was also mentioned in Reference [17].

In the same areas, an increase in the linear trend coefficient for the average annual precipitation was observed based on the ERA5 and GPCC data. At the same time, the highest values of the linear trend were typical for the NCEP dataset. Using the NCEP and APHRO data, we outlined the tendencies for precipitation increase in the eastern part of Western Siberia with a maximum in the southeast (the Altay Mountains) and a decrease in the western and northwestern parts (the Ural Mountains). The conclusion about the closeness of the values of extreme precipitation from observational data to the GPCC data was also confirmed by the consistency of their spatial distribution over the territory of Western Siberia.

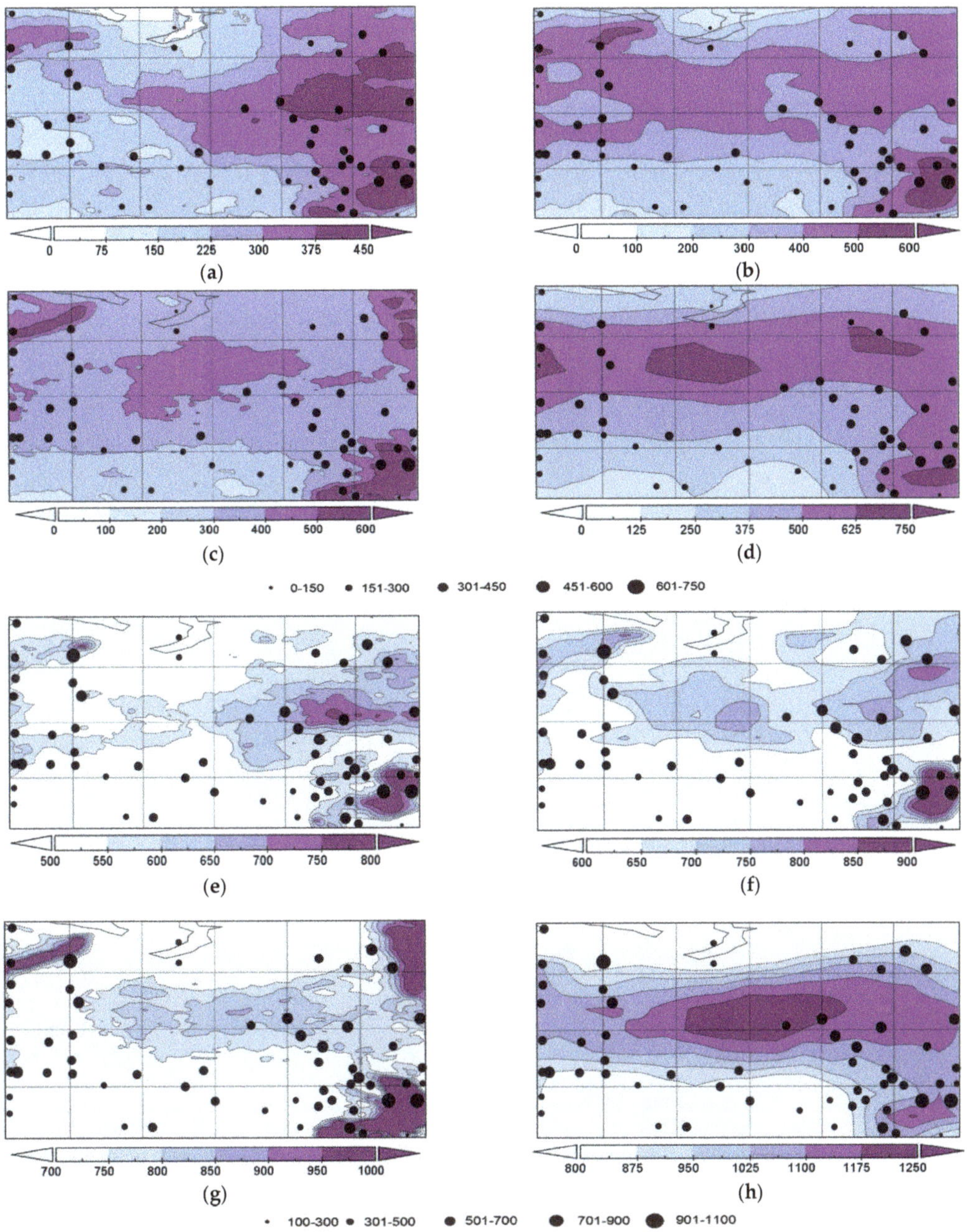

Figure 5. The spatial distribution of precipitation values (mm): (**a**,**e**) APHRO, (**b**,**f**) GPCC, (**c**,**g**) ERA5, and (**d**,**h**) NCEP. Upper panel: (**a**–**d**) extremely low (5%); lower panel: (**e**–**h**) extremely high (95%). Points correspond to the extreme values from the meteorological stations (mm).

A comparative analysis of the amount of precipitation from different datasets was also performed using the Taylor diagrams constructed for the warm and cold seasons for the time interval from 1979 to 2018 (from 1979 to 2007 for the APHRO dataset) (Figure 6).

The data were interpolated from the reanalysis grid nodes to the station coordinates by the bilinear interpolation method. The analysis of the derived correlation coefficients showed a high correlation between the data for extreme precipitation (5% and 95%) at the meteorological stations with the GPCC and ERA data (the values of the correlation coefficients varied from 0.73 to 0.91).

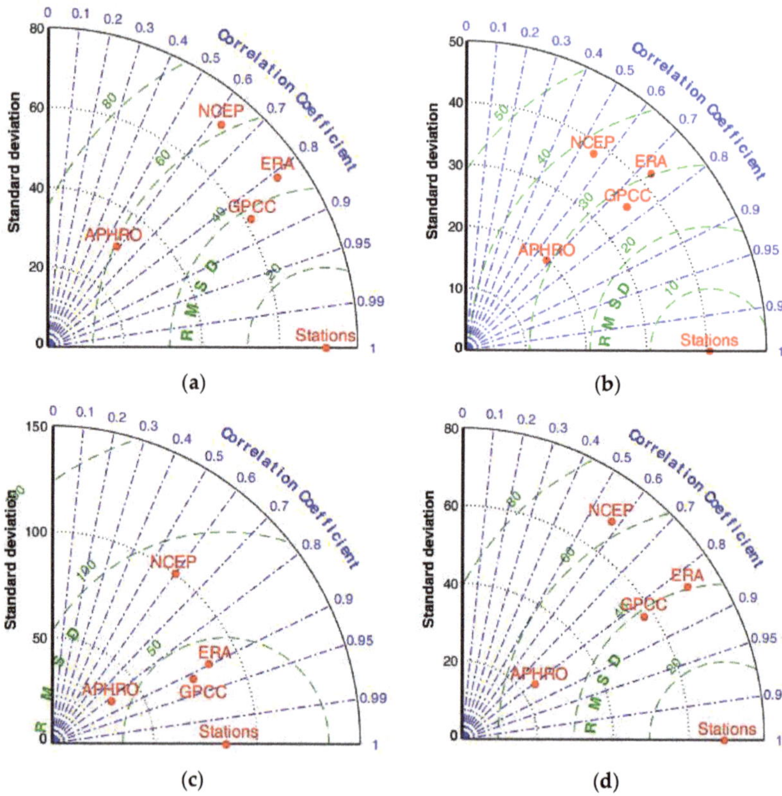

Figure 6. The comparison of the precipitation extremes (using the Taylor diagram) for the period from 1979 to 2018 (from 1979 to 2007 for the APHRO dataset) for the warm (left panel) and cold (right panel) seasons: (**a**,**b**) extremely low (5%) and (**c**,**d**) extremely high (95%). The bilinear interpolation method is used.

These datasets were characterized by variability in the amount of precipitation within the same range. The smallest values of the correlation coefficients (r) were observed in NCEP data ($0.55 < r < 0.62$). At the same time, the precipitation values had less variability than the observational data in the APHRO dataset. However, in both cases, there was a relatively largely centered root mean square error for the precipitation values. We noted that similar trends were typical for the extreme precipitation values of 1% and 99%.

Additionally, the median values in all the datasets were in good agreement with the observational data; particularly, the correlation coefficient varied from 0.76 in the cold season to 0.95 in the warm season compared to 0.65 in the cold season to 0.78 in the warm season for the NCEP dataset.

The results of the comparative analysis of the estimates derived using the bilinear and cubic interpolation methods were generally similar, except for the cold season; the lowest correlation coefficient for the extremely low and high precipitation values was observed for the APHRO dataset ($r < 0.3$) (Figure 7).

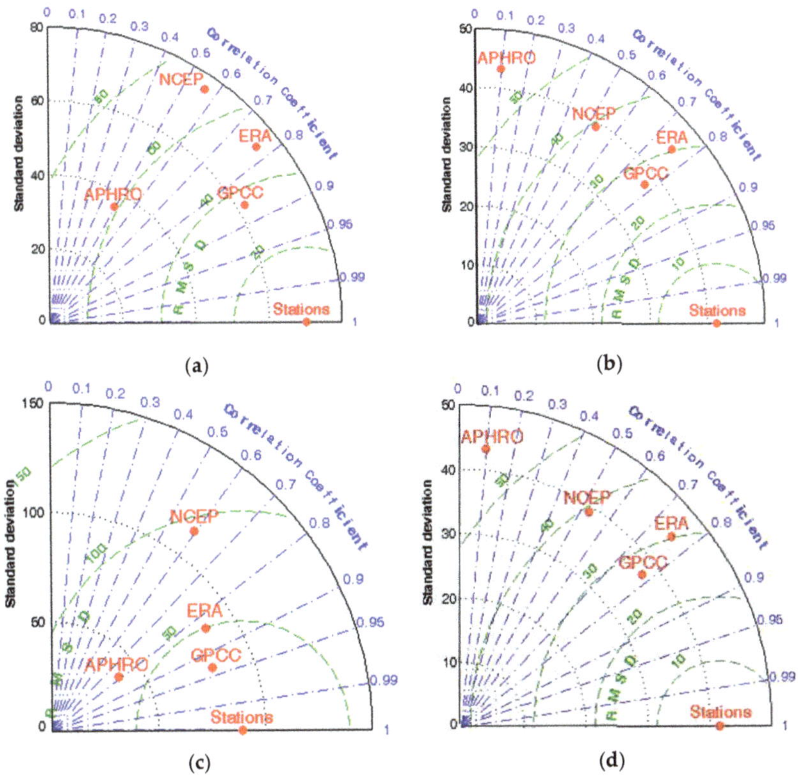

Figure 7. The comparison of the precipitation extremes (using the Taylor diagram) for the period of 1979–2018 (1979–2007 for the APHRO dataset) for the warm (left panel) and cold (right panel) seasons: (**a**,**b**) extremely low (5%) and (**c**,**d**) extremely high (95%). The cubic interpolation method is used.

4. Discussion

We find that the values from the GPCC dataset are closest to those from the observational data. This is probably since the GPCC data represents gridded observational data from stations. However, at the same time, the APHRO archive (which also uses observational data) underestimates values, and the NCEP reanalysis data overestimates ones. The GPCC data also show the best agreement for annually averaged extremely high values. The agreement for other datasets can vary and can depend on the season. For example, the NCEP dataset can reproduce median and extreme values. According to the research for the European continent [32], the NCEP2 dataset also demonstrates the closest to the station data estimates of extreme precipitation.

Applying the spectral analysis to the precipitation time series could allow us to better understand the characteristics of precipitation variability. Moreover, this is a quite new approach to investigating extreme precipitation variability in Western Siberia. For example, the GPCC dataset reveals the periodicities in the time series of observational data from stations in the warm and cold seasons. The ERA5 dataset reproduces the general variability but with a smaller amplitude. Statistically significant fluctuations are mainly distinguished in the warm season at periods of 7 to 8 years, while shorter-period fluctuations also appear in the extreme values series. It should be noted that revealed periodicities can be caused by dynamic processes in the atmosphere described by global atmospheric mechanisms, such as North Atlantic Oscillation (NAO), Atlantic Multidecadal Oscillation (AMO), and

South Oscillation (El Niño), where this periodicity (7–9 years) is also observed in their time series [33]. We suppose that this fact will be useful for the construction of climatic projections. The long-term periodicities systematically (for each characteristic) exist in the NCEP reanalysis data. This is caused by the presence of a trend (Figure 2) in the time series, but a similar trend in the time series derived from meteorological stations could not be observed. However, such periodicities (15–17-year cycles) can come about for other regions of the planet [34]. Additionally, in Reference [35], it was revealed (periodogram-based time series methodology) that the monthly average precipitation has two different periodic structures of six months and twelve months that coincide with the seasonal pattern of the time series. However, the interannual periodicity is not explicit enough. Moreover, there is no information presented about the extreme values variability. The spectral analysis in the framework of this study revealed the periodic structure in the precipitation time series constructed from different datasets, where statistically significant values were mainly observed in the short-period part of the amplitude spectrum (fluctuation scale < 10 years). This result could be useful in the short-term forecasting of both the mean and extreme values of precipitation. For future research, it seems appropriate to apply the methods of multiscale and multivariate statistical analyses (including Wavelet analysis). This will allow us to show the coherency between the components of two time series in the time–frequency domain and to provide better comparison and visualization of the observed periods.

Based on the observations, we see that the precipitation in the northern stations (situated above 60° N) is greater than in the southern ones (situated below the 60° N). The maximum number of precipitations was observed at a station located in the Altay Mountain area, the minimum in the Chuy River Basin and in the Ural Mountains. The spatiotemporal variability of extreme precipitation revealed an increase in precipitation in the northeast and northwest (the arctic zone) and in the central part of the territory (the Siberian Ridges) that was consistent with the observational data. The correlation analysis showed that the GPCC and the ERA5 datasets were in good agreement with the observations (the correlation coefficient was up to 0.91). We obtained quite good agreement between observational data at the stations and GPCC data. The GPCC dataset outcomes can be explained by the fact that GPCC owns the largest and most comprehensive worldwide collection of precipitation data. This is based on daily surface synoptic observations and monthly climate messages [36]. Moreover, it supports regional climate monitoring and climate variability analyses. The ERA5 reanalysis data (replaces the ERA-Interim reanalysis) enhanced the spatial and temporal resolutions in comparison with the other reanalyses, which allowed us to get information that was more detailed. The APHRODITE project develops daily precipitation datasets with high-resolution grids for Asia; however, it has limited time series. The NCEP reanalysis data has a quite coarse grid resolution for the analysis of regional precipitation characteristic variabilities, especially as concerns their extremes.

Thus, we compared different types of precipitation datasets for Western Siberia with different spatial and temporal resolutions, observational data on stations, gridded data, and reanalysis data. The choice of appropriate data source for the research of precipitation characteristic variabilities will firstly depend on the goal of the investigation. Moreover, the regional differences in the long-term tendencies of the precipitation characteristics (means and extremes) will depend on the changes in the used data assimilation and parametrization models in different datasets. The median estimates of the precipitation amount derived from station data and reanalysis data are in better agreement with each other rather than their extreme values. At the same time, in some cases, the temporal variability of the extremes can be quite effectively diagnosed by reanalyses, at least in comparison to the median values of precipitation [32].

We found that some of our results related to the agreement between observational data and reanalysis ones have also been found in similar research [2,8]. However, in Reference [37], it has been found that the APHRO archive is closest to the real observations in comparison with ERA-Interim for the Siberian region. This result is explained by the fact that the validation was made based on a single parameter (RMSE) and did not to take into account other statistical characteristics. The novelty of this study is that we proposed a comparative analysis not only for the mean values of precipitation but also for their extremes. The usage of different statistical methods (descriptive statistics, Fourier spectrum, and Taylor diagrams) makes the results presented in this study more reliable. This is quite important for the arctic part of the region, where an observational grid is significantly sparse.

5. Conclusions

In the framework of this study, we presented a comparative analysis of the atmospheric precipitation characteristics (mean and extremes) in Western Siberia from 1979 to 2018 across different datasets.

The performed analysis was based on data acquired from meteorological stations, global precipitation datasets such as APHRODITE and GPCC, and reanalysis archives, including NCEP-DOE and ERA5. The comparison was based on the methods of descriptive statistics, Fourier spectrum, and Taylor diagrams.

The best agreement of the values from the observational data was observed with the values from GPCC. This archive also represented the periodicities in the time series of observational data from the meteorological stations, especially in the short-period part of the spectrum. Underestimated values were revealed for the APHRODITE archive, while overestimated ones were found for the NCEP reanalysis data. In comparison with GPCC, the ERA5 dataset reproduced the general variability but with a smaller amplitude (the correlation coefficient was up to 0.9). In general, the median estimates of the precipitation amount derived from the meteorological stations' data, as well from the reanalysis data, were in better agreement with each other rather than their extreme values. However, their temporal variability can be effectively described by other datasets.

The results obtained from the validation can be useful in solving various problems in climatology associated with the usage of data on the variable precipitation characteristics and extreme events (when studying the conditions for the formation of droughts, forest fires, degradation of the permafrost zone, etc.), as well as for the development and correction of regional climate models for more accurate climate change projections. In the framework of this study, we focused on a descriptive comparative analysis of the precipitation characteristics, where a spectrum analysis is one of the parts of the research. The novelty of our work is that we made a comparison of the time series amplitude spectrum averaged by the territory values of precipitation, as well as their extreme values.

Thus, we also suppose that the goal of our future work will deal with the application of multiscale and multivariate statistical analyses (including a wavelet analysis) that will allow us to conduct an analysis of the precipitation time series spectrum in more detail and to provide better comparisons and visualization of the results.

Author Contributions: Conceptualization and methodology: E.K.; data resources and curation: Y.M.; analysis: E.K. and S.L.; visualization: Y.M., S.L. and E.K.; writing—original draft preparation: E.K.; writing—review and editing: Y.M., S.L. and I.S.; funding acquisition: I.S.; supervision and project administration: S.L. and I.S. All authors have read and agreed to the published version of the manuscript.

Funding: This research was funded by the Russian Science Foundation (RSF), project # 21-71-10052, https://rscf.ru/en/project/21-71-10052 (accessed on 17 May 2021), Y.M. and S.L. were also supported by the Ministry of Science and Higher Education of the Russian Federation, project # 121031300154-1.

Institutional Review Board Statement: Not applicable.

Informed Consent Statement: Not applicable.

Data Availability Statement: Not applicable.

Conflicts of Interest: The authors declare no conflict of interest.

References

1. Gruza, G.V.; Rankova, E.Y. *Observed and Expected Climate Changes over Russia: Surface Air Temperature*; World Data Center (RIHMI-WDC), Roshydromet: Obninsk, Russia, 2012; p. 194. (In Russian)
2. Available online: http://www.meteorf.ru/press/news/23886/?sphrase_id=667332 (accessed on 17 May 2021). (In Russian).
3. Chernokulsky, A.; Kozlov, F.; Zolina, O.; Bulygina, O.; Mokhov, I.; Semenov, V. Observed changes in convective and stratiform precipitation in Northern Eurasia over the last five decades. *Environ. Res. Lett.* **2019**, *14*, 045001. [CrossRef]
4. Kokorev, V.A.; Sherstiukov, A.B. Meteorological data for studying the current and projected for the future climate change in Russia. *Arct. 21st Century. Environ. Sci.* **2015**, *2*, 5–23, (In Russian with English summary).
5. Leshkevich, N.V. (Ed.) *Assessment Report on Climate Change and Its Consequences in Russian Federation. Part 1: Climate Change*; Roshydromet: Moscow, Russia, 2008; p. 277. (In Russian)
6. Alexander, L.V.; Zhang, X.; Peterson, T.C.; Caesar, J.; Gleason, B.; Klein Tank, A.M.G.; Haylock, M.; Collins, D.; Trewin, B.; Rahimzadeh, F.; et al. Global observed changes in daily climate extremes of temperature and precipitation. *J. Geophys. Res. Atmosph.* **2006**, *111*, D05109. [CrossRef]
7. Zolotokrylin, A.; Cherenkova, E. Seasonal changes in precipitation extremes in Russia for the last several decades and their impact on vital activities of the human population. *Geogr. Environ. Sustain.* **2017**, *10*, 69–82. [CrossRef]
8. Zolina, O.G.; Bulygina, O.N. Current climatic variability of extreme precipitation in Russia. *Fundame. Appl. Clim.* **2016**, *1*, 84–103, (In Russian with English summary). [CrossRef]
9. Korshunova, N.N.; Bulygina, O.N.; Razuvaev, V.N.; Davletshin, S.G. Evaluating extremeness of temperature and precipitation regimes for the territory of the Russian Federation and its regions. *Proc. All-Russ. Res. Inst. Hydrometeorol. Inf.—World Data Cent.* **2018**, *183*, 20–30, (In Russian with English summary).
10. Volkova, M.A.; Cheredko, N.N.; Sokolov, K.I.; Ogurtsov, L.A. The current space-time structure of the extreme precipitation field in Western Siberia. *Tomsk State Univ. J.* **2015**, *390*, 208–216, (In Russian with English summary).
11. Mokhov, I.I.; Semenov, V.A. Weather and Climate Anomalies in Russian Regions Related to Global Climate Change. *Russ. Meteorol. Hydrol.* **2016**, *2*, 16–28. [CrossRef]
12. Kharyutkina, E.V.; Loginov, S.V.; Usova, E.I.; Martynova, Y.V.; Pustovalov, K.N. Tendencies in Changes of Climate Extremality in Western Siberia at the End of the XX Century and the Beginning of the XXI Century. *Fundam. Appl. Climatol.* **2019**, *2*, 45–65. (In Russian) [CrossRef]
13. Sharapova, A.A.; Kuzhevskaya, I.V.; Kashtanova, K.A.; Polyakov, D.V. Characteristics of heavy precipitation in Western Siberia. *Geogr. Bull.* **2017**, *3*, 88–98, (In Russian with English summary). [CrossRef]
14. Khudyakova, T.A.; Martynova, Yu.V. Research of the quality of representation of the Siberian high intensity in reanalysis data. In Proceedings of the International Conference on Environmental Observations, Modelling and Information Systems "ENVIROMIS-2016", Tomsk, Russia, 11–16 July 2016; pp. 67–71.
15. Laptev, A.S.; Zhuravlev, S.A.; Georgievsky, M.V. Estimation of water balance elements of large river basins based on Earth remote sensing data. In Proceedings of the International Scientific Conference in Memory of Outstanding Russian Hydrologist Yury Vinogradov "Third Vinogradov Conference. Facets of Hydrology", Saint Petersburg, Russia, 28–30 March 2018; pp. 204–208.
16. Zolina, O.G. Statistical Modeling of Extreme Precipitation and Their Role in the Regional Hydrological Cycle. Doctoral Thesis, Hydrometeorological Research Center of the Russian Federation, Moscow, Russia, 2018. (In Russian).
17. Voropay, N.; Ryazanova, A.; Dyukarev, E. High-resolution bias-corrected precipitation data over South Siberia, Russia. *Atmos. Res.* **2021**, *254*, 105528. [CrossRef]
18. Allen, M.R.; Ingram, W.J. Constraints on future changes in climate and the hydrologic cycle. *Nature* **2002**, *419*, 224–232. [CrossRef] [PubMed]
19. Available online: http://meteo.ru (accessed on 14 January 2022).
20. Available online: http://www.chikyu.ac.jp/precip (accessed on 14 January 2022).
21. Available online: https://psl.noaa.gov/data/gridded/data.gpcc.html (accessed on 14 January 2022).
22. Available online: https://psl.noaa.gov/data/gridded/data.ncep.reanalysis2.html (accessed on 14 January 2022).
23. Available online: https://reanalyses.org/atmosphere/era5-references (accessed on 14 January 2022).
24. Von Storch, H.; Zwiers, F.W. *Statistical Analysis in Climate Research*; Cambridge University Press: Cambridge, UK, 1999; p. 484. ISBN 0-521-45071-3.
25. Jenkins, G.M.; Watts, D.G. *Spectral Analysis and Its Applications*; Emerson-Adams Press: Boca Raton, FL, USA, 2000; p. 525. ISBN 1892803038.
26. Ghaderpour, E.; Pagiatakis, S.D.; Hassan, Q.K. A survey on Change Detection and Time Series Analysis with Applications. *Appl. Sci.* **2021**, *11*, 6141. [CrossRef]
27. Harris, F. On the Use of Windows for Harmonic Analysis with the Discrete Fourier Transform. *Proc. IEEE* **1978**, *66*, 51–83. [CrossRef]
28. Available online: https://blogs.mathworks.com/cleve/2019/04/29/makima-piecewise-cubic-interpolation (accessed on 14 January 2022).
29. Cressie, N.A.C. The Origins of Kriging. *Math. Geol.* **1990**, *22*, 239–252. [CrossRef]
30. Grimm, J.W.; Lynch, J.A. Statistical analysis of error in estimating wet deposition using five surface estimation algorithms. *Atmos. Environ.* **1991**, *25*, 317–327. [CrossRef]

31. Taylor, K.E. Summarizing multiple aspects of model performance in a single diagram. *J. Geophys. Res.* **2001**, *106*, 7183–7192. [CrossRef]
32. Zolina, O.; Kapala, A.; Simmer, C.; Gulev, S.K. Analysis of extreme precipitation over Europe from different reanalyses: A comparative assessment. *Glob. Planet. Chang.* **2004**, *44*, 129–161. [CrossRef]
33. Moraru, E.I.; Loginov, S.V.; Kharyutkina, E.V. Features of the spatiotemporal variability of the advective heat transfer in the troposphere over the Pacific and Atlantic oceans in the Northern Hemisphere for 1979–2018. *Izv. Atmos. Ocean. Phys.* **2021**, *57*, 606–613. [CrossRef]
34. Joshi, M.K.; Pandey, A.C. Trend and spectral analysis of rainfall over India during 1901–2000. *J. Geophys. Res.* **2011**, *116*, D06104. [CrossRef]
35. Akdi, Y.; Ünlü, K.D. Periodicity in precipitation and temperature for monthly data of Turkey. *Theor. Appl. Climatol.* **2021**, *143*, 957–968. [CrossRef]
36. Bruno, R.; Becker, A.; Schneider, U.; Meyer-Christoffer, A.; Ziese, M. New GPCC Full Data Reanalysis Version 5 Provides High-Quality Gridded Monthly Precipitation Data. *GEWEX News* **2011**, *21*, 4–5.
37. Shulgina, T.; Genina, E.; Gordov, E. Dynamics of climatic characteristics influencing vegetation in Siberia. *Environ. Res. Lett.* **2011**, *6*, 045210. [CrossRef]

atmosphere

MDPI

Article

Variation of High and Low Level Circulation of Meiyu in Jiangsu Province in Recent 30 Years

Ruoxin Hu and Lijuan Wang *

Key Laboratory of Meteorological Disaster, Ministry of Education/Joint International Research Laboratory of Climate and Environment Change/Collaborative Innovation Center on Forecast and Evaluation of Meteorological Disasters, Nanjing University of Information Science and Technology, Nanjing 210044, China; sandy329@126.com
* Correspondence: wljfw@163.com

Abstract: By using the NCEP/NCAR re-analysis data from 1990 to 2019 and the daily precipitation data of CN05.1 gridded observation dataset, the high and low level circulation characteristics and their influence on the onset and precipitation of Meiyu in Jiangsu Province in recent 30 years are studied. Comparing Meiyu in the 2010s with that in the 1990s, it is found that during the 2010s Meiyu was characterized by a late arrival and less precipitation. There were obviously earlier Meiyu years in the 1990s, while no extremely early Meiyu year existed in the 2010s, which was mainly caused by the late northward jump of the upper jet and the ridge line of the western Pacific subtropical high (WPSH hereinafter) in the 2010s. Compared with the 1990s, the 2010s witnessed an eastward position of the South Asia high and a westward position of the subtropical westerly jet during the Meiyu period, which are not conducive to precipitation in the Yangtze-Huaihe region. At the same time, the cold air flowing southward to the Yangtze-Huaihe region was hindered in the 2010s due to the change of blocking in the middle troposphere. In the 2010s, the water vapor transport and the vertical transportation weakened, resulting in the decrease of precipitation in the Yangtze-Huaihe region.

Keywords: Meiyu; high and low level circulation characteristics; the western Pacific subtropical high; the South Asia high; the water vapor transport

Citation: Hu, R.; Wang, L. Variation of High and Low Level Circulation of Meiyu in Jiangsu Province in Recent 30 Years. *Atmosphere* **2021**, *12*, 1258. https://doi.org/10.3390/atmos12101258

Academic Editor: Zhaoxia Pu

Received: 31 August 2021
Accepted: 23 September 2021
Published: 27 September 2021

1. Introduction

Each summer, the northward advancement of the East Asian monsoon effects a rainy season of continuous precipitation in the Yangtze-Huaihe region in China, which is called Meiyu. Frequent occurrences of drought and flood disasters can be seen in the Yangtze-Huaihe region, and those in June and July are mostly related to the abnormal Meiyu in that year [1,2]. Therefore, Meiyu has always been a great concern of meteorological research, and valuable achievements have been made on it. The onset and precipitation of the Meiyu are controlled by large-scale circulation factors, and the variation of the East Asian summer monsoon results in the change of the Meiyu [3,4]. The South Asia high and the western Pacific subtropical high (WPSH hereinafter) can directly affect Meiyu as well [5,6]. Moreover, Meiyu is related to the synoptic-scale systems, such as the blocking high, which can make a significant impact on the situation of Meiyu [7,8]. The Meiyu is also affected by climate change. Under the background of global warming, less precipitation of the Meiyu has occurred in the Yangtze-Huaihe region [9,10].

It has been widely proven that the circulation characteristics of the Meiyu period can influence the Meiyu process. According to an analysis of the observation data from 1961 to 2011, it has been concluded that Meiyu in the Yangtze-Huaihe region started late and ended early in the 21st century, and the length and intensity of the rain season were reduced [11]. At the beginning of this century, the Meiyu in Jiangsu was not typical and it was difficult to determine the start and end date of the Meiyu [12]. In 2018, the Yangtze-Huaihe region had a late Meiyu and less precipitation due to the northerly subtropical upper jet in East Asia, the WPSH, the atypical blocking situation in the middle to high latitudes, and the late onset of the summer monsoon [13,14]. Compared with other years, the year 2017 had a stronger East Asian trough, weaker summer monsoon, and stronger WPSH, resulting in a later Meiyu onset, earlier Meiyu ending, and shorter length and less amount of precipitation in the Yangtze-Huaihe region [15,16]. In addition, the Meiyu lasted less time in 2015 and 2016 [17,18], and was even absent in 2014 [19]. All these facts indicate that new characteristics have emerged for the Meiyu in recent years.

Since the 1990s, the onset of Meiyu over the Yangtze-Huaihe region has been delayed, the temporal features and precipitation process of the Meiyu have changed, and the circulation background shows characteristics different from the past. Based on the atmospheric reanalysis data and the Meiyu time data, this work analyzes the variation characteristics of Meiyu and the changes of high and low level circulation in Jiangsu in recent 30 years. At the same time, standards of Meiyu in each region are not completely consistent, and the performance of Meiyu in the south and north of the Yangtze-Huaihe region is slightly different. As a result, this work focuses on the Meiyu in Jiangsu Province, and uses the onset data of Jiangsu Province. By comparing the characteristics of Meiyu in the 2010s to those in the 1990s, this work discusses the interdecadal evolution of Meiyu in recent 30 years, and further reveals the variation pattern and mechanism of Meiyu.

2. Data and Methods

Data utilized in this work include: (1) the daily precipitation data of CN05.1 gridded observation data from 1990 to 2019 provided by the National Climate Center of China, which is based on the daily observation data of more than 2400 stations nationwide of the National Meteorological Information Center of China, and has a horizontal resolution of $0.25° \times 0.25°$; (2) atmospheric re-analysis data from 1990 to 2019 provided by the National Center for Environmental Prediction/the National Center for Atmospheric Research (NCEP/NCAR) of the U.S., including potential height, meridional wind, zonal wind, vertical velocity, relative humidity and other factors with a horizontal resolution of $0.25° \times 0.25°$; (3) time data of the onset and end of the Meiyu in Jiangsu Province from 1961 to 2019, based on the standards for delimiting Meiyu from Jiangsu Meteorological Observatory.

In this work, the position of the ridge line of the WPSH is determined by the 500-hPa potential height field [20]. The zonal wind meets the geostrophic relationship, and the ridge line meets the following relationship:

$$\begin{cases} u = -\frac{1}{f}\frac{\partial \phi}{\partial y} = 0 \\ \frac{\partial u}{\partial y} > 0 \end{cases} \quad (1)$$

where u is the zonal wind, f is the geostrophic parameter, ϕ is the potential height, and y is the vector of meridional.

3. Results

3.1. Evolution Characteristics of Meiyu Onset and End Time in Jiangsu Province

Based on the onset and end dates of Meiyu in Jiangsu Province, the interdecadal variation characteristics of Meiyu in Jiangsu Province have been discussed. According to the 9-year moving average of the Meiyu onset time of Jiangsu Province from 1961 to 2019 (Figure 1), it can be seen that the overall change trend in the 59 years can be divided into two stages. From 1961 to 1987, the Meiyu onset was gradually advanced over time, and then it was delayed since 1988. The decadal average of the Meiyu onset time shows that Meiyu in Jiangsu was continuously delayed since the 1990s. In this process, the onset and precipitation characteristics of Meiyu and the high and low level circulation background changed. As the initial and end stages of the delay process, the 1990s and the 2010s have obvious differences and can indicate the evolution of Meiyu in recent 30 years. Therefore, the two stages of 1990–1991 and 2010–2019 were analyzed. The 2000s, which can be regarded as a transitional period, will not be specifically discussed in this work.

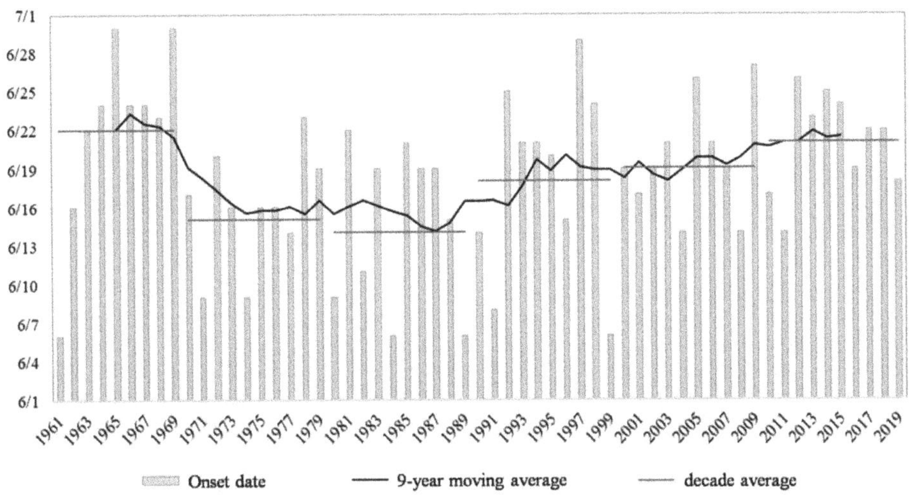

Figure 1. Onset date and interannual and interdecadal sequence of Meiyu in Jiangsu Province from 1961 to 2019.

In order to compare the onset time in the 1990s and 2010s, we standardized the data within each decade respectively, and the standardized sequences of the Meiyu onset time in 1990–1999 (Figure 2a) and 2010–2019 (Figure 2b) were obtained. The positive anomaly standardization means a late onset. In the 1990s, there were great differences in the time of the Meiyu onset in different years. Despite the trend of a delayed Meiyu onset, there were still, though rare, extremely early years in the 59 years. The onsets of Meiyu in the 2010s were more similar than those in the 1990s. Taking 0.6 standard deviations as the indicator, the years whose time of Meiyu onset and length of the Meiyu period exceeding 0.6 positive standard deviations were defined as the early years of the Meiyu onset and the long years of the Meiyu period, respectively. Based on this standard, 1990, 1991 and 1999 were the years with an early arrival of Meiyu in the 1990s, while 1992, 1997 and 1998 were the years with a late arrival of Meiyu in this decade. In the 2010s, 2010, 2011 and 2019 were the years of an early Meiyu arrival, while 2012, 2014 and 2015 witnessed a late arrival. Previous studies have shown that the dates of the Meiyu onset and ending are independent of each other, and there is no significant correlation [2]. Therefore, this work focuses on the difference of abnormal Meiyu onset in different years.

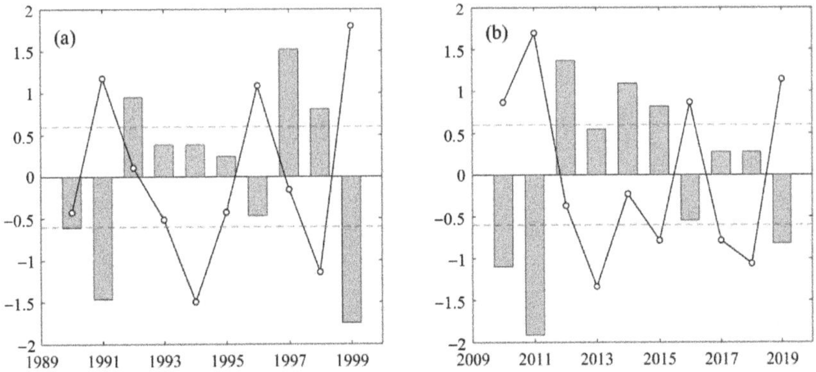

Figure 2. Standardized anomaly of Meiyu onset time from 1990 to 1999 (**a**) and from 2010 to 2019 (**b**).

3.2. Comparison of Meiyu Characteristics and Rainstorms in 1990–1999 and 2010–2019

3.2.1. Meiyu Characteristics

We took the 12 years selected above as the characteristic cases of the 1990s and the 2010s, respectively, and discussed the variation of Meiyu characteristics with time in recent 30 years. We used the averages of the past 30 years (that is, 1990–2019) as the climatological states of Meiyu to compare with the conditions of the characteristic years. It can be seen from Table 1 that Meiyu with an early arrival in the 1990s had a much earlier onset than that in the climatological state, and the end time was close to the climatological situation. In the second decade, the onset time for Meiyu with an early arrival was only slightly earlier than that in the climatological state, while the time of Meiyu's end was much later. Therefore, the duration of the Meiyu period in both the 1990s and 2010s was much longer than that in the climatological state. For the late-arriving Meiyu, the dates of the Meiyu onset and ending, as well as the duration of Meiyu in the 1990s were similar to those in the second decade. Generally speaking, Meiyu started later in the 2010s than in the 1990s, but there was little difference in how long the Meiyu rain period lasted.

Table 1. Temporal characteristics of Meiyu in 1990s and 2010s.

Type	Onset Date	End Date	Days
Meiyu with early arrival in the 1990s	9 June	13 July	35
Meiyu with late arrival in the 1990s	26 June	15 July	20
Meiyu with early arrival in the 2010s	16 June	20 July	35
Meiyu with late arrival in the 2010s	25 June	16 July	22
climatological state of Meiyu	20 June	12 July	23

3.2.2. Rainstorm during Meiyu Period

Rainstorms often occur during the Meiyu period, and the rainfall intensity and range vary under different circulation backgrounds. According to the ground observation specification of the China Meteorological Administration, a rainstorm is defined when the 24-h precipitation reaches more than 50 mm, and a heavy rainstorm is defined when the precipitation reaches more than 100 mm. Based on this standard, this work conducted statistical analysis on the rainstorm process in Jiangsu Province in the 12 characteristic years. In the grid with a horizontal resolution of 0.25°, when rainstorm precipitation occurred at two consecutive points, it was recorded as a rainstorm process, so as to calculate the days of rainstorms or heavy rainstorms. The days with continuous rainstorm in Jiangsu Province were defined as the rainstorm duration days, and the longest record of duration days over the years was regarded as the longest duration of rainstorm. The occurrence of rainstorms in the above years is shown in Table 2.

Table 2. Comparison of rainstorms in the Meiyu period in Jiangsu Province.

Year of Early Meiyu Arrival	Rainstorm Days	Heavy Rainstorm Days	Maximum Rainstorm Duration Days	Year of Late Meiyu Arrival	Rainstorm Days	Heavy Rainstorm Days	Maximum Rainstorm Duration Days
1990	3	0	1	1992	1	0	1
1991	17	8	5	1997	4	2	2
1999	14	3	6	1998	6	0	2
2010	6	1	4	2012	9	1	2
2011	10	2	4	2014	4	0	1
2019	5	0	1	2015	5	1	2

In general, Meiyu with an early arrival had stronger and more precipitation processes, and longer duration days of rainstorm than the late-arriving ones. Meanwhile, it was easier to induce heavy rainstorms in the early-arriving Meiyu. Such a phenomenon was more obvious in the 1990s. During this decade, the rainstorm cases for Meiyu with early and late onsets were significantly different. Early-arriving Meiyu had much more rainstorms and heavy rainstorm days than a late one, and the rainstorm cases lasted longer. In contrast, although there were more rainstorm days in the early-arriving Meiyu periods than in the late ones for the 2010s, the difference was not clear. A relatively significant feature is that the rainstorm duration days in the early cases were generally longer than those in the late ones. The difference of precipitation characteristics in different years was consistent with the difference of the Meiyu onset time. There was a large discrepancy in the Meiyu onset time between the years of the early-arriving and the late-arriving Meiyu in the 1990s, and the number of rainstorm days also varied to a great extent. Particularly, rainstorms occurred frequently in 1991 and 1999. But the 2010s had little difference in rainstorm days between different Meiyu years, while the deviation of the Meiyu onset time was not as much as that in the 1990s, either.

Moreover, looking into the years within the same category, it can be found that the early Meiyu years were more dissimilar between the 1990s and 2010s than the late Meiyu years. For the early-arriving Meiyu years, more rainstorm days, heavy rainstorm days and continuous duration days existed in the 1990s than in the 2010s. On the whole, during the 1990s the rainstorm process in the Meiyu period occurred more frequently, and there was a noteworthy difference between the years of early and late Meiyu arrival.

3.3. Effects of Circulation Variation on Meiyu Onset and Precipitation in 1990–1999 and 2010–2019

3.3.1. South Asia High and Upper-Level Jet

As the strongest and most stable atmospheric circulation system in the upper troposphere of the northern hemisphere in summer, the South Asia high (SAH hereinafter) has a vital impact on summer precipitation in China [21]. Its strength, east-west position and north-south shift are closely linked to the onset and end of Meiyu, and also affects the summer rainfall in the Yangtze-Huaihe region of China. Figure 3 shows the locations of the SAH and the upper-level jet during the Meiyu period in the years with the early and late arrival of Meiyu in the 1990s and 2010s. The height of 16,800 gpm at 100 hPa was taken as the characteristic line of the SAH, and the area with a wind speed over 30 m/s at 200 hPa was considered the upper-level jet region. It can be seen from the figure that in the 2010s the location of the SAH in the years of the early and late Meiyu arrival was relatively similar, and the range was obviously larger than that in the 1990s. The easternmost point of the SAH exceeded 110° E in the 2010s, and the intensity was strong (Figure 3c,d). The location of the SAH in the Meiyu period of the 1990s was slightly westward compared to that in the 2010s, and the range was marginally smaller. The location of the SAH during the years of the late Meiyu arrival in the 1990s was more westward, as its easternmost point was west of 100° E (Figure 3a,b). The SAH extending to the east of 110° E often causes delay of the Meiyu onset [22]. Furthermore, the SAH located further east in June makes it difficult for the rain belt to move northward, and the SAH further east in July leads to the westerly and northerly extension of the WPSH, which is not conducive to precipitation in the Yangtze-Huaihe region [23]. Therefore, the further east location of the SAH is one of the reasons for the later onset and decreased precipitation in the 2010s than in the 1990s.

With the formation of the SAH, the anticyclone circulation strengthens the pressure gradient on the north side, and the subtropical westerly jet in its north enhances [7]. Due to the geostrophic relation, the strength and location of the SAH have a decisive impact on the westerly jet [24]. Taking 120° E as the boundary, the East Asian continental jet and the Western Pacific jet have been proposed according to the location of the upper-level jet axis. The Western Pacific jet plays a significant role in the Meiyu precipitation process in the Yangtze-Huaihe region, as this region is located on the right side of the upper-level jet inlet [25]. The comparison of different decades shows that there was a stable upper-level jet in East Asia in the years with an early Meiyu arrival during the 1990s. The jet had a wide range from east to west, and the Yangtze-Huaihe region was located on the right side of the inlet of the upper-level jet area (Figure 3a). In the years of the late Meiyu arrival during the same decade, the position of the upper-level jet was extremely westward, and the absence of the upper-level jet hindered the occurrence of the rainstorm process (Figure 3b). In the 2010s, the location and intensity of the jet over East Asia were relatively similar between the years of the early and late Meiyu arrival. The upper-level jet was the East Asian continental jet, and the convergence on the right side of the jet stream outlet was not conducive to convection and rainstorms (Figure 3c,d).

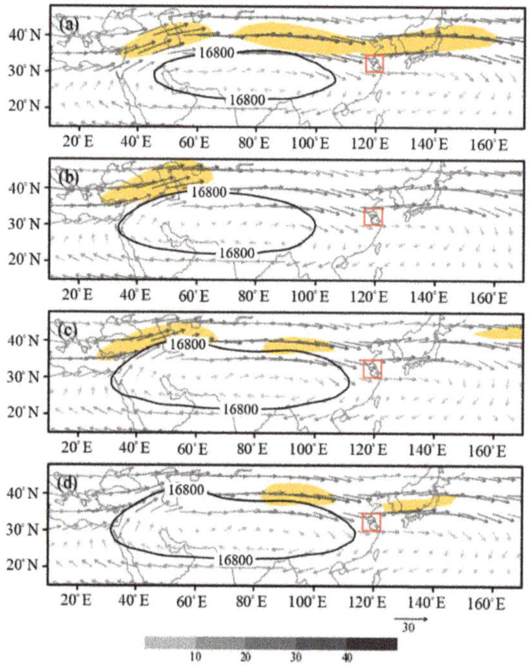

Figure 3. The South Asia high at 100 hPa (the solid line denotes the 16,800-gpm characteristic line), the wind fields at 200 hPa (vector, unit: $m{\cdot}s^{-1}$, the yellow shaded areas denote speed exceeding 30 $m{\cdot}s^{-1}$) and the study region (marked in red) of years of early Meiyu arrival in the 1990s (**a**), late Meiyu arrival in the 1990s (**b**), early Meiyu arrival in the 2010s (**c**), and late Meiyu arrival in the 2010s (**d**).

The north-south position of the upper-level jet before the Meiyu onset and the time of upper-level jet formation are related to the time of the Meiyu onset. As shown in Figure 4, the zonal average of the 200-hPa zonal wind at 110° E to 130° E was used to indicate the intensity of the upper-level jet affecting the Yangtze-Huaihe region, which can reflect the variation of the position and intensity of the upper-level jet for Meiyu of early arrival and late arrival in the two different decades. Combined with the average time of the Meiyu onset, it can be found that the periods with strong jet flow were mostly before the Meiyu onset. There was still a period of strong jet flow after the Meiyu's start in an early Meiyu arrival year in the 1990s, while the jet flow in the range of 110° E to 130° E in the other three types of years was weak in the Meiyu period. Comparing different Meiyu years in the 1990s, the jet axis in the early Meiyu arrival years was basically located at 40° N, which was much more north than that in the late years; the wind speed was higher, and the duration was longer than that in the late years (Figure 4a,b). As pointed out by research, the northerly upper-level jet often corresponds to a stronger East Asian summer monsoon system, and its early northward movement results in the early onset of Meiyu in the Yangtze-Huaihe region [26]. However, an almost opposite phenomenon occurred in the 2010s. The jet intensity in the years of an early Meiyu arrival was weaker, the duration was obviously shorter, and the north-south position deviation was not significant (Figure 4c,d).

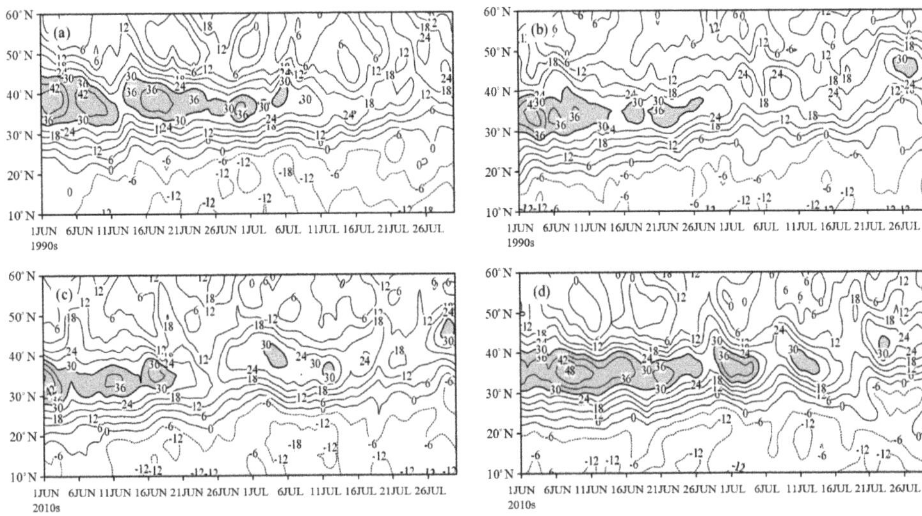

Figure 4. Time-latitude profile along the mean 110~130° E of 200-hPa mean zonal wind from June to July (the solid line denotes wind speed, unit: m·s^{-1}; the shaded areas denotes wind speed exceeding 30 m·s^{-1}) of years of early Meiyu arrival in the 1990s (**a**), late Meiyu arrival in the 1990s (**b**), early Meiyu arrival in the 2010s (**c**), and late Meiyu arrival in the 2010s (**d**).

3.3.2. Western Pacific Subtropical High and Blocking High

The Western Pacific subtropical high (WPSH hereinafter) is a warm anticyclone system, and its location and intensity are closely related to Meiyu. The northward jump of the WPSH ridgeline plays a critical role in the Meiyu's onset and end, and the westward extension position, north-south oscillation and range of the WPSH have an important impact on the rainfall area and intensity of Meiyu's precipitation.

The latitude of the WSPH ridgeline over 120° E at 500 hPa is one of the conditions of the Meiyu onset according to Jiangsu Meteorological Observatory. It stipulates that the ridgeline should be north than 20° N. Therefore, the discussion on the Jiangsu Meiyu onset needs to consider the location of the WPSH and the moving of the ridgeline. Figure 5 compares the range and ridgeline of 500-hPa WPSH before and after the Meiyu onset in the 1990s and 2010s, and takes the average potential height of 110~130° E as an indication to reflect the north-south shift of the WPSH. In the years of the early Meiyu arrival in the 1990s, the range of the WPSH was small, but the ridgeline of the anticyclone system jumped northward earlier and moved to the north of 20° N in the first pentad of June. Although there were short-term fluctuations, the ridgeline basically remained between 20 and 25° N until mid-July. The early northward jump of the WPSH and the early formation of the Meiyu circulation system resulted in the early onset of Meiyu in these years (Figure 5a). In contrast, in the years of the late arrival of Meiyu in the 1990s, the WPSH moved northward much later as it propagated to 20° N in the fifth pentad of June. It moved slowly and stably before the onset, rather than jumping northward rapidly like the case of the early-arriving Meiyu. The WPSH stayed in the south for a long time, which determined the late start of Meiyu in these years (Figure 5b). In the 2010s, the WPSH appeared early in East Asia in the years of the early Meiyu arrival, but its ridgeline was still in a southerly position for some time until a northward shift occurred at the fourth pentad. Therefore, the onset time for the early-arriving Meiyu of the 2010s was later than that of the 1990s (Figure 5c). The WPSH was formed late in the years of the late Meiyu arrival of the 2010s. The ridgeline of the anticyclone system was at the north of 20° N in the first pentad of June, but it remained north for only a few days and then retreated southward. It moved north at the fourth pentad, which was close to the condition of the early Meiyu years of the 2010s. However, although the ridgeline completed the north jump, the clear WPSH characteristic line did not appear until the fifth pentad of June. It means that the high pressure zone did not reach the strength of the WPSH until the end of June (Figure 5d). Zhou have pointed out that for the Meiyu with a late arrival, the ridgeline of the WPSH may push northward quickly in late May, but retreats southward due to instability [27], which is similar to this situation in the years of the late Meiyu arrival in the 2010s.

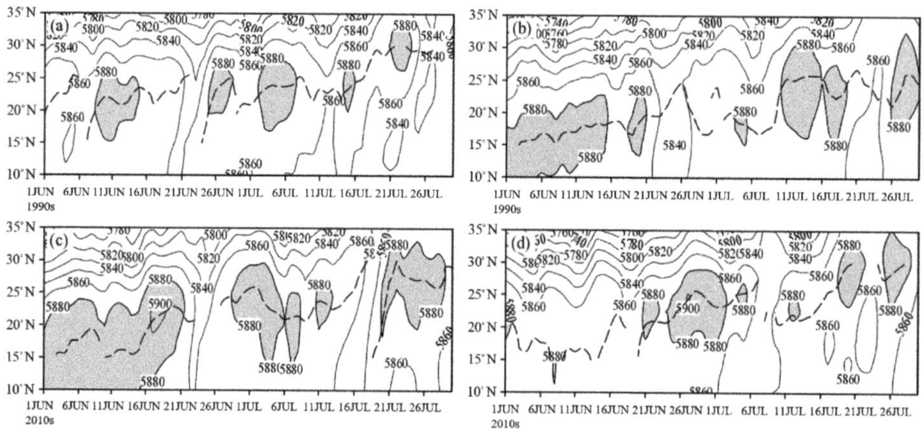

Figure 5. Time-latitude profile of 500-hPa mean geopotential height at 110~130° E (the solid line denotes geopotential height, unit: gpm; the shaded areas denotes geopotential height exceeding 5880 gpm) and western Pacific subtropical high ridgeline (dashed line) from June to July of early-arriving Meiyu in the 1990s (**a**), late-arriving Meiyu in the 1990s (**b**), early-arriving Meiyu in the 2010s (**c**), and late-arriving Meiyu in the 2010s (**d**).

In the middle troposphere, the blocking situation in the middle and high latitudes can also affect Meiyu. In previous studies, the blocking high in the region of 111~150° E is generally regarded as the east-blocking type, that in the region of 81~110° E the Baikal-blocking type, and that in 51~80° E the west-blocking type [28]. The east-blocking high among them, which is stable and less variable, makes more contribution in the years with longer Meiyu periods and has a closer relationship with Meiyu [5]. As can be seen from Figure 6, the blocking situation in the years of the early Meiyu arrival was relatively similar in the 1990s and 2010s, but the blocking situation in the years of the late Meiyu arrival was not typical. For the early-arriving Meiyu in the 1990s, the distribution of blocking high approximated the double-blocking type. Compared with the climatological state, the two high pressures were stronger, and the blocking high was the east-blocking type (Figure 6a). This type of blocking situation can make the cold air invade southward and strengthen the Meiyu front, increase the Meiyu rainstorm, and prolong the Meiyu period by blocking the northward rise of the WPSH [27]. The blocking high for the early-arriving Meiyu in the 2010s was the Baikal-blocking type, which had less effect on cold air transport and Meiyu front maintenance than the east-blocking type (Figure 6c). In contrast, the blocking high of the years with the late Meiyu arrival was weaker. The single-blocking type was shown in the 1990s (Figure 6b), and the blocking situation for the late-arriving Meiyu in the 2010s was not obvious. The westerly belt over middle and high latitudes was relatively flat to the east of the Ural region (Figure 6d). The two kinds of circulation situations can both cause the weak activity of cold air, which is unfavorable to the emergence and maintenance of the Meiyu structure.

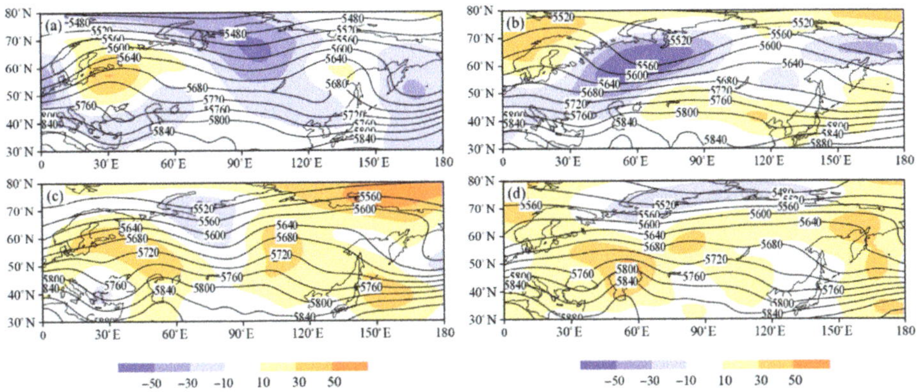

Figure 6. The geopotential height fields at 500 hPa (solid line, units: gpm) and anomaly field compared with climatological state (shaded, units: gpm) of early-arriving Meiyu in the 1990s (**a**), late-arriving Meiyu in the 1990s (**b**), early-arriving Meiyu in the 2010s (**c**), and late-arriving Meiyu in the 2010s (**d**).

3.4. Effects of the Variation of Mesoscale Characteristic Factors on Meiyu Precipitation in 1990–1999 and 2010–2019

3.4.1. Water Vapor Transport

Water vapor transport is one of the key factors in precipitation, and the transport of water vapor by the monsoon circulation system in the lower troposphere has an important impact on the rainfall in the Meiyu period. The average water vapor flux and water vapor flux divergence of 850 hPa were analyzed with composites of the Meiyu periods in the 1990s and 2010s. The results are shown in Figure 7. In the 1990s, the air flow from the South China Sea was more southerly. Apart from the southwest air flow in the northern Indian Ocean, a stream of water vapor was also transported northward to the Yangtze-Huaihe region through the South China Sea, and both were transport routes with strong water vapor flux (Figure 7a). The water vapor flux in the 2010s mainly had a southwest direction. The high value area of the water vapor flux shows that a large amount of water vapor was directly transported from the northern Indian Ocean to the Yangtze-Huaihe region, while the water vapor flux transported from south to north near the South China Sea was not as strong as that in the 1990s (Figure 7b). In addition, comparison of the water vapor flux divergence suggests that the convergence in the Yangtze-Huaihe region in the 2010s was weaker than that in the 1990s, which hindered the rainfall there.

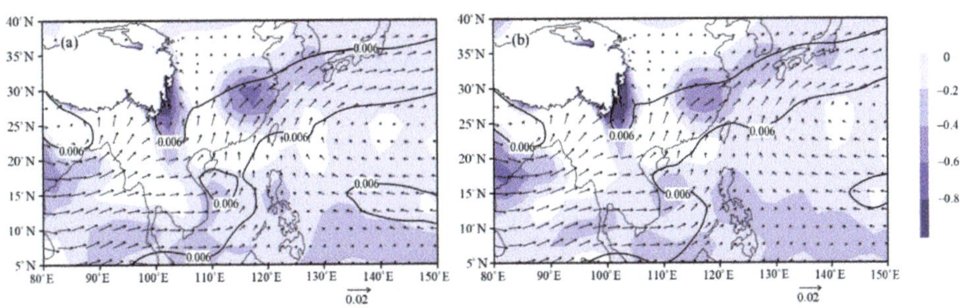

Figure 7. Water vapor flux field at 850 hPa (vector, unit: s^{-1}, the solid line denotes 0.006 s^{-1} line) and its divergence field (shaded, unit: 10^{-8} g/(s·cm^2·hPa)) of Meiyu in the 1990s (**a**) and 2010s (**b**).

The Yangtze-Huaihe region is in the southeast of Asia, and its precipitation during the Meiyu period is affected by air masses from several different sources. The three sources that contribute most to the water vapor transport in the region are the Indian Ocean, the Western Pacific, and the South China Sea [29]. Different water vapor channels and air mass types can have distinct effects on Meiyu precipitation, and this work takes the intensity of the three water vapor transport channels as a standard to compare the water vapor contribution in different decades. Based on the average distribution of the vertical integration of water vapor flux from 1990 to 2019, we recognized three extreme regions of the vector modes as the water vapor transport channels. It was obtained that the southwest channel's range was (85~100° E, 15~22.5° N), the South China Sea channel's range was (107.5~117.5° E, 10~20° N), and the southeast channel's range was (135~150° E, 10~17.5° N) [30]. The intensities of the three channels are shown in Table 3. The average intensity of the three channels in the 1990s was stronger than that in the 2010s, which proves that the water vapor transmission from the three sources to the Yangtze-Huaihe region in the 1990s was stronger. Furthermore, the intensity of the southwest channel and the South China Sea channel increased synchronously in the early-arriving Meiyu years of the 1990s. The years with abnormal precipitation in Meiyu always have significantly higher water vapor transport from the South China Sea than other years [29], which can explain the strong Meiyu precipitation in the early Meiyu arrival years during the 1990s. As revealed by the contrast of water vapor transport between the 1990s and 2010s, stronger water vapor channels and better water vapor income in the Yangtze-Huaihe region were the reasons for more precipitation in the 1990s. The insufficient water vapor acquisition and the transformation from a south water vapor transport channel to a southwest one caused less precipitation in the 2010s.

Table 3. Intensity index of water vapor channel in Meiyu period (unit: $kg{\cdot}m^{-1}{\cdot}s^{-1}$).

Water Vapor Channels	Early-Arriving Meiyu in the 1990s	Late-Arriving Meiyu in the 1990s	Average of the 1990s	Early-Arriving Meiyu in the 2010s	Late-Arriving Meiyu in the 2010s	Average of the 2010s
Southwest channel	371.14	358.43	351.06	326.42	348.21	340.11
Southeast channel	210.35	253.94	240.34	231.74	224.22	230.48
South China sea channel	281.56	273.42	262.28	257.59	284.65	258.64

3.4.2. Vertical Convection

The occurrence of strong rainstorms not only needs an abundant water vapor condition, but also requires the cooperation of a strong upward movement. During the Meiyu period, the low-level jet over the Yangtze-Huaihe region not only provides continuous water vapor transport, but also meets with cold air to form the Meiyu front, thus resulting in vertical convection [31,32]. When comparing the vertical movement in the 1990s and 2010s, the region of 116~122° E was used to characterize the average level of Jiangsu Province and its north and south sides. The longitudinal profile of vertical velocity and radial wind field is shown in Figure 8. In the years of the early Meiyu arrival, there was an upward movement in Jiangsu Province and its south side. In the 1990s, the rising area was wider, and the vertical height of the upper boundary of the rising area was higher (Figure 8a). In the 2010s, the downdraft appeared at the upper troposphere in the south of Jiangsu (Figure 8c). In the 1990s and 2010s, the late-arriving Meiyu experienced a subsidence movement near 20° N in the south of Jiangsu Province. The subsidence in the 1990s was mainly in the middle and low level of the troposphere (Figure 8b), while in the 2010s, there was a subsidence area penetrating the entire height of the troposphere in the south of Jiangsu Province, which was mainly related to the westward extension of the WPSH and the eastward extension of the SAH (Figure 8d). For the precipitation process, the vertical convection is a necessary condition to trigger it. The large-scale upward movement in the years of the early Meiyu arrival in the 1990s was a major factor causing more rainstorms, while the strong downdraft in the south of Jiangsu in the years of the late Meiyu arrival in the 2010s hindered the occurrence of rainstorms.

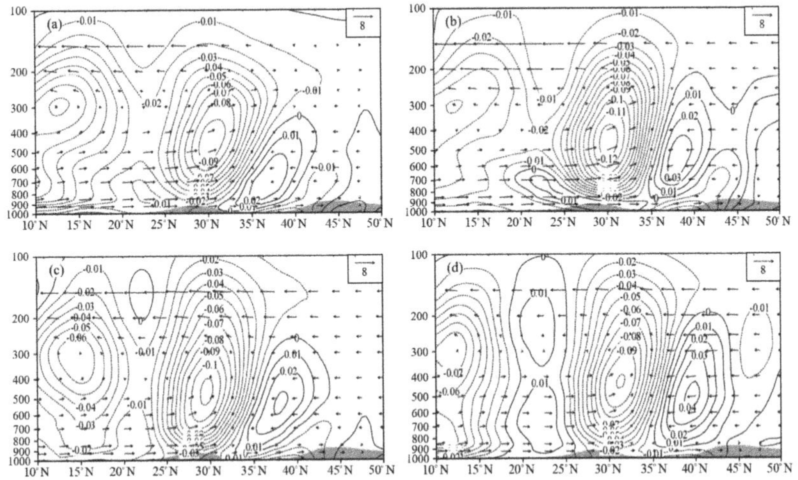

Figure 8. Profile along the mean 116~122° E of vertical velocity field (solid line, unit: m·s^{-1}) and wind field (vector, unit: s^{-1}, the vertical velocity is magnified 10 times); the shaded areas denote the topography of years of early Meiyu arrival in the 1990s (**a**), late Meiyu arrival in the 1990s (**b**), early Meiyu arrival in the 2010s (**c**), and late Meiyu arrival in the 2010s (**d**).

4. Conclusions

This work presents the variation of Meiyu in Jiangsu Province in recent 30 years by comparing the characteristic cases of the 1990s and the 2010s. The main conclusions are as follows:

(1) Since the 1990s, the Meiyu in Jiangsu Province has been delayed, and there was no extremely early Meiyu year in the 2010s. In the 1990s, the precipitation intensity was stronger than that in the 2010s in Jiangsu Province, and the difference between the early and late Meiyu years was obvious. The 2010s had less rainfall, fewer numbers of rainstorm days, and similar precipitation characteristics between the years of the early and late Meiyu arrival.

(2) At the upper troposphere, the SAH and the upper-level jet have a major impact on Meiyu. The position of the SAH in the 2010s was far eastward in comparison with that in the 1990s, which was not conducive to precipitation in the Meiyu period in the Yangtze-Huaihe region. For the early-arriving Meiyu of the 1990s, the effect of the subtropical westerly jet was one of the reasons for multiple rainfall processes, and the strong and northerly upper-level jet in June, which indicated a strong summer monsoon, induced an early Meiyu onset. In the 2010s, the position of the upper-level jet is west and the range is small, and the difference of the upper-level jet was not significant, so the precipitation was weak and similar.

(3) The WPSH and blocking situation in the middle troposphere are important factors affecting Meiyu. The ridgeline of the WPSH in the years of the early Meiyu during the 1990s completed its northward jump at the beginning of June, causing an earlier Meiyu onset. In the late 1990s, the WPSH ridgeline remaining in low latitudes for a long term was the reason for a late Meiyu onset. Although in the 2010s, a large range of WPSH appeared early for the Meiyu of early arrival, the ridgeline did not jump north earlier, so there was no extremely early Meiyu onset. During the 2010s, the WPSH in the late-arriving Meiyu years jumped to the north earlier but retreated to the south soon, and the time it finally completed the north lifting was late, so the Meiyu started late. In addition, the east-blocking high in the early-arriving Meiyu years in the 1990s was more conducive to the southward movement of cold air that strengthened the Meiyu precipitation, while the atypical blocking situation in the years of the late Meiyu arrival of the 1990s and the 2010s hindered the occurrence of Meiyu precipitation.

(4) In the 1990s, there was a stronger water vapor convergence in the Yangtze-Huaihe region, which provided the water vapor condition for more precipitation. The water vapor transport sources are different in different decades, and the change of water vapor channels in the 1990s and the 2010s was also one of the reasons for the variation of precipitation characteristics. On the whole, water vapor transport from the three sources of the Indian Ocean, the South China Sea, and the Western Pacific was stronger in the 1990s than in the 2010s, and less water vapor transport in the 2010s reduced the precipitation in the Yangtze-Huaihe region. In addition, in the 2010s the vertical transportation of Jiangsu and its south side weakened, and there even existed a large subsidence area in the years of the late Meiyu arrival, which had a negative impact on precipitation.

Author Contributions: Supervision, L.W.; Writing—review & editing, R.H. Both authors have read and agreed to the published version of the manuscript.

Funding: This research was jointly supported by the National Natural Science Foundations of China (41975085) and the National Key R&D Program of China (2019YFC1510004).

Institutional Review Board Statement: Not applicable.

Informed Consent Statement: Not applicable.

Data Availability Statement: All data used in this study are available upon request.

Acknowledgments: The authors are grateful to the reviewers and the editor for valuable comments and suggestions.

Conflicts of Interest: The authors declare no conflict of interest.

References

1. Ye, D.; Huang, R. Progress, achievements and problems in the research on the causes and prediction of drought and flood in the Yangtze and yellow river basins in China. *Prog. Geosci.* **1991**, *4*, 24–29.
2. Huang, Q.; Wang, L.; He, J.; Guan, Z. Review of research on Meiyu in Jianghuai. *Zhejiang Meteorol.* **2010**, *31*, 2–7.
3. Tao, S.; Zhu, W.; Zhao, W. On the interannual variation of Mei Yu. *Atmos. Sci.* **1988**, *12*, 13–21.
4. Yang, Z.; Wang, W.; He, L. Anomalous characteristics of tropical atmospheric circulation in rainy and rainy years. *J. Trop. Meteorol.* **1997**, *1*, 47–56.
5. Tao, S.; Xu, S. Circulation characteristics of persistent drought and flood in Jianghuai River Basin in summer. *J. Meteorol.* **1962**, *1*, 1–10.
6. Liu, Y.; Hong, J.; Liu, C.; Zhang, P. Meiyu flood in Huaihe River and seasonal advance anomaly of subtropical high in the Western Pacific. *Atmos. Sci.* **2013**, *37*, 439–450.
7. Zhu, Q.; Lin, J.; Shou, S.; Tang, D. *Synoptic Principles*, 4th ed.; Meteorological Publishing House: Beijing, China, 2007; pp. 354–358.
8. Nie, Y.; Zhou, J.; Gu, X.; Zhou, Y.; Du, X. Diagnosis and analysis of a case of heavy rainstorm in Southeast Guizhou in the western section of "6.18" Meiyu front. *Rainstorm Disaster* **2018**, *37*, 445–454.
9. Xu, H.; Liu, J.; Deng, J. Projections of East Asian summer monsoon change at global warming of 1.5 and 2 °C. In Proceedings of the 35th Annual Meeting of China Meteorological Society S4 2018 National Symposium on Tropical and Marine Meteorology—Symposium on Generation Mechanism and Prediction Technology of Tropical Extreme Weather and Climate Events, Hefei, China, 24 October 2018; China Meteorological Society: Beijing, China, 2018; p. 95.
10. Wu, C.; Liu, Y. Response of Meiyu in Jianghuai river basin to global warming. In Proceedings of the S6 East Asian Climate Multi Time Scale Variation Mechanism and Climate Prediction at the 34th Annual Meeting of China Meteorological Society, Zhengzhou, China, 26–29 September 2017; China Meteorological Society: Beijing, China, 2017; p. 373.
11. Jiang, W.; Gao, H. New characteristics and cause analysis of Meiyu in the middle and lower reaches of the Yangtze River in the 21st century. *Meteorology* **2013**, *39*, 1139–1144.
12. Jiang, W.; Lv, J.; Xiang, Y. Abnormal characteristics and prediction of Meiyu in Jiangsu Province in recent 10 years. In Proceedings of the Subtropical Monsoon and Climate Change Branch of the 27th Annual Meeting of China Meteorological Society, Beijing, China, 21 October 2010.
13. Gu, W.; Chen, L. Characteristics of ocean atmosphere and its impact on China's climate in summer 2018. *Meteorology* **2019**, *45*, 126–134.
14. Zhang, F.; He, L. Atmospheric circulation and weather analysis in June 2018. *Meteorology* **2018**, *44*, 1237–1244.
15. Li, J.; He, L. Atmospheric circulation and weather analysis in June 2017. *Meteorology* **2017**, *43*, 1160–1164.
16. Zheng, Z.; Wang, Y. Characteristics of atmospheric circulation in the northern hemisphere in summer 2017 and its impact on China's weather and climate. *Meteorology* **2018**, *44*, 199–205.
17. Cao, Y.; Zhang, T. Atmospheric circulation and weather analysis in June 2016. *Meteorology* **2016**, *42*, 1154–1160.

18. Si, D.; Liu, Y.; Shao, X.; Wang, Y. Ocean and atmospheric circulation anomalies in 2015 and their impact on China's climate. *Meteorology* **2016**, *42*, 481–488.
19. Guan, Y.; He, L. Atmospheric circulation and weather analysis in June 2014. *Meteorology* **2014**, *40*, 1159–1164.
20. Wen, M.; Shi, X. Relationship between the West Pacific subtropical high activity and condensation latent heat heating in summer 1998. *Plateau Meteorol.* **2006**, *25*, 616–623.
21. Mason, R.B.; Anderson, C.E. The development and decay of the100mb summertime anticyclone over Southern Asia. *Mon. Weather Rev.* **1958**, *91*, 312.
22. Liu, M. Impact of South Asia high on Meiyu characteristics in Jiangsu during flood season. In Proceedings of the Climate Credit Conference of 2007 Annual Meeting of China Meteorological Society, Guangzhou, China, 23–25 November 2007; China Meteorological Society: Beijing, China, 2007; pp. 1340–1348.
23. Lu, M.; Zhu, F.; Huang, W. 100 hPa South Asia High eigenvector and its relationship with summer precipitation in China. *Trop. Meteorol.* **1987**, *1*, 42–49.
24. Liu, J.; Kuang, X.; Zhang, Y. Relationship between subtropical westerly jet and subtropical (South Asian) high in the upper troposphere. *Meteorol. Sci.* **2010**, *30*, 34–41.
25. Du, Y.; Zhang, Y.; Xie, Z. Influence of East-West morphological change of upper westerly jet on spatial distribution of precipitation in Meiyu period. *J. Meteorol.* **2008**, *66*, 566–576.
26. Li, C.; Han, G.; Liu, M.; Sun, H.; Zhou, L. Relationship between East Asian upper jet anomaly and Jianghuai Meiyu. *Meteorol. Sci.* **2015**, *35*, 176–182.
27. Zhou, Z. *Jianghuai Meiyu*; Meteorological Publishing House: Beijing, China, 1996; pp. 134–135.
28. Zhang, C.; Song, L.; Li, Y. Research progress of summer blocking process in East Asia. *J. Meteorol.* **2004**, *1*, 119–127.
29. Jiang, Z.; Ren, W.; Liu, Z.; Yang, H. Analysis of water vapor transport characteristics of Jianghuai Meiyu based on Lagrange method. *J. Meteorol.* **2013**, *71*, 295–304.
30. Wang, Z.; Gao, Q.; Hu, B.; Sun, Y. Study on water vapor transport characteristics during Meiyu period in Jianghuai region in recent 50 years. *J. Atmos. Sci.* **2017**, *40*, 48–60.
31. Guo, Y.; Wang, J.; Li, C.; Long, L. Effect of frontogenesis on Rainstorm in Hubei during the plum flood season in 2011. *Meteorology* **2014**, *40*, 86–93.
32. Yang, S.; Lu, Y.; Yu, C. Mesoscale convective system and low-level wind field impact analysis of a Meiyu Front Rainstorm. *Meteorology* **2017**, *43*, 21–33.

atmosphere

Article

A 7-Year Climatology of Warm-Sector Heavy Rainfall over South China during the Pre-Summer Months

Tao Chen [1] **and Da-Lin Zhang** [2,*]

[1] National Meteorological Center, China Meteorological Administration, Beijing 100081, China; chentao@cma.gov.cn
[2] Department of Atmospheric and Oceanic Science, University of Maryland, College Park, MD 20742-2425, USA
* Correspondence: dalin@umd.edu

Abstract: In view of the limited predictability of heavy rainfall (HR) events and the limited understanding of the physical mechanisms governing the initiation and organization of the associated mesoscale convective systems (MCSs), a composite analysis of 58 HR events over the warm sector (i.e., far ahead of the surface cold front), referred to as WSHR events, over South China during the months of April to June 2008~2014 is performed in terms of precipitation, large-scale circulations, pre-storm environmental conditions, and MCS types. Results show that the large-scale circulations of the WSHR events can be categorized into pre-frontal, southwesterly warm and moist ascending airflow, and low-level vortex types, with higher frequency occurrences of the former two types. Their pre-storm environments are characterized by a deep moist layer with >50 mm column-integrated precipitable water, high convective available potential energy with the equivalent potential temperature of ≥340 K at 850 hPa, weak vertical wind shear below 400 hPa, and a low-level jet near 925 hPa with weak warm advection, based on atmospheric parameter composite. Three classes of the corresponding MCSs, exhibiting peak convective activity in the afternoon and the early morning hours, can be identified as linear-shaped, a leading convective line adjoined with trailing stratiform rainfall, and comma-shaped, respectively. It is found that many linear-shaped MCSs in coastal regions are triggered by local topography, enhanced by sea breezes, whereas the latter two classes of MCSs experience isentropic lifting in the southwesterly warm and moist flows. They all develop in large-scale environments with favorable quasi-geostrophic forcing, albeit weak. Conceptual models are finally developed to facilitate our understanding and prediction of the WSHR events over South China.

Keywords: warm-sector heavy rainfall; mesoscale convective systems; statistical analysis; South China

Citation: Chen, T.; Zhang, D.-L. A 7-Year Climatology of Warm-Sector Heavy Rainfall over South China during the Pre-Summer Months. *Atmosphere* **2021**, *12*, 914. https://doi.org/10.3390/atmos12070914

Academic Editor: Roberto Fraile

Received: 13 June 2021
Accepted: 8 July 2021
Published: 15 July 2021

Publisher's Note: MDPI stays neutral with regard to jurisdictional claims in published maps and institutional affiliations.

1. Introduction

Previous studies have shown that more than 40% of annual rainfall in South China occurs during the pre-summer months (April to June) [1–4]. In contrast to relatively weak and broad rainfall occurring behind slowly moving surface cold fronts, many of the growing-season rainfall events over the region are heavy and highly localized, where heavy rainfall is defined herein as the daily rainfall amount of greater than 50 mm, following that defined by the China Meteorological Administration. They are typically generated in mesoscale convective systems (MCSs) that develop in the warm sector, i.e., far ahead of surface fronts, where southwesterly warm and moist flows prevail with weak thermal gradients. Hence, they have been often regarded as warm-sector heavy rainfall (WSHR) events, although some of them may be initiated near surface fronts and then propagate rapidly ahead [5–8].

Because of their high impacts on societal and economic activities, there has been considerable interest in studying WSHR events over South China during the past few decades. For example, high-resolution mesoscale observing networks were established in late 1990s in South China to monitor the development of heavy rainfall (HR) and other severe convective weather events. In addition, several field experiments were carried out, such as the South China Sea Monsoon Experiment (SCSMEX, 1996–2000) [9,10], Huanan Area Mesoscale Experiment (HUAMEX, 1998) [11,12] and the Southern China Monsoon Rainfall Experiment (SCMREX, 2014–2017) [13], to examine the development of HR-producing MCSs. A few climatological conceptual models were developed for some WSHR events, in which a quasi-stationary frontal system, formed in the southerly monsoonal air from the South China Sea (SCS)-Indochina Peninsula interacting with northerly cold air, provides a favorable condition for the generation of WSHR events over South China during the pre-summer months. After analyzing two high-impact WSHR events, Sun and Zhao [14–16], revealed the importance of surface heating, topography and sea breeze circulations in the associated HR-producing MCSs, which differ from those associated with dynamical forcing such as frontal lifting and upper-level troughs. Both observational and numerical modeling studies have shown the effects of topographical lifting on the development of HR-producing MCSs [17–20]. The importance of low-level jets (LLJs), and southwesterly moisture supply has also been considered in the growth and organization of HR-producing MCSs [21–24]. However, HR events are often associated with long-lived MCSs in weak southwesterly wind environments [25–27]. These rainfall events also exhibit significant diurnal variations [28,29], with high-frequency HR occurring during midnight to the early morning hours that are closely related to LLJs, pre-existing cool pool, orographic lifting and the planetary boundary layer (PBL) processes including urban heat island effects [30–32].

Although considerable progress has been made in understanding the development of WSHR events in South China, many scientific issues are still elusive. In particular, current operational (global and regional) numerical weather prediction (NWP) models still remain either hit or miss in predicting the timing and location of convective initiation and the development of WSHR-producing MCSs, thus showing little skill in the associated quantitative precipitation forecasts [13,33–35]. Unlike rainstorms under the influences of synoptic dynamic forcing [36], quasi-stationary fronts [37], shearlines or low-level vortices and southwesterly LLJ [38–41] are often the necessary but not sufficient conditions for determining the exact location, timing and amount of HR [42]. Therefore, it could be useful to apply the ingredient-based analysis of environmental conditions for WSHR-producing MCSs. Moreover, from the perspective of operational forecasts, it is desirable to examine climatologically the geographical and temporal distributions of HR in South China, develop conceptual models for the formation of different WSHR-producing MCSs under typical environmental conditions, and refine empirical WSHR forecast techniques.

On the other hand, HR amounts are closely related to the structures, organization and propagation of MCSs. In this regard, several three-dimensional morphological models of MCSs were developed, based on Doppler radar observations [43,44]. Doswell et al. [45] showed the relationship between localized rainfall, convective organization and movement by demonstrating that flash floods could be produced by passing through a series of convective elements aligned linearly, like an advancing train, the so-called echo-training process. Schumacher and Johnson [46] classify Two archetypes of heavy-rain-producing linear MCSs for multiple flash flood events in the Midwest of the United States: A line of training convective elements with an adjoint trailing stratiform region (TL/AS), and an area of back-building convection with a trailing stratiform region. Numerous HR events in South China and elsewhere in China were also found to be directly linked to multiple mesoscale echo- and rainband-training processes [47]. Recently, Liu et al. [48] classified three types of persistent HR events in South China, based on their correlations with geographical locations of the HR event occurrences.

Evidently, few statistical studies have been systematically performed to classify WSHR-producing MCSs during the pre-summer months in accordance with synoptic environments and topographic forcing in South China. Given our limited understanding of the large-scale conditions associated with WSHR and the morphologies of HR-producing MCSs during the pre-summer months over South China, the objectives of this study are to (i) document the spatiotemporal characteristics of WSHR and classify a few major types of WSHR events during the years of 2008–2014; (ii) identify the large-scale environmental conditions associated with the different types of WSHR events; and (iii) analyze the structures, organization and evolution of typical MCSs that produce the WSHR events.

The next section describes the data and methodologies used for this study. Section 3 presents the spatiotemporal characteristics of WSHR and describes the classification of major WSHR events, based on their large-scale flow patterns. Their associated environmental parameters, i.e., total precipitable water (PWAT), convective available potential energy (CAPE), and the level of free convection (LFC) are also shown. Section 4 presents some typical MCS morphologies using radar observations. A summary and concluding remarks are given in the final section.

2. Data Source and Methodology

In this study, observations from the conventional (2418) surface weather stations and upper-air network during the pre-summer months of 2008–2014, archived by the National Meteorological Center of the China Meteorological Administration (CMA/NMC), are used to describe the proximity of environmental conditions in which WSHR events occurred. The 6-hourly National Centers for Environmental Prediction (NCEP) Global Forecast System (GFS) $0.5^\circ \times 0.5^\circ$ gridded model dataset is used to reveal the large-scale flows and the environmental conditions under which WSHR-producing MCSs develop.

In general, South China is geographically referred to as the vast territory on the south of Mt. Nanling, including most portion of Guangxi (GX) Province, all of Guangdong (GD) and Hainan (HN) provinces, and part of Fujian (FJ) Province (see Figure 1). In this study, we focus on the area ranging within 20–27° N and 105–120° E, which includes almost all of South China, except for Hainan Province. The mean annual precipitation in South China, obtained from 16 climatological stations, is 1614 mm, according to Year 1981–2010 precipitation statistics, over 45% of which occurred during the pre-summer months [2]. For the present study, out of 16 national surface stations, nine received top annual rainfall amounts over South China, as highlighted with red dots in Figure 1. The top annual rainfall amounts ranged from 1453 mm at Wuzhou, GX (No. 59265) to 2221 mm at Yangjiang, GD (No. 59663).

A composite analysis and subjective classification of 58 WSHR events will be performed to develop conceptual models of different MCSs for various types of WSHR events, based on the given large-scale background flows. This approach has been used by Maddox et al. [49] to classify four categories of large-scale flow patterns that are favorable for the generation of flooding rainfall in the central and eastern United States, and by Tao [1], and Huang [2] to characterize various large-scale circulation features, such as surface fronts, LLJs, and low-level vortices, that are involved in the generation of WSHR events in South China. In this study, a surface front is defined as a warm/moist—cold/dry air boundary, accompanied with a height trough aloft. An LLJ is simply defined as the peak horizontal wind speed of at least 10 m s^{-1} below 850 hPa, though occurring mostly near 925 hPa, while a low-level vortex (LV) is just a closed circulation in the lowest 150 hPa. Given the weak-gradient environments in which the WSHR events took place, an objective classification scheme, e.g., through regional-scale vertical wind shear, CAPE, PWAT, and relative vorticity, could be developed to identify some unique dynamical and thermodynamic characteristics of the WSHR events.

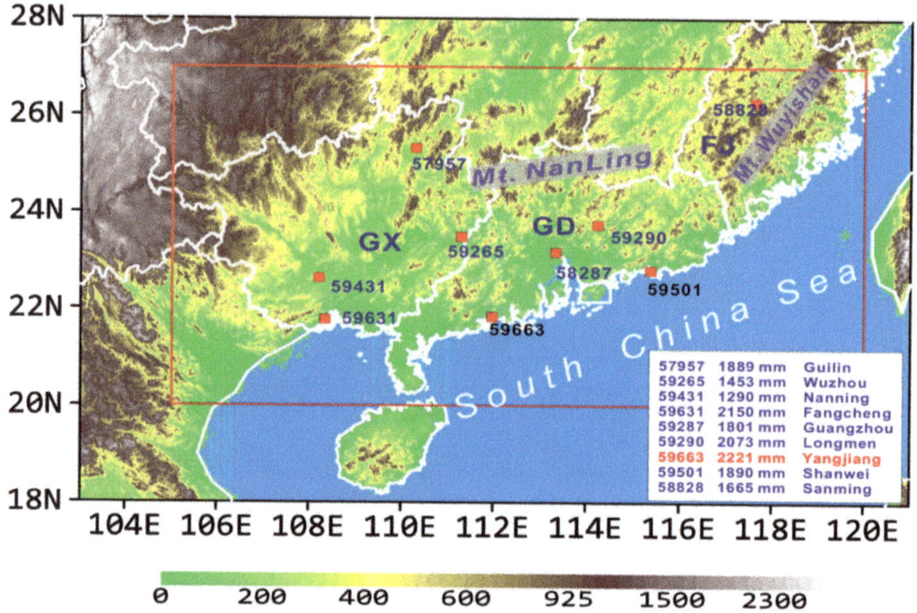

Figure 1. Distribution of topography (shaded, m) and some related national surface stations (red dots with identifications) over South China. The inner red frame denotes the region of interest (20–27° N, 105–120° E) for the present study. The top nine annual precipitation stations (with their identifiers) averaged during 1981–2010 are given with their corresponding amounts listed in the bottom right box. Letters, "GX", "GD", and "FJ" denote the province of Guangxi, Guangdong and Fujian, respectively, similarly for the rest of figures.

3. Spatiotemporal Characteristics of WSHR and Large-Scale Mean Flows

In this section, we examine first the temporal distribution of WSHR events, and classify them into different categories, based on their common background characteristics. Then, the corresponding composite fields will be analyzed to gain insight into different characteristics in large-scale flows and energy supply, HR generation mechanisms, and their spatial distributions in South China.

3.1. Characteristic of WSHR

In this study, an HR event is defined when 24-h (i.e., between 0800–0800 BST: Beijing Standard Time = UTC + 8 h) accumulated precipitation at 3 and more adjacent (less than 200 km apart) or at five scattered (at least 300 km apart) national stations over South China equals to or exceeds 50 mm. Clearly, this definition will filter out HR events generated by local or random deep convection. Table 1 shows, on average, 37.7 HR days per year during the pre-summer months of 2008–2014, 65% of which take place after the onset of summer monsoon in East Asia. There are a total of 58 WSHR days or events (see Appendix A, Appendix B, Appendix C for their individual occurrence dates), which give 8.3 WSHR events per year or 22.0% of the total HR events during the pre-summer months of the 7-year period, and a total of 145 typical WSHR-producing MCSs move across South China. Higher-frequency WSHR events occur during the pre-summer months of 2008 and 2009, as compared to the lowest-frequency WSHR events in 2011.

Horizontal distribution of the mean daily rainfall amount from all 58 WSHR events over South China is displayed in Figure 2, showing the top five rainfall maxima, denoted as "C_1"–"C_5", four of which are more than 100 mm day^{-1}. They are located near surface stations in Guilin of GX; Yangjiang of GD, Shanwei of GD, and the Pearl River Estuary of GD, respectively. A single-day rainfall amount reaches 445.7 mm, named as the top one

(C_1) WSHR event, which occurs in Yunxiao of FJ, on 14 June 2008 as a result of a low-level vortex interacting with strong monsoonal flows. The top 1–10 WSHR maxima correspond more or less to climatological precipitation maxima.

Table 1. The number of HR versus WSHR days, and the ratio of HR days to WSHR days over South China during the pre-summer months of 2008–2014.

Year	2008	2009	2010	2011	2012	2013	2014	Mean
HR days	40	25	40	32	46	40	41	37.7
WSHR days	14	9	8	5	6	10	6	8.3
Ratio	35%	36%	20%	15.6%	13.0%	25%	14.6%	22.0%

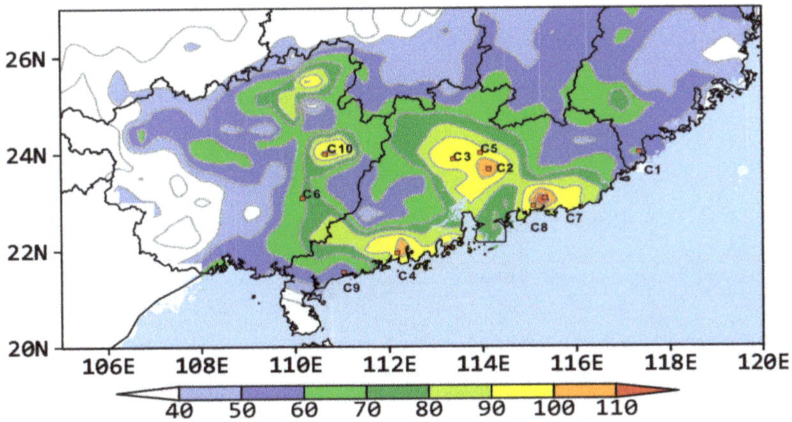

Figure 2. Horizontal distribution of the mean daily precipitation amount composite (shaded, mm) from the 58 WSHR events over South China during the pre-summer months of 2008–2014. Symbols, "C_1"–"C_{10}", denote the top 10 daily rainfall maxima, which are also listed in Table 2.

Table 2. The locations (and station numbers), dates (day/month/year), amounts (mm), and classified weather types of the top ten daily-averaged WSHR events during the pre-summer months of 2008–2014 (also see Figure 2).

Index	Station ID/Name	Date	Amount(mm)	Type
C_1	59322/Yunxiao	13 June 2008	445.7	LLV
C_2	59290/Longmen	14 June 2008	327.0	PSF
C_3	59087/Fogang	15 May 2013	292.4	WMF
C_4	59663/Yangjiang	28 June 2011	256.5	WMF
C_5	59094/Wongyuan	6 May 2010	240.1	PSF
C_6	59254/Guiping	12 May 2012	237.2	LLV
C_7	59500/Haifeng	21 June 2012	226.4	WMF
C_8	59501/Shanwei	7 June 2008	216.9	WMF
C_9	59664/Dianbai	5 May 2008	209.8	WMF
C_{10}	59069/Zhaoping	22 June 2012	206.9	LLV

3.2. Classification of WSHR Events

Based on the previous studies of HR-producing MCSs, it is convenient to use the surface front and low-level vortex as the two key weather system identifiers to classify the large-scale environments in which all the above WSHR events occur. The following three major types of flow configuration could be identified: (i) a pre-surface-frontal (PSF) type, (ii) a warm-moist airflow (WMF) type, and (iii) a low-level vortex (LLV) type. The remaining HR events appear to be more associated with typical frontal rainbelts, tropical cyclones or randomly generated local thunderstorms. Monthly distribution of each type of WSHR events is given in Figure 3, showing the least occurrences of (10) WSHR events in April and the frequent occurrences in May and June (24 in each month). The frequencies of the individual types of WSHR events appear to follow closely the seasonal transition from more surface frontal passages in April to more influences of southwesterly monsoonal flows in May and June. In particular, few LLV-type WSHR events could occur in April and May due to the presence of more baroclinicity on the lee side of the Tibet Plateau, but they become the dominant HR producer in June as more moist monsoonal air is processed. After describing briefly their general characteristics below, more detailed rainfall characteristics and mean environmental conditions associated with the three types of WSHR events are presented in subsections 3b-d that follow, respectively.

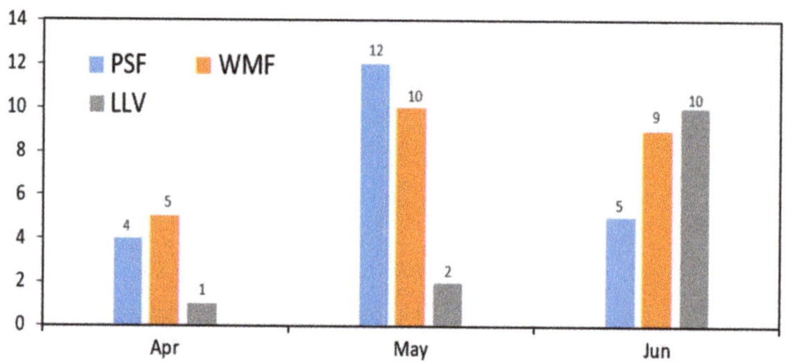

Figure 3. Monthly number distribution of the three major types (i.e., PSF, WMF, and LLV) of the 58 WSHR events over South China during the pre-summer months of 2008–2014.

(i) The PSF type consists of 21 HR events (see Appendix A for their occurrence dates), accounting for 36.2% of the total WSHR events. It occurs in the moist southwesterly flows far ahead of a surface front (i.e., in a warm sector), and the associated WSHR belt is well separated from those rainbelts with much weak rainfall intensity along or behind the frontal zone.

(ii) The WMF type consists of 24 HR events (see Appendix B for their occurrence dates), accounting for 41.4% of the total WSHR events, which is of the highest frequency among the three types. Its large-scale flow pattern is dominated by a warm-moist southwesterly airstream with little evidence of a surface front, but experiencing northeastward isentropic lifting [50–54]. It is often accompanied by a southwesterly LLJ extending from the SCS, with some typical monsoonal rainfall characteristics.

(iii) The LLV type consists of 13 HR events (see Appendix C for their occurrence dates), accounting for 22.4% of the total WSHR events. An LLV is often generated on the lee side of the Yunnan-Guizhou Plateau and then moves eastward across GX, during which course HR takes place in its southern and eastern quadrants. This type of WSHR events tends to be a HR producer. One example is the record-breaking HR event of 13–15 June 2008, listed as the top one HR producer in Table 2, that is associated with a slow-moving LLV across GX with high-θ_e air masses fed by southwesterly monsoonal flows.

In addition to the above three-types of large-scale flow patterns, some WSHR events may be produced by the other types of MCSs, such as LLVs moving from the SCS that are similar to landfalling tropical depressions. Two such examples are the WSHR events of 5–6 June 2008 and 23–24 May 2009 leading to torrential rainfall over the coastal regions of South China. Nevertheless, few such WSHR events occur during the pre-summer months of 2008–2014, and so they will not be investigated herein.

3.3. *Large-Scale Mean Flows of the WSHR Events*

3.3.1. The PSF-Type Events

The relative contribution (%) of the PSF-type rainfall to the total WSHR amount during the pre-summer months of the study period is given in Figure 4, showing a narrow small-percentage (i.e., <12.5%) zone of about 80-km width along the southern coastal region, followed by an elongated high-percentage (i.e., >15%) belt of about 120-km width inland, and a secondary rainbelt (i.e., <20%) in the southern Hunan Province. Several distinct high percentage (i.e., >20%) maxima are located near Nanning (No. 59431) and Wuzhou (No. 59265) of GX, Longmen (No. 59290) of GD, and Sanming (No. 58828) of FJ, which coincide well with several top HR centers shown in Figure 2.

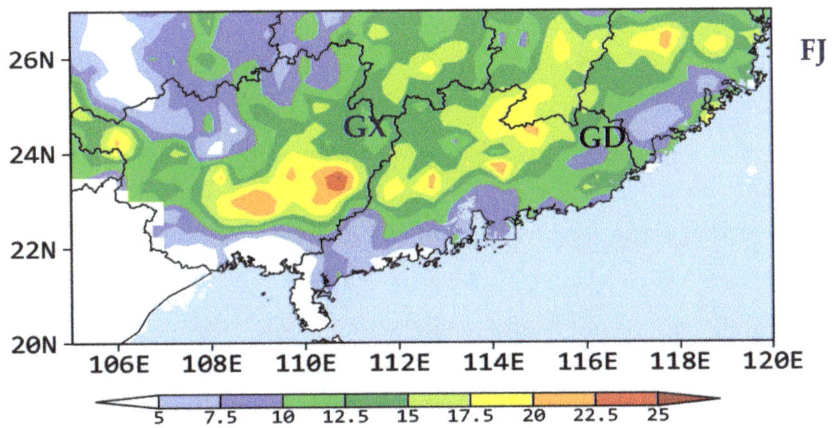

Figure 4. The percentage (%, shadings) of the 21 PSF-type WSHR events with respect to the total WSHR amount during the pre-summer months of 2008–2014.

The composite 500-hPa height field for the PSF events is given in Figure 5a, showing a weak trough located to the north of 25 °N, offshore of East China, with northwesterly flows of cold and dry air. In contrast, South China, i.e., on the south of 25° N, is featured by a nearly zonal flow pattern. This is a typical circulation pattern during the pre-summer months in South China, in which a quasi-stationary front sustains between a dry-cold air mass from the midlatitudes and a warm-moist air mass from tropical oceans.

The composite 850-hPa wind field, also given in Figure 5a, shows a well-defined cold front, as characterized by an arc-shaped shearline, just approaching to Mt. Nanling, as the cold-dry airmass from aloft descends anticyclonically southwestward. The cold front becomes shallower and weaker to the southwest, due partly to the blocking effects of Mt. Nanling, and it moves slower (and even nearly quasi-stationary), after passing the mountains. Thus, its warm sector is prevailed by the southwesterly winds of 8~10 m s^{-1} with a high equivalent potential temperature (θ_e) tongue of greater than 344 K extending from Indochina Peninsula and the SCS. In addition, air masses of strong and weak-θ_e gradients are distributed over the northern and southern portion of South China, respectively, implying the presence of little thermal advection in the warm sector. Since WSHR occurs in the high-θ_e tongue region with high moisture content, its amount is much greater than that along the surface front. In particular, the continuous supply of high-θ_e air by the prevailing southwesterly flow provides not only the necessary moisture content for the WSHR production, but also helps maintain conditional instability, that is removed by deep convection, for the persistent convective overturning in the warm sector. In this regard, the low-level high-θ_e airstream could be considered as one of the most important thermodynamic forcing for producing WSHR events over South China.

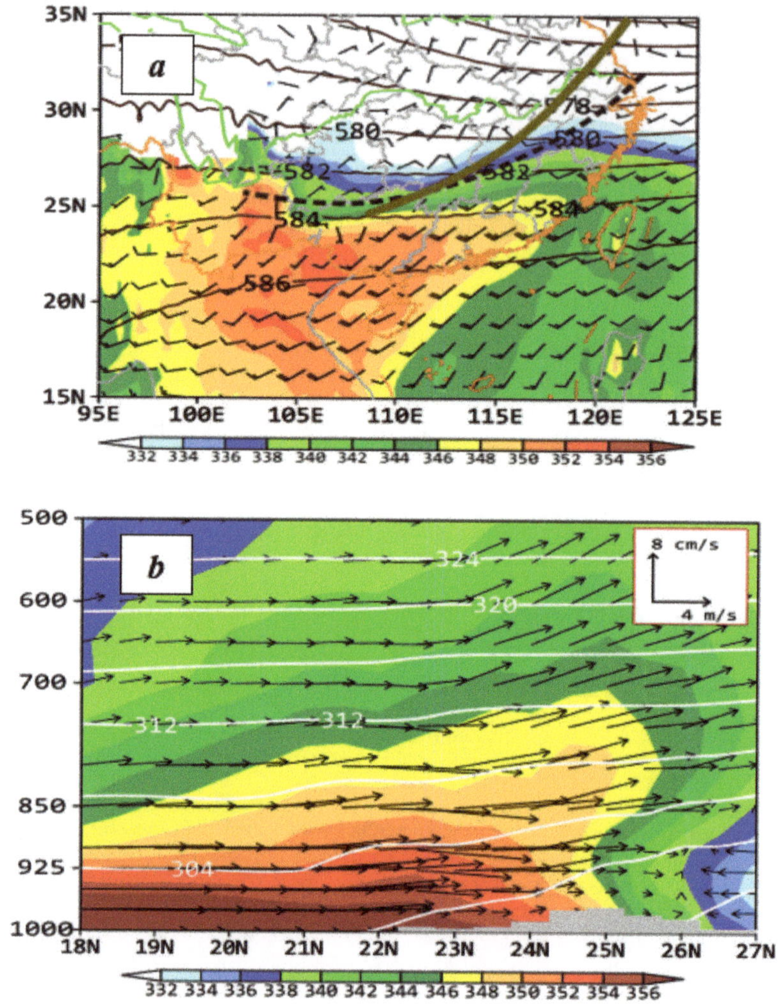

Figure 5. Composite analysis of the 21 PSF-type WSHR events from the NCEP's 0800 BST GFS data: (**a**) Geostrophic height (contoured at 20-m intervals) at 500 hPa, horizontal wind barbs (a full barb is 4 m s^{-1}) and equivalent potential temperature (shaded, θ_e, K) at 850 hPa; and (**b**) South-north vertical cross section of potential temperature (contoured at 4-K intervals) and θ_e (shaded), superimposed with in-plane flow vectors, along 112° E (near Yangjiang with the top HR amount) during the pre-summer months of 2008–2014. Dashed lines and the green thick line in (**a**) denote a shearline and roughly a trough axis, respectively. Grey shadings in (**b**) denote topography.

A south-north vertical cross section of secondary circulation along 112° E (near Yangjiang) is given in Figure 5b, showing that more pronounced low-level isentropic lifting of about 10 cm s^{-1} occurs to the north of 24° N, which is more associated with the shallow cold frontal system. However, the PSF-type HR occurs in the 21–24° N range (cf. Figures 4 and 5b), whose maxima are 100–150 km ahead of the cold front. A detailed analysis of a few PSF-type HR events reveals that their initial convective activity could be traced back to the leading edge of the surface front, and then its organization into MCSs leads to the occurrences of the PSF-type of WSHR events as they propagate rapidly eastward (not shown). Nevertheless, the vertical lapse rate of θ_e from the surface to 700 hPa on the south of 23° N indicates the presence of more significant convective instability than that on its north. Furthermore, given the presence of very humid air during this season in South China, the southwesterly airstream of high-θ_e (>354 K) airmass in the PBL would allow deep convection to be triggered along convectively generated outflow boundaries. Convective triggering along the cold outflow boundaries accounts partly for the eastward propagation of MCSs and explains the generation of an elongated HR belt of about 120-km width with a distance of about 80 km inland from the southern coastline (Figure 4).

3.3.2. The WMF-Type Events

The WMF-type of WSHR events is of the highest frequency among the three classified types, as shown in Figure 3, and it is also more localized in the coastal region of South China (Figure 6). Hence, this type of WSHR events is distinct from the PSF-type events. Figure 6 exhibits three pronounced WMF-type of WSHR maxima: the most intense one at Yangjiang of GD (No. 59663), and then Shanwei of GD (No. 59501) and Fangcheng of GX (No. 59631) in this order, which are all located near the coastline with some hills, albeit less than 500 m altitude; they are also three of the top nine HR stations shown in Figure 1. This appears to indicate the possible roles of topographical forcing and sea breezes in generating the WSHR maxima. Of significance is that the WMF-type events account for more than 20% of the total pre-summer months rainfall amount in the Yangjiang area.

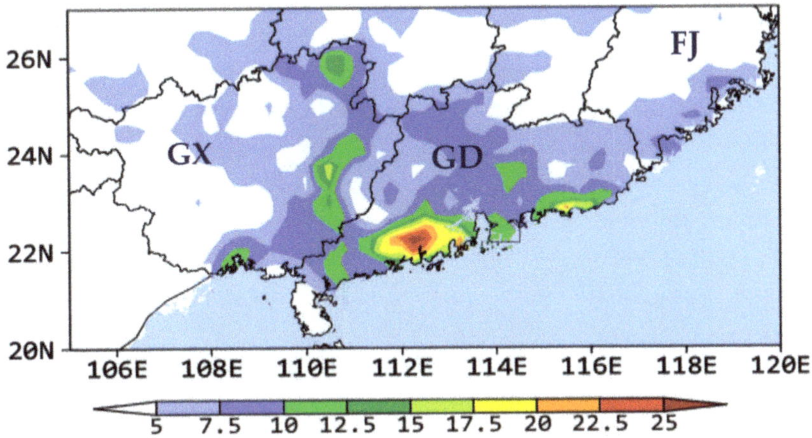

Figure 6. As in Figure 4, except for the 24 WMF-type WSHR events.

As compared to the PSF-type of WSHR events, the WMF-type composite height field at 500 hPa is characterized by a distinct short-wave trough with the lower-tropospheric warm advection on the lee side of the Tibet Plateau with southwesterly flows over South China (Figure 7a). This would facilitate quasi-geostrophic ascent that helps bring the moist southwesterly monsoonal air underneath to saturation, thereby preconditioning the large-scale environment for HR production. The 850-hPa composite fields show the presence of south-to-southwesterly winds of up to 10 m s^{-1} with θ_e > 340 K over a vast area on the south of the Yangtze River Basin, including South China. Unlike the PSF-type events, there is little evidence of a low-level shearline or a front-like zone. Thus, it is often hard to trace directly the initiation of MCSs associated with the WMF-type of WSHR events. An examination of Figures 1, 6 and 7a shows that the above-mentioned WSHR maxima coincide reasonably well with the land-ocean contrast and local topography. This indicates the importance of orographic lifting in triggering deep convection, and of the echo-training process in producing the subsequent HR. In addition, we may assume that convectively generated moist downdrafts, and old thermal boundaries from the previous dissipated MCSs as well as the surface heating could also provide favorable triggering of deep convection leading to the WSHR-producing MCSs.

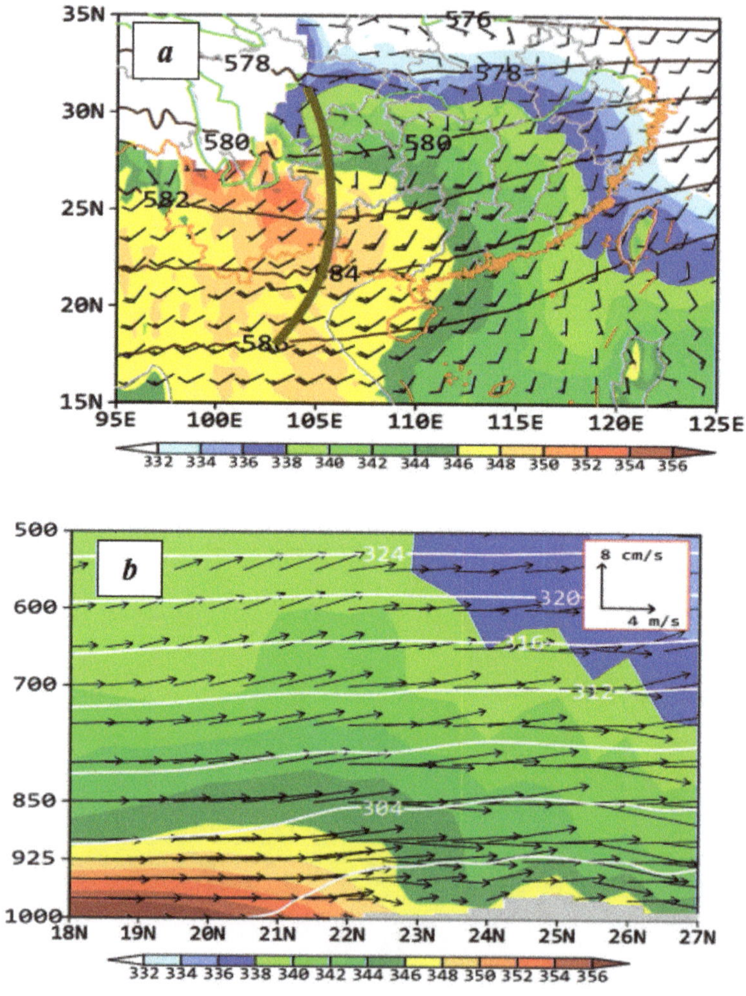

Figure 7. **(a,b)** As in Figure 5a,b, respectively, except for the 24 WMF-type WSHR events.

A south–north vertical cross section through the Yangjiang station with the top HR amount reveals favorable isentropic uplifting of southwesterly flows below 850 hPa, in association with the above-mentioned quasi-geostrophic ascent, which is more pronounced across the coastline (Figure 7b). Unlike the PSF-type events that take place in an environment with a deep-layer high-θ_e air over a wide area in the warm sector, the horizontal extent and depth of high-θ_e air associated with the WMF-type events are much limited to the west of 110° E and the south of 22° N (cf. Figure 5a,b and Figure 7a,b). This different three-dimensional distribution of high-θ_e air appears to explain why the latter exhibits more localized HR compared to more widespread HR in the former (cf. Figures 4 and 6).

3.3.3. The LLV-Type Events

Figure 8 shows more zonally distributed high-percentage rainfall associated with the LLV-type WSHR events, covering a zonal belt of 50–80 km width with several high-percentage rainfall maxima across GX around 24° N. This rainbelt is located to the north of the high-percentage rainbelt associated with the PSF-type events and on the southern side of Mt. Nanling (cf. Figures 4 and 8). A composite wind and θ_e analysis shows clearly the presence of an elliptic-shaped LLV with $\theta_e > 348$ K at 850 hPa around 24° N, with a corresponding mesotrough at 500 hPa that is more pronounced than that associated with the PSF- and WMF-type events (cf. Figures 5a, 7a and 9a). An analysis of several LLV cases indicates that the LLVs are usually formed on the lee side of the Yunnan-Guizhou Plateau in close association with the passage of midlevel troughs [3,52,54], and HR begins to develop as they move into GX's northwestern border (Figure 8). The high-percentage rainbelt just corresponds to the paths of LLVs that are collocated with the midlevel trough (cf. Figures 8 and 9a). Note that the LLVs under study differ from southwest vortices discussed by Kuo et al. [55], and Li et al. [56], which are more topographically related to their origins over the Sichuan Basin. Because of the conservative property of cyclonic vorticity in the presence of weak vertical wind shear, the associated MCSs are longer lived than those associated with the other two-type HR events. Normally, it takes 2–3 days for them to move across GX and GD due to the presence of weak-gradient flows, especially in June. Such slow movements allow the MCSs to drop more rainwater along their paths, thus often producing HR and regional flash floods over South China.

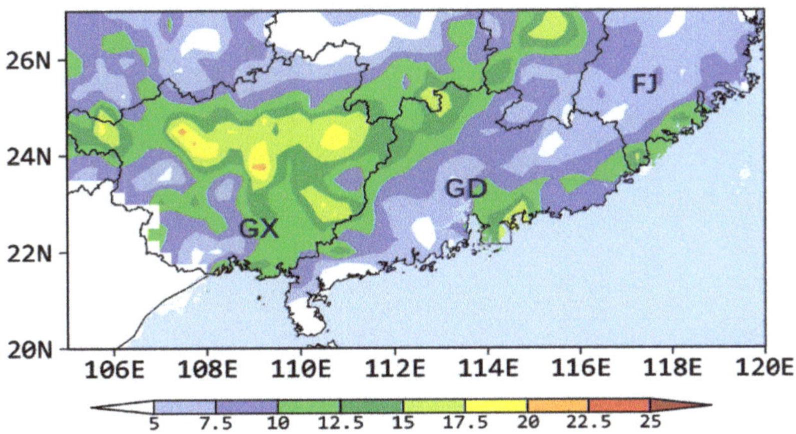

Figure 8. As in Figure 4, except for the 13 LLV-type WSHR events.

As compared to the other two types of WSHR events, the predictability of the LLV-type events by NWP models appears to be superior due again to the conservative property of LLVs [54,57]. LLVs are typically of 300–400 km in diameter, and more evident below 700 hPa, even in the composite fields (Figure 9a). HR usually occurs in the eastern semicircle of LLVs, where is also the favorable region for isentropic uplifting from the potential vorticity viewpoint, given westerly vertical wind shears in the pre-storm environments. This can also be seen from Figure 9b, showing isentropic ascent of high-θ_e air into the LLV region. When interacting with a warm-moist airstream in the PBL, deep convection can form successively at the southern to southeastern periphery of LLVs. The resulting latent heat release would in turn help enhance the intensity of LLVs. The three-dimensional extent of high-θ_e air is greater than that associated with the PSF-type events (cf. Figures 5 and 9). In addition, the topographical lifting of the high-θ_e air over Mt. Yunkai and Mt. Yunwu near the GD-GX border appears to be more pronounced in the LLV-type events than that in the other two types, due to their developments over different geographical locations. This can be seen from the sloping terrain shown in Figures 5b, 7b and 9b.

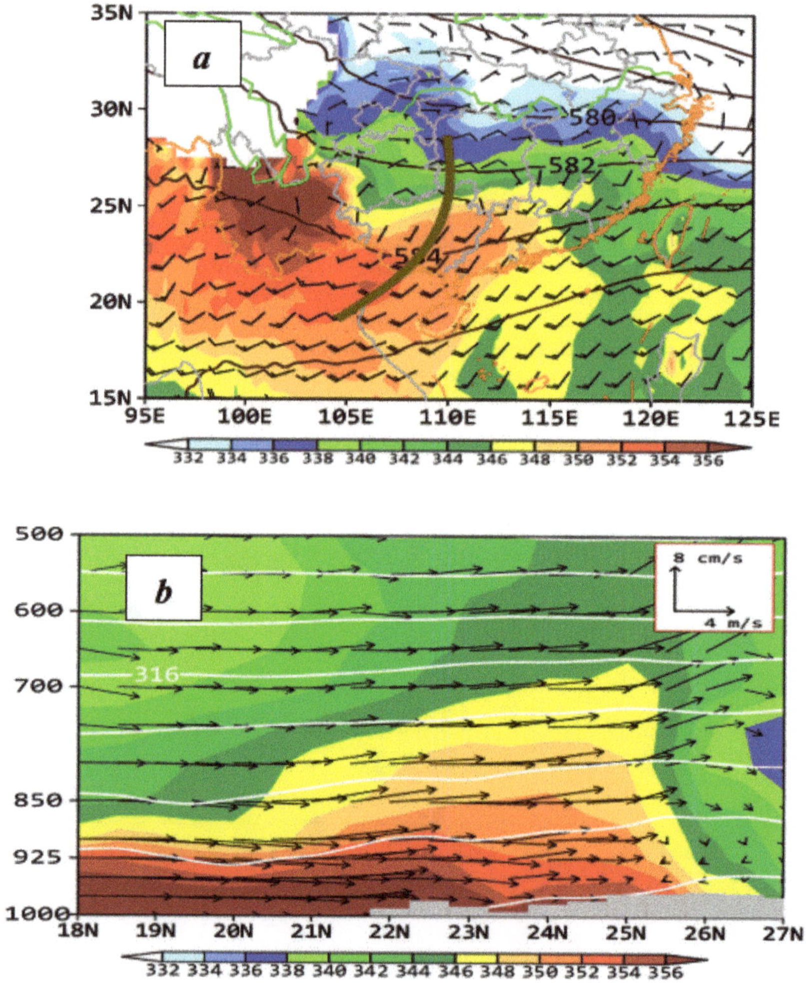

Figure 9. As in Figure 5a,b,respectively, except for the 13 LLV-type WSHR events, with (**b**) along 108 °E through roughly the LLV core region indicated in (**a**) and the HR region indicated in Figure 8.

4. Environmental Thermodynamical Parameters

After seeing different statistical characteristics of rainfall and large-scale flows associated with the three types of WSHR events, it is desirable to examine their corresponding pre-storm environmental thermodynamical parameters in terms of stability, PWAT and CAPE [58–60]. Accurate characterizing the pre-storm environmental soundings is often limited due partly to lacking observations in pre-storm environments with respect to approaching MCSs, and partly to the coarse spatial and temporal resolutions of conventional observations. Given the general weak-gradient warm sector environments in which the MCSs of interest develop, soundings at the Yangjiang station taken at 0800 BST, ahead of approaching MCSs are synthesized in Figure 10 for each type of WSHR events. We acknowledge that this approach of using single-station soundings may be more biased somewhat for those HR events occurring far from Yangjiang, e.g., the LLV-type WSHR events. However, composite single-station soundings are shown in Figure 10 to reveal some common features in pre-storm environmental parameters as described below.

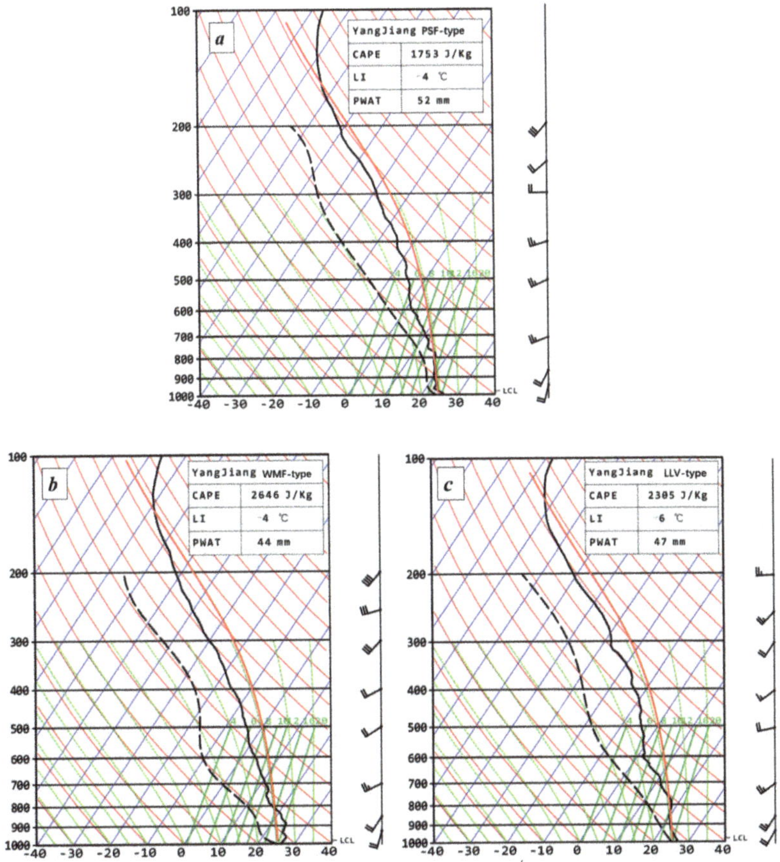

Figure 10. Composite Skew-T/Log P diagrams for the (**a**) PSF-, (**b**) WMF-, and (**c**) LLV-type WSHR events, which are taken at the Yangjiang station at 0800 BST that are far ahead of the approaching MCSs in both the location and timing. A full barb is 4 m s^{-1}.

- The level of free convection (LFC) appears in the lowest 50 hPa, with the lifted index (LI) of less than −4 °C and CAPE ranging between 1753 and 2646 J kg^{-1}. All these indicate the presence of pronounced conditional instability, relatively easy triggering of deep convection in the PBL, and pronounced updrafts in MCSs.
- A deep moist layer is present below 700 hPa with relative humidity (RH) of greater than 85%, and the melting (i.e., 0 °C) level and the tropopause located near 550 and 150 hPa, respectively. They imply the availability of considerable moisture for HR production, and higher precipitation efficiency with more dominant warm cloud microphysics processes.
- A warm, moist airflow of 8–10 m s^{-1} is evident below 850 hPa, with little vertical wind shear from the PBL top to 400 hPa, and slight cyclonic rotation in wind vectors above. They suggest the presence of sustained supply of high-θ_e air for continued convective overturning, and the existence of little thermal advection below 400 hPa but weak cold advection above.

Figure 11 shows that the composite PWAT is generally higher than 50 mm over South China, with the peak value of 60 mm over the coastal regions. This is more or less consistent with the generation of more rainfall with several HR maxima in the regions (cf. Figures 2 and 11). The box and whisker plots of PWAT, obtained in the coastal regions, show that the peak value of more than 65 mm appears in all the three types of WSHR events, with the minimized 45 mm for the WMF-type WSHR events. In general, the percentiles between 25–75% probability distribution in the WMF-type events show relatively wider spreads, with a minimum PWAT value of 44 mm, compared to the other two types. On average, one may take the PWAT of 50–55 mm as the typical value for most WSHR events in South China, which is much higher than 40 mm for the flash flood events occurring in the United States [49].

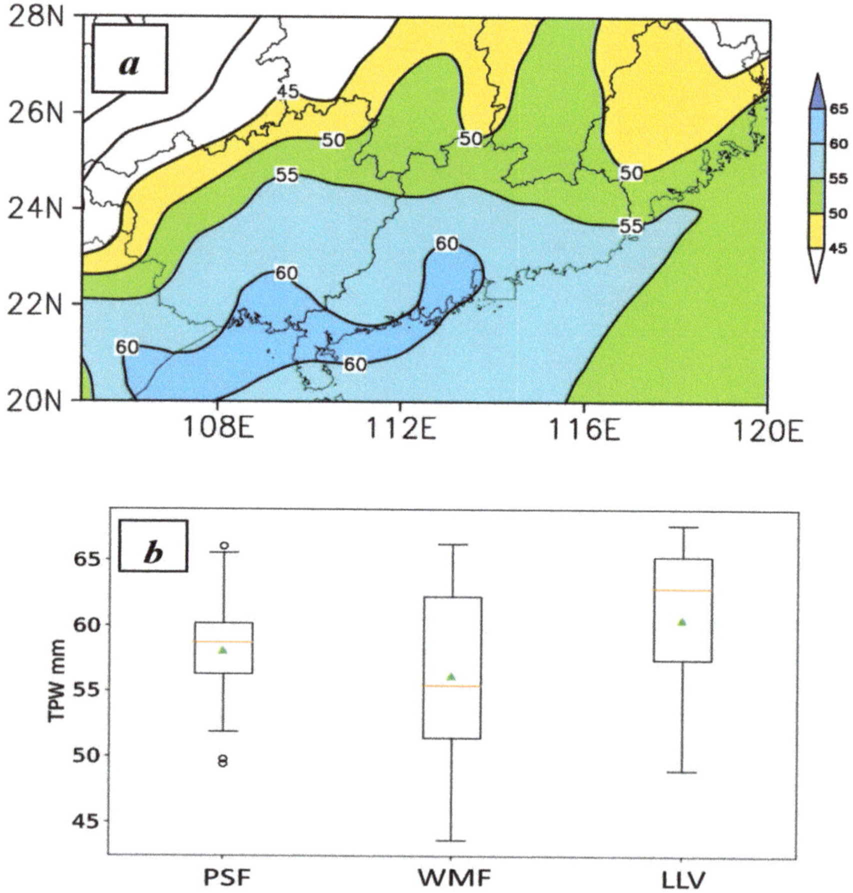

Figure 11. (**a**) Distribution of the daily (at 0800 BST) mean PWAT (mm) that is averaged for the 58 WSHR events under study; and (**b**) the box-whisker plots of the PWAT distribution, taken at the Yangjiang Station, for each type of WSHR events.

Figure 12a shows the distribution of the composite CAPE at 1400 BST, when the daily major convective outbreaks occur. We see that CAPE decreases northward, which is more or less consistent with the distribution of the low-level moisture field. Nevertheless, the magnitudes of 500–1500 J kg^{-1} can be seen over South China, indicating the general existence of convective instability during early afternoon hours. The box and whisker plots of CAPE show the presence of CAPE beyond 2500 J kg^{-1} in all the three types of WSHR events (Figure 12b). The percentiles between 25 and 75% occur mainly in the range of 1000–2000 J kg^{-1}, with the peak CAPE median number for the LLV-type events.

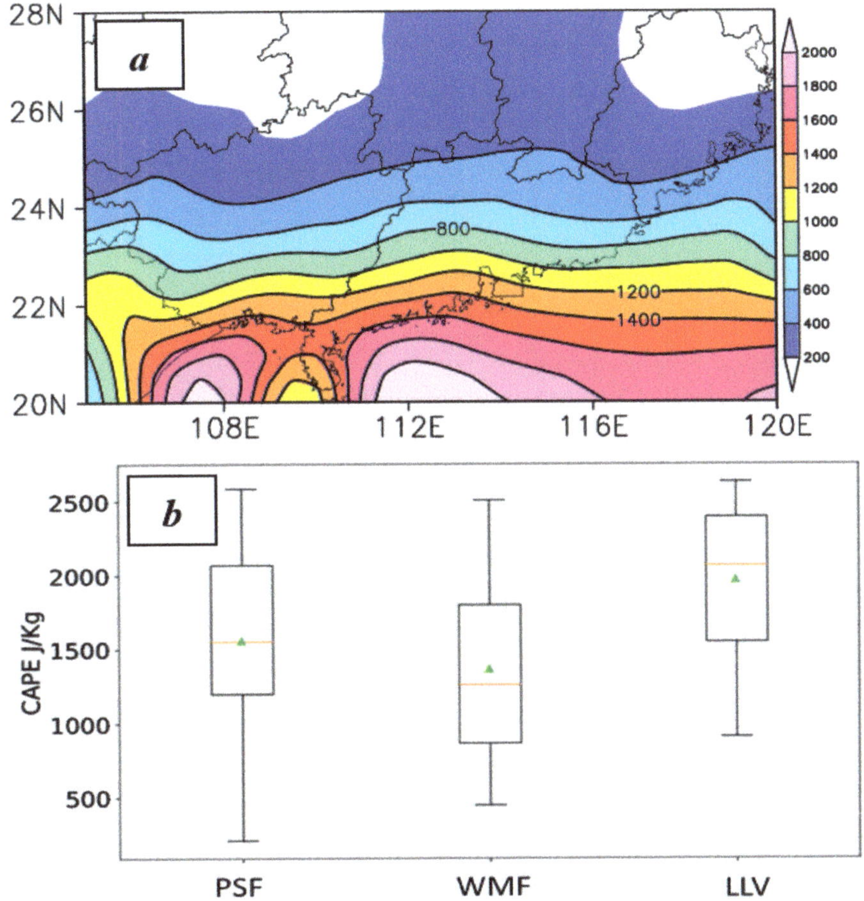

Figure 12. (**a**) As in Figure 11a,b, respectively, except for the mean CAPE in (**b**) from the 1400 BST soundings.

5. Representative Radar Echo Characteristics of WSHR-Producing MCSs

Since all the WSHR events are produced by MCSs, it is of interest to examine if different types of WSHR events classified in Section 3 correspond to different types of MCSs. This can be achieved by analyzing the organizational characteristics of all the MCSs occurring over South China during the pre-summer months in the context of radar echo morphologies. Radar observations have been used to characterize the rainfall structure and intensity of various types of MCSs [25,43,47]. Using radar observations, Schumacher and Johnson [61] documented two types of MCSs that are responsible for HR events in the United States: training line/adjoining stratiform (TL/AS), and back-building/quasi-stationary (BB).

After carefully analyzing all the WSHR events in South China, based on their radar echo morphologies, the following three major types of MCSs are identified: (a) linear-shaped, echo and band training, (b) TL/AS-like organization, and (c) a spirally shaped MCS, i.e., with mesoscale rainbands adjoined with a slowly moving mesovortex or LLV. A total of 145 MCSs are analyzed subjectively from the 58 WSHR events (Table 3). It is found that nearly half of the MCSs can be characterized with echo- and band-training or quasi-stationary linear echoes, and 31.7% of them have the similar form of TL/AS organization with long lifecycles and large spatial extent. The comma-shaped MCSs have temporal and spatial scales that are large and smaller than the linear-shaped echoes and TL/AS

MCSs, respectively. Randomly formed afternoon convection, sea-breeze convection and the other less-organized MCSs have also been found to produce some localized HR events, but they are not examined in this study. These different types of WSHR-producing MCSs just reflect the presence of variable and complicated mesoscale convective processes involved in the generation of WSHR events in South China. In the next three subsections, some representative cases for the above three major types of MCSs are presented, respectively.

Table 3. The number, percentage (%), mean lifespan (hours) and mean maximum length scale (km) of three MCS types associated with the 58 WSHR events.

MCS Type	Number	Ratio (%)	Mean Life Span (h)	Mean Maximum Length Scale (km)
Linear-shaped	71	48.9	3.2	175
TL/AS-like	46	31.7	5.2	315
Comma-shaped	28	19.3	3.7	225

5.1. Linear-Shaped MCSs

Linearly shaped MCSs are the primary type of MCSs leading to WSHR events in South China. HR is generated when a series of convective cells repeatedly forms upstream of their predecessors (i.e., back building) and then moves along the same linear path, i.e., the echo-training. Samples of such MCSs are shown in composite radar mosaic in Figure 13, showing that convective cells are successively initiated in the coastal region and then they move northeastward along the coastline, leading to a mesoscale rainbelt of about 100~200 km with HR in a few hours. The organizational process of echo training cannot fully be attributed to the back-building effect due to the presence of weak downdrafts in deep moist environments. The best documented echo-training process in South China is related to local topographic features, e.g., near Yangjiang [13], where convective initiation occurs as a result of topographical lifting, and the subsequent growth and downstream propagation of convective cells along the same path account for the HR production near Yangjiang. Figure 14 illustrates a schematic of how such linear-shaped MCSs develop in South China.

A further analysis of the composite radar echoes indicates a higher frequency of linear-shaped MCSs near the stations of Fangcheng, Yangjiang and Shanwei along the southern coast of South China (see Figure 1 for their locations), where isolated topography plays an important role in convective initiation. The latter two stations coincide with the climatological rainfall maxima in WSHR events, as discussed in Section 2. Since these rainfall maxima are all close to the coastline, we may expect the sea breeze induced convergence to contribute to the initiation and subsequent organization of the associated linear MCSs, as mentioned before.

5.2. TL/AS MCSs

The TL/AS type of MCSs, as described by Schumacher and Johnson [61], is the No. 2 contributor of the WSHR events in South China. Figure 15 shows the TL/AS type of MCSs but with the following three different configurations: (i) a southwest-northeast elongated MCS moving northeastward with the sustained leading convective line generating HR along GD's coastal regions and a near-symmetric stratiform region to the north (Figure 15a); and (ii) an MCS followed by another MCS along the mean flows moving southeastward with successive newborn convective cells building at the end of their leading lines (Figure 15b). The MCS in Figure 15a, and the MCSs in Figure 15b are associated with the PSF- and WMF-type of WSHR events, respectively.

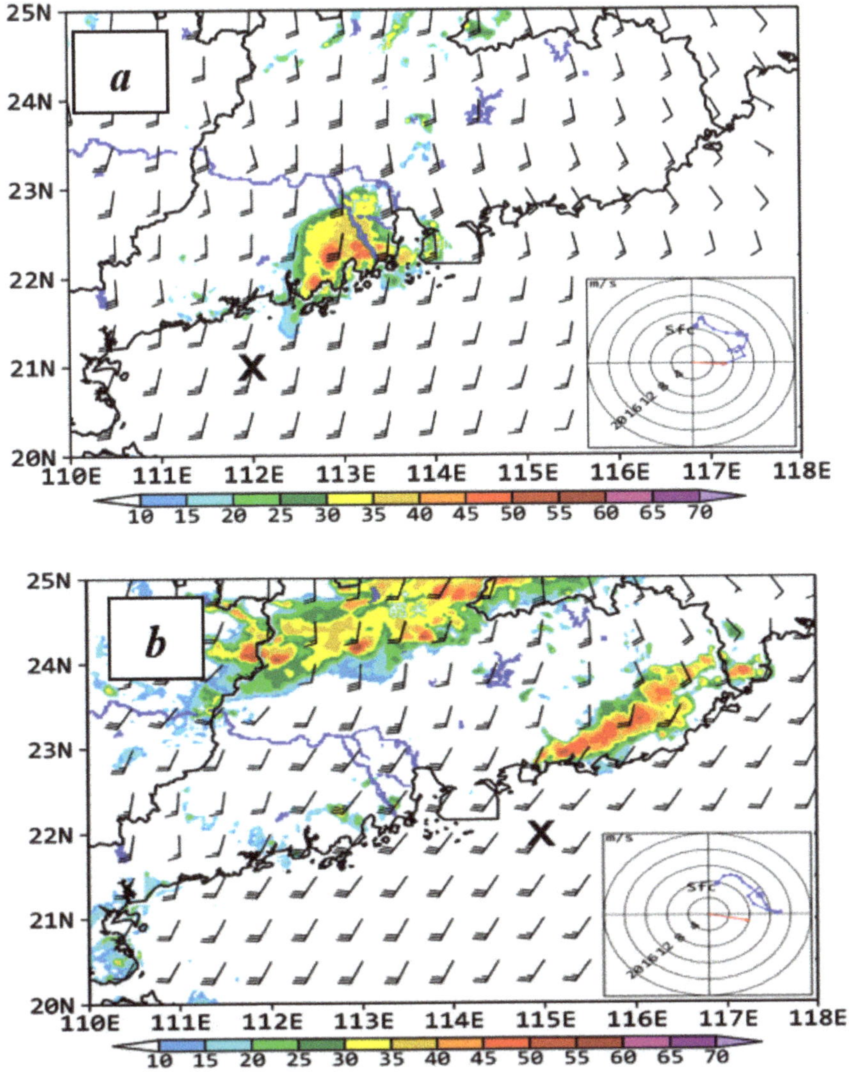

Figure 13. The NCEP GFS analysis of horizontal winds (a full barb is 4 m s^{-1}) at 925 hPa and the composite radar reflectivity (RR, shadings) mosaic at z = 1 km for the linear-shaped MCSs with hodographs taken at '×', and a bold arrow showing the moving direction of convective cells: (**a**) horizontal winds at 0800 BST, and RR at 0600 BST 26 May 2013; (**b**) horizontal winds at 2000 BST, and RR at 2340 BST 26 May 2013.

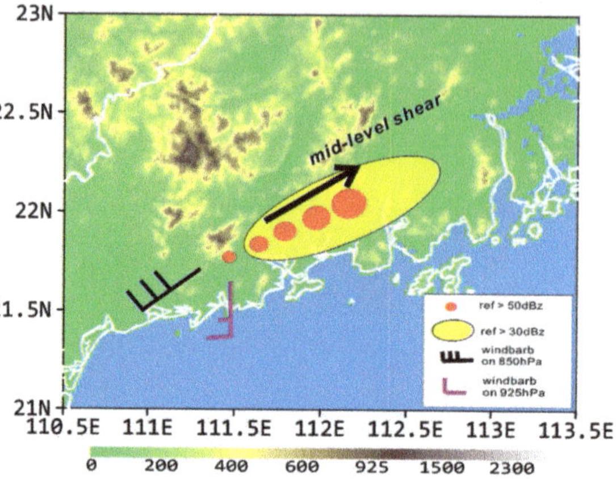

Figure 14. A conceptual model for the linear-shaped MCSs with background topography of South China.

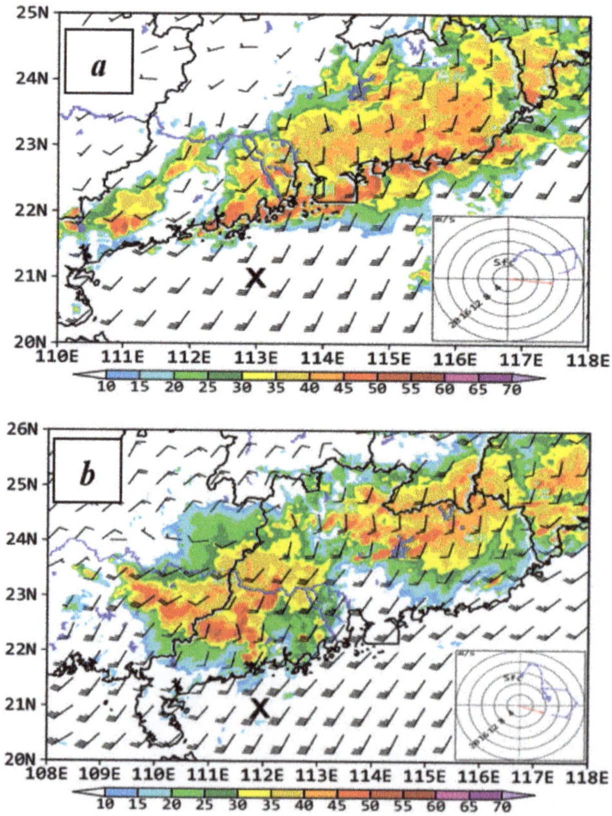

Figure 15. As in Figure 13, but the TL/AS type of MCSs: (**a**) horizontal winds (HW) at 925 hPa at 0800 BST and the composite radar reflectivity (RR) at z =1 km at 0450 BST 22 May 2013; (**b**) horizontal winds at 925 hPa at 0800 BST and RR at z = 1 km at 0900 BST 16 May 2013.

Hodographs in the pre-storm environment, as given in Figures 13 and 15, exhibit clockwise rotation in the lower troposphere (but above the PBL), implying the presence of weak warm advection in the southwesterly monsoonal flows ahead of the midlevel troughs shown in Figures 5a, 7a and 9a. In both types of MCSs, the weak midlevel vertical shear sampled in pre-storm environments is in agreement with the composite wind profiles at Yangjiang station (Figure 10). However, the low-level (weak) warm advection and midlevel vertical shear are more prominent for the TL/AS type of than those of linear-shaped MCSs. The presence of warm advection plus the quasi-geostrophic lifting ahead of the trough, albeit on the order of magnitude of a few cm s^{-1}, and sustained moisture supply in the PBL ensure the preconditioning of convectively favorable environments. Developing into organized MCSs will depend on their interaction, e.g., with local topography, convectively generated cold outflows, and a favorable larger-scale environment such as an LLJ and an ample moisture source.

The above features are summarized by a conceptual model for a PSF-type WSHR event in Figure 16. Given the presence of a slow-moving, weak surface front in South China, as shown in Figure 5b, an MCS initiated in the warm sector due to topographic lifting and prefrontal low-level convergence would grow more rapidly than a frontal MCS, leading to a PSF-type WSHR event. The warm sector is often characterized with ample moisture content and conditional instability, and prevailing southwesterly flows with high CAPE, low CIN and θ_e> 340 K in a deep-moist environment.

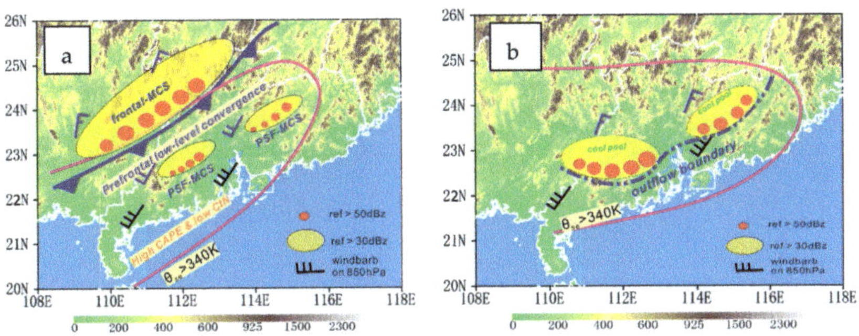

Figure 16. A conceptual model for a typical PSF-type WSHR event with the development of TL/AS MCSs, (**a**) T = T_0, (**b**) T = T_0 + 6 h.

5.3. Comma-Shaped MCSs

In the LLV-type of WSHR events, HR is usually produced by slow-moving MCSs comprising of a series of convective cells with a large area of trailing stratiform clouds in the eastern semicircle of a low-level vortex. Figure 17a shows such an MCS with radar echoes of more than 30 dBz covering the foreside of a low-level vortex. Given the favorable pre-storm environmental conditions, HR results from the long-lived MCS having several rainbelts due to the inertial stability of the mesovortex, in which the MCS is embedded [35,52–54]. Although the hourly rainfall rate is generally weaker than the other two types of MCSs described in the preceding two subsections, the long-lived lifecycle and the slow-moving nature of the low-level vortex are conducive for HR production. Furthermore, the associated multiple rainbands distributed along an LLJ could sustain for hours, leading to large rainfall accumulation. The MCSs of 23 June 2012 is an example in which rainbands are triggered by local topography, and they grow in coverage and rainfall intensity in a convectively unstable environment, and become organized by a low-level mesovortex and an LLJ (Figure 9). A conceptual model, given in Figure 17b, shows the presence of moist southwesterly flows with weak westerly vertical wind shear, and favorable upward motion in the eastern semicircle of a low-level vortex that facilitate the initiation of deep convection in the southeastern quadrant of high-θe flows, and the subsequent upscale growth into an MCS. Sustained rainfall occurs as long as moisture supply continues, leading to the generation of large rainfall totals over South China despite its weak rainfall rates.

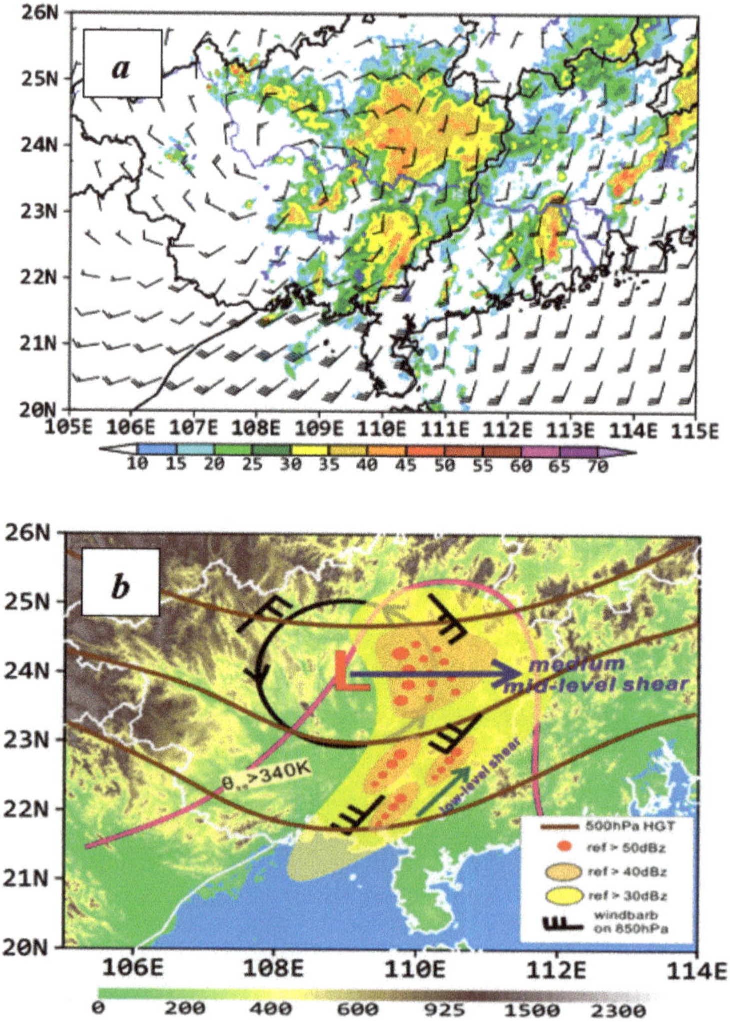

Figure 17. As in Figure 13 but for the comma-shaped MCSs: (**a**) horizontal winds at 925 hPa at 1400 BST and the composite radar reflectivity at z = 1 km at 1240 BST 23 Jun 2012; and (**b**) A conceptual model for the LLV-type WSHR event, with thick arrows in black and green denoting the mid-level and low-level environmental vertical wind shear, respectively.

6. Summary and Concluding Remarks

In this study, a composite analysis of 58 WSHR events occurring in South China during the pre-summer months of 2008~2014 is performed in the context of precipitation, large-scale circulations, pre-storm environmental conditions, and the morphological characteristics of the associated HR-producing MCSs, using the NCEP reanalysis, conventional surface and upper-air observations as well as radar observations. Some major results are summarized as follows.

- Results show that the large-scale circulations governing the development of the 58 WSHR events can be categorized into the following three types: PSF, WMF, and LLV, with higher-frequency occurrences for the former two types. Although the LLV-type WSHR events occur less frequently, they are often the HR producers influencing much larger areas due to the inertial stability or long life cycle of their associated mesoscale vortical flows.
- A statistical analysis of HR-producing pre-storm environments indicates that PWAT of more than 50 mm, and θ_e at 850 hPa of higher than 340 K can be used as the necessary conditions to predict the possible development of WSHR events over South China. A composite analysis of pre-storm soundings shows that most WSHR events occur in an environment with a deep moist layer, pronounced CAPE, weak vertical wind shear below 400 hPa, and an LLJ near 925 hPa with weak warm advection in the lower troposphere.
- Radar echo analyses indicate that the three different types of WSHR events correspond well to the following three organizational modes of HR-producing MCSs: linear-shaped, a leading convective line adjoined with trailing stratiform, and comma-shaped with a low-level vortex. Many line-shaped MCSs over the coastal regions appear to be triggered by land-sea contrasts that are enhanced by sea breezes in South China, whereas the latter two classes of MCSs tend to experience isentropic lifting in the southwesterly warm and moist flows. They all develop in large-scale environments with favorable quasi-geostrophic forcing, including (weak) warm advection in the lower troposphere.

In conclusion, we may state that the WSHR events over South China tend to develop in the following three types of large-scale flows with convectively favorable conditions: pre-frontal, ascending southwesterly flows, and low-level vortical flows, and that the corresponding HR-producing MCSs exhibit mainly linearly shaped, comma-shaped, and a leading convective line with a trailing stratiform region. In particular, the classification of several types of WSHR events and the categorization of the corresponding MCSs appear to add some new understanding of flow configurations and storm morphologies associated with those pre-summer WSHR events. It should be mentioned, however, that due to lacking high-resolution observations, it is not possible to examine herein how deep convection under the influences of the above three different types of flow regimes is triggered and then organized into HR-producing MCSs with the above three different morphologies. In this regard, high-resolution numerical simulations should be performed to explore the roles of various processes, such as topography, surface heating and land-surface conditions, and cold outflow boundaries in convective initiation, and verify conceptual models that are developed herein just through limited coarse-resolution observations.

Author Contributions: Writing, T.C.; Supervision, D.-L.Z. All authors have read and agreed to the published version of the manuscript.

Funding: This work was supported by the National (Key) Basic Research and Development Program of China (2017YFC150210 and 2019YFC1510400), and the Development Foundation of the Chinese Academy of Meteorological Sciences (2020KJ022).

Institutional Review Board Statement: This study was approved and guided by the Ethics Committee of the National Meteorological Center, China Meteorological Administration.

Informed Consent Statement: Informed consent was obtained from all subjects involved in the study.

Acknowledgments: This work was completed during the first author's one-year visitation to the Department of Atmospheric and Oceanic Science, University of Maryland.

Conflicts of Interest: The authors declare no conflict of interest.

Appendix A. A List of the Occurrence Dates (Date/Month/Year) of the 21 PSF-Type of WSHR Events Identified in This Study

12/04/2008 04/05/2008 28/05/2008 29/05/2008 24/04/2009 03/06/2009 09/06/2009 06/05/2010
09/05/2010 28/05/2010 09/06/2010 14/06/2010 15/06/2010 18/04/2012 12/08/2012 29/04/2013
15/05/2013 16/05/2013 19/05/2013 20/05/2013 11/05/2014

Appendix B. As in Appendix A but for the 24 WMF-Type of WSHR Events

05/05/2008 30/05/2008 07/06/2008 08/06/2008 15/04/2009 07/06/2009 08/06/2009 27/06/2009
28/06/2009 08/06/2010 06/05/2011 07/05/2011 08/05/2011 28/06/2011 29/06/2011 03/05/2012
03/04/2013 04/04/2013 28/04/2013 07/05/2013 08/05/2013 07/05/2014 08/05/2014 09/05/2014

Appendix C. As in Appendix A but for the 13 LLV-Type of WSHR Events

11/06/2008 12/06/2008 13/06/2008 27/06/2008 28/06/2008 19/05/2009 31/05/2010 12/05/2012
21/06/2012 19/04/2012 22/06/2012 04/06/2014 05/06/2014

References

1. Tao, S. *Heavy Rainstorms in China*; Science Press: Beijing, China, 1980; p. 225. (In Chinese)
2. Huang, S. *The Presummer Heavy Rainfall over South. China*; Guangdong Science and Technology Press: Guangzhou, China, 1986; p. 243. (In Chinese)
3. Ding, Y.H. Summer monsoon rainfalls in China. *J. Meteorol. Soc. Jpn.* **1992**, *70*, 373–396. [CrossRef]
4. Ding, Y.H. *Monsoons over China*; Kluwer Academic: Beijing, China, 1994; p. 419.
5. Nozumi, Y.; Arakawa, H. Prefrontal rain bands located in the warm sector of subtropical cyclones over the ocean. *J. Geophys. Res. Space Phys.* **1968**, *73*, 487–492. [CrossRef]
6. Moore, J.T.; Glass, F.H.; Graves, C.E.; Rochette, S.M.; Singer, M.J. The environment of warm-season elevated thunderstorms associated with heavy rainfall over the central United States. *Weather Forecast.* **2003**, *18*, 861–878. [CrossRef]
7. Xia, R.; Zhao, S.; Sun, J. A study of circumstances of meso-β-scale systems of strong heavy rainfall in warm sector ahead of fronts in South China. *Chin. J. Atmos. Sci.* **2005**, *6*, 988–1008. (In Chinese)
8. Zhong, L.; Mu, R.; Zhang, D.-L.; Zhao, P.; Zhang, Z.; Wang, N. An observational analysis of warm-sector rainfall characteristics associated with the 21 July 2012 Beijing extreme rainfall event. *J. Geophys. Res. Atmos.* **2015**, *120*, 3274–3291. [CrossRef]
9. Ding, Y.H.; Cao, Q.; Zhang, Y.; Gan, Z.; Zhang, X. The South China Sea monsoon experiment (SCSMEX) implementation plan. *Adv. Atmos. Sci.* **1997**, *14*, 255–270.
10. Lau, K.M.; Ding, Y.; Wang, J.T.; Johnson, R.; Keenan, T.; Cifelli, R.; Gerlach, J.; Thiele, O.; Rickenbach, T.; Tsay, S.-C.; et al. A report of the field operations and early results of the South China Sea Monsoon Experiment (SCSMEX). *Bull. Amer. Meteorol. Soc.* **2000**, *81*, 1261–1270. [CrossRef]
11. Zhou, X.; Xue, J.; Tao, Z. *Experiment of Heavy Rainfall in South China during Pre-Rainy Season (HUAMEX), 1998*; China Meteor. Press: Beijing, China, 2003; p. 220.
12. Zhang, R.; Ni, Y.; Liu, L.; Luo, Y.; Wang, Y. South China Heavy Rainfall Experiments (SCHeREX). *J. Meteorol. Soc. Jpn.* **2011**, *89*, 153–166. [CrossRef]
13. Luo, Y.; Zhang, R.; Wan, Q.; Wang, B.; Wong, W.K.; Hu, Z.; Jou, B.J.; Lin, Y.; Johnson, R.H.; Chang, C.P.; et al. The Southern China Monsoon Rainfall Experiment (SCMREX). *Bull. Amer. Meteorol. Soc.* **2017**, *98*, 999–1013. [CrossRef]
14. Sun, J.; Zhao, S. A diagnosis and simulation study of a strong heavy rainfall in South China. *Chin. J. Atmos. Sci.* **2000**, *24*, 382–391. (In Chinese)
15. Sun, J.; Zhao, S. A study of mesoscale convective systems and its environmental fields during the June 1994 record heavy rainfall of South China. Part I. A numerical simulation study of meso-convective system inducing heavy rainfall. *Chin. J. Atmos. Sci.* **2002**, *26*, 541–557. (In Chinese)
16. Sun, J.; Zhao, S. A study of mesoscale convective systems and its environmental fields during the June 1994 record heavy rainfall in South China. Part II. Effect of physical processes, initial environmental fields and topography on meso-convective system. *Chin. J. Atmos. Sci.* **2002**, *26*, 633–646. (In Chinese)
17. Maddox, R.A.; Chappell, C.F.; Hoxit, L.R. Synoptic and meso-α scale aspects of flash flood events1. *Bull. Amer. Meteorol. Soc.* **1979**, *60*, 115–123. [CrossRef]
18. Wang, C.-C.; Chen, G.T.-J.; Chen, T.-C.; Tsuboki, K. A numerical study on the effects of Taiwan topography on a convective line during the Mei-Yu season. *Mon. Weather Rev.* **2005**, *133*, 3217–3242. [CrossRef]
19. Xu, W.; Zipser, E.J.; Chen, Y.L.; Liu, C.; Liou, Y.C.; Lee, W.C.; Jong-Dao Jou, B. An orography-associated extreme rainfall event during TiMREX: Initiation, storm evolution, and maintenance. *Mon. Weather Rev.* **2012**, *140*, 2555–2574. [CrossRef]
20. Geerts, B.; Parsons, D.; Ziegler, C.L.; Weckwerth, T.M.; Biggerstaff, M.I.; Clark, R.D.; Coniglio, M.C.; Demoz, B.B.; Ferrare, R.A.; Gallus, W.A., Jr.; et al. The 2015 Plains Elevated Convection at Night (PECAN) field project. *Bull. Amer. Meteorol. Soc.* **2017**, *98*, 767–786. [CrossRef]

21. Zhang, D.-L.; Fritsch, J.M. Numerical simulation of the meso-scale structure and evolution of the 1977 Johnstown flood. Part I: Model description and verification. *J. Atmos. Sci.* **1986**, *43*, 1913–1943. [CrossRef]
22. Chou, L.C.; Chang, C.-P.; Williams, R.T. A numerical simulation of the Mei-Yu front and the associated low-level jet. *Mon. Weather Rev.* **1990**, *118*, 1408–1428. [CrossRef]
23. Chen, S.J.; Kuo, Y.-H.; Wang, W.; Tao, Z.Y.; Cui, B. A modeling case study of heavy rainstorms along the Mei-Yu front. *Mon. Weather Rev.* **1998**, *126*, 2330–2351. [CrossRef]
24. Chen, G.T.J.; Wang, C.C.; Lin, L.F. A diagnostic study of a retreating Mei-Yu front and the accompanying low-level jet formation and intensification. *Mon. Weather Rev.* **2006**, *134*, 874–896. [CrossRef]
25. Luo, Y.; Gong, Y.; Zhang, D.-L. Initiation and organizational modes of an extreme-rain-producing mesoscale convective system along a Mei-Yu front in east China. *Mon. Weather Rev.* **2014**, *142*, 203–221. [CrossRef]
26. Wu, M.; Luo, Y. Mesoscale observational analysis of lifting mechanism of a warm-sector convective system producing the maximal daily precipitation in China mainland during pre-summer rainy season of 2015. *J. Meteorol. Res.* **2016**, *30*, 719–736. [CrossRef]
27. Liu, X.; Luo, Y.; Guan, Z.; Zhang, D.-L. An extreme rainfall event in coastal South China during SCMREX-2014: Formation and roles of Rainband and Echo trainings. *J. Geophys. Res. Atmos.* **2018**, *123*, 9256–9278. [CrossRef]
28. Yu, R.; Zhou, T.; Xiong, A.; Zhu, Y.; Li, J. Diurnal variations of summer precipitation over contiguous China. *Geophys. Res. Lett.* **2007**, *34*, L01704. [CrossRef]
29. Jiang, Z.; Zhang, D.-L.; Xia, R.; Qian, T. Diurnal variations of presummer rainfall over southern China. *J. Climatol.* **2017**, *30*, 755–773. [CrossRef]
30. Huang, S.; Li, Z.; Liang, B.; Bao, C. *Mechanism and Prediction of Pre-Rainy Season Heavy Rain Fall in South China*; Collected Papers of Pre-Rainy Season Heavy Rainfall in South China; Beijing China Meteorological Press: Beijing, China, 1981; pp. 1–8. (In Chinese)
31. Wang, C.-C.; Chen, G.T.-J.; Huang, S.-Y. Remote trigger of deep convection by cold outflow over the Taiwan Strait in the Mei-Yu season: A modeling study of the 8 June 2007 case. *Mon. Weather Rev.* **2011**, *139*, 2854–2875. [CrossRef]
32. Yin, J.; Zhang, D.-L.; Luo, Y.; Ma, R. On the extreme rainfall event of 7 May 2017 over the coastal city of Guangzhou. Part I: Impacts of urbanization and orography. *Mon. Weather Rev.* **2020**, *148*, 955–979. [CrossRef]
33. Davis, R.S. Flash Flood Forecast and Detection Methods. In *Severe Convective Storms*; American Meteorological Society: Boston, MA, USA, 2001; Volume 50, pp. 481–525.
34. Fritsch, J.M.; Kane, R.J.; Chelius, C.R. The contribution of mesoscale convective weather systems to the warm-season precipitation in the United States. *J. Appl. Meteor. Climatol.* **1986**, *25*, 1333–1345. [CrossRef]
35. Fritsch, J.M.; Murphy, J.D.; Kain, J.S. Warm core vortex amplification over land. *J. Atmos. Sci.* **1994**, *51*, 1780–1807. [CrossRef]
36. Maddox, R.A. Mesoscale convective complexes. *Bull. Amer. Meteorol. Soc.* **1980**, *61*, 1374–1387. [CrossRef]
37. Chen, G.T.J. Observational aspects of the Mei-Yu phenomena in subtropical China. *J. Meteorol. Soc. Jpn.* **1983**, *61*, 306–312. [CrossRef]
38. Tao, S.; Chen, L. A review of recent research on the East Asian summer monsoon in China. In *Monsoon Meteorology*; Chang, C.P., Krishnamurti, T.N., Eds.; Oxford University Press: London, UK, 1987; pp. 60–92.
39. Yao, X.; Ma, J.; Zhang, D.-L.; Yan, L. A 33-yr Meiyu-season climatology of shear lines over the Yangtze-Huai River Basin in Eastern China. *J. Appl. Meteorol. Climatol.* **2020**, *59*, 1125–1137. [CrossRef]
40. Chen, G.T.J.; Yu, C.C. Study of low-level jet and extremely heavy rainfall over northern Taiwan in the Mei-Yu season. *Mon. Weather Rev.* **1988**, *116*, 884–891. [CrossRef]
41. Wang, C.-C.; Hsu, J.C.-S.; Chen, G.T.-J.; Lee, D.-I. A study of two propagating heavy-rainfall episodes near Taiwan during SoWMEX/TiMREX IOP-8 in June 2008. Part I: Synoptic evolution, episode propagation, and model control simulation. *Mon. Weather Rev.* **2014**, *142*, 2619–2643. [CrossRef]
42. Lin, P.F.; Chang, P.L.; Jou, B.J.; Wilson, J.W.; Roberts, R.D. Warm season afternoon thunderstorm characteristics under weak synoptic-scale forcing over Taiwan Island. *Weather Forecast.* **2011**, *26*, 44–60. [CrossRef]
43. Bluestein, H.B.; Jain, M.H. Formation of mesoscale lines of precipitation: Severe squall lines in Oklahoma during the spring. *J. Atmos. Sci.* **1985**, *42*, 1711–1732. [CrossRef]
44. Parker, M.D.; Johnson, R.H. Organizational modes of midlatitude mesoscale convective systems. *Mon. Weather Rev.* **2000**, *128*, 3413–3436. [CrossRef]
45. Doswell, C.A., III; Brooks, H.E.; Maddox, R.A. Flash-flood forecasting: An ingredients-based methodology. *Weather Forecast.* **1996**, *11*, 360–381. [CrossRef]
46. Schumacher, R.S.; Johnson, R.H. Organization and environmental properties of extreme-rain-producing mesoscale convective systems. *Mon. Weather Rev.* **2005**, *133*, 961–976. [CrossRef]
47. Luo, Y.; Qian, W.; Zhang, R.; Zhang, D.-L. Gridded hourly precipitation analysis from high-density rain-gauge network over the Yangtze-Huai Rivers Basin during the 2007 Meiyu season and comparison with CMORPH. *J. Hydrometeor.* **2013**, *14*, 1243–1258. [CrossRef]
48. Liu, R.; Sun, J.; Wei, J.; Fu, S. Classification of persistent heavy rainfall events over South China and associated moisture source analysis. *J. Meteorol. Res.* **2016**, *30*, 678–693. [CrossRef]
49. Maddox, R.A.; Canova, F.; Hoxit, L.R. Meteorological characteristics of flash flood events over the western United States. *Mon. Weather Rev.* **1980**, *108*, 1866–1877. [CrossRef]

50. Raymond, D.J.; Jiang, H. A theory for long-lived mesoscale convective systems. *J. Atmos. Sci.* **1990**, *47*, 3067–3077. [CrossRef]
51. Zhang, M.; Zhang, D.-L. Sub-kilometer simulation of a torrential-rain-producing mesoscale convective system in East China. Part I: Model verification and convective organization. *Mon. Weather Rev.* **2012**, *140*, 184–201. [CrossRef]
52. Shu, Y.; Sun, J.; Pan, Y. Characteristics of mesoscale vortices over China in 2015. *J. Meteorol. Res.* **2017**, *31*, 1149–1160. [CrossRef]
53. Chen, T.; Zhang, F.H.; Duan, Y.H. A study of relationship between a southwest vortex and the mesoscale convective systems during the severe "6.12" rainstorm event in Guangxi province. *Acta Meteorol. Sin.* **2011**, *69*, 472–485. (In Chinese)
54. Zhang, D.-L.; Fritsch, J.M. Numerical simulation of the meso-β scale structure and evolution of the 1977 Johnstown flood. Part II: Inertially stable warm-core vortex and the mesoscale convective complex. *J. Atmos. Sci.* **1987**, *44*, 2593–2612. [CrossRef]
55. Kuo, Y.-H.; Cheng, L.; Bao, J.-W. Numerical simulation of the 1981 Sichuan flood. Part I: Evolution of a mesoscale southwest vortex. *Mon. Weather Rev.* **1988**, *116*, 2481–2504. [CrossRef]
56. Li, J.; Du, J.; Zhang, D.-L.; Cui, C.; Liao, Y. Ensemble-based analysis and sensitivity of mesoscale forecasts of a vortex over southwest China. *Q. J. R. Meteorol. Soc.* **2014**, *140*, 766–782. [CrossRef]
57. Zhang, D.-L.; Fritsch, J.M. A numerical investigation of a convectively generated, inertially stable, extratropical warm-core mesovortex over land. Part I: Structure and evolution. *Mon. Weather Rev.* **1988**, *116*, 2660–2687. [CrossRef]
58. Tian, F.; Zheng, Y.; Zhang, T.; Zhang, X.; Mao, D.; Sun, J.; Zhao, S. Statistical characteristics of environmental parameters for warm season short-duration heavy rainfall over central and eastern China. *J. Meteorol. Res.* **2015**, *29*, 370–384. [CrossRef]
59. Xia, R.; Zhang, D.-L.; Zhang, C.; Wang, Y. Synoptic control of convective rainfall rates and cloud-to-ground lightning frequencies in warm-season mesoscale convective systems over North China. *Mon. Weather Rev.* **2018**, *146*, 813–831. [CrossRef]
60. Wang, H.; Luo, Y.; Jou, B.J.D. Initiation, maintenance, and properties of convection in an extreme rainfall event during SCMREX: Observational analysis. *J. Geophys. Res. Atmos.* **2014**, *119*, 13206–13232. [CrossRef]
61. Schumacher, R.S.; Johnson, R.H. Characteristics of U.S. extreme rain events during 1999–2003. *Weather Forecast.* **2006**, *21*, 69–85. [CrossRef]

atmosphere

MDPI

Article

Characteristics of Precipitation Diurnal Cycle over a Mountainous Area of Sumatra Island including MJO and Seasonal Signatures Based on the 15-Year Optical Rain Gauge Data, WRF Model and IMERG

Marzuki Marzuki [1,*], Helmi Yusnaini [1], Ravidho Ramadhan [1,2], Fredolin Tangang [3], Abdul Azim Bin Amirudin [3], Hiroyuki Hashiguchi [4], Toyoshi Shimomai [5] and Mutya Vonnisa [1]

1 Department of Physics, Universitas Andalas, Padang 25163, Indonesia; helmiyusnaini@gmail.com (H.Y.); ravidhoramadhan20@gmail.com (R.R.); mutyavonnisa@sci.unand.ac.id (M.V.)
2 Department of Physics, Universitas Gadjah Mada, Yogyakarta 55281, Indonesia
3 Department of Earth Sciences and Environment, Faculty of Science and Technology, Universiti Kebangsaan Malaysia, Bangi 43600, Selangor, Malaysia; tangang@ukm.edu.my (F.T.); p94165@siswa.ukm.edu.my (A.A.B.A.)
4 Research Institute for Sustainable Humanosphere (RISH), Kyoto University, Gokasho, Uji, Kyoto 611-0011, Japan; hasiguti@rish.kyoto-u.ac.jp
5 Department of Electronic and Control Systems Engineering, Shimane University, Shimane 690-0823, Japan; shimomai@ecs.shimane-u.ac.jp
* Correspondence: marzuki@sci.unand.ac.id

Citation: Marzuki, M.; Yusnaini, H.; Ramadhan, R.; Tangang, F.; Amirudin, A.A.B.; Hashiguchi, H.; Shimomai, T.; Vonnisa, M. Characteristics of Precipitation Diurnal Cycle over a Mountainous Area of Sumatra Island including MJO and Seasonal Signatures Based on the 15-Year Optical Rain Gauge Data, WRF Model and IMERG. *Atmosphere* **2022**, *13*, 63. https://doi.org/10.3390/atmos13010063

Academic Editors: Tomeu Rigo, Zuohao Cao, Huaqing Cai and Xiaofan Li

Received: 15 November 2021
Accepted: 28 December 2021
Published: 30 December 2021

Abstract: In this study we investigate the characteristics of the diurnal precipitation cycle including the Madden–Julian oscillation (MJO) and seasonal influences over a mountainous area in Sumatra Island based on the in situ measurement of precipitation using the optical rain gauge (ORG). For comparison with ORG data, the characteristics based on the Global Precipitation Measurement (GPM) mission (IMERG) and Weather Research and Forecasting (WRF) simulations were also investigated. Fifteen years of ORG data over a mountainous area of Sumatra, namely, at Kototabang (100.32° E, 0.20° S), were analyzed to obtain the characteristics of the diurnal cycle of precipitation in this region. The diurnal cycle of precipitation presented a single peak in the late afternoon, and the peak time difference was closely related to the rain event duration. The MJO acts to modulate the diurnal amplitude but not the diurnal phase. A high precipitation amount (PA) and frequency (PF) were observed during phases 2, 3, and 4, along with an increase in the number of longer-duration rain events, but the diurnal phase was similar in all MJO phases. In terms of season, the highest PA and PF values were observed during pre-southwest and pre-northeast monsoon seasons. WRF simulation reproduced the diurnal phase correctly and more realistically than the IMERG products. However, it largely overestimated the amplitude of the diurnal cycle in comparison with ORG. These disagreements could be related to the resolution and quality of IMERG and WRF data.

Keywords: diurnal cycle; Kototabang; Sumatra; optical rain gauge; IMERG; WRF

1. Introduction

The estimation of precipitation at high altitudes in Indonesia, including Sumatra, is challenging because in situ measurements remain scarce. Sumatra is one of the largest islands in Indonesia, and is directly adjacent to the Indian Ocean (see Figure 1). It is considered important in global atmospheric circulation because its position is almost perpendicular to the propagation of winds and clouds from the Indian Ocean [1–3]. The combination of mesoscale variability with topography and the coastline controls diabatic heating in Indonesia, including Sumatra [4]. The Barisan mountain range in Sumatra, with an average altitude of 2000 m, plays an essential role in the convection process in this region [5,6]. Sumatra's topography induces convection in Sumatra and the surrounding area [7–9]. Interestingly, rainfall characteristics on both sides of the Barisan mountains are different [10], with more frequent rainfall in the western part of the Barisan mountains, resulting in a greater amount of rain in this area than on the eastern side. However, the precipitation intensity in the western part of the Barisan mountains is smaller than that on the eastern side [11].

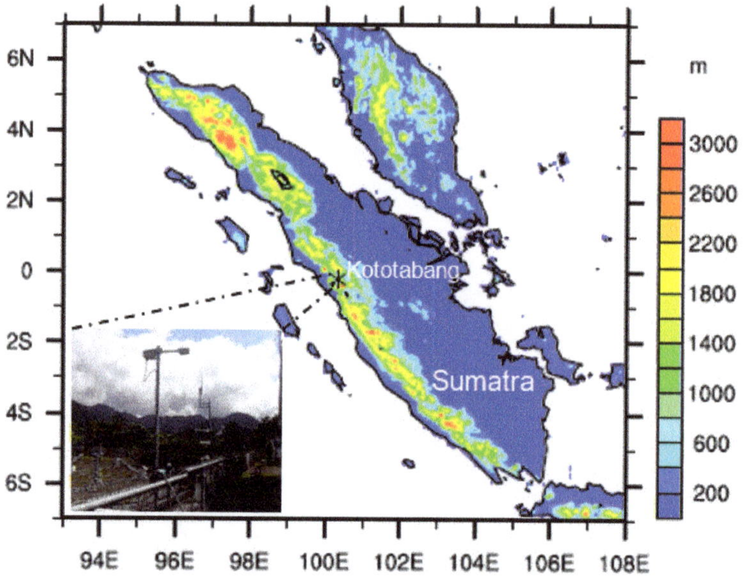

Figure 1. The topography of Sumatra and the location of the Equatorial Atmosphere Observatory (EAO) at Kototabang.

In 2001, an atmospheric observation center called the Equatorial Atmosphere Observatory (EAO) was established in Sumatra, precisely at Kototabang (100.32° E, 0.20° S, 865 m above sea level). Kototabang is located in the mountainous area of the Barisan mountains (Figure 1). The main instrument at EAO was the equatorial atmosphere radar (EAR), which was a large atmospheric radar for atmospheric observations [12]. Supporting instruments at EAO included a boundary layer radar, an X-band weather radar, a radiometer, a ceilometer, lidar, an optical rain gauge (ORG), and a disdrometer. With the use of this instrument, several research projects have been carried out, such as the Coupling Processes in the Equatorial Atmosphere (CPEA) study [13]. Although many research results have been reported using the data in the EAO, some observations have yet to be analyzed. This paper presents the results of ORG observations during 15 years of operation in Kototabang. These data are used to characterize the diurnal variation of rainfall in Kototabang, including its Madden–Julian oscillation (MJO) and seasonal signatures.

Diurnal variation is the most fundamental mode of rainfall variability in the tropics [14]. Hence, an in-depth study of the diurnal cycle in rainfall is essential in helping to understand the relationship between the rainfall processes and related factors [15]. However, studies on the diurnal cycles of rainfall remain challenging and they are often hampered by the limited observational data on an hourly scale in many parts of Indonesia, including in Sumatra. Wu et al. [8] analyzed precipitation data during March 2001 in Kototabang and demonstrated a relationship between diurnal variations and water vapor. Water vapor in Kototabang increases during the day and reaches its maximum in the late afternoon. Mori et al. [5] observed the migration of rainfall peaks in Sumatra using Tropical Rainfall Measuring Mission (TRMM) precipitation radar (PR) data for three years

(1998–2000). They found that the rainfall peaks on mainland Sumatra were observed at 15:00 and 20:00 local standard time (LST) and were dominated by convective rain. In the early morning, stratiform and convective types of rain appeared but they were dominant in the surrounding ocean. Many other studies have investigated the diurnal variations of rainfall in Sumatra by utilizing satellite data or in situ observations over a limited period. To overcome these limitations, Marzuki et al. [11] and Suryani et al. [16] analyzed the diurnal variations of rainfall in Sumatra by analyzing rain gauge data from 186 stations throughout Sumatra, which were operated by the Meteorology, Climatology, and Geophysical Agency (BMKG). The analysis was based on rain gauge data during 2015–2019, and did not include Kototabang ORG data. These studies enriched the understanding of the rainfall in Sumatra because they examined diurnal variations from the perspectives of precipitation amount (PA), frequency (PF), and intensity (PI), which have not been studied for the Sumatra area before.

This paper presents a follow-up of the work of Marzuki et al. [11], focusing on a study area in a mountainous area of Sumatra where in situ measurements remain scarce. There are several factors affecting the diurnal cycle of rainfall, namely, surface temperature, moist convection, the formation of clouds, boundary layer development [14], regional and synoptic-scale dynamical, and thermal conditions [17]. Over a mountainous area, the interaction of these factors with local-scale phenomena of land–sea and mountain–valley breezes may produce different characteristics in terms of the diurnal cycles of precipitation [8]. Marzuki et al. [18] showed the prominent role of the island's mountainous areas in the development and propagation of the precipitation system over Sumatra island.

In addition to the climatology of the diurnal cycle, the MJO and seasonal signature of the diurnal cycle in rainfall have also been investigated, and over mountainous areas these effects can be unique. Several studies have shown the effect of seasons [19,20] and MJO [21] on the diurnal cycle of precipitation. MJO modulates the diurnal precipitation cycle over the Indonesian maritime continent (IMC) due to an increase in the low-level moisture background associated with MJO [21]. Furthermore, the highest amplitude of diurnal variation in precipitation was observed during the active phase of the monsoon season, as observed in the Bay of Bengal [20]. The effect of MJO and seasons is not discussed by Marzuki et al. [11] due to the limited duration of the available data. Recently, Marzuki et al. [18] investigated seasons' effects on Sumatra's diurnal cycle in relation to the land/sea contrast in precipitation, but they also did not discuss the MJO. Thus, using the 15-year data record of ORG in Kototabang will enhance our understanding of diurnal variations in Sumatra, including its MJO and seasonal signature.

The diurnal cycle of rainfall is also a critical test of many aspects of the physical parameterizations in weather and climate models [15]. Global climate models usually fail to simulate properly the regional processes and their spatial variability for precipitation in mountainous areas [22]. Hara et al. [23] compared the precipitation simulated by a 20 km-grid Meteorological Research Institute General Circulation Model (MRI-GCM) and the near-surface rain data of TRMM 2A25. They failed to simulate the diurnal cycle over islands of which the horizontal scale was larger than 200 km, such as Sumatra and Borneo. There was a difference in peak time between MRI-GCM and TRMM 2A25 because the cumulus convective parameterization in the 20 km grid spacing did not adequately represent the coupling of convection and local circulations [23]. In this study, we examined the ability of the Weather Research and Forecasting (WRF) model to reproduce the diurnal cycle of the precipitation in the mountains of Sumatra, using a smaller simulation grid of 5 km. ORG data are very useful in evaluating the ability of the WRF to capture the diurnal cycle of precipitation over a mountainous area of Sumatra. Although there are differences in the sampling area between ORG and WRF, comparisons between point rain gauges such as ORG with grid data (WRF) are widely used in the validation of precipitation from WRF models, especially with regard to diurnal cycles, for which rain gauge data are always used as a reference value. In addition to point rain gauge observations, the simulation results are also compared with gridded precipitation data from the Integrated

Multi-satellite Retrievals for GPM (IMERG). The IMERG was chosen because these data have better temporal and spatial resolution than others [24] and are believed to be the most accurate data at present.

2. Materials and Methods

Precipitation data were collected on an 815 Optical Rain Gauge (ORG) during 2002–2016. ORG data in Kototabang have been used for various studies, such as telecommunication [25–27], weather radar [28], and precipitation variability studies [3,29]. This instrument was based on scintillation technology, in which raindrops fall through the beam, leading to variations in the intensity of the infrared light. These irregularities, known as scintillations, are detected by the sensor and converted to the rain rate. ORG reports the rain rate in mm h^{-1} with the dynamic range of 0.1 to 500 mm h^{-1}. The detailed specifications of this instrument can be found in the company website [30]. This study used a sampling time of 10 min, which is the same as that of previous studies using rain gauge data in Sumatra [11]. In general, the availability of data was high, except for a few months (Figure 2). In 2002, the percentage of data availability was around 76% because there were no observations in January and February. In 2007, data availability was about 70% due to the absence of observations in January and February and the lack of observations in October and November (<60%).

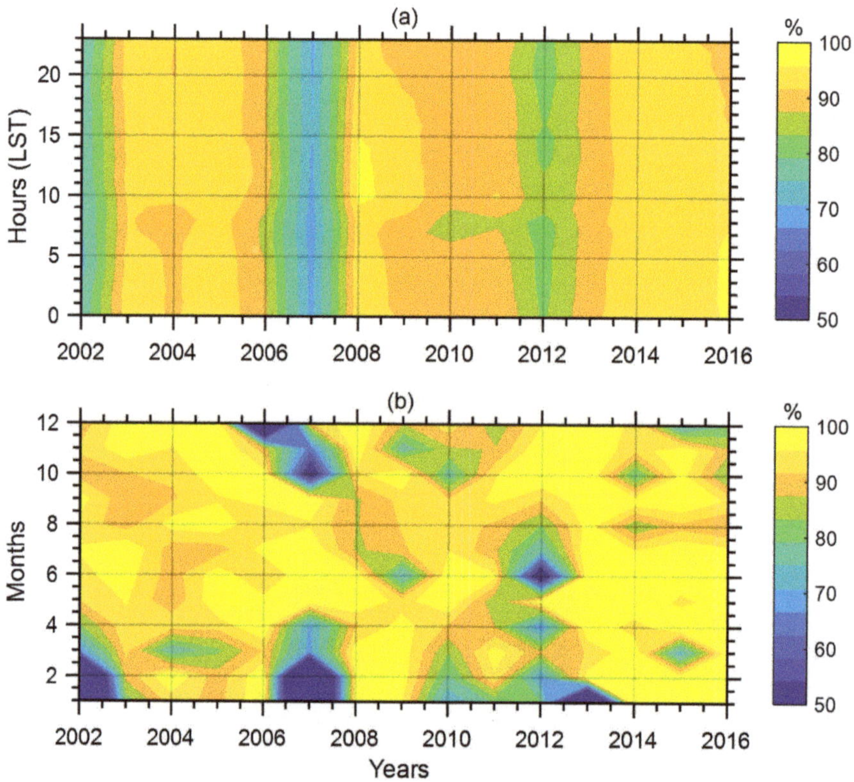

Figure 2. Availability of the data on an hourly (**a**) and monthly (**b**) basis. The percentage indicates the ratio of hourly (monthly) recorded samples to the total number of samples that would have been recorded had no measurement failures occurred during a given year.

The diurnal variation was analyzed from three parameters, namely, the precipitation amount (PA), precipitation frequency (PF), and precipitation intensity (PI). PA is defined as the total accumulated rainfall divided by the total hours of observation. Furthermore, the PF is the total 10-min data with rain above 0.1 mm h^{-1} divided by the total 10-min observation data, and PI is defined as the total accumulated rainfall divided by the number of rainy hours ($R \geq 0.1$ mm h^{-1}). The definitions of PA, PF, and PI, are the same as those presented in the previous study [11]. Calculations were carried out for each hour in local standard time (LST) units. The precipitation at Kototabang is influenced by seasons and Madden–Julian oscillation (MJO) [3,31–34]. Therefore, the effect of seasons on the diurnal variation of rainfall was investigated by calculating PA, PF, and PI for each month. In addition, the effect of the MJO on the diurnal variation of rainfall in Kototabang was also investigated. The MJO index was downloaded from the Australian Meteorological Agency website [35]. This study only considered strong MJO, which is indicated by an MJO index more than one. The effect of MJO was observed for the period of December–January–February (DJF), March–April–May (MAM), June–July–August (JJA), and September–October–November (SON), which was based on the convection pattern in the IMC [2,34,36]. The duration of a rain event can also affect the peak time of the rain [37]. To investigate such an effect, we classified rain events into three durations, following Marzuki et al. [11], namely, <3 h (short-duration rain events), 3–6 h (medium-duration rain events), and >6 h (long-duration rain events), as also used by Marzuki et al. [11]. A rain event was defined as a continuous event that was not interrupted by an hourly rainfall of less than 0.1 mm.

The diurnal variations of precipitation from the ORG data were compared with satellite-derived (Integrated Multi-satellite Retrievals for the Global Precipitation Measurement (GPM) mission (IMERG)) and atmospheric model outputs (WRF, Weather Research and Forecasting). IMERG data were available in 0.1° grids every half hour. The IMERG precipitation product is generated from the GPM mission, which unifies observations from a network of partner satellites in the GPM constellation [38]. In this study, IMERG final-run product V06 (IMERG-F) was used, since it is recommended for research activities due to its better accuracy ([39]. The WRF simulations were carried out with a nest domain with a horizontal resolution of 25 km in the outer domain and 5 km in the inner domain (Figure 3). The model was configured with 32 vertical layers with 50 hPa at the top level. The WRF model was setup in a non-hydrostatic mode and was initialized and forced with the European Centre for medium-range weather forecasting (ECMWF) reanalysis (ERA-Interim). The time scale for this experiment was 3 years of simulation with a 1 year spin-up time from 2013 to 2017. The physics options for this experiment followed those of Ratna et al. [40]. These included the WSM 3-class simple ice scheme for microphysics schemes [41], the Unified NOAH scheme for land surface schemes [42] and Betts–Miller–Janjic scheme for cumulus analysis [43,44]. Short-wave and long-wave schemes were based on the Dudhia scheme and the Yonsei University model (RRTM) scheme, respectively [45,46].

The difference in sampling area between ORG point observations and IMERG-based and WRF-based grid values can cause uncertainty when these three forms of data are directly compared. To quantitatively evaluate the performance of WRF and IMERG-F in estimating hourly rainfall at Kototabang, several statistical parameters were calculated as in Table 1. The equation to calculate these parameters can be seen in some references, e.g., [47]. The observation accuracy of IMERG-F and WRF data is still low compared to that of ORG, as can be seen from the low CC and high RMSE and RB values. The RB values of IMERG and WRF, when compared with ORG, are positive. WRF has the largest RB value when compared to ORG. Although the accuracy of IMERG and WRF is still not good when observing hourly rainfall, the ability of IMERG data and the WRF model to detect hourly rainfall is good enough, as can be seen from the POD value. Although the POD value is good enough, IMERG and WRF were often wrong in detecting hourly rainfall in Kototabang, as seen from the high FAR value. This result may indicate the inability of one observation point to represent rainfall variations in one WRF grid (5 km × 5 km) and IMERG-F grid (0.1° × 0.1°) due to significant small-scale spatial rainfall variability in the mountainous area of Sumatra. However, the comparison between WRF and IMERG, which are grids of data, does not produce better results even though the POD value between ORG and IMERG is better than that between IMERG and WRF (Table 1). The test results show that the POD values of the three data are good enough, which indicates their potential to detect the same rain. In addition, a time series of hourly IMERG-F data shows that IMERG-F can capture the temporal trends of hourly precipitation, in comparison with point rain gauge observations [48,49]. Therefore, although caution is needed, the comparison of these three data will provide valuable information regarding the performance of each data in describing the diurnal cycle of precipitation.

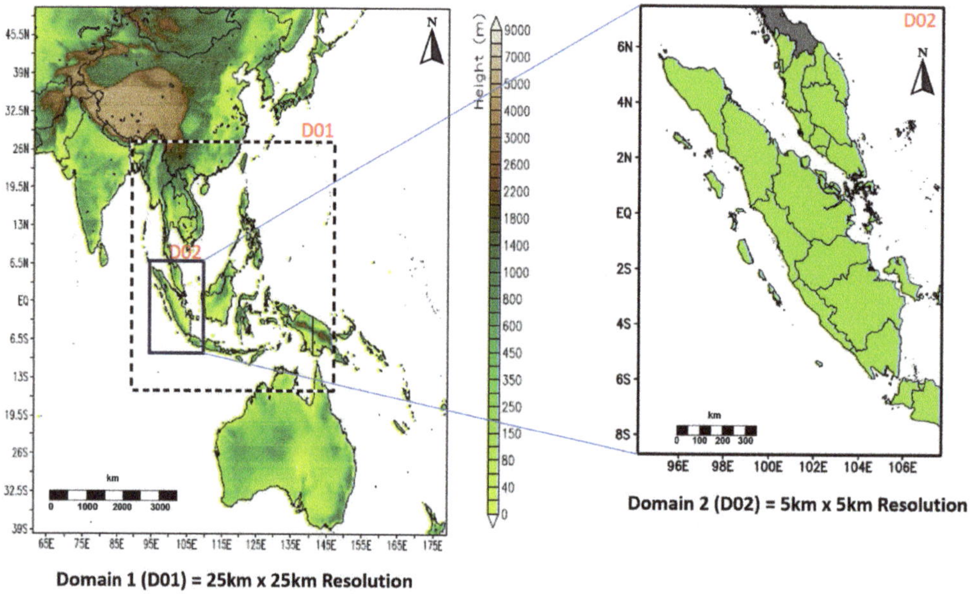

Figure 3. Spatial configuration of the domains used in the WRF simulations.

Table 1. Evaluation metrics calculated for ORG, WRF, and IMERG-F data at Kototabang for two rainfall rate (R) thresholds.

Parameters	ORG-WRF	IMERG-WRF	ORG-IMERG
Threshold: $R \geq 0.1$ Correlation coefficient (CC)	0.091	0.125	0.286
Root-mean-square error (RMSE)	5.255	2.807	3.678
Relative bias (RB)	0.313	0.179	0.123
Probability of detection (POD)	0.480	0.418	0.767
False alarm ratio (FAR)	0.799	0.554	0.698
Critical success index (CSI)	0.165	0.275	0.276
Threshold: $R \geq 0.5$ Correlation coefficient (CC)	0.088	0.117	0.279
Root-mean-square error (RMSE)	5.284	2.853	3.713
Relative bias (RB)	0.270	0.225	0.046
Probability of detection (POD)	0.394	0.336	0.642
False alarm ratio (FAR)	0.835	0.699	0.701
Critical success index (CSI)	0.132	0.189	0.256

3. Results and Discussion

3.1. Climatology of Diurnal Cycles of PA, PF, and PI

Figure 4 shows the diurnal variations in PA, PF, and PI from the ORG observation in Kototabang during 2002–2016. The average climatological PA, PF, and PI were 0.26 mm h^{-1}, 7.40%, and 3.16 mmh^{-1}, respectively, which are similar to those obtained from rain gauge data in Sumatra, as described in Marzuki et al. [11], especially for PA. About 98% of the rain gauge stations operated by the Meteorology, Climatology, and Geophysical Agency (BMKG) showed an average PA value of <0.5 mm h^{-1} [11]. Both PA, PF, and PI showed a single peak, in contrast to those found in the UK, which show two peaks [15]. Rain is more frequent in the late afternoon, which is indicated by a larger PF, but the intensity (PI) is larger in the early afternoon. The peak time of PI was observed at 13:00–14:00 LST with a peak value of 6.55 mm h^{-1}. On the other hand, peaks of PF and PA were observed at 17:00 LST, with values of 13.3% and 0.68 mm h^{-1}, respectively. The diurnal peak exhibited high PI/low PF, suggesting that both PI and PF are important indicators of PA, as is found in the Philippines [50]. However, the contribution of PF to PA is much greater than that of PI because the peak time of PA is always similar to that of PF (Figure 4).

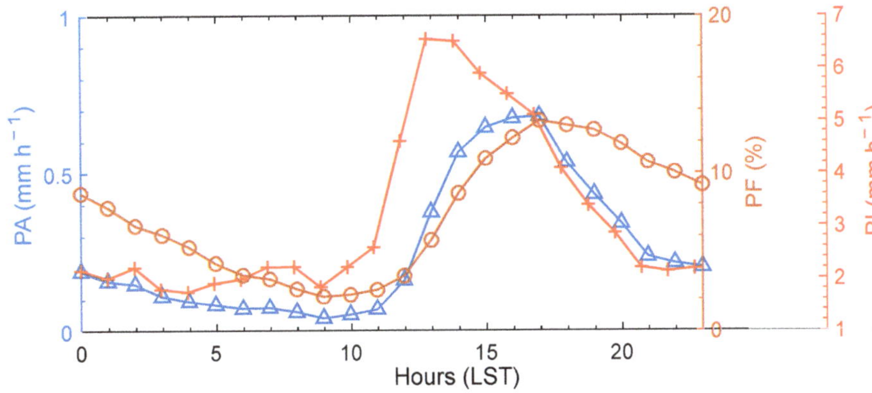

Figure 4. Climatology of diurnal variations of PA (blue), PF (merigold), and PI (red) at Kototabang during 2002–2016.

3.2. Diurnal Cycles of PA, PF, and PI with Different Durations

Rain in Kototabang is dominated by short-duration rain events (<3 h). During 2002–2016, 3996 rain events were observed, in which 3089 events (77.3%) were short-duration rain events. Furthermore, 670 (16.8%) and 237 (5.9%) were rain events of 3–6 h and >6 h in duration. Previous studies in Sumatra that were based on the BMKG rain gauge and IMERG observation data also showed the dominance of short-duration rain events, with percentages of 50–80% [18].

Figure 5 shows the diurnal variations in PA, PF, and PI for events with different durations. The peak time of short-duration rain events came earlier than that of long-duration events. For PA, rain peaks were observed at 14:00, 17:00, and 17:00 LST for <3 h, 3–6 h, and >6 h rain event durations, respectively. Moreover, for PF, peaks were observed at 15:00, 18:00, and 22:00 LST. Significant differences in peak time were not observed for PI, where, for events with durations < 3 h, peaks were observed at 13:00–14:00 LST, whereas peaks for events with durations > 6 h were observed at 15:00 LST. The dependence of the rain peak on the duration of the rain, as observed in this study, is consistent with previous studies in Sumatra [18].

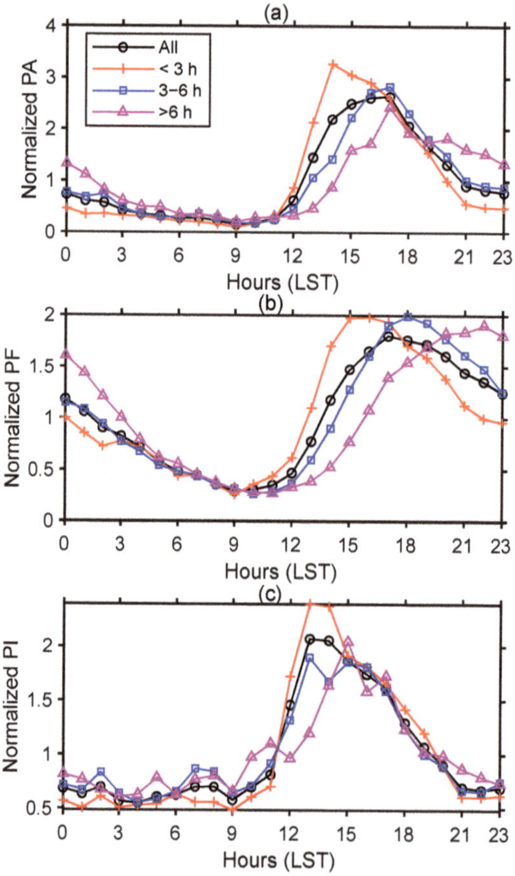

Figure 5. Diurnal variations in PA (**a**), PF (**b**), and PI (**c**) for different rain event durations. PA, PF, and PI values for each hour were divided by the average value.

The difference in rainfall peaks for each duration is consistent with the evolution of MCS clouds. Based on radar observations at Kototabang, short durations of heavy convective rain are often followed by longer durations of light rain from the stratiform portion of the cloud system [7]. The occurrence ratio between stratiform and deep convective rain is about 3:2, which contributes to the 2:3 ratio of total rainfall [7,29]. This ratio is comparable to that found in this study. The stratiform peak (long-duration rain) comes later than convective precipitation (short-duration rain), which is consistent with Figure 5.

3.3. Seasonal Changes in Diurnal Cycles of PA, PF, and PI

Figure 6 shows the PA, PF, and PI values for each month. PA and PF values in April and November were larger than those in other months. The highest PA value was observed in April at 15:00 LST with a value of 1.08 mm h^{-1}, followed by November with a PA of 1.00 mm h^{-1} at 16:00 LST. The minimum PA value was observed in July with a value of 0.50 mm h^{-1} at 15:00 LST. For PF, the highest value was observed in November with 21.7% at 22:00 LST, followed by April with 18.42% at 17:00 LST. Similarly to PA, the minimum value of PF was also observed in July, with a value of 8.66% at 17:00 LST. On the other hand, for PI, the highest value was observed in February with 9.73 mm h^{-1} at 13:00 LST, followed by August with a value of 9.62 mm h^{-1} at 13:00 LST. The two peaks of PA and PF found in this study were consistent with the monsoon season. In terms of season, the

Kototabang area can be divided into four seasons; pre-southwest (April–May), southwest (June–September), pre-northeast (October–November), and northeast (December–March). Pre-southwest and pre-northeast monsoons are wet seasons at Kototabang [34]. The two peaks of PA and PF were also associated with the southward and northward movement of the inter-tropical convergence zone (ITCZ) [51]. Consistently with a previous study [36], the peaks of PA and PF in November were much higher than those in April.

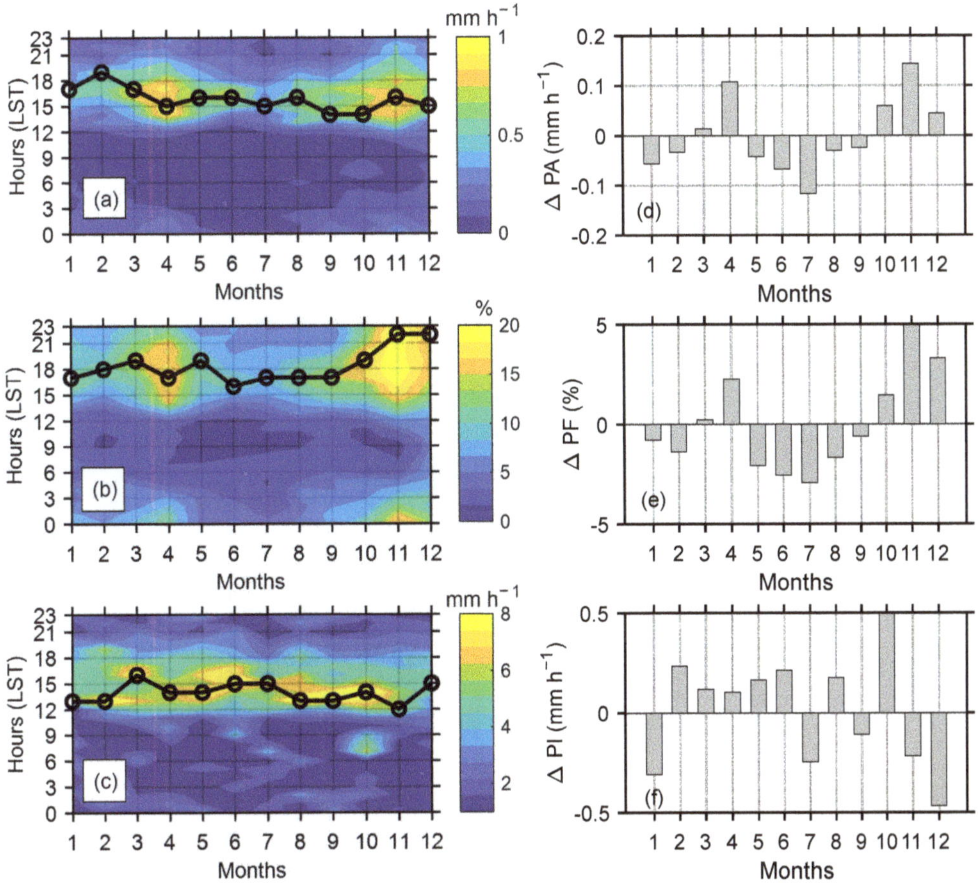

Figure 6. Monthly variations in diurnal cycles of PA (**a**), PF (**b**), and PI (**c**), as well as their differences (Δ) with average climatological values (**d**–**f**) described in Section 3.1. Lines with circle markers indicate the peak times of PA (**a**), PF (**b**), and PI (**c**).

The minimum PI was observed in the month when the PA and PF values were at the maximum. Thus, PA was more influenced by PF than PI, as also observed in Figure 4. This condition can be seen more clearly in the differences (Δ) of PA, PF, and PI values with the averages of the climatological values as described in Section 3.1. In general, when ΔPA and ΔPF are positive, the value of ΔPI is negative (Figure 6d–f).

The number of rain events also showed monthly variations, with the highest number in April and November (Figure 7). During April, 369 short-duration rain events (<3 h) were observed, accounting for 79% of the total events. Furthermore, 346 (70%) short-duration rain events were observed during November. In June and July, the number of rain events was much smaller than in April and November, but the percentages of short-duration rain events were higher, namely, 84% and 82%, respectively. The lack of medium- and long-duration run events during this period caused small PA and PF values (Figure 6d,e). The increase in the number of medium- and long-duration rain events during April and November is partially due to the favorable environmental shear conditions necessary to sustain a long-lived cloud system [2]. Seasonal variations in long-duration rainfall events had also been observed in central-eastern China [52].

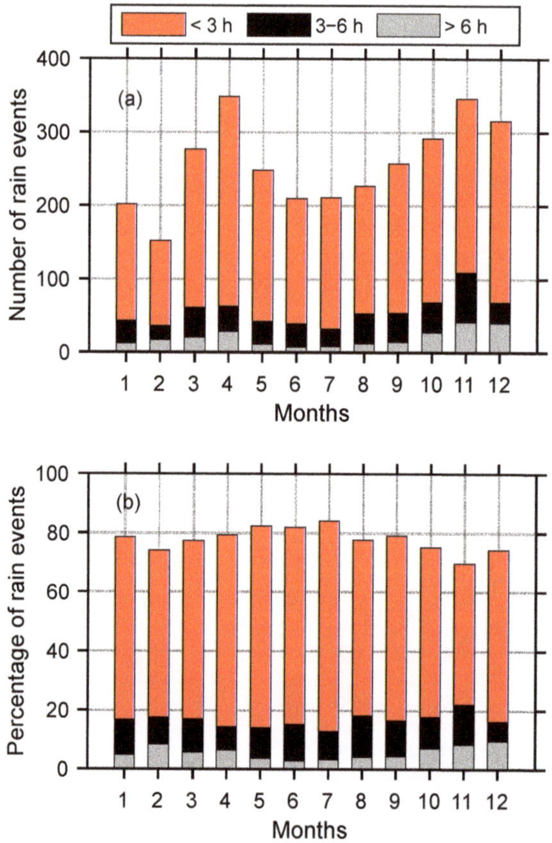

Figure 7. Monthly variations in the number (**a**) and percentage (**b**) of rain events for different rain event durations. The percentages shown in Figure 7b were calculated by dividing the number of events for each duration by the total number of events in that month.

Although there was an increase in PA and PF (Figure 6a,b), as well as the number of rain events (Figure 7) during the pre-southwest (April–May) and pre-northeast (October–November), the peak times for PA and PF did not shift significantly. PA and PF with high values were observed longer (15:00–03:00 LST the next day) (Figure 6a,b), but later peak times were only evident for PF during the pre-northeast monsoon, which is consistent with the BMKG rain gauge observations [18].

3.4. *Effect of MJO on Diurnal Cycles of PA, PF, and PI*

Figure 8 shows the effect of the MJO on the diurnal variation of precipitation in Kototabang. There was an increase in PA and PF values in phases 2, 3, and 4, with the maximum values observed in phase 3, namely, 1.21 mm h^{-1} and 21.39%, respectively. A smaller PA value was observed in phases 5 to 7, around 0.33, 0.49, and 0.40 mm h^{-1}. Such a condition was also observed for PF, with values for phases 5, 6, and 7 being 8.91%, 7.10%, and 7.48%. The average PA and PF values during phases 2, 3, and 4 were larger than the average climatological PA and PF values in Kototabang. In contrast, the average values in phases 5, 6, and 7 were smaller than the climatological values (Figure 8d,e). Although the highest PA and PF values were observed in phase 3, the PI value was relatively small in this phase (~7.15 mm h^{-1}), smaller than the climatological PA value in Kototabang (Figure 7f). During the active phase of the MJO (phases 2–4), cloud clusters (CCs), which developed in the convective envelope of a super cloud cluster (SCC) with a period of several days, mainly induced the formation of convective activities over Sumatera [3,33,53]. On the other hand, during the inactive phase of MJO (5–8), convective activities caused by local circulation were prominent at Kototabang [53].

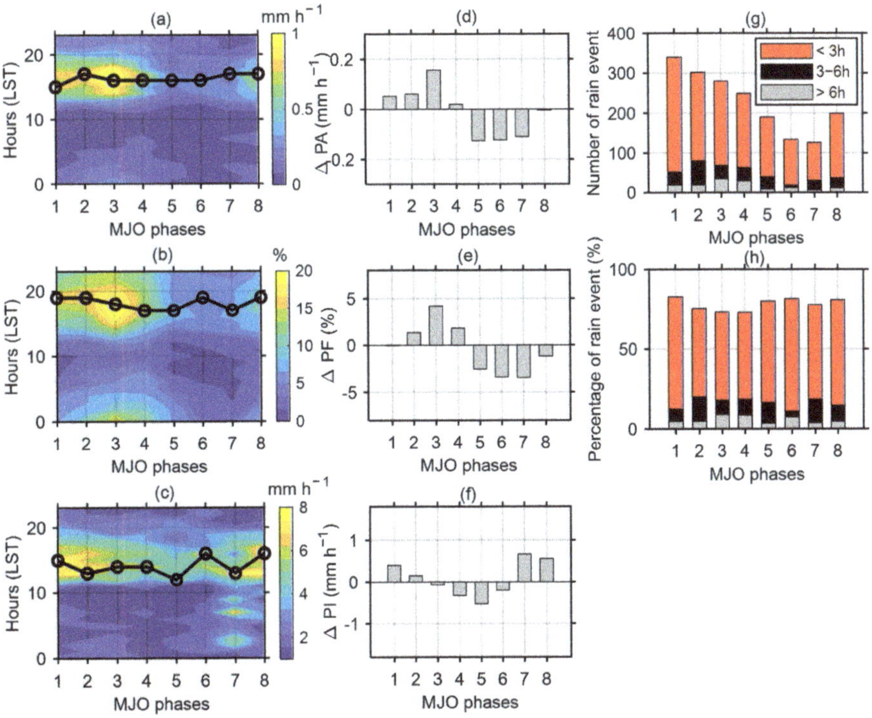

Figure 8. Intraseasonal variations in terms of MJO of the diurnal cycle of PA (**a**), PF (**b**), and PI (**c**), as well as their differences (**d**–**f**) with the average climatological values described in Section 3.1, along with the number (**g**) and percentage (**h**) of rain events for different rain event durations. Lines with circle markers indicate the peak time of PA (**a**), PF (**b**), and PI (**c**). The percentage shown in Figure 8h is calculated by dividing the number of events for each duration by the total number of events in that MJO phase.

The number of rain events in Kototabang was highest in phase 1 and decreased as the MJO phase increased (Figure 8g). The number of rain events increased again in phase 8. Although the number of rain events was huge in phase 1, rain in this phase was dominated by short-duration rain events (83%), followed by medium-duration rain events (12.44%) and long-duration rain (4.63%). The number of medium- and long-duration rain events increased in phases 2, 3, and 4, along with the increase in the convective system induced by CCs over Sumatra [3,33,53]. The percentage of short-duration rain events (<3 h) during phases 2, 3, and 4 were 75.31%, 73.11%, 73.02%, respectively. Furthermore, the percentage of rain events with durations from 3–6 h were 19.95%, 18.02%, and 18.48%. For rain events with durations > 6 h, the values were 4.74%, 8.88%, and 8.50%, respectively (Figure 8h). This feature is consistent with Figure 8b, where relatively high PF values in phases 2, 3 and 4, were observed from 12:00 LST to 5:00 LST (the next day), indicating a relatively high number of medium- and long-duration rain events. On an annual basis, MJO phases 1 occurred more often than the other phases throughout the year [54], so the number of rain events in this phase was also greater (Figure 8g).

The peak times of PA and PF were slightly different for each MJO phase. The PA peak in phase 2 was 17:00 LST, whereas for phases 3–6 the peak was 16:00 LST. In phases 7 and 8, the peak was 17:00 LST (Figure 8a). The peak time of PF varied between 17:00 and 19:00 LST. During phases 2, 3, and 4, PF peaks were observed at 19:00, 18:00, and 17:00 LST, respectively (Figure 7b). The PI peaks fluctuated more and peaks were observed at 13:00, 14:00, and 14:00 LST during phases 2, 3, and 4. During phase 8, the peak time of PI was observed at 16:00 LST (Figure 8c). Thus, the diurnal cycle of precipitation at Kototabang increased during the active MJO (phases 2–4) compared with the suppressed MJO (phases 5–8), but the diurnal phase was similar in both regimes, which was consistent with some previous studies [21,55].

Figure 9 shows the effect of MJO on PA, PF, and PI on a seasonal basis. Due to the limited number of rain events for each month, the result is displayed for the DJF, MAM, JJA, and SON periods. In general, the effect of MJO for each season was similar, with the PA and PF values during phases 1–4 being larger than phases 5–8. Furthermore, the PA and PF values during phases 1–4 were larger than the average climatological PA, in contrast to phases 5–8, in which the values were smaller than the climatological values. Slightly different conditions were observed during JJA, where, in general, PA and PF were negative, indicating a suppressed convection period (Figure 9g,h). During JJA, the number of strong MJO events (index ≥ 1) was much smaller than in other periods [54].

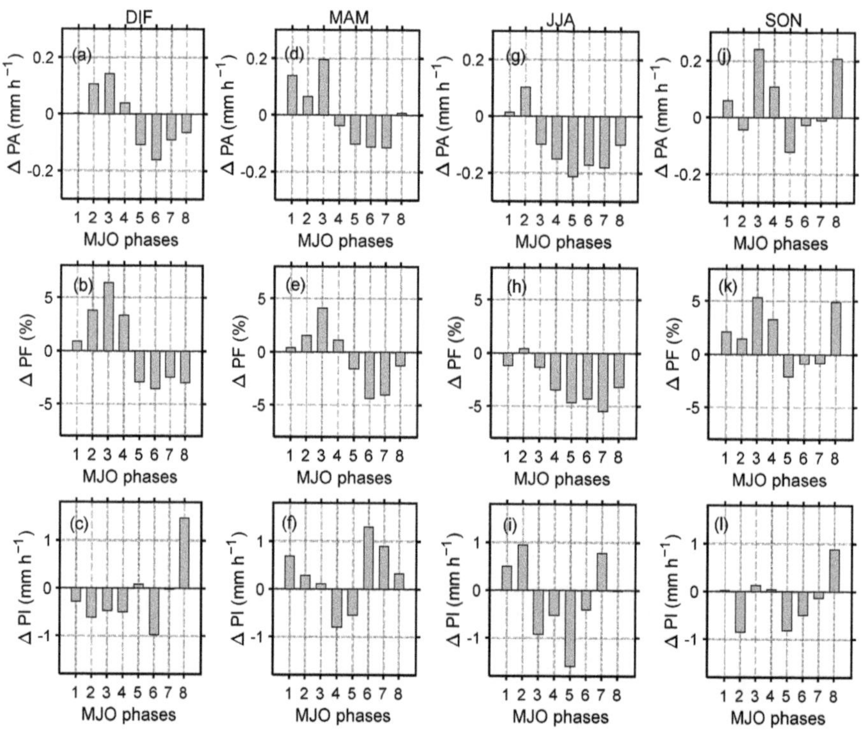

Figure 9. Same as Figure 7(**d–f**), but for DJF (**a–c**), MAM (**d–f**), JJA (**g–i**), and SON (**j–l**) periods.

3.5. Comparison of Diurnal Cycle from ORG with IMERG and WRF Model

Figure 10 shows the comparison of the diurnal cycle from the ORG data with the WRF model IMERG-F data. Since the WRF simulations were carried out only for the period of 2013–2015, the comparison to the ORG and IMERG-F data was carried out only for this limited period. The results of the WRF model were averaged for the grids of 100.2965–100.3417° E and −0.2304–0.1851° S, whereas the IMERG-F data were taken from the closest grids to Kototabang (−0.25° S and 100.35° E). The WRF model produced the same diurnal cycle phase as the ORG data. Both PA and PF peaks were observed at 16:00 LST (Figure 10a,b,d,e). The PA and PF peaks of IMERG-F were observed at 18:00 LST, which differed by two hours from those of ORG and WRF. Such differences have also been found in Sumatra using BMKG rain gauge data [18] and in China [49]. The PI values of both the WRF and IMERG-F models did not show a significant diurnal pattern, in contrast to the ORG data, which showed dominant peaks at 14:00 (Figure 10c) and at 14:00-16:00 LST (Figure 10f).

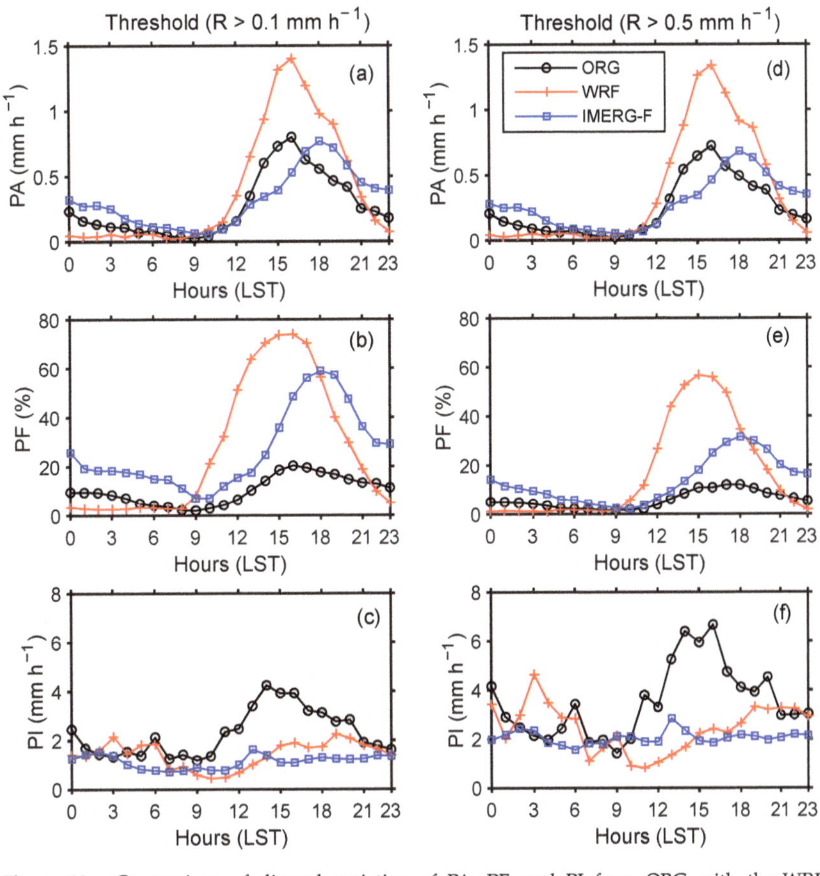

Figure 10. Comparison of diurnal variation of PA, PF, and PI from ORG with the WRF model and IMER-F data during 2013–2015. Calculations used two rainfall rate (R) thresholds: $R > 0.1$ mm h^{-1} (**a–c**) and $R > 0.5$ mm h^{-1} (**d–f**).

Although the phase of the diurnal cycle in the WRF and ORG data is the same, the amplitudes of the two are different. The PA and PF peaks of the WRF model are much larger than those of the ORG. The peaks of PA (PF) of the ORG, WRF, and IMERG data for a threshold $R > 0.1$ mm h^{-1} were 0.80 (20.36), 1.40 (73.92), and 0.77 mm h^{-1} (58.92%), respectively. The PF amplitude from the WRF model is three times larger than that of the ORG. The PA peak of ORG was close to the PA peak of IMERG. WRF's success in modeling diurnal cycle phases, as well as its failure to represent precipitation frequency and intensity, for mountainous areas have also been reported in Peru [22].

The significant difference between the diurnal cycle amplitude between WRF and ORG may be caused by several reasons. First, it may have come from a high FAR value. Although the POD value was good enough, WRF was often wrong in detecting hourly rainfall in Kototabang, as seen from the high FAR value (Table 1). This may also be the cause for the high PF value. Second, the high value of FAR may be caused by the inability of one observation point to represent rainfall variations in one WRF grid (5 km $\times$ 5 km) due to significant small-scale spatial rainfall variability in the mountainous area of Sumatra. However, WRF had a better resolution than the IMERG-F grid (0.1° $\times$ 0.1°), but the POD value from ORG vs. IMERG was better than IMERG vs. WRF (Table 1). Even the peak PA values between ORG and IMERG-F were almost the same. Thus, this difference is not solely due to the difference in sampling area between ORG and WRF but is also influenced by the performance of WRF in modeling diurnal cycles in mountainous areas.

4. Conclusions

The diurnal precipitation cycle at Kototabang presented a single peak in the late afternoon, and the peak time was closely related to the duration of rain events. The peak time of short-duration rain events came earlier than those of medium- and long-duration events, which is consistent with the evolution of the mesoscale convective system (MCS). MJO and the seasons influence the diurnal cycle of precipitation at Kototabang. This influence is more dominant on the diurnal amplitude than the diurnal phase. The mean and diurnal maximum values of precipitation amount (PA) and frequency (PF) increased significantly during the active MJO stage (phases 2–4) compared with the suppressed MJO stage (phases 5–8). Increases in PA and PF were also observed during the pre-southwest and pre-northeast monsoons. In addition to PA and PF, the number of medium- and long-duration rain events also increased. However, the phase of the diurnal cycle was similar in all MJO phases and seasons, indicating the dominant late afternoon peak of rainfall events from convective activities caused by local circulation. WRF simulation reproduced the phases of the diurnal cycle correctly, but it largely overestimated the amplitude of the diurnal cycle compared to ORG. A complex interaction between local circulation and other factors in mountainous areas may be the cause of this shortcoming. MJO and seasons may modify the local circulation at Kototabang. This issue is being investigated, and the results will be published in another journal. In addition, the WRF model testing in this study was still limited to one point of observation. More extensive testing using more rain gauge stations will also need to be carried out in the future.

Author Contributions: Conceptualization, M.M.; methodology, M.M.; software, H.Y., R.R. and A.A.B.A.; formal analysis, M.M.; investigation, M.M., H.Y., and R.R.; resources, M.M. and F.T.; data curation, M.M. and H.H.; writing—original draft preparation, M.M.; writing—review and editing, M.M., M.V., F.T., H.H., and T.S.; visualization, M.M., H.Y., R.R., and A.A.B.A.; supervision, M.M. and F.T.; project administration, H.Y.; funding acquisition, M.M. and F.T. All authors have read and agreed to the published version of the manuscript.

Funding: This study was supported by 2019–2021 Basic Research Grants from Ministry of Research, Technology and Higher Education/Ministry of Education, Culture, Research, and Technology (Contract no: T/3/UN.16.17/PT.01.03/PD-Kebencanaan/2019; T/18/UN.16.17/PT.01.03/AMD/PD-Kebencanaan/2020, and 021/E4.1/AK.04.PT/2021), Universiti Kebangsaan Malaysia grant (GUP-2019-035) and Malaysian Ministry of Higher Education Grant (LRGS/1/2020/UKM-UKM/01/6/1).

Institutional Review Board Statement: Not applicable.

Informed Consent Statement: Not applicable.

Data Availability Statement: All data used in this study are available upon request.

Acknowledgments: Optical Rain Gauges (ORGs) were operated by National Institute of Aeronautics and Space (LAPAN), Shimane University, and Kyoto University under financial support from the Ministry of Education, Culture, Sports, Science, and Technology (MEXT) of Japan. We also thank the National Aeronautics and Space Administration (NASA) for providing IMERG data.

Conflicts of Interest: The authors declare no conflict of interest.

References

1. Nitta, T.; Mizuno, T.; Takahashi, K. Multi-Scale Convective Systems during the Initial Phase of the 1986/87 El Niño. *J. Meteorol. Soc. Japan Ser. II* **1992**, *70*, 447–466. [CrossRef]
2. Marzuki, M.; Hashiguchi, H.; Yamamoto, M.K.; Yamamoto, M.; Mori, S.; Yamanaka, M.D.; Carbone, R.E.; Tuttle, J.D. Cloud episode propagation over the Indonesian Maritime Continent from 10years of infrared brightness temperature observations. *Atmos. Res.* **2013**, *120–121*, 268–286. [CrossRef]
3. Marzuki; Hashiguchi, H.; Kozu, T.; Shimomai, T.; Shibagaki, Y.; Takahashi, Y. Precipitation microstructure in different Madden-Julian Oscillation phases over Sumatra. *Atmos. Res.* **2016**, *168*, 121–138. [CrossRef]
4. Vincent, C.L.; Lane, T.P. Mesoscale variation in diabatic heating around Sumatra, and its modulation with the Madden–Julian oscillation. *Mon. Weather Rev.* **2018**, *146*, 2599–2614. [CrossRef]
5. Mori, S.; Jun-Ichi, H.; Tauhid, Y.I.; Yamanaka, M.D.; Okamoto, N.; Murata, F.; Sakurai, N.; Hashiguchi, H.; Sribimawati, T. Diurnal land-sea rainfall peak migration over Sumatera Island, Indonesian Maritime Continent, observed by TRMM satellite and intensive rawinsonde soundings. *Mon. Weather Rev.* **2004**, *132*, 2021–2039. [CrossRef]
6. Yamanaka, M.D. Physical climatology of Indonesian maritime continent: An outline to comprehend observational studies. *Atmos. Res.* **2016**, *178–179*, 231–259. [CrossRef]
7. Renggono, F.; Hashiguchi, H.; Fukao, S.; Yamanaka, M.D.; Ogino, S.-Y.; Okamoto, N.; Murata, F.; Sitorus, B.P.; Kudsy, M.; Kartasasmita, M. Precipitating clouds observed by 1.3-GHz boundary layer radars in equatorial Indonesia. *Ann. Geophys.* **2001**, *19*, 889–897. [CrossRef]
8. Wu, P.; Hamada, J.I.; Mori, S.; Tauhid, Y.I.; Yamanaka, M.D.; Kimura, F. Diurnal variation of precipitable water over a mountainous area of Sumatra Island. *J. Appl. Meteorol.* **2003**, *42*, 1107–1115. [CrossRef]
9. Wu, P.; Hara, M.; Hamada, J.I.; Yamanaka, M.D.; Kimura, F. Why a large amount of rain falls over the sea in the vicinity of western Sumatra Island during nighttime. *J. Appl. Meteorol. Climatol.* **2009**, *48*, 1345–1361. [CrossRef]
10. Tan, H.; Ray, P.; Barrett, B.; Dudhia, J.; Moncrieff, M.W. Systematic patterns in land precipitation due to convection in neighboring islands in the Maritime Continent during MJO propagation. *J. Geophys. Res. Atmos.* **2021**, *126*, e2020JD033465. [CrossRef]
11. Marzuki, M.; Suryanti, K.; Yusnaini, H.; Tangang, F.; Muharsyah, R.; Vonnisa, M.; Devianto, D. Diurnal variation of precipitation from the perspectives of precipitation amount, intensity and duration over Sumatra from rain gauge observations. *Int. J. Climatol.* **2021**, *41*, 4386–4397. [CrossRef]
12. Fukao, S.; Hashiguchi, H.; Yamamoto, M.; Tsuda, T.; Nakamura, T.; Yamamoto, M.K.; Sato, T.; Hagio, M.; Yabugaki, Y. Equatorial Atmosphere Radar (EAR): System description and first results. *Radio Sci.* **2003**, *38*, 1053. [CrossRef]
13. Fukao, S. Coupling processes in the equatorial atmosphere (CPEA): A project overview. *J. Meteorol. Soc. Japan Ser. II* **2006**, *84*, 1–18. [CrossRef]
14. Yang, G.Y.; Slingo, J. The diurnal cycle in the tropics. *Mon. Weather Rev.* **2001**, *129*, 784–801. [CrossRef]
15. Xiao, C.; Yuan, W.; Yu, R. Diurnal cycle of rainfall in amount, frequency, intensity, duration, and the seasonality over the UK. *Int. J. Climatol.* **2018**, *38*, 4967–4978. [CrossRef]
16. Suryanti, K.; Marzuki; Muharsyah, R.; Vonnisa, M.; Tangang, F. Diurnal rainfall variability in West Sumatra from rain gauge observation. *AIP Conf. Proc.* **2020**, *2221*, 090003.
17. Dai, A.; Giorgi, F.; Trenberth, K.E. Observed and model-simulasted diurnal cycles of precipitation over the contiguous United States. *J. Geophys. Res. Atmos.* **1999**, *104*, 6377–6402. [CrossRef]
18. Marzuki, M.; Helmi, Y.; Fredolin, T.; Robi, M.; Mutya, V.; Harmadi, H. Land-Sea Contrast of Diurnal Cycle Characteristics and Rain Event Propagations over Sumatra According to Different Rain Duration and Seasons. *Atmos. Res.* 2021; *final rev.*
19. Yin, S.; Chen, D.; Xie, Y. Diurnal variations of precipitation during the warm season over China. *Int. J. Climatol.* **2009**, *29*, 1154–1170. [CrossRef]
20. Rajan, D.; Pattnaik, S.; Karunasagar, S. Mean diurnal cycle of monsoon precipitation: Unified model NCUM. *Model. Earth Syst. Environ.* **2020**, *7*, 2847–2857. [CrossRef]
21. Tian, B.; Waliser, D.E.; Fetzer, E.J. Modulation of the diurnal cycle of tropical deep convective clouds by the MJO. *Geophys. Res. Lett.* **2006**, *33*, L20704. [CrossRef]
22. Mourre, L.; Condom, T.; Junquas, C.; Lebel, T.; Sicart, J.E.; Figueroa, R.; Cochachin, A. Spatio-temporal assessment of WRF, TRMM and in situ precipitation data in a tropical mountain environment (Cordillera Blanca, Peru). *Hydrol. Earth Syst. Sci.* **2016**, *20*, 125–141. [CrossRef]
23. Hara, M.; Yoshikane, T.; Takahashi, H.G.; Kimura, F.; Noda, A.; Tokioka, T. Assessment of the diurnal cycle of precipitation over the Maritime Continent simulated by a 20 km mesh GCM using TRMM PR data. *J. Meteorol. Soc. Japan Ser. II* **2009**, *87*, 413–424. [CrossRef]

24. Tan, J.; Huffman, G.J.; Bolvin, D.T.; Nelkin, E.J. Diurnal Cycle of IMERG V06 Precipitation. *Geophys. Res. Lett.* **2019**, *46*, 13584–13592. [CrossRef]
25. Marzuki, M.; Kozu, T.; Shimomai, T.; Randeu, W.L.; Hashiguchi, H.; Shibagaki, Y. Diurnal variation of rain attenuation obtained from measurement of raindrop size distribution in equatorial Indonesia. *IEEE Trans. Antennas Propag.* **2009**, *57*, 1191–1196. [CrossRef]
26. Marzuki; Hashiguchi, H.; Shimomai, T.; Randeu, W.L. Cumulative distributions of rainfall rate over Sumatra. *Prog. Electromagn. Res. M* **2016**, *49*, 1–8. [CrossRef]
27. Marzuki; Yoseva, M.; Hashiguchi, H.; Vonnisa, M.; Harmadi; Luini, L.; Nugroho, S.; Muzirwan; Shafii, M.A. Characteristics of Rain Attenuation for Microwave-to-terahertz Waveband from Raindrop Size Distribution Observation in Indonesia. In Proceedings of the Progress in Electromagnetics Research Symposium, Rome, Italy, 17–20 June 2019; Volume 2019.
28. Marzuki; Hashiguchi, H.; Vonnisa, M.; Harmadi; Muzirwan; Nugroho, S.; Yoseva, M. Z-R Relationships for Weather Radar in Indonesia from the Particle Size and Velocity (Parsivel) Optical Disdrometer. In Proceedings of the Progress in Electromagnetics Research Symposium, Toyama, Japan, 1–4 August 2018; Volume 2018.
29. Marzuki; Randeu, W.L.; Kozu, T.; Shimomai, T.; Hashiguchi, H.; Schönhuber, M. Raindrop axis ratios, fall velocities and size distribution over Sumatra from 2D-Video Disdrometer measurement. *Atmos. Res.* **2013**, *119*, 23–37. [CrossRef]
30. Optical Rain Gauge. Available online: http://www.opticalscientific (accessed on 11 November 2021).
31. Kozu, T.; Shimomai, T.; Akramin, Z.; Marzuki; Shibagaki, Y.; Hashiguchi, H. Intraseasonal variation of raindrop size distribution at Koto Tabang, West Sumatra, Indonesia. *Geophys. Res. Lett.* **2005**, *32*, L07803. [CrossRef]
32. Marzuki, F.; Kozu, T.; Shimomai, T.; Hashiguchi, H.; Randeu, W.L.; Vonnisa, M. Raindrop size distributions of convective rain over equatorial Indonesia during the first CPEA campaign. *Atmos. Res.* **2010**, *96*, 645–655. [CrossRef]
33. Shibagaki, Y.; Kozu, T.; Shimomai, T.; Mori, S.; Murata, F.; Fujiyoshi, Y.; Hashiguchi, H.; Fukao, S. Evolution of a super cloud cluster and the associated wind fields observed over the Indonesian maritime continent during the first CPEA campaign. *J. Meteorol. Soc. Japan Ser. II* **2006**, *84*, 19–31. [CrossRef]
34. Kozu, T.; Reddy, K.K.; Mori, S.; Thurai, M.; Ong, J.T.; Rao, D.N.; Shimomai, T. Seasonal and diurnal variations of raindrop size distribution in Asian monsoon region. *J. Meteorol. Soc. Japan* **2006**, *84*, 195–209. [CrossRef]
35. Madden-Julian Oscillation (MJO). Available online: http://www.bom.gov.au/climate/mjo/ (accessed on 11 November 2021).
36. Aldrian, E.; Susanto, R.D. Identification of three dominant rainfall regions within Indonesia and their relationship to sea surface temperature. *Int. J. Climatol.* **2003**, *23*, 1435–1452. [CrossRef]
37. Yu, R.; Xu, Y.; Zhou, T.; Li, J. Relation between rainfall duration and diurnal variation in the warm season precipitation over central eastern China. *Geophys. Res. Lett.* **2007**, *34*, L13703. [CrossRef]
38. Huffman, G.J.; Bolvin, D.T.; Braithwaite, D.; Hsu, K.-L.; Joyce, R.J.; Kidd, C.; Nelkin, E.J.; Sorooshian, S.; Stocker, E.F.; Tan, J. Integrated multi-satellite retrievals for the Global Precipitation Measurement (GPM) mission (IMERG). In *Satellite Precipitation Measurement*; Springer: Cham, Germany, 2020; pp. 343–353.
39. Wang, J.; Petersen, W.A.; Wolff, D.B. Validation of satellite-based precipitation products from TRMM to GPM. *Remote Sens.* **2021**, *13*, 1745. [CrossRef]
40. Ratna, S.B.; Ratnam, J.V.; Behera, S.K.; Tangang, F.T.; Yamagata, T. Validation of the WRF regional climate model over the subregions of Southeast Asia: Climatology and interannual variability. *Clim. Res.* **2017**, *71*, 263–280. [CrossRef]
41. Hong, S.-Y.; Dudhia, J.; Chen, S.-H. A revised approach to ice microphysical processes for the bulk parameterization of clouds and precipitation. *Mon. Weather Rev.* **2004**, *132*, 103–120. [CrossRef]
42. Chen, F.; Dudhia, J. Coupling an advanced land surface–hydrology model with the Penn State–NCAR MM5 modeling system. Part I: Model implementation and sensitivity. *Mon. Weather Rev.* **2001**, *129*, 569–585. [CrossRef]
43. Betts, A.K.; Miller, M.J. A new convective adjustment scheme. Part II: Single column tests using GATE wave, BOMEX, ATEX and arctic air-mass data sets. *Q. J. R. Meteorol. Soc.* **1986**, *112*, 693–709.
44. Janjić, Z.I. The step-mountain eta coordinate model: Further developments of the convection, viscous sublayer, and turbulence closure schemes. *Mon. Weather Rev.* **1994**, *122*, 927–945. [CrossRef]
45. Mlawer, E.J.; Taubman, S.J.; Brown, P.D.; Iacono, M.J.; Clough, S.A. Radiative transfer for inhomogeneous atmospheres: RRTM, a validated correlated-k model for the longwave. *J. Geophys. Res. Atmos.* **1997**, *102*, 16663–16682. [CrossRef]
46. Dudhia, J. Numerical study of convection observed during the winter monsoon experiment using a mesoscale two-dimensional model. *J. Atmos. Sci.* **1989**, *46*, 3077–3107. [CrossRef]
47. Xu, R.; Tian, F.; Yang, L.; Hu, H.; Lu, H.; Hou, A. Ground validation of GPM IMERG and TRMM 3B42V7 rainfall products over southern Tibetan Plateau based on a high-density rain gauge network. *J. Geophys. Res. Atmos.* **2017**, *122*, 910–924. [CrossRef]
48. Yusnaini, H.; Ramadhan, R.; Marzuki, M.; Ningsih, A.P.; Hashiguchi, H.; Shimomai, T.; Vonnisa, M.; Harmadi, H.; Suryanto, W.; Sholihun, S. Statistical Comparison of IMERG Precipitation Products with Optical Rain Gauge Observations over Kototabang, Indonesia. *J. Ilmu Fis. Univ. Andalas* **2022**, *14*, 10–20.
49. Tang, S.; Li, R.; He, J.; Wang, H.; Fan, X.; Yao, S. Comparative evaluation of the GPM IMERG early, late, and final hourly precipitation products using the CMPA data over Sichuan Basin of China. *Water* **2020**, *12*, 554. [CrossRef]
50. Hilario, M.R.A.; Olaguera, L.M.; Narisma, G.T.; Matsumoto, J. Diurnal characteristics of summer precipitation over Luzon Island, Philippines. *Asia-Pac. J. Atmos. Sci.* **2020**, *57*, 573–585. [CrossRef]
51. Waliser, D.E.; Gautier, C. A satellite-derived climatology of the ITCZ. *J. Clim.* **1993**, *6*, 2162–2174. [CrossRef]

52. Yuan, W.; Yu, R.; Chen, H.; Li, J.; Zhang, M. Subseasonal characteristics of diurnal variation in summer monsoon rainfall over central eastern China. *J. Clim.* **2010**, *23*, 6684–6695. [CrossRef]
53. Seto, T.H.; Yamamoto, M.K.; Hashiguchi, H.; Fukao, S. Convective activities associated with intraseasonal variation over Sumatera, Indonesia, observed with the equatorial atmosphere radar. *Ann. Geophys.* **2004**, *22*, 3899–3916. [CrossRef]
54. Lafleur, D.M.; Barrett, B.S.; Henderson, G.R. Some climatological aspects of the Madden–Julian Oscillation (MJO). *J. Clim.* **2015**, *28*, 6039–6053. [CrossRef]
55. Suzuki, T. Diurnal cycle of deep convection in super clusters embedded in the Madden-Julian Oscillation. *J. Geophys. Res. Atmos.* **2009**, *114*, D22102. [CrossRef]

atmosphere

Article

Variability of Precipitation Recycling and Moisture Sources over the Colombian Pacific Region: A Precipitationshed Approach

Angelica M. Enciso [1,2,*], Olga Lucia Baquero [3], Daniel Escobar-Carbonari [1], Jeimar Tapasco [1] and Wilmar L. Cerón [4]

1 Climate Action, Alliance of Bioversity International and the International Center for Tropical Agriculture (CIAT), Palmira 763537, Colombia; d.escobar@cgiar.org (D.E.-C.); j.tapasco@cgiar.org (J.T.)
2 Postgraduate Program on Integrated Water Resources Management, Faculty of Engineering, Universidad del Valle, Cali 760032, Colombia
3 Cinara Institute, Faculty of Engineering, Universidad del Valle, Cali 760032, Colombia; olga.baquero@correounivalle.edu.co
4 Department of Geography, Faculty of Humanities, Universidad del Valle, Cali 760032, Colombia; wilmar.ceron@correounivalle.edu.co
* Correspondence: a.enciso@cgiar.org; Tel.: +57-312-766-02-35

Abstract: This study assessed the precipitation recycling and moisture sources in the Colombian Pacific region between 1980–2017, based on the monitoring of moisture in the atmosphere through the Eulerian Water Accounting Model-2 layer (WAM2 layer) and the delimitation of the area contributing to terrestrial and oceanic moisture in the region is performed using the "precipitationshed" approach. The results indicate a unimodal precipitation recycling ratio for the North and Central Pacific and Patía-Mira regions, with the highest percentages between March and April, reaching 30% and 34%, respectively, and the lowest between September and October (between 19% and 21%). Moreover, monthly changes in the circulation of the region promote a remarkable variability of the sources that contribute to the precipitation of the study area and the spatial dynamics of the precipitationshed. From December to April, the main contributions come from continental sources in eastern Colombia and Venezuela, the tropical North Atlantic, and the Caribbean Sea, a period of high activity of the Orinoco Low-Level jet. In September, the moisture source region is located over the Pacific Ocean, where a southwesterly cross-equatorial circulation predominates, converging in western Colombia, known as the Choco Jet (CJ), decreasing the continental contribution. An intensified Caribbean Low-Level Jet inhibits moisture sources from the north between June and August, strengthening a southerly cross-equatorial flow from the Amazon River basin and the southeastern tropical Pacific. The March–April (September–October) season of higher (lower) recycling of continental precipitation is related to the weakening (strengthening) of the CJ in the first (second) half of the year, which decreases (increases) the contribution of moisture from the Pacific Ocean to the region, increasing (decreasing) the influence of land-based sources in the study area.

Keywords: moisture transport; evaporation; precipitation; Pacific region; WAM-2 layer; precipitationshed

Citation: Enciso, A.M.; Baquero, O.L.; Escobar-Carbonari, D.; Tapasco, J.; Cerón, W.L. Variability of Precipitation Recycling and Moisture Sources over the Colombian Pacific Region: A Precipitationshed Approach. *Atmosphere* **2022**, *13*, 1202. https://doi.org/10.3390/atmos13081202

Academic Editors: Zuohao Cao, Huaqing Cai and Xiaofan Li

Received: 11 June 2022
Accepted: 23 July 2022
Published: 30 July 2022

1. Introduction

As a result of the interactions between the atmosphere and the Earth's surface, precipitation and evaporation processes are generated. Precipitation over the land surface originates mainly from two mechanisms, advection of water vapor from the oceans and evaporation of surface moisture, whose contribution to the total precipitation in a region is called recycled precipitation [1–3]. Most surface hydrology studies are based on analyzing the division of precipitation between runoff and evaporation, describing how water molecules fall from the atmosphere to the surface. The concept of recycled precipitation describes a similar division. Still, in this case, it does not analyze where precipitation falls but rather the origin of the water vapor molecules that form it [4]. Although the source of precipitation is difficult to establish due to the variability of the processes that produce it, it is a fundamental factor in understanding the role of the hydrological cycle in the climate system. According to Van Der Ent [5], it is estimated that, on average, 40% of terrestrial precipitation comes from terrestrial evaporation, and 57% of this returns as precipitation over land.

In recent years, several researchers have developed new definitions of moisture recycling for studies on continental moisture feedback processes [5–8]. Among which is the one developed by Keys [8], which introduces the concept of "precipitationshed" or "atmospheric basins", defined as the surface area of both water and land that provides evaporation to the precipitation of a specific region. This concept has been used to highlight areas where livelihoods depend on rainfed agriculture and where changes in the land use in precipitationshed could have significant consequences on society. The importance of the "atmospheric basins" approach is the inclusion of evaporation input in distant areas from the precipitation of a specific region, thus establishing hydroclimatological connections between remotely separated areas, which often overwhelm political–administrative divisions [9,10]. Therefore, terrestrial sources and water vapor sinks represent a particular interest in the hydrological analysis, as well as the understanding of moisture fluxes responsible for the transport of water vapor, a phenomenon that occurs on several temporal and spatial scales that generates a spatiotemporal redistribution of precipitation over the planet [11]. For example, in some regions of South America, precipitation recycling has been explored finding out that varies substantially in dry and wet years and under the influence of phenomena such as the North Atlantic Oscillation (NAO) or El Niño Southern Oscillation (ENSO) [4,10,12–20].

Otherwise, Van Der Ent [7] indicates that in the tropics and mountainous lands, the scale of the recycling process can be from 500 to 2000 km, and the timescale ranges vary from 3 to 20 days, except for deserts, where it is much longer. In this regard, Cuartas et al. [2], evidenced that in the Amazon the moisture recycling can represent between 35% and 50% of the total precipitation where the terrestrial recycling of moisture is essential in sustaining regional precipitation [13,21], whereas Satyamurti et al. [12] identified that wet years show about 55% more moisture convergence than dry years in the Amazon basin. A reduction in moisture inflow across the eastern and northern boundaries of the basin and an increase in outflow across the southern border at 15° S led to drier conditions. For his part, Martinez et al. [20] estimated the variability of moisture sources in the La Plata River Basin (LPB) using an extended version of the Dynamic Recycling Model, finding that 37% of the mean annual precipitation over the LPB comes from the South Pacific and Atlantic tropical oceans. The remaining 63% comes from South America, including 23% from local sources in the LPB and 20% from the southern Amazon.

In Colombia, several researchers have tried to explain some of the climatic and hydrological phenomena related to the interaction between the atmosphere and the surface, finding that the climate of the country, mainly in the center and western part, is strongly influenced by the physical interactions that occur in the Pacific Ocean, where moisture influx occurs at the lowest pressure levels (below 800 hPa), mainly due to surface winds or jets such as the Choco Jet (CJ) located at 925 hPa [22–24], which transports large amounts of moisture to the region. According to Jaramillo et al. [25], this jet contributes approximately 57% of the total precipitation in western Colombia and the Gulf of Panama. In addition, Gallego et al. [18] indicate that the CJ is deeply related to the dynamics of the Intertropical Convergence Zone (ITCZ) in the eastern equatorial Pacific and is responsible for up to 30% of the total precipitation in central and northern South America. Furthermore, in addition to the significant influence of the humid oceanic masses on the Pacific region's climate, it has been determined that the Andes Mountain range is the main determining factor of the geographic, physical, biological, and hydrological configuration of the Pacific sub ecosystem. According to Velásquez-Restrepo and Poveda [26], these characteristics make the Pacific an ideal space for understanding the functioning and integrity of the hydrological cycle, mainly on atmospheric interactions with oceanic and surface processes.

Cuartas [27] and Cuartas and Poveda [2] indicated that the average value of atmospheric moisture input in Colombia is 5716 mm/year, with an important variability during ENSO phases, coming mainly from easterly and westerly trade winds. On the other hand, they found that, on average, 31% of the total precipitation in the country is due to evaporated moisture from the surface, and for the Pacific region, it is about 18%. Meanwhile, Hoyos et al. [10] evaluated the sources and processes of moisture transport for Colombia, finding that moisture from the Atlantic Ocean and terrestrial recycling are the main sources of moisture in the country; they also found that the CJ plays an important role in the convergence of moisture over western Colombia. Arias et al. [16] examined the primary sources of moisture during La Niña 2010–2012, finding that the main sources were the Pacific Ocean through the CJ and the Caribbean Sea, through the weakening of the Caribbean Low-Level Jet (CLLJ) and the development of southward anomalies toward northern South America.

Quantifying the terrestrial evaporation which feeds precipitation over land, i.e., the magnitude of recycled moisture, is of great importance in the analysis of the interactions and feedbacks between surface and atmospheric hydrology, being a possible indicator of climate sensitivity to land use changes [1,4,6,28], which is essential to understand the impact of anthropogenic activities on climate, as they can modify terrestrial moisture fluxes through changes in land use and water management.

The understanding of the interactions between precipitation and evaporation has changed over time. There are several models for calculating recycled precipitation; however, most of them are modifications of the generalized one-dimensional model of Budyko [29], which expresses the percentage of recycling in the direction of a single stream of velocity u and length l and as a function of regional evaporation (E) and atmospheric moisture transport (Q). This model was modified by Brubaker et al. [1] by considering the flow input to a region in two directions, for his part Eltahir and Bras [4] proposed a model that calculates the local recycling ratio on a spatially distributed grid for monthly or longer time scales, and Dominguez et al. [30] developed a model derived from the two-dimensional atmospheric water balance equation, which estimates the local recycling ratio for daily or longer time scales and only uses the hypothesis of a well-mixed atmosphere. Finally, Van Der Ent et al. [6] proposed a numerical model based on the atmospheric water balance, estimating the recycling ratio daily. The latter model labels each water particle to be traced back to the origin, determining the spatio-temporal distribution of the moisture origin rather than simply estimating the recycling ratio on a large spatial and temporal scale [7].

Among the most widely used moisture tracking models are Lagrangian models such as the 2D Dynamic Recycling Model (DRM) developed by Dominguez et al. [30], the Quasi-Isentropic 3D Backward Trajectory (QIBT) method by Dirmeyer and Brubaker [31], and others such as FLEXPART and HYSPLIT [7]. Furthermore, the Eulerian models allow the tracking of moisture on a global scale [6,7] and are highlighted by the speed of the calculation due to their simplicity, but also due to their Eulerian grid, which allows them to track the origin of moisture quickly in both large and small areas. An example of such an Eulerian trajectory model is the WAM-2 layers (Water Accounting Model-2 layers) [7].

Therefore, considering the complexity of the Colombian Pacific region, this study aims to identify the recycled precipitation and moisture sources in the study area by tracking atmospheric moisture through the Eulerian Water Accounting Model-2 layer (WAM-2 layer). This method estimates the recycling ratio of continental precipitation and the contribution of terrestrial moisture sources to the study area's precipitation. Additionally, it delimits the area that contributes both terrestrial and oceanic moisture to the precipitation of the Colombian Pacific under the "precipitationshed" approach implemented by Keys [8]. This study contributes to the appropriate management and conservation of water resources in the Colombian Pacific region, increasing knowledge on land–atmosphere feedback and ocean–atmosphere feedback.

2. Materials and Methods

The percentage of recycled precipitation might be local (ρ), regional (ρ_r) and continental (ρ_c) [4,6,13,21,28,32]. Here, we define surface precipitation as a combination of oceanic and terrestrial moisture sources; this approach makes it possible to analyze the influence of evaporation from remote locations on the precipitation in a particular geographical area. Precipitation is defined as:

$$P(t,\ x,\ y) = P_c(t,\ x,\ y) + P_o(t,\ x,\ y) \tag{1}$$

where P_c is continental-sourced precipitation (evaporated from a continent region) and P_o is ocean-sourced precipitation (evaporated from the ocean). The recycling ratio of continental precipitation, which shows the dependence of precipitation at a given location relative to continental evaporation, is provided as:

$$\rho_c(t,\ x,\ y) = \frac{P_c(t,\ x,\ y)}{P\ (t,\ x,\ y)} \tag{2}$$

WAM-2 layers perform 2D (x, y) moisture tracking, globally and regionally, following the moisture backward (evaporation that will be precipitated in a predefined region) and forwards (precipitation that evaporated from a predefined region) in time. The model splits the atmosphere into two layers making moisture tracking more reliable than vertically integrated moisture fluxes [7]. Van Der Ent et al. [7] showed that the WAM-2 layer model provides similar results to a complex and highly detailed moisture tracking scheme in a regional climate model (RCM-tag) [33] but with a lower computational effort [34]. The WAM single-layer model struggled to estimate the correct moisture flux direction in this case study. However, by adding another layer, the results were closer to the RCM-tag method that directly uses highly accurate three-dimensional water tracking (including phase transitions) within a regional climate model.

The shear layer is approximately at the sigma level, corresponding to around 800 hPa. The horizontal moisture fluxes in the lower layer are calculated between the surface and the sigma level, while the horizontal moisture fluxes in the upper layer are calculated from sigma level to 175 hPa. Moreover, the vertical velocity given at the sigma level of around 800 hPa was used to calculate the moisture transport between the lower and upper layers. On the preceding basis, moisture is tracked from where it enters to where it leaves the atmosphere as evaporation and precipitation. Therefore, it is possible to identify when and where precipitation from a specific region entered the atmosphere as evaporation as time progresses.

The WAM-2 layer model is open access and requires input data with a daily resolution specified in Table 1. These input data were taken from the ERA-Interim climate reanalysis project dataset provided by the European Centre for Medium-Range Weather Forecasts Interim Reanalysis (ECMWF/ERA-I) [35,36]. According to Hoyos [10], the ECMWF/ERA-I dataset has a good qualitative, and quantitative representation of the Colombian climate features; depicting more realistically the regional orography when compared to the first-generation reanalysis data (NCEP/NCAR and ERA-40), allowing a better representation of regional humidity and atmospheric transport [10,35,37]. Furthermore, previous studies have used this dataset, presenting results suitable for the analysis of climate variability at interannual and decadal scales which affect Colombia [16,38,39]. All variables were obtained for the 1980–2017 period in the region 20° N–15° S and 120° W–60° W, with a horizontal resolution of 0.75° × 0.75°. The moisture source and the average recycling ratio were estimated over the same period. Precipitation and evaporation data at 3 h intervals, specific humidity, zonal and meridional wind velocity at 24 pressure levels (175–1000 hPa) (as required by WAM-2 layer), and surface pressure were used to calculate the vertically integrated moisture flux and precipitable water every 6 h.

Table 1. Variables are included in the WAM-2 layer, with the units described in ERA-I.

	Surface Variables	
Symbol	**Variable**	**Unit**
P	Precipitation	m·day^{-1}
E	Evaporation	m·day^{-1}
TCW	Total column water	kg·m^{-2}
TCWV	Total column of water vapor	kg·m^{-2}
EWVF	Vertical integral of eastward water vapor flux	kg·m^{-1}·s^{-1}
NWVF	Vertical integral of northward water vapor flux	kg·m^{-1}·s^{-1}
ECLWF	Vertical integral of eastward cloud liquid water flux	kg·m^{-1}·s^{-1}
NCLWF	Vertical integral of northward cloud liquid water flux	kg·m^{-1}·s^{-1}
ECFWF	Vertical integral of eastward cloud solid water flux	kg·m^{-1}·s^{-1}
NCFWF	Vertical integral of northward cloud solid water flux	kg·m^{-1}·s^{-1}
p_s	Surface pressure	Pa
	Pressure levels variables	
q	Specific humidity	kg·kg^{-1}
u	Zonal component of wind	m·s^{-1}
v	Meridional component of wind	m·s^{-1}

The fundamental principle for the WAM-2 layers numerical model is the atmospheric water balance:

$$\frac{\partial W_k}{\partial t} = \frac{\partial (W_k u)}{\partial x} + \frac{\partial (W_k v)}{\partial y} + E_k - P_k + \xi_k \pm Q_V \quad (3)$$

where W_k is the atmospheric moisture storage (precipitable water) in layer k (either the upper or lower layer), E_k is evaporation flowing into layer k, P_k is precipitation leaving layer k, ξ is an essential residual to correct the precipitation or evaporation, Q_V is the vertical moisture transport between the lower and upper layers, x is the longitudinal direction (zonal component) and y is the latitudinal direction (meridional component). Moisture transport is calculated over grid cell boundaries. The change in atmospheric moisture due to horizontal transport is defined by:

$$\frac{\Delta(Wu)}{\Delta x} = F_{k,x}^{-} - F_{k,x}^{+}, \ \frac{\Delta(Wv)}{\Delta x} = F_{k,y}^{-} - F_{k,y}^{+} \quad (4)$$

where F_k is the moisture flow over the boundary of a grid cell in the upper or lower layer, which is positive from west to east and from south to north. The superscript "−" represents the west and south boundaries of the grid cell and "+" represents the east and north boundaries. The vertically integrated moisture flux (F_k) is calculated as follows:

$$F_k = \frac{L}{g\rho} \int_{ptop}^{p_{bottom}} q u_h dp \quad (5)$$

where L is the length of perpendicular cell to the moisture flow direction, g is the gravity, ρ is the density of liquid water (1000 kg·m^{-3}), p is the pressure, q specific humidity and u_h is the horizontal component in either x or y direction. For the top layer, applies $ptop = 0$ and $pbottom = pdivide$. For the bottom layer applies $ptop = pdivide$ and $pbottom = psurface$. Where $pdivide$ is the pressure between the upper and lower layer, which corresponds to 81,283 Pa at a standard surface pressure of 101,325 Pa and can be calculated as follows [5]:

$$p_{divide} = 7438.803 + 0.728786 \times p_{surface}\ [Pa] \quad (6)$$

Over the surface, the bottom layer represents about 40–80% of the total moisture storage in the column, and 30–70% of the total horizontal moisture flow [5]. Evaporation E enters only the lower layer, so $E_k = E$ in this layer while $E_k = 0$ in the upper layer.

Precipitation is assumed to be removed immediately from moisture storage (there is no downward precipitation exchange between the upper and lower layer), and the "well-mixed atmosphere" condition is assumed for precipitation:

$$P_k = P\frac{W_k}{W} \tag{7}$$

where P is the total precipitation and W is the total atmospheric storage in the vertical.

The residual ξ is the result of the assimilation of the ERA-Interim dataset and the fact that the decoupled tracking scheme calculates the water balance at gross spatial and temporal resolution. The vertical moisture transport Q_V is difficult to calculate because, in addition to the transport by the mean wind velocity in the vertical, there is a scattering moisture exchange due to the convective scheme in ERA-Interim. Therefore, it is assumed that the vertical exchange is the closing term of the water balance. However, due to the residual ξ the water balance cannot be completely closed. Therefore, the closure is defined by the ratio of upper and lower layer residuals, which is proportional to the moisture content of the layers:

$$\frac{\xi_{top}}{W_{top}} = \frac{\xi_{bottom}}{W_{bottom}} \tag{8}$$

From the above equation, the vertical moisture transport can be calculated as:

$$Q_V = \frac{W_{bottom}}{W}\left(\xi^*_{bottom} + \xi^*_{top}\right) - \xi^*_{bottom} \tag{9}$$

where ξ^*_{bottom} and ξ^*_{top} are the residuals to be considered before vertical transport.

The same water balance is applied for tracking moisture from a given origin (continental, regional or local) in WAM-2 layers. For example, the water balance of the evaporation to be tracked (identified by the subscript Ω) in the lower layer of the atmosphere for forwarding tracking (trajectory of moisture from where it evaporates to where it precipitates) is described by:

$$\frac{\partial W_{\Omega,\ bottom}}{\partial t} = \frac{\partial(W_{\Omega,bottom}u)}{\partial x} + \frac{\partial(W_{\Omega,bottom}v)}{\partial y} + E_\Omega - P_\Omega + \xi_\Omega \pm Q_{V,\Omega} \tag{10}$$

For backward tracking (trajectory of moisture from where it precipitates to where it evaporates) and upper layer calculation, equations similar to the above are used. Van Der Ent [7] found that the vertical flow was too small to adequately represent the vertical transport of the monitored water, which is attributed to the turbulent moisture exchange between the upper and lower layer. Based on Van Der Ent's trial-and-error tests [7], this situation was resolved by keeping Q_V as the net vertical moisture flux and using a vertical flux of $4Q_V$ in the net flow direction and $3Q_V$ in the opposite direction during the tagging experiments. Despite simplifying turbulent moisture exchange, the authors considered it a suitable parameterization for the study, and their results were not very sensitive to turbulent moisture exchange. For further details on the application of this model, refer to https://github.com/ruudvdent/WAM2layersPython (accessed on 22 July 2022). The specific structure of commands applied for the model execution and the division of the variables into atmospheric layers can be explicitly found at https://github.com/ruudvdent/WAM2layersPython/blob/master/Fluxes_and_States_Masterscript.py (accessed on 22 July 2022).

For this study, the Colombian Pacific region was selected as the unit of analysis based on the seasonality of the monthly mean precipitation in Colombia as defined by the Institute of Hydrology, Meteorology and Environmental Studies (IDEAM) [40] (Figure 1a,b). Out of the 12 regions defined by IDEAM, we used the North and Central Pacific (NCP) and Patía-Mira (P-M) regions covering the Colombian Pacific, as shown in Figure 1a. To determine the recycled precipitation in the Colombian Pacific region, moisture was tracked forward in time, which allows for determining the amount (percentage) of the total precipitation from the region that originated from the evaporation of continental sources.

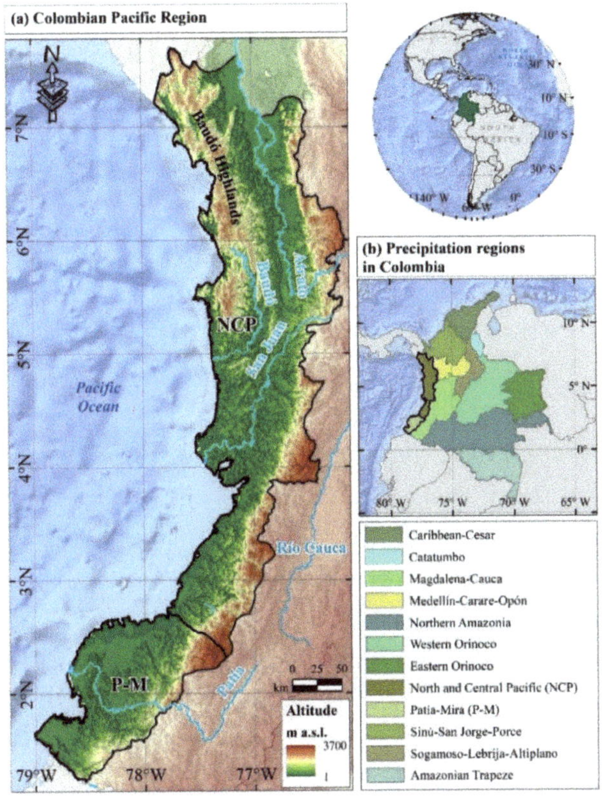

Figure 1. (**a**) Study area and (**b**) precipitation regions in Colombia provided by IDEAM [40] file number 20199050007812 from 11 February 2019.

In this case, the continental precipitation recycling ratio (ρ_c) was estimated, as described in Equation (2), whose numerical implementation in the model is given by:

$$\rho_c = \frac{\sum_{t=t_0}^{t_{final}} P \frac{W_{con}}{W}}{\sum_{t=t_0}^{t_{final}} P} \tag{11}$$

where, W_{con} is the atmospheric moisture storage of continental origin. The storage of atmospheric moisture of continental origin is the humidity coming from the evapotranspiration of the continental (terrestrial) zone. In the model, a layer or mask is added to define the oceanic and continental region of the study area, which in this case is slightly lower than the region where the information is downloaded and corresponds to the 15° N–6.75° S and 98.25° W–60.75° W. The region where the forward trajectory of the moisture is calculated and where the spatial distribution of the evaporated precipitation is obtained corresponds to the above-mentioned region. For the backward trajectory (where the spatial distribution of the evaporation that will be precipitated (precipitationshed) is obtained from the model and based on the loaded layer, the location of the study region (Colombian Pacific region) is specified.

To calculate the precipitationshed or source region of precipitation in the Pacific, we use the adaptation carried out by Keys et al. [41] to the WAM-2 layer, which allows backtracking of precipitation from the source region (in our case, the Pacific region), to identify the sources of evaporation or humidity. The backtracking method is based on the following approach:

$$P_{\Omega}(t,\ x_{\Omega}, y_{\Omega},\ A_{\Omega},\ \Delta_{\Omega}) = \int_{i=0}^{p}\int_{j=0}^{m} E_{\Omega}(t,\ x_i, y_i) \quad (12)$$

where P_{Ω} is the precipitation in the sink region Ω (defined by longitude x_{Ω}, latitude y_{Ω}, area A_{Ω} and shape Δ_{Ω}). Specifically, the amount of evaporation E_{Ω} reaching the region Ω, which traveled through the atmosphere, and ends up as precipitation in that region, is calculated for each cell. E_{Ω} is integrated over all grid cells, where i and j are the cell indices and p and m are the cell numbers along the parallel and meridian, respectively. Its numerical implementation in the model is given by:

$$E_{\Omega} = \sum_{t=t_0}^{t_{final}} E\frac{W_{bottom}}{W} \quad (13)$$

where, W_o is the moisture storage tracked or labeled in the lower layer, and W is the total moisture in all layers. To find the precipitationshed with WAM-2 layer, a backward tracking in time is performed from the region defined as a moisture sink, which in this study corresponds to the Colombian Pacific region, whose result allows identifying the area contributing the highest percentage to the moisture that precipitates in this region. Nevertheless, the results of the model on their own do not correspond to the precipitationshed, because each cell, even if it is quite small or distant from the sink area, can contribute to the evaporation estimate; therefore, a threshold must be established to define a spatially explicit boundary of the precipitationshed, based on the evaporation contribution. In this case, the threshold was defined as the cells that contribute 80% of evaporation or humidity to the precipitation over the Colombian Pacific; this is based on the topographic limits of the country that directly affect the behavior of the airflow and on the usefulness of these limits for future analysis of vulnerability to changes in land use and land cover.

Moreover, indices of water balance variables, precipitation, evapotranspiration, and moisture convergence (moisture divergence multiplied by −1, hereinafter ConvQ), corresponding to the mean for each region, NCP and P-M separately, were constructed. To examine the variability of the continental recycled precipitation and its relationship with the winds coming from the Pacific Ocean, the CJ index was used as the mean of the zonal winds at 925 hPa in the region between 2° N and 7° N along the 80° W meridian. Several authors used the zonal winds at 925 hPa in northern South America to characterize the CJ and the associated moisture transport on monthly and seasonal scales [15,16,24,38,42].

3. Results and Discussion

3.1. Water Balance

The spatio-temporal variability of water balance variables, including precipitation (P), evapotranspiration (E), moisture convergence (ConvQ), and the ratio between the precipitation and evapotranspiration (E/P) at annual and monthly scales for the NCP and P-M regions between 1980–2017 are presented below.

Figure 2a shows the total annual precipitation over the Colombian Pacific, with mean values between 10,000 and 12,000 mm·yr^{-1} in the northeast of the study area and precipitation close to 9000 mm·yr^{-1} at the border of the NCP and P-M regions, while in the south the annual precipitation oscillates between 3000 and 7000 mm. The mean annual precipitation in the NCP region was 9611 mm·yr^{-1}, and in P-M, it was 5397 mm·yr^{-1}, exceeding the country's annual average of 3189 mm·yr^{-1} reported by Vallejo [11] and consistent with Cerón et al. [43]. They observed three cores of high precipitation over the Colombian Biogeographic Chocó region.

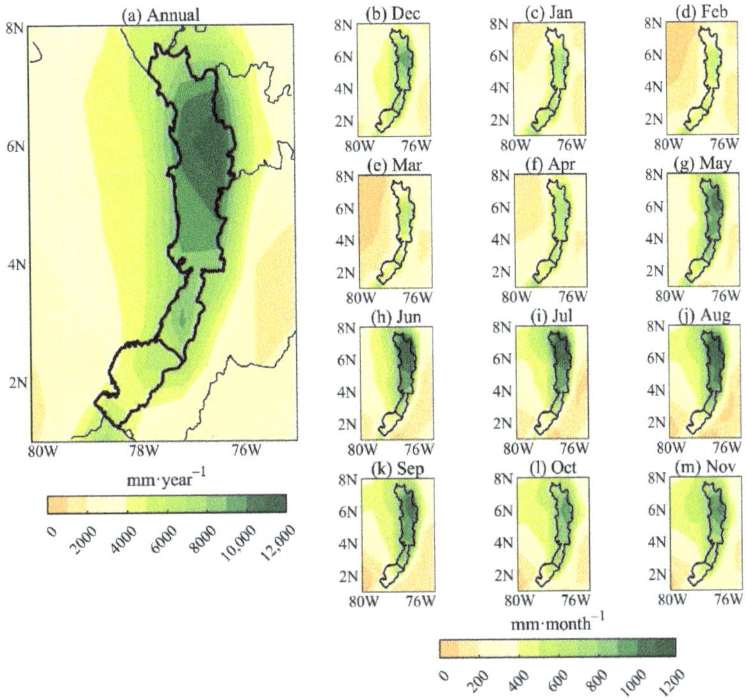

Figure 2. Annual and monthly means precipitation in the Colombian Pacific regions for the 1980–2017 period. (**a**) Annual; (**b**) Dec; (**c**) Jan; (**d**) Feb; (**e**) Mar; (**f**) Apr; (**g**) May; (**h**) Jun; (**i**) Jul; (**j**) Aug; (**k**) Sep; (**l**) Oct; (**m**) Nov.

On a monthly scale, precipitation processes are conditioned by the latitudinal migration of the ITCZ, which, combined with the orographic effects created by the Serranía del Baudó and the foothills of the Western Cordillera, serve as a natural barrier for the air masses coming from the Pacific Ocean, discharging their moisture on the western slopes of the mountain ranges as orographic precipitation [15,43–46]. In the NCP region, there is no defined dry season, and the distribution during the year is relatively uniform (Figure 2b–m); however, from December to April, precipitation is lower (Figure 2b–f), when the ITCZ reaches its southernmost position, whereas, between May and November (Figure 2g–m) the highest precipitation is observed, the moment in which the ITCZ and in general the atmospheric systems record their greatest northward displacement [43,47–49]. Furthermore, in the P-M region, the least rainy season is observed from June to November (Figure 2h–m), when the country's ITCZ is located in its most northerly position.

The analysis of evapotranspiration, which is another important component of the water balance, shows slightly pronounced differences in the study area (Figure 3a), reaching similar annual values in both regions, 921 mm·yr^{-1} for P-M and 855 mm·yr^{-1} for NCP; the lowest annual evapotranspiration (<650 mm·yr^{-1}) occurs towards the center of the region, including the hydrographic zone of the San Juan River and the Pacific coast of the department of Valle del Cauca. The results are consistent with those reported by Vallejo [11], who indicates mean values between 600 and 800 mm·yr^{-1}. According to Vallejo [11], the low evapotranspiration corresponds to the low amount of solar radiation that reaches the area due to the large-scale convective processes and a constant saturation of the atmosphere, coherent with the high convergence of annual moisture observed in the region (Figure 4a), in agreement with the studies of Velasco and Frischt [50], Zipser et al. [51], Zuluaga and Houze [52] and Jaramillo et al. [25]. On a monthly scale (Figure 3b–m), over the NCP region, evapotranspiration decreases while precipitation increases from May to November (<80 mm·month^{-1}; Figures 2g–m and 3g–m), and in coherence with the intensification of moisture convergence in the same period (Figure 4g–m), at the same time that higher evapotranspiration values occur in the P-M region (70–100 mm·month^{-1}) when precipitation and moisture convergence decrease in this region (Figures 2g–m and 4g–m). It should be noted that the magnitude of evapotranspiration concerning precipitation and moisture convergence does not show high variability in the two regions, with values below 90 mm·month^{-1} (standard deviation of 8 mm in NCP and 7 mm in P-M).

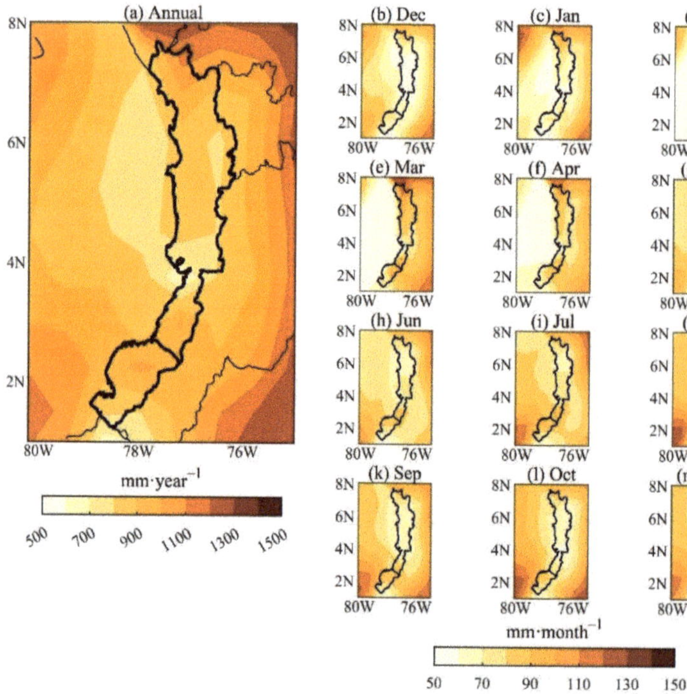

Figure 3. Annual and monthly means evapotranspiration in the Colombian Pacific regions for the 1980–2017 period. (**a**) Annual; (**b**) Dec; (**c**) Jan; (**d**) Feb; (**e**) Mar; (**f**) Apr; (**g**) May; (**h**) Jun; (**i**) Jul; (**j**) Aug; (**k**) Sep; (**l**) Oct; (**m**) Nov.

According to Cuartas and Poveda [2] and Marengo [53], a region acts as a source (sink) of moisture to the atmosphere when evaporation is greater (less) than precipitation. Additionally, Satyamurty et al. [12] and Do Nascimento et al. [54] explain that when there is moisture divergence (convergence) in a given region, it behaves as a source (sink) of moisture to neighboring regions. The relationship between evapotranspiration and precipitation (Figure 5), as an approximation to the regional moisture recycling ratio, shows that between January and April, the NCP has the greatest contribution of moisture to the atmosphere, between 11% and 17% of the total precipitation (Figure 5a), related to the reduction in the moisture convergence towards the region and with the increase in regional recycling (E/P); the rest of the year, the E/P ratio observed corresponds to 7% of the total precipitation, which indicates that the moisture recycled during most of the year comes from in other regions of the continent. In the case of the P-M region (Figure 5b), the region makes the most outstanding moisture to the atmosphere from June to November (dry season), when the E/P ratio varies between 18% and 23%, while moisture convergence decreases; this regional recycling represents almost 100% of the total, highlighting the regional processes of evapotranspiration over precipitation during the driest season, while during the period of higher precipitation, moisture arriving from other regions becomes more important. This is an indicator of the low rate of regional moisture recycling in the Colombian Pacific, which suggests that the region behaves as a moisture sink (E < P).

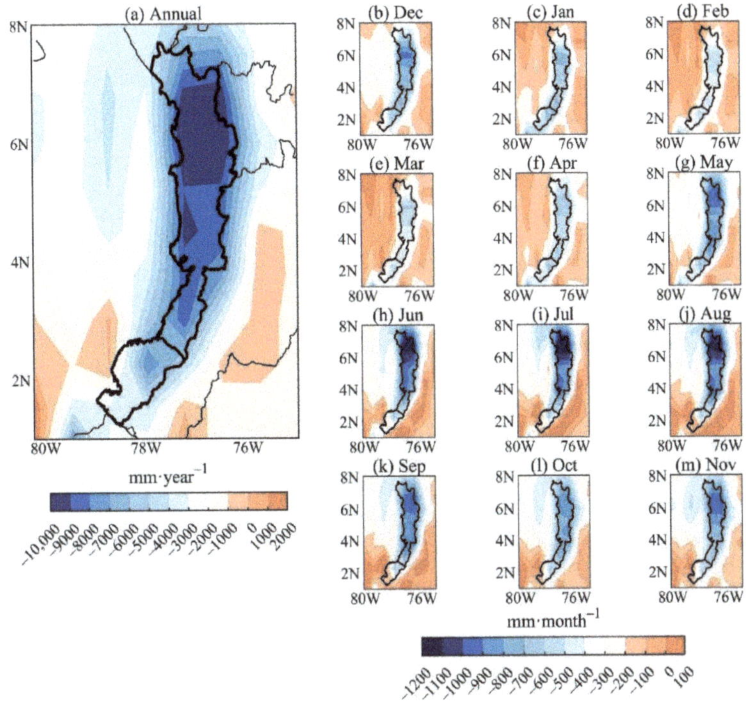

Figure 4. Annual and monthly means moisture divergence in the Colombian Pacific regions for the 1980–2017 period. Negative values reflect rising motions, and positive values reflect sinking motions. (**a**) Annual; (**b**) Dec; (**c**) Jan; (**d**) Feb; (**e**) Mar; (**f**) Apr; (**g**) May; (**h**) Jun; (**i**) Jul; (**j**) Aug; (**k**) Sep; (**l**) Oct; (**m**) Nov.

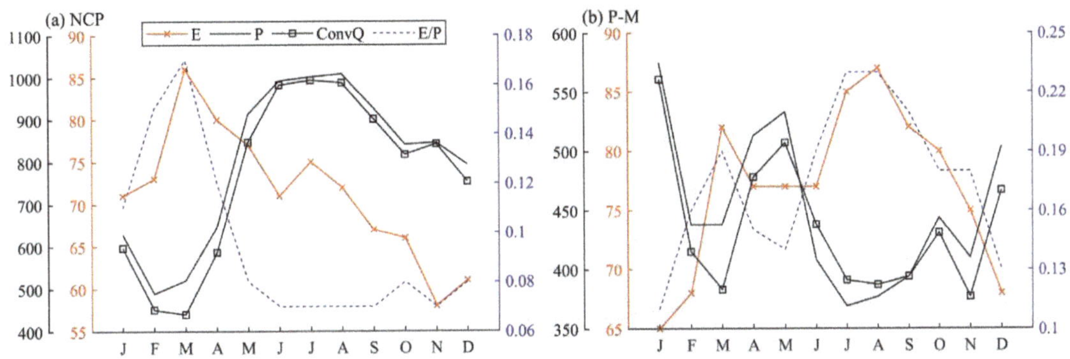

Figure 5. Water balance variables (mm month^{-1}) over the (**a**) North and Central Pacific (NCP) and (**b**) Patia-Maria (P-M) regions from 1980 to 2017. The left y-axis is orange for evapotranspiration (E), black for precipitation (P) and moisture convergence (ConvQ), and the right y-axis for E/P ratio (blue).

3.2. Continental Precipitation Recycling Ratio

Based on the results of the WAM-2 layer model implementation and ERA-I reanalysis data, results concerning the continental precipitation recycling ratio and the delimitation of the precipitationshed of the Colombian Pacific region are presented. Figure 6a shows the spatial distribution of the annual continental precipitation recycling ratio (ρ_c), being on average higher in P-M with 26% per year and lower in NCP with 23%; these low values ratify the dominance of oceanic moisture sources over continental moisture recycling. Likewise, the annual recycling ratio increases from north to south and from west to east (coast to the mountain range). On a monthly scale (Figure 6b–m), the first half of the year shows a marked latitudinal orientation of the recycling ratio, with the highest values in the P-M region (south) and decreasing towards the NCP region (north), which may be related to the migration of the ITCZ from its southernmost position (December–February; DJF) to the north in the March–May (MAM) quarter. During the second half of the year, the recycling ratio shows a marked longitudinal orientation, with the highest values near the foothills of the western cordillera and the lowest near the Pacific Ocean, highlighting the role of the orographic barrier of the Colombian Andes and the moisture transport associated with the CJ during the second half of the year [23,42]. The CJ intensifies from June and reaches its highest velocity from September to November (SON) with values at its core (centered at 5° N along 80° W) of 5–8 m·s^{-1}, in agreement with the maximum precipitation observed in the region during the second half of the year [16,24,55], and weakens from December to May with velocities between 2 and 3 m·s^{-1}. The interaction of the CJ with the topography of the western Andes and the trade winds from the east favors deep convection producing large amounts of precipitation [15,22,23,56,57].

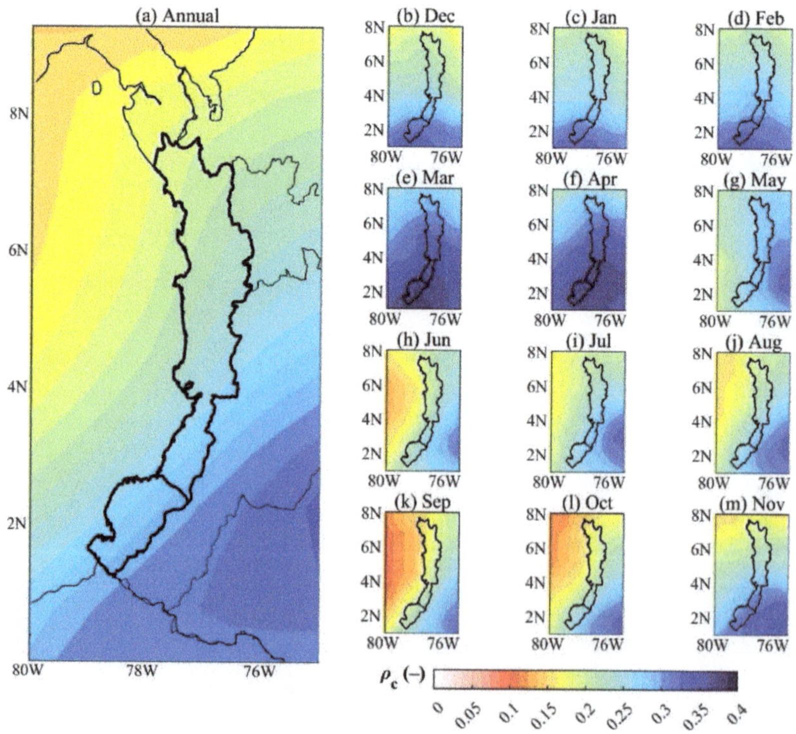

Figure 6. Annual and monthly mean continental precipitation recycling ratio (ρ_c). (**a**) Annual; (**b**) Dec; (**c**) Jan; (**d**) Feb; (**e**) Mar; (**f**) Apr; (**g**) May; (**h**) Jun; (**i**) Jul; (**j**) Aug; (**k**) Sep; (**l**) Oct; (**m**) Nov.

Figure 7a shows the monthly variability of the precipitation recycling ratio for the NCP and P-M regions. A unimodal recycling ratio pattern is observed in both areas, with the highest percentages between March and April, reaching 30% and 34% for NCP and P-M, respectively, and consistent with the spatial pattern shown in Figure 6e,f. Moreover, the lowest recycling season is between June and October, with the lowest percentage of NCP recycling occurring between September–October, with a value between 19% and 20%. In contrast, P-M has its lowest recycling ratio in September, corresponding to 21%. Figure 7b presents the climatological zonal wind values associated with the CJ, showing a greater wind intensity between September and November, and the interquartile distance (3rd quartile minus 1st quartile) is greater than during the rest of the year, while the winds associated with the CJ are less intense from January to April. In this sense, the season of higher (lower) continental precipitation recycling ratio is related to the weakening (strengthening) of the CJ in the first (second) half of the year, which decreases (increases) the contribution of moisture from the Pacific Ocean to the region, increasing (decreasing) the influence of the contribution of land-based sources in the study area. Furthermore, the higher continental precipitation recycling ratio in the P-M region can be related to its higher evapotranspiration concerning NCP and its southern location in the zone of greatest influence of the CJ, which allows the influence of other phenomena such as moisture feedback [7,8,49,58].

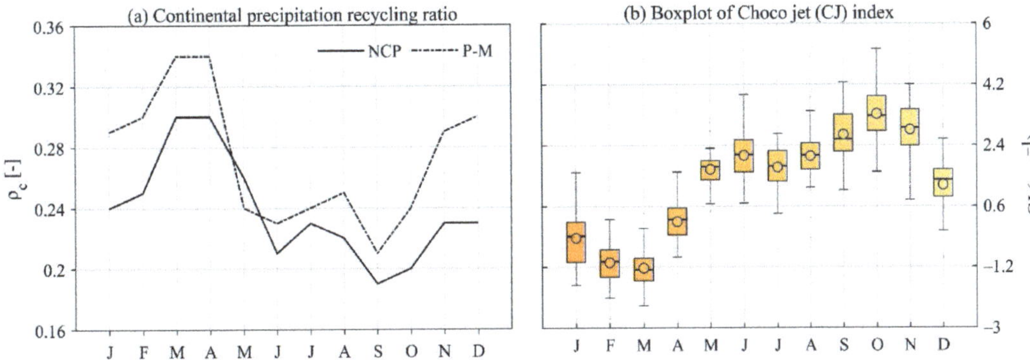

Figure 7. (**a**) Continental precipitation recycling ratio (ρ_c) for the North and Central Pacific (NCP) and Patía-Mira (P-M) regions. (**b**) Boxplots of the Choco jet index (m·s^{-1}). In (**b**) each box represents the range between the first and third quartiles, divided by the median of the sample, with maximum and minimum values (whiskers) shown by vertical stems.

3.3. *Precipitationshed*

Figure 8 shows the monthly variability of the main sources that contribute to precipitation in the Colombian Pacific region and the changes in monthly precipitationshed. The contributions of these different sources change throughout the year, driven by seasonal changes in circulation. From December to April (Figure 8a–e), the largest contributions come from northern sources, mostly the tropical North Atlantic (TNA) and the Caribbean Sea from October to February, and from eastern Colombia and Venezuela; a period when the Orinoco Low-Level jet (OLLJ) exhibits its maximum wind speed, with values around 8–10 m·s^{-1} during DJF [59]. According to Builes-Jaramillo et al. [59], the OLLJ transports atmospheric moisture from TNA, linked to an area of moisture flux divergence located over northeastern South America. During June to August (Figure 8g–i), the ITCZ migration to its northernmost position results in an area of moisture flux convergence over TNA and the Caribbean Sea, which strengthens the CLLJ and inhibits the entrance of moisture from northerlies; thus, the southerly cross-equatorial flow from the Amazon River basin and the southeastern tropical Pacific predominates, as has been documented by Arias et al. [16] and Builes-Jaramillo et al. [59]. In September (Figure 8j), a southwesterly cross-equatorial circulation predominates, converging over the eastern Pacific and western Colombia, consistent with the spatial pattern associated with the easterly low-level jet, also known as the CJ [15,18,23,42]. The CJ transports moisture in the lower troposphere, interacting with the Colombian orography, inducing moisture convergence and convection in western Colombia, which leads to high precipitation over the Colombian Pacific [10,16], with greater intensity during the September to November quarter [23].

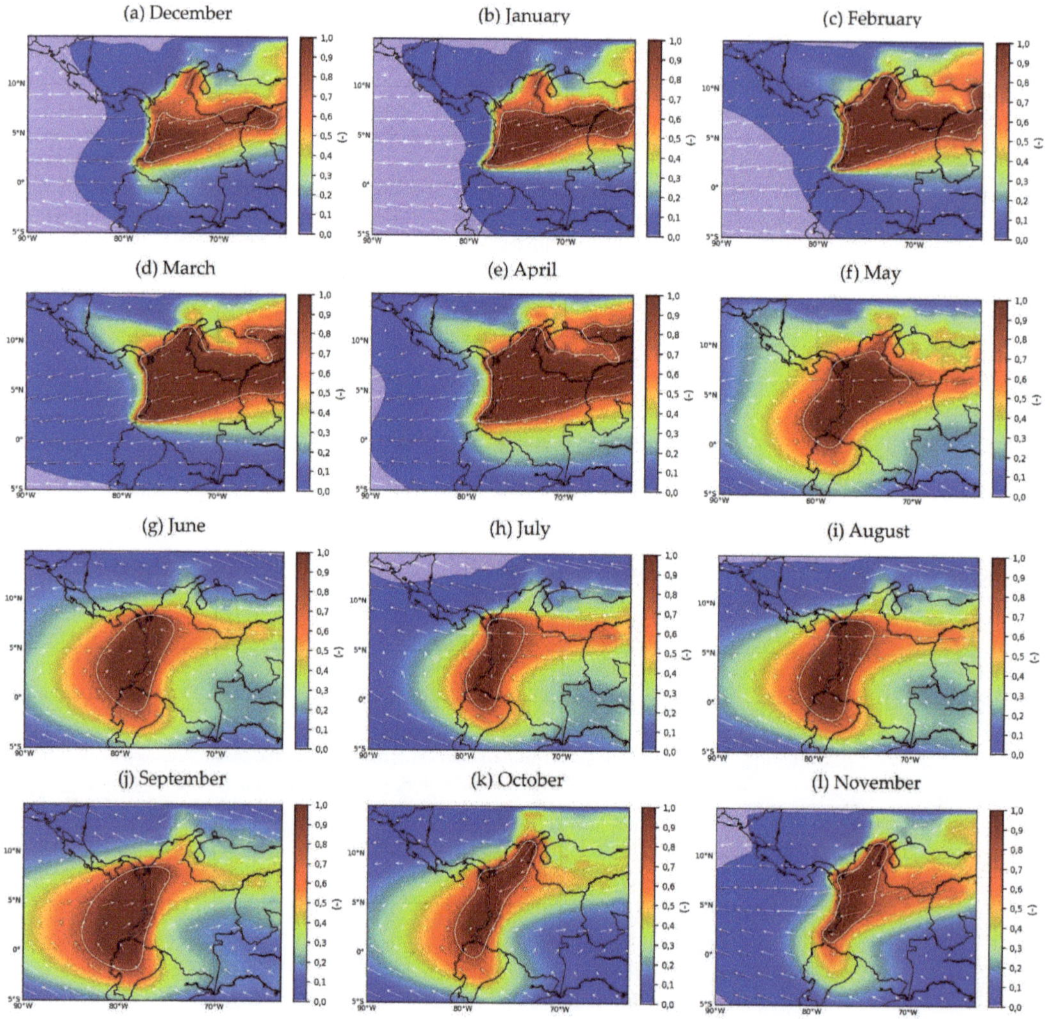

Figure 8. Seasonal variability of precipitationshed in the Colombian Pacific region and vertically integrated moisture fluxes (vectors) from 1980 to 2017 for (**a**) December, (**b**) January, (**c**) February, (**d**) March, (**e**) April, (**f**) May, (**g**) June, (**h**) July, (**i**) August, (**j**) September, (**k**) October, and (**l**) November. The white line represents the threshold whereby the grids with the higher values account for 80% of the tracked moisture. The grids compose the Primary Source region, also known as the precipitationshed. The color bar represents the contribution ratio of the moisture sources to the precipitation. The arrows represent the 1000–175 hPa vertically integrated moisture flux ($Kg{\cdot}m^{-1}{\cdot}s^{-1}$) over the study area.

These seasonal changes in the region's circulation drive a remarkable spatio-temporal variability of precipitationshed (Figure 8), which is significantly more dynamic than the relatively static limits of the surface watersheds since they depend on a defined threshold and the variability of climatic phenomena at multiple scales. Figure 8 highlights the large spatio-temporal variability of the 80% threshold of tracked moisture. The most significant contrast of moisture sources occurs during April and September (Figure 8e,j), representing the highest and lowest continental precipitation recycling ratio (Figure 7a). During April, the largest source region is located on the continent, reflecting the strong influence of northeastern Colombia and Venezuela and the transport of moisture from the Atlantic Ocean; while in September, the continental recycling ratio is reduced, and the moisture source region is located mainly in the Pacific Ocean, and the continental contribution decreases.

Even though the ocean is highlighted as the main source of moisture for the delimited atmospheric basin, these regions are located near the continent, which coincides with the results of Van Der Ent and Savenije [14], who found that ocean moisture source areas are more intense closer to the land surface. As shown in Figure 8, terrestrial sources of moisture encompass a larger area than oceanic sources; however, this is not directly related to the amount of moisture input [14]. This finding can also be related to those found for the precipitation recycling ratio, which indicates that in the NCP, this recycling comes from other regions, which are identified according to the precipitationshed as neighboring and in greater extension towards the northeastern part of the study region, covering almost all of the Andean region, the Orinoquia and the Colombian Caribbean.

4. Conclusions

The objective of this paper was to assess precipitation recycling and moisture sources in the Colombian Pacific region during the 1980–2017 period. We have used the atmospheric moisture tracking model WAM-2 layers to track moisture fluxes, as well as used the precipitationshed approach for the delimitation of the area contributing to terrestrial and oceanic moisture in the study area. We summarize our findings as follows: (1) The results show that on average, the Colombian Pacific region acts as a moisture sink ($E < P$), where convergence is mainly from the Pacific Ocean and represents the largest contribution to the precipitation production, in contrast to the regional (E/P) and neighboring areas. (2) Precipitationshed and its moisture sources present a significant monthly variability, with continental sources from eastern Colombia and Venezuela, and the tropical North Atlantic from December to April; the Pacific Ocean takes preponderance in these contributions during September–October, when the CJ intensifies and the continental contribution decreases; while the Amazon basin and the southeastern tropical Pacific make their greatest contributions during June and August. The importance of the "precipitationshed" approach in the analysis of the hydroclimatology of the Pacific region is the inclusion of the contribution of evaporation from remote land surfaces to the region's precipitation, a factor that, to our knowledge, had not been considered before. This could become an input for analyzing the impacts that changes in land cover and land use may have on evapotranspiration ratios in this region. Furthermore, the alterations that climate change could represent, such as variations in moisture transport, could affect the interactions between the source regions and the continental Pacific.

Further research could address the analysis of moisture recycling by considering other atmospheric monitoring models, such as those developed under a Lagrangian approach, the comparison of both results will improve the accuracy of the results. In addition, the analysis should be carried out considering the impacts of climate change and climate variability events such as the ENSO phases, which affect the region's climatic conditions and the country to a great extent. Research involving land cover and land use scenarios is recommended to explore the vulnerability of moisture recycling in the precipitationshed of the Pacific region.

Author Contributions: Conceptualization, A.M.E. and O.L.B.; methodology, A.M.E. and O.L.B.; software, A.M.E. and W.L.C.; validation, A.M.E., O.L.B. and W.L.C.; formal analysis, A.M.E., O.L.B. and W.L.C.; investigation, A.M.E., O.L.B. and W.L.C.; resources, D.E.-C. and J.T.; data curation, A.M.E. and W.L.C.; writing—original draft preparation, A.M.E. and W.L.C.; writing—review and editing, A.M.E., O.L.B., D.E.-C., J.T. and W.L.C.; visualization, A.M.E. and W.L.C.; supervision, O.L.B. and J.T.; funding acquisition, J.T. All authors have read and agreed to the published version of the manuscript.

Funding: This work was implemented as part of the Climate Action, Alliance of Bioversity International, and the International Center for Tropical Agriculture, Palmira, Colombia, under grants AEC D103D014OOP2.

Institutional Review Board Statement: Not applicable.

Informed Consent Statement: Not applicable.

Data Availability Statement: Not applicable.

Acknowledgments: The authors are grateful to the International Center for Tropical Agriculture (CIAT) and Universidad del Valle.

Conflicts of Interest: The authors declare no conflict of interest. The founding sponsors had no role in the design, analysis, and interpretation of data, the writing manuscript, or the decision to publish the results.

References

1. Brubaker, K.L.; Entekhabi, D.; Eagleson, P.S. Estimation of continental precipitation recycling. *J. Clim.* **1993**, *6*, 1077–1089. [CrossRef]
2. Cuartas, A.; Poveda, G. Balance atmosférico de humedad y estimación de la precipitación reciclada en colombia según el reanális ncep/ncar. *Meteorol. Colomb.* **2002**, *5*, 57–65.
3. Sánchez, C. *Estimation of the Precipitation Recycling Ratio*; Delft University of Technology—National University of Singapore: Delft, The Netherlands, 2016.
4. Eltahir, E.; Bras, R. Precipitation Recycling. *Rev. Geophys.* **1996**, *34*, 367–378. [CrossRef]
5. Van Der Ent, R. A New View on the Hydrological Cycle over Continents. Ph.D. Thesis, Delft University of Technology, Delft, The Netherlands, 2014.
6. Van Der Ent, R.J.; Savenije, H.H.G.; Schaefli, B.; Steele-Dunne, S.C. Origin and fate of atmospheric moisture over continents. *Water Resour. Res.* **2010**, *46*, 1–12. [CrossRef]
7. Van Der Ent, R.J.; Wang-Erlandsson, L.; Keys, P.W.; Savenije, H.H.G. Contrasting roles of interception and transpiration in the hydrological cycle—Part 2: Moisture recycling. *Earth Syst. Dyn.* **2014**, *5*, 471–489. [CrossRef]
8. Keys, P.W. *The Precipitationshed: Concepts, Methods, and Applications*; Stockholm University, Faculty of Science, Stockholm Resilience Centre: Stockholm, Sweden, 2016; ISBN 9789176494646.
9. Hoyos, I. *Transporte de Humedad Atmosférica en Colombia: Origen, Variabilidad y Acople con Fenómenos Climáticos Globales*; Universidad de Antioquia: Medellín, Colombia, 2017.
10. Hoyos, I.; Dominguez, F.; Cañón-Barriga, J.; Martínez, J.A.; Nieto, R.; Gimeno, L.; Dirmeyer, P.A. Moisture origin and transport processes in Colombia, northern South America. *Clim. Dyn.* **2018**, *50*, 971–990. [CrossRef]
11. Vallejo, L. *Dinámica Espacio-Temporal de Los Ríos Aéreos en el Norte de sur América y Posibles Efectos del Cambio Climático*; Universidad Nacional de Colombia: Medellín, Colombia, 2014.
12. Satyamurty, P.; da Costa, C.P.W.; Manzi, A.O. Moisture source for the Amazon Basin: A study of contrasting years. *Theor. Appl. Climatol.* **2013**, *111*, 195–209. [CrossRef]
13. Eltahir, E.A.B.; Bras, R.L. Precipitation recycling in the Amazon basin. *Q. J. R. Meteorol. Soc.* **1994**, *120*, 861–880. [CrossRef]
14. Van Der Ent, R.J.; Savenije, H.H.G. Oceanic sources of continental precipitation and the correlation with sea surface temperature. *Water Resour. Res.* **2013**, *49*, 3993–4004. [CrossRef]
15. Poveda, G.; Jaramillo, L.; Vallejo, L.F. Seasonal precipitation patterns along pathways of South American low-level jets and aerial rivers. *Water Resour. Res.* **2014**, *50*, 98–118. [CrossRef]
16. Arias, P.A.; Martínez, J.A.; Vieira, S.C. Moisture sources to the 2010–2012 anomalous wet season in northern South America. *Clim. Dyn.* **2015**, *45*, 2861–2884. [CrossRef]
17. Morales, J.; Arias, P.; Martínez, J. Role of Caribbean low-level jet and Choco jet in the transport of moisture patterns towards Central America. *Conf. Proc. Pap.* **2017**, *1*, 4861.
18. Gallego, D.; García-Herrera, R.; Gómez-Delgado, F.D.P.; Ordoñez-Perez, P.; Ribera, P. Tracking the moisture transport from the Pacific towards Central and northern South America since the late 19th century. *Earth Syst. Dyn.* **2019**, *10*, 319–331. [CrossRef]
19. Drumond, A.; Nieto, R.; Gimeno, L.; Ambrizzi, T. A Lagrangian identification of major sources of moisture over Central Brazil and la Plata Basin. *J. Geophys. Res. Atmos.* **2008**, *113*, 1–9. [CrossRef]
20. Martinez, A.J.; Dominguez, F. Sources of atmospheric moisture for the La Plata River Basin. *J. Clim.* **2014**, *27*, 6737–6753. [CrossRef]

21. Gimeno, L.; Stohl, A.; Trigo, R.M.; Dominguez, F.; Yoshimura, K.; Yu, L.; Drumond, A.; Durán-quesada, A.M.; Nieto, R. Oceanic and terrestrial sources of continental precipitation. *Rev. Geophys.* **2012**, *50*, 1–41. [CrossRef]
22. Poveda, G.; Mesa, O. La corriente de chorro superficial del oeste ("del Chocó") y otras dos corrientes del chorro en Colombia: Climatología y variabilidad durante las fases ENSO. *Rev. Académica Colomb. Ciencias Tierra* **1999**, *23*, 517–528.
23. Yepes, J.; Poveda, G.; Mejía, J.F.; Moreno, L.; Rueda, C. Choco-jex: A research experiment focused on the Chocó low-level jet over the far eastern Pacific and western Colombia. *Bull. Am. Meteorol. Soc.* **2019**, *100*, 779–796. [CrossRef]
24. Cerón, W.L.; Kayano, M.T.; Andreoli, R.V.; Avila-Diaz, A.; de Souza, I.P.; Souza, R.A.F. Pacific and atlantic multidecadal variability relations with the choco and caribbean low-level jets during the 1900–2015 period. *Atmosphere* **2021**, *12*, 1120. [CrossRef]
25. Jaramillo, L.; Poveda, G.; Mejía, J.F. Mesoscale convective systems and other precipitation features over the tropical Americas and surrounding seas as seen by TRMM. *Int. J. Climatol.* **2017**, *37*, 380–397. [CrossRef]
26. Velásquez-Restrepo, M.; Poveda, G. Estimación del balance hídrico de la región Pacífica. *Dyna* **2019**, *86*, 297–306. [CrossRef]
27. Cuartas, L. *Balance Atmosférico de Humedad Para Colombia (Tesis de Maestría)*; Universidad Nacional de Colombia: Medellín, Columbia, 2001.
28. Savenije, H.H.G. New definitions for moisture recycling and the relationship with land-use changes in the Sahel. *J. Hydrol.* **1995**, *167*, 57–78. [CrossRef]
29. Budyko, M.I. *Climate and life*; Budyko, M.I., Miller, D.H., Eds.; Academic Press: London, UK; New York, NY, USA, 1974; ISBN 978-0-12-139450-9. Available online: https://www.sciencedirect.com/bookseries/international-geophysics/vol/18/suppl/C (accessed on 22 July 2022).
30. Dominguez, F.; Liang, X.; Ting, M. Impact of Atmospheric Moisture Storage on Precipitation Recycling. *J. Clim.* **2006**, *19*, 1513–1530. [CrossRef]
31. Dirmeyer, P.; Brubaker, K. Characterization of the Global Hydrologic Cycle from a Back-Trajectory Analysis of Atmospheric Water Vapor. *J. Hydrometeorol.* **2007**, *8*, 20–37. [CrossRef]
32. Burde, G.; Zangvil, A. The estimation of regional precipitation recycling. Part I: Review of recycling models. *J. Clim.* **2001**, *14*, 2509–2527. [CrossRef]
33. Knoche, H.R.; Kunstmann, H. Tracking atmospheric water pathways by direct evaporation tagging: A case study for West Africa. *J. Geophys. Res. Atmos.* **2013**, *118*, 12345–12358. [CrossRef]
34. Van Der Ent, R.J.; Tuinenburg, O.A.; Knoche, H.R.; Kunstmann, H.; Savenije, H.H.G. Should we use a simple or complex model for moisture recycling and atmospheric moisture tracking? *Hydrol. Earth Syst. Sci.* **2013**, *17*, 4869–4884. [CrossRef]
35. Dee, D.P.; Uppala, S.M.; Simmons, A.J.; Berrisford, P.; Poli, P.; Kobayashi, S.; Andrae, U.; Balmaseda, M.A.; Balsamo, G.; Bauer, P.; et al. The ERA-Interim reanalysis: Configuration and performance of the data assimilation system. *Q. J. R. Meteorol. Soc.* **2011**, *137*, 553–597. [CrossRef]
36. European Centre for Medium-Range Weather Forecasts Interim Reanalysis (ECMWF/ERA-I) Atmospheric Data. Available online: http://apps.ecmwf.int/datasets/data/interim-full-moda/levtype=pl/ (accessed on 6 June 2019).
37. Hoyos, I.; Baquero-Bernal, A.; Jacob, D.; Rodríguez, B. Variability of extreme events in the Colombian Pacific and Caribbean catchment basins. *Clim. Dyn.* **2013**, *40*, 1985–2003. [CrossRef]
38. Cerón, W.L.; Andreoli, R.V.; Kayano, M.T.; Souza, R.A.F.; Jones, C.; Carvalho, L.M.V. The Influence of the Atlantic Multidecadal Oscillation on the Choco Low-Level Jet and Precipitation in Colombia. *Atmosphere* **2020**, *11*, 174. [CrossRef]
39. Cerón, W.L.; Kayano, M.T.; Andreoli, R.V.; Avila, A.; Canchala, T.; Francés, F.; Ayes Rivera, I.; Alfonso-Morales, W.; Ferreira de Souza, R.A.; Carvajal-Escobar, Y. Streamflow Intensification Driven by the Atlantic Multidecadal Oscillation (AMO) in the Atrato River Basin, Northwestern Colombia. *Water* **2020**, *12*, 216. [CrossRef]
40. Institute of Hydrology, Meteorology and Environmental Studies (IDEAM). *Atlas climatológico de Colombia*; IDEAM: Bogotá, Colombia, 2017.
41. Keys, P.; Van Der Ent, R.; Gordon, L.; Hoff, H.; Nikoli, R.; Savenije, H. Analyzing precipitationsheds to understand the vulnerability of rainfall dependent regions. *Biogeosciences* **2012**, *9*, 733–746. [CrossRef]
42. Cerón, W.L.; Andreoli, R.V.; Kayano, M.T.; Avila-Diaz, A. Role of the eastern Pacific-Caribbean Sea SST gradient in the Choco low-level jet variations from 1900–2015. *Clim. Res.* **2021**, *83*, 61–74. [CrossRef]
43. Cerón, W.L.; Andreoli, R.V.; Kayano, M.T.; Ferreria, D.S.R.; Canchala, N.T.; Carvajal-Escobar, Y. Comparison of spatial interpolation methods for annual and seasonal rainfall in two hotspots of biodiversity in South America. *An. Acad. Bras. Ciencias* **2021**, *93*, 1–22. [CrossRef] [PubMed]
44. Guzmán, D.; Ruíz, J.F.; Cadena, M. *Regionalización de Colombia Según la Estacionalidad de la Precipitación Media Mensual*; A Través Análisis de Componentes Principales (ACP): Bogotá, Colombia, 2014.
45. Estupiñan, A.R.C. Estudio de la Variabilidad Espacio Temporal de la Precipitación en Colombia. Ph.D. Thesis, Universidad Nacional de Colombia, Medellín, Colombia, 2016. Available online: http://bdigital.unal.edu.co/54014/1/1110490004.2016.pdf (accessed on 5 October 2019).
46. Cerón, W.L.; Andreoli, R.V.; Kayano, M.T.; Canchala, T.; Ocampo-marulanda, C.; Avila-diaz, A.; Antunes, J. Trend Pattern of Heavy and Intense Rainfall Events in Colombia from 1981–2018: A Trend-EOF Approach. *Atmosphere* **2022**, *13*, 156. [CrossRef]
47. Canchala, T.; Cerón, W.L.; Francés, F.; Carvajal-Escobar, Y.; Andreoli, R.V.; Kayano, M.T.; Alfonso-Morales, W.; Caicedo-Bravo, E.; de Souza, R.A.F. Streamflow variability in colombian pacific basins and their teleconnections with climate indices. *Water* **2020**, *12*, 526. [CrossRef]

48. Canchala, T.; Alfonso-Morales, W.; Cerón, W.L.; Carvajal-Escobar, Y.; Caicedo-Bravo, E. Teleconnections between monthly rainfall variability and large-scale climate indices in Southwestern Colombia. *Water* **2020**, *12*, 1863. [CrossRef]
49. Poveda, G. La hidroclimatología de Colombia: Una síntesis desde la escala inter-decadal hasta la escala diurna. *Rev. Acad. Colomb. Cienc.* **2004**, *28*, 201–222.
50. Velasco, I.; Fritsch, J.M. Mesoscale convective complexes in the Americas. *J. Geophys. Res. Atmos.* **1987**, *92*, 9591–9613. [CrossRef]
51. Zipser, E.J.; Cecil, D.J.; Liu, C.; Nesbitt, S.W.; Yorty, D.P. Where are the most: Intense thunderstorms on Earth? *Bull. Am. Meteorol. Soc.* **2006**, *87*, 1057–1071. [CrossRef]
52. Zuluaga, M.D.; Houze, R.A. Extreme convection of the near-equatorial Americas, Africa, and adjoining oceans as seen by TRMM. *Mon. Weather Rev.* **2015**, *143*, 298–316. [CrossRef]
53. Marengo, J.A. Characteristics and spatio-temporal variability of the Amazon river basin water budget. *Clim. Dyn.* **2005**, *24*, 11–22. [CrossRef]
54. Do Nascimento, M.G.; Herdies, D.L.; De Souza, D.O. The south American water balance: The influence of low-level jets. *J. Clim.* **2016**, *29*, 1429–1449. [CrossRef]
55. Poveda, G.; Álvarez, D.M.; Rueda, Ó.A. Hydro-climatic variability over the Andes of Colombia associated with ENSO: A review of climatic processes and their impact on one of the Earth's most important biodiversity hotspots. *Clim. Dyn.* **2011**, *36*, 2233–2249. [CrossRef]
56. Sakamoto, M.S.; Ambrizzi, T.; Poveda, G. Moisture Sources and Life Cycle of Convective Systems over Western Colombia. *Adv. Meteorol.* **2011**, *2011*, 1–11. [CrossRef]
57. Durán-Quesada, A.M.; Gimeno, L.; Amador, J. Role of moisture transport for Central American precipitation. *Earth Syst. Dyn.* **2017**, *8*, 147–161. [CrossRef]
58. Bosilovich, M.G. On the vertical distribution of local and remote sources of water for precipitation. *Meteorol. Atmos. Phys.* **2002**, *80*, 31–41. [CrossRef]
59. Builes-Jaramillo, A.; Yepes, J.; Salas, H.D. The Orinoco Low-Level Jet and Its Association with the Hydroclimatology of Northern South America. *J. Hydrometeorol.* **2022**, *23*, 209–223. [CrossRef]

MDPI
St. Alban-Anlage 66
4052 Basel
Switzerland
Tel. +41 61 683 77 34
Fax +41 61 302 89 18
www.mdpi.com

Atmosphere Editorial Office
E-mail: atmosphere@mdpi.com
www.mdpi.com/journal/atmosphere

www.ingramcontent.com/pod-product-compliance
Lightning Source LLC
Chambersburg PA
CBHW061819210926
51850CB00006B/754
9783036576077